Antoine Henrot (Ed.)

Shape optimization and spectral theory

Contributors

Pedro R.S. Antunes
Mark Ashbaugh
Virginie Bonnaillie-Noël
Lorenzo Brasco
Dorin Bucur
Giuseppe Buttazzo
Guido De Philippis
Pedro Freitas
Alexandre Girouard
Bernard Helffer
James Kennedy
Jimmy Lamboley
Richard S. Laugesen
Edouard Oudet
Michel Pierre
Iosif Polterovich
Bartłomiej A. Siudeja
Bozhidar Velichkov

Antoine Henrot (Ed.)

Shape optimization and spectral theory

Managing Editor: Agnieszka Bednarczyk-Drąg
Associate Editor: Filippo A. E. Nuccio Mortarino Majno di Capriglio
Language Editor: Nick Rogers

DE GRUYTER
OPEN

ISBN 978-3-11-055085-6
e-ISBN 978-3-11-055088-7

Library of Congress Cataloging-in-Publication Data
A CIP catalog record for this book has been applied for at the Library of Congress.

Published by De Gruyter Open Ltd, Warsaw/Berlin
Part of Walter de Gruyter GmbH, Berlin/Boston
The book is published with open access at www.degruyter.com.
Cover illustration: © Bartlomiej Siudeja

Managing Editor: Agnieszka Bednarczyk-Drąg
Associate Editor: Filippo A. E. Nuccio Mortarino Majno di Capriglio
Language Editor: Nick Rogers

www.degruyteropen.com

Contents

Antoine Henrot

1 Introduction

1.1 General introduction

This collective book has the ambition to give an overview of recent results in spectral geometry and its links with shape optimization. The questions we are interested in are:

- Does there exist a set which minimizes (or maximizes) the k-th eigenvalue of a given elliptic operator (mainly the Laplacian or the Schrödinger operator) with given boundary conditions (mainly Dirichlet, Robin or Steklov boundary conditions in this book) among sets of given volume?
- If existence is proved, what can be said about the regularity of the optimal set?
- Can one compute the optimal set? Can one give some geometric properties of it?
- Is it possible to prove some stability results for the well-known isoperimetric inequalities involving the first or the second eigenvalue, namely some quantitative isoperimetric inequalities for eigenvalues?
- More generally, can one write some universal inequalities related to the eigenvalues of elliptic operators?
- For some more specific shapes like triangles, can one prove more precise bounds and observe some patterns of the eigenfunctions?

A first survey on these kinds of extremum problems appeared in 2006 in the book "Extremum Problems for eigenvalues of elliptic operators", see [505]. It turns out that much progress has been made in the last ten years which leads us to think that an update and an extension would be valuable for a wide community who is interested in spectral theory and its links with geometry, calculus of variations, free boundary problems and shape optimization.

1.2 Content of the book

We now describe in more detail the content of each chapter. For most of the notation used here and in this book, we refer to the end of this Introduction chapter (Section 1.4).

Antoine Henrot: Institut Élie Cartan, UMR 7502, Université de Lorraine, CNRS BP 70239, 54506 Vandoeuvre-lès-Nancy, France, E-mail: antoine.henrot@univ-lorraine.fr

Existence results **by D. Bucur**

In Chapter 2, existence of a minimizer for a problem like

$$\min\{\lambda_k(\Omega), \Omega \subset \mathbb{R}^d, |\Omega| = c\} \tag{1.1}$$

or more generally like

$$\min\{F(\lambda_1(\Omega), \dots \lambda_k(\Omega)), \Omega \subset \mathbb{R}^d, |\Omega| = c\} \tag{1.2}$$

is studied. It is necessary to enlarge the class of admissible sets to *quasi-open* sets which is actually the largest, and more natural, class for which eigenvalues of the Dirichlet-Laplacian are well defined. A famous existence result due to G. Buttazzo and G. Dal Maso in [238] was previously available but with the supplementary assumption that the sets have to lie in some fixed bounded "box" D. A lot of effort has been done to remove this assumption and to get a general existence result. These efforts eventually led to the papers by D. Bucur [206] and A. Pratelli-D. Mazzoleni [700] where this open problem of existence was solved. These two papers use a completely different strategy. In [206], the notion of *shape subsolution* for the torsion energy was introduced, and it was proved that every such subsolution has to be bounded and has finite perimeter. A second argument, showed that minimizers for (1.1) are shape subsolutions, so they are bounded. The author finished the proof by using a concentration-compactness argument, like in [219]. The approach of [700] is different: a surgery result proved that some parts (like long and tiny tentacles), can be cut out from every set such that, after small modifications and rescaling, the new set has a diameter uniformly bounded and its first k eigenvalues are smaller. In this way, the existence problem in $\mathbb{R}^d$ can be reduced to the local case of Buttazzo and Dal Maso. In Chapter 2, the main ideas of the proof of the global existence result are provided, using a combination of the two methods: shape subsolutions and surgery. Moreover, it is proved that the optimal set is bounded and has finite perimeter.

Regularity of optimal spectral domains **by J. Lamboley and M. Pierre**

Once existence has been proved (for example for Problem (1.1)), an important and difficult issue is to study the regularity of the optimal set $\Omega^\star$. As we have seen, the existence theorem provides a solution which is only quasi-open and this is a very weak regularity. Now it seems reasonable to expect much more regularity such as Lipschitz or C^2 boundary, even analytic (at least in two dimensions). We would like to have this kind of regularity to be able to write *optimality conditions* thanks to the shape derivative, see [510, chapter 5]. For computing such shape derivatives, a minimal regularity is required (e.g. C^2 regularity if we want to consider the trace of the gradient on the boundary which occurs in the classical Hadamard's formula). It turns out that the main difficulty in this regularity issue is the first step: to be able to reach weak but consistent regularity, for example that the boundary of the optimal domain is locally a graph. Once this is done, it is often possible to use powerful tools developed in the theory of regularity for free boundary problems (e.g. by L. Caffarelli and co-authors) to reach the desired properties. Chapter 3 is devoted to these questions and presents the known results. The authors also point out many open problems which remain to be solved. In a first section, they consider the problem

$$\min\{\lambda_1(\Omega), \Omega \subset D, |\Omega| = c\} \tag{1.3}$$

where D is a given bounded domain. Of course, if D is large enough (or c small enough) in order that it contains the ball of volume c, this one is the solution and there is nothing to prove. Otherwise, the regularity of the optimal domain is studied in detail and is well understood. The fact that the state function (the first eigenfunction u_1) is positive plays an important role in this analysis. Actually, the first step is to study the global regularity of u_1 naturally extended by zero outside $\Omega^\star$. For that purpose, the authors show that Problem (1.3) is actually equivalent to a penalized version

$$\min\{\lambda_1(\Omega) + \mu[|\Omega| - c]^+, \Omega \subset D\} \tag{1.4}$$

(where $[x]^+$ denotes the positive part of x) for μ large enough. It turns out that it is much more convenient to work with this penalized version, in particular it is easier to perform variations and exploit the minimality of the domain which is used to prove that the eigenfunction is globally Lipschitz continuous. This idea of the penalized version works as well for any eigenvalue λ_k but without a box constraint: Problem (1.1) is equivalent to

$$\min\{\lambda_k(\Omega) + \mu|\Omega|, \Omega \subset \mathbb{R}^d\} \tag{1.5}$$

for a particular value of μ. Knowing that the state function is Lipschitz continuous is a first (important) step in the study of the regularity of the boundary of the optimal set, but obviously not sufficient. The next step is to prove that the gradient of the state function does not degenerate at the boundary, in order to be able to use some "implicit function theorem" to deduce regularity of the boundary itself. This is done in this chapter for λ_1 where analyticity of the boundary is proved in dimension $d = 2$, while regularity of the *reduced boundary* is proved in higher dimension. Then, for Problem (1.3) the regularity of $\partial\Omega^\star$ up to the boundary of the box D is studied. At last, some problems with a *convexity* constraint are considered.

The Robin problem **by D. Bucur, P. Freitas and J. Kennedy**

While the two previous chapters deal with Dirichlet boundary conditions, the next one deals with Robin boundary conditions: $\frac{\partial u}{\partial n} + \alpha u = 0$ on $\partial\Omega$. This is a very important case since it can be seen as a generalization (or interpolation) of the Dirichlet and

Neumann cases. Neumann boundary conditions correspond to $\alpha = 0$ while Dirichlet ones correspond to $\alpha \to +\infty$. Chapter 4 presents a very complete overview on qualitative properties for the Robin eigenvalues. Among many results shown in that chapter we can find a study of the curves $\alpha \mapsto \lambda_k(\alpha, \Omega)$ together with a precise asymptotic expansion of $\lambda_k(\alpha, \Omega)$ for both $\alpha \to \pm\infty$. Concerning isoperimetric inequalities, while the minimization problem for the first two eigenvalues is now well understood when α is positive, the corresponding maximization problems for negative α remains open. This is the last problem for which an isoperimetric inequality for the first eigenvalue of the Laplace operator has not yet been solved. Chapter 4 begins with a clear presentation of the minimization of $\lambda_1(\alpha, \Omega)$ for $\alpha > 0$ recalling the arguments used by M.H. Bossel and D. Daners. Then, this result is applied to solve the minimization problem for $\lambda_2(\alpha, \Omega)$ (result due to J. Kennedy). The situation for higher eigenvalues seems to be much more complex, since the optimizers are expected to depend on the boundary parameter α as suggested by numerical simulations. Nevertheless some properties are also given in that case. The case $\alpha < 0$ is still more intriguing. The long-standing conjecture that the ball should be the maximizer for every value of α has been very recently disproved by P. Freitas and D. Krejčiřík. More precisely, they proved that the disk is indeed the maximizer for *small* values of α (in dimension 2). But the ball cannot be the maximizer for *large* (negative) values of α since an annulus (for $d = 2$) or a spherical shell (for $d \geq 3$) gives a larger value. This can be seen thanks to the precise evaluation of the asymptotic expansion of $\lambda_1(\alpha)$ when $\alpha \to -\infty$ that was obtained previously.

Spectral geometry of the Steklov problem **by A. Girouard and I. Polterovich**

Chapter 5 presents an overview of the geometric properties of the Steklov eigenvalues and eigenfunctions. One can consider two natural constraints for an isoperimetric inequality for the first non trivial Steklov eigenvalue. Under the volume constraint, the fact that the ball is the extremal domain has been proved by F. Brock. A perimeter constraint appears to be more natural from the following viewpoint: the Steklov eigenvalues are the eigenvalues of the Dirichlet-to-Neumann operator, which is an operator defined on the boundary of the domain. In that case, Weinstock has proved that the disk is the maximizer in the class of simply-connected planar domains. However, this topological assumption cannot be removed, as can be seen from the example of an annulus. Moreover, for general Euclidean domains the question of existence of a maximizer remains open. Nevertheless, it is known that for simply connected planar domains, the k-th normalized Steklov eigenvalue is maximized in the limit by a disjoint union of $k - 1$ identical disks for any $k \geq 2$. Some geometric bounds are also obtained in higher dimensions, but they are more complicated, as they involve other geometric quantities, such as the isoperimetric ratio. Other interesting questions are also discussed in this chapter, in particular, isospectrality (Can one hear the shape of a drum whose mass is concentrated on the boundary?). It is largely open, since the usual techniques applied in the Dirichlet case, such as the transplantation technique, do not work for planar domains with the Steklov boundary condition. Another topic covered in the chapter is the study of nodal lines and nodal domains. Bounds for the multiplicity of the Steklov eigenvalues, as well as the asymptotic distribution of Steklov eigenvalues are also considered. It turns out that spectral asymptotics in the Steklov case strongly depend on the regularity of the domain.

Triangles and other special domains **by R.S. Laugesen and B. Siudeja**

Chapter 6 reports on known and conjectured spectral properties of the Laplacian on special domains, like triangles, rectangles or rhombi. Topics include sharp lower bounds and sharp upper bounds, as well as inverse problems, hot spots, and nodal domains. The authors consider both Dirichlet and Neumann boundary conditions (and sometimes mixed Dirichlet–Neumann conditions). This chapter begins with classical applications of symmetrization techniques for finding optimal domains, for example in the class of triangles. Then the method of unknown trial functions is presented and used to obtain sharp lower bounds for triangles, including a sharp Poincaré inequality. The method consists in transplanting the (unknown) eigenfunction of an arbitrary triangle to yield trial functions for the (known) eigenvalues of certain equilateral and right triangles. The method of choosing clever trial functions allows one to get sharp upper bounds. For isosceles or right triangles, explicit expressions for the eigenvalues are not available, but nevertheless, the authors are able to get some monotonicity formulas in these cases. The chapter also discusses the inverse problem "Can one hear the shape of a triangular drum?" and the positive answer by Durso and Grieser–Maronna. Lastly, qualitative properties of eigenfunctions and their nodal regions are investigated:

- simplicity of the low eigenvalues λ_2 and μ_2 for triangles,
- the hot spots conjecture for acute triangles,
- the Courant-sharp property (which means that the number of nodal domains is exactly the rank of the eigenvalue),
- the sign of the Neumann eigenfunctions on the boundary of the domain.

Spectral inequalities in quantitative form **by L. Brasco and G. de Philippis**

When an isoperimetric inequality, like Faber-Krahn inequality, is proved, a very natural question is the stability issue. Namely: assume that a domain Ω has a first Dirichlet eigenvalue very close to the first Dirichlet eigenvalue of a ball of same volume, to what extent can we claim that Ω itself is close to a ball? Moreover, can we quantify it? This kind of question has a long history for the classical (geometric) isoperimetric inequality. Concerning the eigenvalues, it started in the 1990's and enjoyed a renewed success during the ten last years particularly thanks to the Italian school. The aim of this chapter is to give a complete picture on recent results about quantitative improvements of sharp inequalities for eigenvalues of the Laplacian with all the classical boundary conditions. The authors begin by the case of the Faber-Krahn inequality. The distance to the ball (or more generally to the optimal domain) can be expressed in different ways. A popular choice consists in using the so-called *Fraenkel asymmetry*, which is a L^1 distance between the characteristic functions:

$$\mathcal{A}(\Omega) := \inf\left\{\frac{|\Omega\Delta B|}{|\Omega|}\ B \text{ ball such that } |B| = |\Omega|\right\}. \tag{1.6}$$

A discussion is undertaken so as to compare it to other measures of asymmetry. Then, the stability of the Szegő-Weinberger and Brock-Weinstock inequalities are treated. For each of these situations, the authors present the relevant stability result and then discuss its sharpness, in particular the sharpness of the exponent on the Fraenkel asymmetry which occurs in the quantitative inequality and this is not always as simple as one may think. For the quantitative Faber-Krahn inequality, the sharp exponent is 2 and several weaker results were available in the literature before being able to get this exponent. The proof, whose main ideas are given here, consists in obtaining a quantitative estimate for the torsional rigidity by some selection principle. Some interesting applications of the quantitative Faber-Krahn inequality to estimates of the so called harmonic radius are also considered. Another Section is devoted to presenting the proofs of other spectral inequalities, involving the second Dirichlet eigenvalue λ_2 as well, as the Hong-Krahn-Szego inequality for λ_2 and the Ashbaugh-Benguria inequality for the ratio λ_2/λ_1. For the Hong-Krahn-Szego inequality, it is important to notice that it is the first example of quantitative isoperimetric inequality for which the target set is no longer the ball, since the optimal domain is the disjoint union of two identical balls. Obviously, this requires a different version for the "asymmetry" adapted to this situation.

Universal inequalities **by M. Ashbaugh**

Inequalities involving eigenvalues are called *universal* when they hold in complete generality, requiring no hypotheses on the domain (other than that it is of dimension d). It contrasts with most of the isoperimetric inequalities of other chapters which are obtained with a volume (or perimeter) constraint. Famous examples of such inequalities (for the Dirichlet-Laplacian) are

$$\frac{\lambda_2}{\lambda_1} \le 1 + \frac{4}{d} \quad \text{or more generally} \quad \lambda_{m+1} - \lambda_m \le \frac{4}{md}\sum_{i=1}^{m}\lambda_i \tag{1.7}$$

which were proved for (planar) domains by Payne, Pólya and Weinberger in [746] and which may be considered as the starting point of that study. Other more general results have been obtained by Hile and Protter on the one hand and Yang on the other, and Chapter 8 shows in a very clear way all these results and the hierarchy between them. Let me also mention that the first inequality in (1.7) has been later improved as $\lambda_2/\lambda_1 \le j^2_{d/2,1}/j^2_{d/2-1,1}$ by the author and R. Benguria in [62], [64] giving here the sharp inequality. Chapter 8 presents in an unified way these universal inequalities by choosing the more general approach. All the proofs are given and the author often chooses to present simpler proofs or more general results than the original ones. In particular, several statements are given with the Laplacian replaced by a Schrödinger operator with both a scalar and vectorial potential. The authors considers whether it is possible to deduce from the previous universal inequalities some explicit bounds for a given eigenvalue, say λ_{m+1}. Moreover, we would like these bounds to be in good accordance with Weyl's law. Several interesting results in that direction are presented. This chapter aims to put in perspective all of this material, by giving historical markers and references.

Spectral optimization problems for Schrödinger operators **by G. Buttazzo and B. Velichkov**

In this chapter Schrödinger operators of the form $-\Delta + V(x)$ are considered, with Dirichlet boundary conditions on a bounded open set $D \subset \mathbb{R}^d$. The question is now to find optimal potentials for some suitable optimization criteria. In general, the optimization problems studied here can be written as $\min\{F(V), V \in \mathcal{V}\}$ where F is a suitable cost functional and $\mathcal{V}$ is a suitable class of admissible potentials. The case of spectral functionals $\min\{\Phi(\lambda(V)), V \in \mathcal{V}\}$, where $\lambda(V)$ is the spectrum of the Schrödinger operator, which will be assumed to be discrete, are also considered. Here, the interest is not only in potentials V which are bounded functions, but also in their natural extension: *the capacitary measures* which are nonnegative Borel measures on D, possibly taking the value $+\infty$ and vanishing on all sets of capacity zero. The class of capacitary measures is very large and contains both the cases of standard potentials $V(x)$, in which $\mu = V dx$, as well as the case of classical domains, where we set $\mu = +\infty_{D\setminus\Omega}$ which is the measure defined by

$$\mu(E) = \begin{cases} 0 & \text{if } \operatorname{cap}(\Omega \setminus E) = 0 \\ +\infty & \text{if } \operatorname{cap}(\Omega \setminus E) > 0. \end{cases} \tag{1.8}$$

In that sense, this can be seen as a unified presentation of these two classical problems. The authors are mainly interested in existence theorems. They first prove a very general existence result in the class of capacitary measures. Then they consider the more specific case of integrable potentials: $V \in L^p(D)$ for which other general existence theorems are proved. Some examples where the optimal potential can be explicitly determined are also presented. In the last section, they consider another class of admissible potentials, namely the function $V(x)$ such that $\int_D \Psi(V)\,dx \le 1$ where Ψ satisfies some assumptions allowing possible large potentials.

Nodal and spectral minimal partitions **by V. Bonnaillie-Noël and B. Helffer**

This chapter is devoted to the analysis of *minimal partitions* and their relations with the nodal domains of eigenfunctions. Let Ω be a fixed domain, then a k-partition $\mathcal{D}$ of Ω is a family of k disjoint sub-domains of Ω: $D_1, D_2, \dots, D_k$. It is a natural and popular question to ask what possible k-partition $\mathcal{D}$ minimizes

$$\Lambda(\mathcal{D}) := \max\left(\lambda_1(D_1), \dots, \lambda_1(D_k)\right), \tag{1.9}$$

where $\lambda_1(D_i)$ denotes the first Dirichlet eigenvalue of the sub-domain D_i? In particular, is this minimal partition related to the nodal domains of a given eigenfunction, for example the k-th eigenfunction associated to $\lambda_k(\Omega)$. As the reader will discover, this is always the case for $k = 2$, but this is true for higher values of k if and only if $\lambda_k(\Omega)$ is *Courant-sharp* which means that it has a corresponding eigenfunction with exactly k nodal domains, saturating in this way the famous Courant nodal Theorem. This minimal partition problem has strong links with models in mathematical ecology where the sub-domains represent the strong competition limit of segregating species in population dynamics. A complete overview of what is known about Courant-sharp eigenvalues is presented here. The famous Pleijel's Theorem states that there are only a finite number of such eigenvalues and even gives an upper bound of the number of nodal domains. For some particular domains (not many), it is possible to give explicitly the eigenvalues which are Courant-sharp. As explained, this occurs in particular the case for thin domains, square, some rectangles, some torus, disk and circular sectors. Then, the minimal partitions are studied. Existence and regularity of minimal partitions is stated. The case of 3-partitions and their possible topologies are investigated in more detail. Some explicit results and conjectures, supported by numerical simulations are given. A generalization to p-minimal k-partitions where the ℓ^∞ norm defining Λ in (1.9) is replaced by the ℓ^p norm is also considered. Then the authors introduce the Aharonov-Bohm operators. It turns out that minimal partitions can be recognized as nodal partitions of eigenfunctions of these operators. This gives interesting necessary conditions for candidates to be minimal partitions. At last the asymptotic behavior of minimal partitions when $k \to +\infty$ is discussed. In particular the hexagonal conjecture and some other qualitative properties are presented.

Numerical results for spectral optimization problems **by P. Antunes and E. Oudet**

This chapter is devoted to numerical methods which have been introduced to solve the previous problems. In the first two sections two of these approaches which have been successful in recent years on spectral problems are explained. The first one consists in introducing some global optimization tools to provide a good initial guess of the optimal profile. This step does not require any topological information on the set but is restricted to a small class of shapes. Then the method of fundamental solutions (to compute the eigenvalues) is described. It allows, in a second stage, the identification and precise evaluation of shapes which are locally optimal. A constant preoccupation is to decrease the complexity of the optimization problem by introducing a reduction of the number of parameters which still allows a precise computation of the cost function. For example, the parametrization of the boundaries of the open sets as level set functions, for example level sets of truncated Fourier series can be very efficient. The chapter ends with a presentation of the best domains obtained numerically for both Dirichlet and Neumann eigenvalues λ_k and μ_k (for $k = 1$ to 10 or 15) and some conjectures inspired by these numerical results.

1.3 Balls and union of balls

One of the most important topics discussed in this book is the determination of which domain minimizes or maximizes a given eigenvalue. For low eigenvalues, actually the two first eigenvalues, and for most boundary conditions, the optimal domains are known and it turns out that they are the same: the ball for the first eigenvalue and the union of two identical balls for the second. These results are recalled in different chapters of this book, but let us sum up it here.

First eigenvalue-Dirichlet The ball minimizes $\lambda_1(\Omega)$ among sets of given volume (Faber-Krahn inequality), see [377] and [603].

First (non-trivial) eigenvalue-Neumann The ball maximizes $\mu_2(\Omega)$ among sets of given volume (Szegő-Weinberger inequality) , see [838] for Lipschitz simply connected planar domains and [871] for the general case.

First eigenvalue-Robin ($\alpha > 0$) The ball minimizes $\lambda_1(\Omega, \alpha)$ among sets of given volume (Bossel-Daners inequality), see [169] for the two-dimensional case and [319] for the general case.

First eigenvalue-Steklov The ball maximizes $\sigma_2(\Omega)$ among sets of given volume (Brock-Weinstock inequality), see [873] for the two-dimensional case and [192] for the general case.

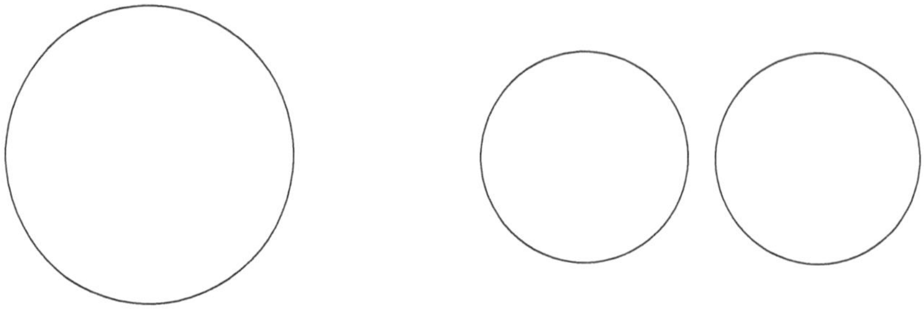

Fig. 1.1. Left: the disk minimizes the first eigenvalue (Dirichlet or Robin) and maximizes the first non trivial eigenvalue (Neumann or Steklov). Right: two disks minimizes the second eigenvalue (Dirichlet or Robin) and maximizes the second non trivial eigenvalue (Neumann or Steklov)

Second eigenvalue-Dirichlet The union of two identical balls minimizes $\lambda_2(\Omega)$ among sets of given volume (Hong-Krahn-Szego inequality), see [604], [534].

Second (non-trivial) eigenvalue-Neumann The union of two identical disks maximizes $\mu_3(\Omega)$ among simply connected bounded planar domains of given volume, see [428].

Second eigenvalue-Robin ($\alpha > 0$) The union of two identical balls minimizes $\lambda_1(\Omega, \alpha)$ among sets of given volume, see [586] and Theorem 4.36 in Chapter 4.

Second eigenvalue-Steklov The union of two identical disks maximizes $\sigma_3(\Omega)$ among simply connected bounded planar domains of given volume, see [513], [430] and Chapter 5.

In view of the previous results, it is a natural question to ask whether there are other eigenvalues for which balls or union of balls could be the optimal domain. For Dirichlet eigenvalues, this question has been recently investigated in the PhD thesis of A. Berger, see [137]. She proves that for $d = 2$, only λ_1 and λ_3 can be minimized by the disk (it is still a conjecture for λ_3). Moreover, only λ_2 and λ_4 can be minimized by union of disks (it is still a conjecture for λ_4).

Let us finish this section by giving the eigenvalues of the ball. In dimension 2, the eigenvalues of the disk B_R of radius R for the Laplacian with Dirichlet boundary conditions and the corresponding eigenfunctions (not normalized) are given by

$$
\begin{aligned}
&\lambda_{0,k} = \frac{j_{0,k}^2}{R^2}, \; k \geq 1, \\
&u_{0,k}(r,\theta) = J_0(j_{0,k} r/R), \; k \geq 1, \\
&\lambda_{n,k} = \frac{j_{n,k}^2}{R^2}, \; n, k \geq 1, \text{ double eigenvalue} \\
&u_{n,k}(r,\theta) = \begin{cases} J_n(j_{n,k} r/R) \cos n\theta \\ J_n(j_{n,k} r/R) \sin n\theta \end{cases}, \; n, k \geq 1,
\end{aligned}
\tag{1.10}
$$

where $j_{n,k}$ is the k-th zero of the Bessel function J_n.
For the Laplacian with Neumann boundary conditions, the eigenvalues and eigen-

functions of the disk B_R are:

$$\begin{aligned}
&\mu_{0,k} = \frac{j_{0,k}^{'2}}{R^2}, \; k \geq 1, \\
&v_{0,k}(r,\theta) = J_0(j_{0,k}^{'} r/R), \; k \geq 1, \\
&\mu_{n,k} = \frac{j_{n,k}^{'2}}{R^2}, \; n, k \geq 1, \text{ double eigenvalue} \\
&v_{n,k}(r,\theta) = \begin{cases} J_n(j_{n,k}^{'} r/R) \cos n\theta \\ J_n(j_{n,k}^{'} r/R) \sin n\theta \end{cases}, \; n, k \geq 1,
\end{aligned} \tag{1.11}$$

where $j_{n,k}^{'}$ is the k-th zero of $J_n^{'}$ (the derivative of the Bessel function J_n).

In dimension three, for the ball B_R, the eigenvalues and eigenfunctions of the Dirichlet-Laplacian are given by $\lambda_{n,k} = j^2_{n+\frac{1}{2},k}/R^2$, $n \in \mathbb{N}, k \in \mathbb{N}$ which is of multiplicity $2n+1$ and is associated to the eigenfunctions

$$v_{n,k}(r,\theta,\phi) = \begin{cases}
\dfrac{J_{n+\frac{1}{2}}\left(\frac{j_{n+\frac{1}{2},k}}{R} r\right)}{\sqrt{r}} P_n^0(\cos\theta), \\
\dfrac{J_{n+\frac{1}{2}}\left(\frac{j_{n+\frac{1}{2},k}}{R} r\right)}{\sqrt{r}} P_n^1(\cos\theta)\cos\phi, \\
\dfrac{J_{n+\frac{1}{2}}\left(\frac{j_{n+\frac{1}{2},k}}{R} r\right)}{\sqrt{r}} P_n^1(\cos\theta)\sin\phi, \\
\vdots \\
\dfrac{J_{n+\frac{1}{2}}\left(\frac{j_{n+\frac{1}{2},k}}{R} r\right)}{\sqrt{r}} P_n^n(\cos\theta)\cos(n\phi), \\
\dfrac{J_{n+\frac{1}{2}}\left(\frac{j_{n+\frac{1}{2},k}}{R} r\right)}{\sqrt{r}} P_n^n(\cos\theta)\sin(n\phi)
\end{cases} \tag{1.12}$$

where P_n^q denote the associated Legendre polynomial, see [4]. Similar formulae hold for the Neumann eigenvalues and eigenfunctions where the eigenvalues are the roots of some transcendental equations involving Bessel functions. In higher dimension d, the eigenvalues of the ball B_R still involve the zeros of the Bessel functions $J_{d/2-1}, J_{d/2}, \ldots$. For example

$$\lambda_1(B_R) = \frac{j^2_{d/2-1,1}}{R^2} \qquad \lambda_2(B_R) = \lambda_3(B_R) = \ldots = \lambda_{N+1}(B_R) = \frac{j^2_{d/2,1}}{R^2} \tag{1.13}$$

while the eigenfunctions combine Bessel functions for the radial part and spherical harmonics for the angular part, see [292]

1.4 Notation

Ω is an open set (or a quasi-open set, see Chapter 2) in $\mathbb{R}^d$. We will denote by $H^1(\Omega)$ the classical Sobolev space:

$$H^1(\Omega) = \left\{ u \in L^2(\Omega), \text{ such that } \frac{\partial u}{\partial x_i} \in L^2(\Omega), i = 1, \ldots, d \right\}$$

and $H_0^1(\Omega)$ is defined as the closure in $H^1(\Omega)$ of C^∞ functions with compact support in Ω.

Eigenvalues and eigenfunctions of the Laplace-Dirichlet operator. We will denote by $\lambda_k(\Omega)$, $k \geq 1$ (or more simply λ_k when the context makes the domain clear) the k-th eigenvalue of the Laplacian with Dirichlet boundary conditions, counted with multiplicity:

$$\begin{cases} -\Delta u &= \lambda u \quad \text{in } \Omega, \\ u &= 0 \quad \text{on } \partial\Omega. \end{cases} \tag{1.14}$$

The corresponding eigenfunction is usually normalized by $\int_\Omega u^2\,dx = 1$.

Eigenvalues and eigenfunctions of the Laplace-Neumann operator. We will denote by $\mu_k(\Omega)$, $k \geq 1$ (or more simply μ_k when the context makes the domain clear) the k-th eigenvalue of the Laplacian with Neumann boundary conditions, counted with multiplicity:

$$\begin{cases} -\Delta u &= \mu u \quad \text{in } \Omega, \\ \frac{\partial u}{\partial n} &= 0 \quad \text{on } \partial\Omega. \end{cases} \tag{1.15}$$

Therefore, by convention, $\mu_1(\Omega) = 0$. The corresponding eigenfunction is usually normalized by $\int_\Omega u^2\,dx = 1$.

Eigenvalues and eigenfunctions of the Laplace-Robin operator. Let α a real number, we will denote by $\lambda_k(\Omega, \alpha)$ (or more simply $\lambda_k(\alpha)$ or λ_k when no confusion can occur) the k-th eigenvalue of the Laplacian with Robin boundary conditions, counted with multiplicity:

$$\begin{cases} -\Delta u &= \lambda u \quad \text{in } \Omega, \\ \frac{\partial u}{\partial n} + \alpha u &= 0 \quad \text{on } \partial\Omega. \end{cases} \tag{1.16}$$

The corresponding eigenfunction is usually normalized by $\int_\Omega u^2\,dx = 1$.

Eigenvalues and eigenfunctions of the Laplace-Steklov operator. We will denote by $\sigma_k(\Omega)$ (or more simply σ_k when the context makes the domain clear) the k-th eigenvalue of the Laplacian with Steklov boundary conditions, counted with multiplicity:

$$\begin{cases} \Delta u &= 0 \quad \text{in } \Omega, \\ \frac{\partial u}{\partial n} &= \sigma u \quad \text{on } \partial\Omega. \end{cases} \tag{1.17}$$

Therefore, by convention, $\sigma_1(\Omega) = 0$. Here Ω can be a compact Riemannian manifold of dimension $n \geq 2$ and in that case Δ is the Laplace-Beltrami operator. The corresponding eigenfunction can be normalized either by $\int_\Omega u^2\,dx = 1$ or more frequently by $\int_{\partial\Omega} u^2\,d\sigma = 1$.

Dorin Bucur

2 Existence results

2.1 Setting the problem

In this chapter, we denote by d the dimension of the space. Let $d \geq 2$ and $\Omega \subseteq \mathbb{R}^d$ be an open set of finite measure. Then the spectrum of the Laplace operator with Dirichlet boundary conditions of Ω consists only on eigenvalues which can be ordered (counting the multiplicity)

$$0 < \lambda_1(\Omega) \leq \lambda_2(\Omega) \leq \cdots \to +\infty.$$

For every $k \in \mathbb{N}$ there exist non-zero functions u (eigenfunctions) that saisfy the equation

$$\begin{cases} -\Delta u &= \lambda_k(\Omega)u \quad \text{in } \Omega \\ u &= 0 \qquad\qquad \text{on } \partial\Omega. \end{cases}$$

If Ω is smooth, then the function $u \in C^2(\Omega) \cap C(\overline{\Omega})$ satisfies the equation in a classical sense. If Ω is just an open set without any regularity, then u satisfies the equation in the following weak sense

$$u \in H_0^1(\Omega), \quad \forall v \in H_0^1(\Omega), \int_\Omega \nabla u \nabla v dx = \lambda_k(\Omega) \int_\Omega uv dx.$$

Let $c > 0$, $k \in \mathbb{N}$ be given, and $F : \mathbb{R}^k \to \mathbb{R}$. The generic spectral optimization problems we discuss in this chapter is

$$\min \left\{ F\big(\lambda_1(\Omega), .., \lambda_k(\Omega)\big) : \Omega \subset \mathbb{R}^d, \Omega \text{ open}, \ |\Omega| = c \right\}. \tag{2.1}$$

Some particular problems have been studied intensively in the last century. We refer the reader to [505] for a recent survey of the topic. Here is a short list of results.

- The Faber-Krahn inequality asserts that the solution of

$$\min \left\{ \lambda_1(\Omega) : \Omega \subset \mathbb{R}^d, \Omega \text{ open}, |\Omega| = c \right\} \tag{2.2}$$

  is the ball of volume c.
- The solution of

$$\min \left\{ \lambda_2(\Omega) : \Omega \subset \mathbb{R}^d, \Omega \text{ open}, |\Omega| = c \right\} \tag{2.3}$$

  consists of two equal and disjoint balls of volume $\frac{c}{2}$ (Krahn-Szegő).

Dorin Bucur: Institut Universitaire de France, Laboratoire de Mathématiques, CNRS UMR 5127, Université de Savoie, Campus Scientifique, 73376 Le-Bourget-Du-Lac, France, E-mail: dorin.bucur@univ-savoie.fr

- Ashbaugh and Benguria proved in [64] that the solution of

$$\max\left\{\frac{\lambda_2(\Omega)}{\lambda_1(\Omega)} : \Omega \subset \mathbb{R}^d, \Omega \text{ open and of finite measure}\right\} \tag{2.4}$$

  is the ball.

An intriguing question is to find the solution of

$$\min\left\{\lambda_k(\Omega) : \Omega \subset \mathbb{R}^d, \Omega \text{ open}, |\Omega| = c\right\} \tag{2.5}$$

for every $k \in \mathbb{N}$. Unfortunately, starting with $k \geq 3$ very few answers are available. In two dimensions, for $k = 3$ it is conjectured that the minimizer is the disc, while in dimension 3 it has been observed numerically by Oudet that the minimizer is not the ball, cf Chapter 11.

Wolf and Keller proved that for $k = 13$, in $\mathbb{R}^2$, the minimizer is not a union of discs, but again Oudet [737] numerically observed that for $k = 5$ to 15 the minimizer is not the disc. In [137], it was rigorously proved by Berger, that for any $k \geq 5$ in $\mathbb{R}^2$, the ball cannot be minimizer.

Several computations were carried out ([42, 737]) providing evidence that the optimal shapes are close to those presented in Chapter 11.

At this point, when no analytical solutions of those problems can be expected, the question is of a qualitative nature. One would like to prove that problem (2.1), or more precisely problem (2.5), has a solution and to gather some information about it. *Does the optimal set have finite perimeter ? Is it bounded ? Is its boundary smooth ? Does it have any symmetry ? Is it convex ? Is it the ball ?*

Problem (2.1) may have or not a solution (for general functions F, the complete answer is not known), but a negative answer may have at least three meanings:

- a solution of problem (2.1) does not exist (i.e. in the class of open sets), but there exists a solution provided the family of open sets is enlarged to a class of Borel subsets of $\mathbb{R}^d$ where the eigenvalue problem is still well posed (i.e. the family of *quasi-open* sets, see Section 2.2 below). This issue is very similar to a "classical" existence result, only that the solution is a *quasi-open* set. In fact the class of quasi-open sets is the largest class of Borel subsets of $\mathbb{R}^d$, where the Dirichlet Laplacian is well defined and inherits a strong maximum principle.
- a solution of problem (2.1) does not exist, even if the class of sets is enlarged, but there exists a solution in a larger class of *relaxed objects* where the eigenvalue problem is well posed. This class consists of *positive Borel measures*, absolutely continuous with respect to capacity (see Section 2.2 below for definition and properties of capacity). Roughly speaking, those measures are limits of sequences of open sets in some suitable sense, and account for the asymptotic behavior of the oscillating boundaries in the sense of capacities (see Remark 2.5 below).
- a solution of problem (2.1) does not exist, in the sense that the infimum is not attained by any "geometrical" object.

In this chapter, we shall analyze the existence of a solution for problem (2.1), and we shall prove that it indeed exists (in the enlarged class of quasi-open sets), provided some assumptions are satisfied by the functional F. Of course, one expects to have smooth open sets as minimizers, at least for problem (2.5), but this has not yet been proved in general. The question of proving existence in the class of open sets is in fact a regularity problem for which we refer the reader to Chapter 3. Some qualitative properties will be proved in this chapter, e.g. the boundedness of the optimal sets and the fact that they have a finite perimeter, since they play a crucial role in the existence question.

2.2 The spectrum of the Dirichlet-Laplacian on open and quasi-open sets

Since the existence question requires us to work in a class of sets larger than the class of open sets, in this section we recall basic facts about capacity and quasi-open sets. We also list some properties of the eigenvalues of the Dirichlet-Laplacian and of the eigenfunctions.

Capacity and quasi-open sets. Let $E \subseteq \mathbb{R}^d$. The *capacity* of E is defined by

$$cap(E) = \inf\left\{ \int |\nabla u|^2 + |u|^2 dx, \quad u \in \mathcal{U}_E \right\}$$

where $\mathcal{U}_E$ is the class of all functions $u \in H^1(\mathbb{R}^d)$ such that $u \geq 1$ almost everywhere (shortly a.e.) in an open neighborhood of E.

A property $p(x)$ is said to hold *quasi everywhere* on E (shortly q.e. on E) if the set of all points $x \in E$ for which the property $p(x)$ does not hold has capacity zero.

A set $\Omega \subseteq \mathbb{R}^d$ is called *quasi-open* if for every $\epsilon > 0$ there exists an open set U_ϵ such that $\Omega \cup U_\epsilon$ is open and $cap(U_\epsilon) < \epsilon$. Clearly, every open set is quasi-open. A function $u : \mathbb{R}^d \mapsto \mathbb{R}$ is said to be *quasi-continuous* if for all $\epsilon > 0$ there exists an open set U_ϵ with $cap(U_\epsilon) < \epsilon$ such that $u_{|U_\epsilon^c}$ is continuous (see [472]).

Every function $u \in H^1(\mathbb{R}^d)$ has a quasi-continuous representative, $\tilde{u}$, such that $\tilde{u}(x) = u(x)$ a.e. This representative is unique up to a set of zero capacity and can be computed by

$$\tilde{u}(x) = \lim_{r\to 0} \frac{\int_{B_r(x)} u(y)dy}{|B_r(x)|}, \quad \text{q.e. } x \in \mathbb{R}^d.$$

The limit above exists quasi everywhere. In particular, the level set $\{\tilde{u} > 0\}$ is a quasi-open set. From now on, every time we speak about the pointwise behavior of a Sobolev function, we refer to a quasi-continuous representative.

The Sobolev spaces. If $\Omega \subseteq \mathbb{R}^d$ is an open set, the Sobolev space $H_0^1(\Omega)$ is defined as $cl_{H^1(\mathbb{R}^d)} C_0^\infty(\Omega)$, the closure of the space of C^∞ functions with compact support in Ω, in the H^1-norm. For a quasi-open set $\Omega \subseteq \mathbb{R}^d$, the Sobolev space $H_0^1(\Omega)$ is defined as a subspace of $H^1(\mathbb{R}^d)$ by:

$$H_0^1(\Omega) = \{u \in H^1(\mathbb{R}^d) \ : \quad u = 0 \quad \text{q.e. on} \quad \mathbb{R}^d \setminus \Omega\}.$$

If Ω is open, the space $H_0^1(\Omega)$ defined above coincides with the usual Sobolev space (see [472]).

From this perspective, for every open or quasi-open set, $H_0^1(\Omega)$ is a subspace of $H^1(\mathbb{R}^d)$, as long as every function of $H_0^1(\Omega)$ is understood as being extended by 0 on $\mathbb{R}^d \setminus \Omega$.

If $\Omega \subseteq \mathbb{R}^d$ is a quasi-open set of finite measure (not necessarily bounded), the injection $H_0^1(\Omega) \hookrightarrow L^2(\Omega)$ is compact.

The spectrum of the Dirichlet-Laplacian. For every quasi-open set of finite measure $\Omega \subseteq \mathbb{R}^d$, we introduce the resolvent operator $R_\Omega : L^2(\mathbb{R}^d) \to L^2(\mathbb{R}^d)$, by $R_\Omega(f) = u$, where u solves the equation

$$\begin{cases} -\Delta u &= f \text{ in } \Omega \\ \quad u &\in H_0^1(\Omega) \end{cases}$$

in the weak sense

$$\forall \phi \in H_0^1(\Omega) \quad \int_\Omega \nabla u \nabla \phi dx = \int_\Omega f\phi dx. \tag{2.6}$$

R_Ω is a compact, self-adjoint, positive operator having a sequence of eigenvalues converging to 0. The inverses of its eigenvalues are the eigenvalues of the Dirichlet-Laplacian on Ω and are denoted (multiplicity being counted) by

$$0 < \lambda_1(\Omega) \le \lambda_2(\Omega) \le \cdots \le \lambda_k(\Omega) \le \cdots \to +\infty.$$

These values can be defined by the min-max formula

$$\lambda_k(\Omega) = \min_{S \in \mathcal{S}_k} \max_{u \in S \setminus \{0\}} \frac{\int_\Omega |\nabla u|^2 dx}{\int_\Omega u^2 dx}, \tag{2.7}$$

where $\mathcal{S}_k$ stands for the family of all subspaces of dimension k in $H_0^1(\Omega)$. A function $u \in H_0^1(\Omega)$ for which equality holds, is called an eigenfunction and satisfies the equation

$$\begin{cases} -\Delta u &= \lambda_k(\Omega) u & \text{in } \Omega \\ \quad u &= 0 & \text{on } \partial\Omega \end{cases}$$

in the weak sense

$$\forall \phi \in H_0^1(\Omega) \quad \int_\Omega \nabla u \nabla \phi dx = \lambda_k(\Omega) \int_\Omega u\phi dx.$$

The system of L^2-normalized eigenfunctions is a Hilbert basis of $H_0^1(\Omega)$.

We list below some properties of the eigenvalues. Let $\Omega, \Omega_1, \Omega_2 \subseteq \mathbb{R}^d$ be quasi-open sets of finite measure.

- (Rescaling) $\forall t > 0, \forall k \in \mathbb{N}, \lambda_k(t\Omega) = \frac{1}{t^2}\lambda_k(\Omega)$.
- (Spectrum of the union) If Ω_1, Ω_2 are disjoint, then the eigenvalues of $\Omega_1 \cup \Omega_2$ are the union of the sets of eigenvalues of Ω_1 and Ω_2 with multiplicities being counted.
- (Monotonicity) Assume that $\Omega_1 \subseteq \Omega_2$. Then

$$\forall k \in \mathbb{N}, \lambda_k(\Omega_2) \leq \lambda_k(\Omega_1).$$

- (Control of the variation)

$$\left|\frac{1}{\lambda_k(\Omega_1)} - \frac{1}{\lambda_k(\Omega_2)}\right| \leq \|R_{\Omega_1} - R_{\Omega_2}\|_{\mathcal{L}(L^2(\mathbb{R}^d))}.$$

Moreover, if $\Omega_1 \subseteq \Omega_2$ (see Bucur [206]), for every $k \in \mathbb{N}$

$$\left|\frac{1}{\lambda_k(\Omega_1)} - \frac{1}{\lambda_k(\Omega_2)}\right| \leq 4k^2 e^{1/4\pi}\lambda_k(\Omega_2)^{d/2}(E(\Omega_1) - E(\Omega_2)). \tag{2.8}$$

where

$$E(\Omega) := \min_{u\in H_0^1(\Omega)} \frac{1}{2}\int_{\mathbb{R}^d} |\nabla u|^2 dx - \int_{\mathbb{R}^d} u dx. \tag{2.9}$$

is the torsion energy. The unique function which minimizes $E(\Omega)$ is called the torsion function and is denoted w_Ω and satisfies in a weak sense

$$-\Delta w_\Omega = 1 \quad \text{in } \Omega, \qquad w_\Omega \in H_0^1(\Omega).$$

The torsion function plays a key role in understanding the behavior of the spectrum of the Dirichlet Laplacian for small geometric domain perturbations. Assume that $\lambda_k(\Omega_2) = K$ and $\Omega_1 \subseteq \Omega_2$ such that

$$E(\Omega_1) - E(\Omega_2) = \frac{1}{2}\int_{\Omega_2} (w_{\Omega_2} - w_{\Omega_1})dx \leq \frac{1}{8k^2 e^{1/4\pi} K^{d/2+1}}. \tag{2.10}$$

Then we get a control on the magnitude of $\lambda_k(\Omega_1)$. Precisely, if (2.10) holds, we get

$$\lambda_k(\Omega_1) \leq 2\lambda_k(\Omega_2). \tag{2.11}$$

As $\int_{\Omega_2} (w_{\Omega_2} - w_{\Omega_1})dx$ becomes smaller, the eigenvalues $\lambda_k(\Omega_1)$ and $\lambda_k(\Omega_2)$ become closer.
Roughly speaking, this property asserts that the variation of the eigenvalues for inner perturbations of a quasi-open set is controlled by the variation of the L^1-norm of the torsion functions (see Section 2.5).
- (Control by the torsion function, see Van den Berg [860])

$$\frac{1}{\lambda_1(\Omega)} \leq \|w_\Omega\|_\infty \leq \frac{4 + 3d\log 2}{\lambda_1(\Omega)}. \tag{2.12}$$

- (Ratio of eigenvalues) For all $k \in \mathbb{N}$ there exists a constant M_k, depending *only* on k and the dimension d, such that (see for instance [57])

$$1 \leq \frac{\lambda_k(\Omega)}{\lambda_1(\Omega)} \leq M_k. \tag{2.13}$$

- (L^∞-bound of the eigenfunctions) If u_k is an L^2-normalized eigenfunction of $\lambda_k(\Omega)$, then

$$\|u_k\|_\infty \leq C_d \lambda_k(\Omega)^{\frac{d}{4}}. \tag{2.14}$$

2.3 Existence results: bounded design region

Let $D \subseteq \mathbb{R}^d$ be a bounded open set. In this section we shall prove an existence result for a local version of problem (2.1), i.e.

$$\min\left\{F\big(\lambda_1(\Omega), .., \lambda_k(\Omega)\big) : \Omega \subset D, \Omega \text{ quasi-open}, |\Omega| = c\right\} \tag{2.15}$$

Theorem 2.1. (Buttazzo-Dal Maso) *Let $F : \mathbb{R}^k \to \mathbb{R}$ be non-decreasing in each variable and lower semicontinuous. Then problem (2.15) has at least one solution.*

The first proof of this theorem, given by Buttazzo and Dal Maso in [238], involved quite technical results describing the so called relaxation phenomenon and covered a more general situation than just a functional depending on the eigenvalues. We give below a direct proof, which does not require the knowledge of the relaxed problem or properties of the weak gamma convergence (see Remark 2.6), since we are concerned only with functionals depending on eigenvalues.

A series of remarks at the end of the proof will explain the necessity of the monotonicity hypothesis on F and the boundedness of the design region D.

Proof. Assume that $(\Omega_n)_n$ is a minimizing sequence for problem (2.15) and that $u_n^1, \dots, u_n^k$ are L^2-normalized eigenfunctions corresponding to $\lambda_1(\Omega_n), \dots, \lambda_k(\Omega_n)$, two by two orthogonal in $L^2(D)$. We can assume that $(\lambda_k(\Omega_n))_n$ is a bounded sequence, otherwise existence occurs trivially as a consequence of the monotonicity of F (from some rank on, F will be constant on the sets Ω_n). After extracting a subsequence we can assume that $(u_n^i)_n$ converges weakly in $H_0^1(D)$, strongly in $L^2(D)$ and pointwise a.e. to a function $u^i \in H_0^1(D)$, for $i = 1, \dots, k$. We can also assume that u^i are quasi-continuous, so that defining

$$\Omega := \bigcup_{i=1}^k \{u^i \neq 0\},$$

we built a quasi-open set $\Omega \subseteq D$ such that $u^i \in H_0^1(\Omega)$. From the pointwise a.e. convergence, we get

$$1_{\{u^i \neq 0\}} \leq \liminf_{n\to\infty} 1_{\Omega_n} \text{ a.e.}$$

so that $1_\Omega \le \liminf_{n\to\infty} 1_{\Omega_n}$ a.e. This implies $|\Omega| \le c$.

On the other hand

$$\lambda_i(\Omega) \le \liminf_{n\to\infty} \lambda_i(\Omega_n), \forall i = 1, \dots, k.$$

Indeed, this is a consequence of the definition of the eigenvalues. Let us denote $S_i = span\{u^1, \dots, u^i\}$. We have, for some $(\alpha_j)_{j=1}^i \in \mathbb{R}^i$

$$\lambda_i(\Omega) \le \max_{u\in S_i} \frac{\int_\Omega |\nabla u|^2 dx}{\int_\Omega |u|^2 dx} = \frac{\int_\Omega |\sum_{j=1}^i \alpha_j \nabla u^j|^2 dx}{\int_\Omega |\sum_{j=1}^i \alpha_j u^j|^2 dx}.$$

Hence

$$\lambda_i(\Omega) \le \liminf_{n\to\infty} \frac{\int_{\Omega_n} |\sum_{j=1}^i \alpha_j \nabla u_n^j|^2 dx}{\int_{\Omega_n} |\sum_{j=1}^i \alpha_j u_n^j|^2 dx} \le \liminf_{n\to\infty} \lambda_i(\Omega_n),$$

the last inequality being a consequence of the min-max formula on Ω_n.

Using the lower semincontinuity and the monotonicity of F, this gives

$$F(\lambda_1(\Omega), \dots, \lambda_k(\Omega)) \le \liminf_{n\to\infty} F(\lambda_1(\Omega_n), \dots, \lambda_k(\Omega_n)).$$

If $|\Omega| < c$, then adding an open set U to Ω such that $U \subseteq D$ and $|\Omega \cup U| = c$, we get a solution for (2.15). This is again a consequence of the monotonicity of F and of the eigenvalues on inclusions of sets. □

Remark 2.2. The monotonicity of the functional F is crucial. From a technical point of view, this hypothesis is used above in the construction of the optimal set: the limit set Ω may have a measure strictly lower than c, and so by enlarging it, we get the solution. When enlarging the set Ω, the eigenvalues do not increase! The functional F is tailored to behave well during this procedure. Nevertheless, there are situations in which existence of a solution holds for functionals F that are not required to satisfy this property. This is the case of functionals depending only on $\lambda_1(\Omega)$ and $\lambda_2(\Omega)$ (see [207, Section 6.4]). It is not known whether a general existence result as Theorem 2.1 may hold if the monotonicity assumption on F is dropped and replaced by a weaker hypothesis.

Remark 2.3. The boundedness of the design region D plays an important role for compactness, in the construction of the optimal set Ω. The *only* fact we used in the proof was that $H_0^1(D)$ is compactly embedded in $L^2(D)$, so Theorem 2.1 holds with this hypothesis, instead the stricter hypothesis D bounded. Nevertheless, if D is not bounded, the existence of a solution may fail. For example, in $\mathbb{R}^2$ we consider

$$D = \bigcup_{i=3}^{+\infty} B_{\frac{1}{2}-\frac{1}{i}}(i, 0), \quad c = \frac{\pi}{4}, \quad F(\Omega) = \lambda_1(\Omega),$$

where $B_r(x)$ is the ball centered at x of radius r.

Then the infimum of the functional is $\lambda_1(B_{\frac{1}{2}})$, which is not attained.

For general unbounded design regions D, Theorem 2.1 may not apply. Nevertheless, in the particular, and very important, case $D = \mathbb{R}^d$, the existence result holds. This issue is discussed in the next section.

Remark 2.4. Depending on F, the optimal set Ω may satisfy some regularity properties. We refer to Theorem 2.10 in the next section and to Chapter 3.

Remark 2.5. Given an arbitrary sequence of quasi-open subsets of D, the full behavior of the spectrum for at least one subsequence is completely understood. In fact, it was proved (see Dal Maso and Mosco [314]), that there exists a subsequence (still denoted using the same index) and a positive Borel measure μ, absolutely continuous with respect to the capacity, such that the sequence of resolvent operators $R_{\Omega_n} : L^2(D) \to L^2(D)$ converges in the operator norm on $L^2(D)$ to R_μ, defined by $R_\mu(f) = u_{\mu,f}$

$$\begin{cases} -\Delta u_{\mu,f} + \mu u_{\mu,f} & = f \text{ in } D \\ u_{\mu,f} & \in H_0^1(D) \cap L^2(D,\mu) \end{cases}$$

in the sense

$$\forall \phi \in H_0^1(D) \cap L^2(D,\mu) \quad \int_D \nabla u_{\mu,f} \nabla \phi dx + \int_D u_{\mu,f} \phi d\mu = \int_D f \phi dx. \tag{2.16}$$

The function $u_{\mu,f}$ is also the unique minimizer in $H_0^1(D) \cap L^2(D,\mu)$ of the functional

$$u \mapsto \frac{1}{2} \int_D |\nabla u|^2 dx + \frac{1}{2} \int_D u^2 d\mu - \int_D f u dx.$$

Then, $\lambda_k(\mu)$ are the inverses of the eigenvalues of the positive, self-adjoint and compact operator R_μ and are defined via the min-max formula

$$\lambda_k(\mu) = \min_{S \in \mathcal{S}_k} \max_{u \in S \setminus \{0\}} \frac{\int_D |\nabla u|^2 dx + \int_D u^2 d\mu}{\int_D u^2 dx}, \tag{2.17}$$

where $\mathcal{S}_k$ stands for the family of all subspaces of dimension k in $H_0^1(D) \cap L^2(D,\mu)$.

As a consequence of the convergence of the resolvent operators, for every $k \in \mathbb{N}$, $\lambda_k(\Omega_n) \to \lambda_k(\mu)$. Of course, from the point of view of the existence theorem 2.1, the measure μ does not provide a solution. The original proof of Buttazzo-Dal Maso (done for a larger class of functionals) consisted in replacing the measure μ, obtained from a minimizing sequence, by a quasi-open set which is, roughly speaking, the union of all sets of μ-finite measure. This set was proved to be optimal, thanks to the monotonicity of F.

Remark 2.6. There is a second proof of Theorem 2.1, given in [210], which is based on the so called *weak gamma convergence*. It is said that Ω_n weakly gamma converges to Ω if $(w_{\Omega_n})_n$ converges weakly in $H_0^1(D)$ to some function w, and $\Omega = \{w > 0\}$. This convergence is compact in the family of quasi-open subsets of D, and the Lebesgue measure is lower semicontinuous. The key consequence of this convergence is the following property.

– *Assume Ω_n weakly gamma converges to Ω. For all sequences $(u_{n_k})_k$, such that $u_{n_k} \in H_0^1(\Omega_{n_k})$ and u_{n_k} converges weakly in $H_0^1(D)$ to some function u, then u has to belong to $H_0^1(\Omega)$.*

Following the proof of Theorem 2.1, this property implies immediately that

$$\forall k \in \mathbb{N}, \lambda_k(\Omega) \le \liminf_{n\to\infty} \lambda_k(\Omega_n).$$

As a consequence, existence in Theorem 2.1 comes by a compactness-semicontinuity argument.

Remark 2.7. More importantly, the argument of the previous remark works for every $k \in \mathbb{N}$, since the set Ω is built from the torsion functions and not from the first eigenfunctions. As a consequence, the existence result can be extended to functionals depending on the full spectrum. So, if $F : \mathbb{R}^{\mathbb{N}} \to \mathbb{R}$ is lower semicontinuous in a suitable sense and non decreasing in each variable, then the existence result proved in Theorem 2.1 could apply. An example would be

$$\Omega \mapsto \left(\sum_{k=1}^{\infty} e^{-\lambda_k(\Omega)}\right)^{-1}.$$

2.4 Global existence results: the design region is $\mathbb{R}^d$

In this section we deal with the problem

$$\min\left\{F\big(\lambda_1(\Omega), \dots, \lambda_k(\Omega)\big) : \Omega \text{ quasi-open}, \Omega \subseteq \mathbb{R}^d, |\Omega| = c\right\} \tag{2.18}$$

The passage from a bounded design region D to $\mathbb{R}^d$ is not trivial. In fact, the compact embedding of $H_0^1(D)$ in $L^2(D)$, which played a crucial role in the proof, fails to be true in $\mathbb{R}^d$. For example, in the proof of Theorem 2.1, if $D = \mathbb{R}^d$, the limit functions u^i may all be equal to zero. This occurs, for instance, if the sets Ω_n have distances to the origin tending towards $+\infty$.

From this perspective, a first attempt to solve the existence problem in $\mathbb{R}^d$ was done in [219] and was based on the concentration compactness principle of Pierre-Louis Lions. The following result holds (see [204, 207]).

Theorem 2.8. **(Bucur)** *Let $(\Omega_n)_n$ be a sequence of quasi-open sets of $\mathbb{R}^d$ of measure equal to c. One of the following situations holds.*

Compactness: *There exists a subsequence* $(\Omega_{n_k})_{k\in\mathbb{N}}$*, a sequence of vectors* $(y_k)_{k\in\mathbb{N}} \subseteq \mathbb{R}^d$ *and a positive Borel measure* μ*, vanishing on sets of zero capacity, such that*

$$\|R_{y_k+\Omega_{n_k}} - R_\mu\|_{\mathcal{L}(L^2(\mathbb{R}^d))} \to 0 \text{ as } k \to +\infty$$

and $\bigcup_{k\in\mathbb{N}} H_0^1(\Omega_{n_k})$ *is collectively compactly embedded in* $L^2(\mathbb{R}^d)$.
Dichotomy: *There exists a subsequence* $(\Omega_{n_k})_{k\in\mathbb{N}}$ *and a sequence of subsets* $\tilde{\Omega}_k \subseteq \Omega_{n_k}$*, such that*

$$\|R_{\Omega_{n_k}} - R_{\tilde{\Omega}_k}\|_{\mathcal{L}(L^2(\mathbb{R}^d))} \to 0, \quad and \quad \tilde{\Omega}_k = \Omega_k^1 \cup \Omega_k^2$$

with $d(\Omega_k^1, \Omega_k^2) \to \infty$ *and* $\liminf_{n\to\infty} |\Omega_k^i| > 0$ *for* $i = 1, 2$.

In [219], this result was used to prove the existence of a minimizer for λ_3. Following Theorem 2.8, a minimizing sequence $(\Omega_n)_n$ can be either in the compactness case, in which we get existence, or in the dichotomy case, in which case, the minimizing sequence can be chosen such that it consists of disconnected sets. In this situation, the problem is reduced to finding the minimizers of λ_1 and λ_2 which are known.

In order to use this argument to prove the existence of a minimizer for λ_4, it would be enough to prove that the minimizer for λ_3 is, for instance, bounded. In this case, the dichotomy would lead to a combination of a minimizer of λ_3 and a ball. If the minimizer of λ_3 (that we know its existence) was not bounded, then the union of the minimizer and a ball may always have a non-trivial intersection. This would be the case if the minimizer of λ_3 was a dense set in $\mathbb{R}^d$. Of course, this situation is not expected, but to exclude it one has to understand some qualitative properties of the minimizers.

The global existence result in $\mathbb{R}^d$, was proved independently in [206] and [700], by completely different methods. In [206], the notion of *shape subsolution* for the torsion energy (see the next section) was introduced, and it was proved that every such subsolution has to be a bounded set with finite perimeter. A second argument showed that minimizers for (2.18) are shape subsolutions, so they are bounded, hence the concentration-compactness theorem 2.8 can be used.

The proof in [700] used a surgery result which states that from every set with a diameter large enough, some parts can be cut out such that after small modifications and rescaling, the new set has a diameter below some treshold and not larger (low) eigenvalues. In this way, replacing the minimizing sequence, the existence problem in $\mathbb{R}^d$ was reduced to the local case of Buttazzo and Dal Maso.

In the next section, we shall give the main ideas of the proof of the global existence result, using a combination of the two methods: shape subsolutions and surgery. The following result was proved in [221], as an extension of the surgery result of [700] and using the subsolution method of [206].

Lemma 2.9. (surgery) *For every* $K, c > 0$*, there exists* $D, C > 0$ *depending only on* K, c *and the dimension* d *such that for every quasi-open set* $\Omega \subset \mathbb{R}^d$ *with* $|\Omega| = c$ *there exists a quasi-open set* $\tilde{\Omega}$ *with* $|\tilde{\Omega}| = c$*,* $\text{diam}\,(\tilde{\Omega}) \le D$*,* $Per(\tilde{\Omega}) \le C$ *and, if for some* $k \in \mathbb{N}$ *it* *holds* $\lambda_k(\Omega) \le K$, *then*

$$\forall 1 \le i \le k, \quad \lambda_i(\tilde{\Omega}) \le \lambda_i(\Omega). \tag{2.19}$$

Moreover, if $Per(\Omega) > C$ *the inequalities (2.19) are strict.*

We shall give the lines of the proof of this lemma in the next section. Here is the main consequence.

Theorem 2.10. (Bucur, Mazzoleni, Pratelli) *Let* $F : \mathbb{R}^k \to \mathbb{R}$ *be non-decreasing in each variable and lower semicontinuous. Then problem (2.18) has at least one solution. If F is strictly increasing in at least one variable, then every solution of (2.18) is a bounded set with finite perimeter.*

Proof. Let $(\Omega_n)_n$ be a minimizing sequence for (2.18). One can choose Ω_n such that $\lambda_k(\Omega_n) \le K$ for all k and some suitable value K. Otherwise, the minimum of F would be formally achieved at $(+\infty, \dots, +\infty)$, which from the monotonicity assumption implies that F has to be constant. As a consequence, we can use the surgery Lemma 2.10 and find a new sequence $(\tilde{\Omega}_n)$ which has a uniformly bounded diameter and satisfies the measure constraint. Since $\forall 1 \le i \le k$, $\lambda_i(\tilde{\Omega}) \le \lambda_i(\Omega)$, the monotonicity of F implies that this new sequence is also minimizing. At this point, we can use the Buttazzo-Dal Maso Theorem 2.1, up to possible translations of all $\tilde{\Omega}$ in the ball centered at the origin of radius D, and get the existence of an optimal set.

Assume now that Ω is optimal and F is strictly increasing in at least one variable. If $Per(\Omega) > C$, then we have that $\tilde{\Omega}$ is a new minimizer with the first k eigenvalues strictly smaller than those in Ω. Since F is strictly increasing, we are in contradiction with the optimality of Ω.

□

2.5 Subsolutions for the torsion energy

In order to explain the main lines of the proof of Lemma 2.9, we recall the notion of *shape subsolutions* introduced in [206], which allows us to replace the study of a general spectral functional with the study of the torsion energy. Roughly speaking, if a shape is optimal for a general spectral functional, it may satisfy some sub-optimality conditions for the torsion energy. From this last condition, one could deduce interesting qualitative properties on the optimal shape, like information on the perimeter and outer density, which is related to the boundedness. The key argument is that the variation of an eigenvalue for an arbitrary geometric perturbation of the domain can be controlled by the variation of the torsion energy, for the same perturbation (see inequality (2.8)).

Definition 2.11. *We say that a quasi-open set $\Omega \subset \mathbb{R}^d$ is a local shape subsolution for the torsion energy if there exists $\eta, \delta > 0$ such that for all quasi-open sets $A \subseteq \Omega$ with the property that $\int (w_\Omega - w_A)dx < \delta$ we have*

$$E(\Omega) + \eta|\Omega| \le E(A) + \eta|A|.$$

The main result proved by Bucur in [206] is the following.

Theorem 2.12. *Assume Ω is a local shape subsolution for the torsion energy. Then Ω is bounded and has finite perimeter.*

In the result above, the diameter is understood as the maximal length of the orthogonal projection of Ω on lines. Both the diameter and the perimeter depend only on $|\Omega|, \eta, \delta, d$.

Proof. (of Theorem 2.12) The proof of the boundedness is a consequence of the following Lemma, inspired from the seminal paper of Alt and Caffarelli [25].

Lemma 2.13. *Assume Ω is a local shape subsolution for the torsion energy. There exists $r_0 > 0$, $C_0 > 0$ such that for every $x_0 \in \mathbb{R}^d$ and $r \in (0, r_0)$*

$$\text{if } \sup_{x \in B_r(x_0)} w_\Omega(x) \le C_0 r \text{ then } w_\Omega = 0 \text{ on } B_{\frac{1}{2}r}(x_0). \tag{2.20}$$

The proof of this lemma is classical. We refer the reader to [25], or to [206], for the specific situation of the torsion function.

In order to gather information on the boundedness of a shape subsolution, one observes that for every $\theta > 0$, there exists $\delta_0 > 0$ depending only on N, θ such that if $w_\Omega(x_0) \ge \theta$ for some $x_0 \in \mathbb{R}^d$, then

$$\int_{B_\delta(x_0)} w_\Omega dx \ge \frac{w_\Omega(x_0)\omega_d}{2}\delta^d, \qquad \forall \delta \in (0, \delta_0).$$

Indeed, for every $x_0 \in \mathbb{R}^d$ the function $x \mapsto w_\Omega(x) + \frac{|x-x_0|^2}{2d}$ is subharmonic in $\mathbb{R}^d$. Consequently, for every $\delta > 0$

$$\theta \le w_\Omega(x_0) \le \frac{1}{|B_\delta|}\int_{B_\delta(x_0)} (w(x) + \frac{|x-x_0|^2}{2d})\,dx = \frac{1}{|B_\delta|}\int_{B_\delta(x_0)} w dx + \frac{\delta^2}{2(d+2)}.$$

For δ_0 sufficiently small, we have $\forall\ 0 < \delta \le \delta_0$

$$\int_{B_\delta(x_0)} w_\Omega dx \ge \frac{w_\Omega(x_0)\omega_d}{2}\delta^d.$$

As a consequence, if Ω is unbounded (or has a large diameter, in the sense that the Hausdorff measure of the projection of the set Ω on one line is large) we get that the measure is larger than any constant (depending on the length of the diameter).

In order to prove that Ω has finite perimeter, we consider for every $\varepsilon > 0$, the test function $w_\varepsilon = (w_\Omega - \varepsilon)^+$, which is the torsion function on the set $\{w_\Omega > \varepsilon\}$. We get

$$\frac{1}{2}\int |\nabla w_\Omega|^2 dx - \int w_\Omega dx + \eta|\Omega| \le \frac{1}{2}\int |\nabla w_\varepsilon|^2 dx - \int w_\varepsilon dx + \eta|\{u_\varepsilon > 0\}|.$$

Consequently

$$\frac{1}{2}\int_{0\le w_\Omega\le\varepsilon} |\nabla w_\Omega|^2 dx + \eta|\{0 \le w_\Omega \le \varepsilon\}| \le$$

$$\le \int w_\Omega - w_\varepsilon dx = \int_{0\le w_\Omega\le\varepsilon} w_\Omega + \varepsilon|\{w_\Omega > \varepsilon\}| \le \varepsilon|\Omega|.$$

By Cauchy-Schwarz

$$\Big(\int_{0\le w_\Omega\le\varepsilon} |\nabla w_\Omega| dx\Big)^2 \le \int_{0\le w_\Omega\le\varepsilon} |\nabla w_\Omega|^2 dx |\{0 \le w_\Omega \le \varepsilon\}| \le 2\varepsilon^2/\eta|\Omega|^2,$$

so that

$$\int_{0\le w_\Omega\le\varepsilon} |\nabla w_\Omega| dx \le \varepsilon\sqrt{\frac{2}{\eta}}|\Omega|.$$

Using the co-area formula and the average theorem, we find $\varepsilon_n > 0$, $\varepsilon_n \to 0$ such that

$$\mathcal{H}^{d-1}(\partial^*\{w_\Omega > \varepsilon_n\}) \le \sqrt{\frac{2}{\eta}}|\Omega|,$$

where $\partial^*\Omega$ denotes the measure theoretic boundary. Passing to the limit, we get

$$\mathcal{H}^{d-1}(\partial^*\Omega) \le \sqrt{\frac{2}{\eta}}|\Omega|.$$

This last inequality implies that Ω has a finite perimeter in the geometric measure theoretical sense.

□

Sketch of the proof of Lemma 2.9. Given $\Omega \subseteq \mathbb{R}^d$, a quasi-open set of volume c, we find first a solution of the following problem

$$\min\{E(A) + \eta|A| : A \subseteq \Omega\},$$

for a suitably chosen value $\eta > 0$, which will be fixed later.

For every $\eta > 0$, this problem has a solution. The existence can be proved by the direct method of the calculus of variations, as a consequence of the compact embedding $H_0^1(\Omega)$ in $L^2(\Omega)$. In fact any weak gamma limit of a minimizing sequence $(A_n)_n$ is a solution.

Let us denote Ω_η a solution. We define the set

$$\tilde{\Omega} = \Big(\frac{|\Omega|}{|\Omega_\eta|}\Big)^{\frac{1}{d}}\Omega_\eta.$$

Since Ω_η is a subsolution for the torsion energy, the diameter and perimeter of $\tilde{\Omega}$ are controlled only by η, c and d.

It is easy to notice that if η is small enough (the precise value will be fixed at the end), then the first k eigenvalues of $\tilde{\Omega}$ are not larger than the corresponding eigenvalues on Ω. This is essentially a consequence of inequality (2.8)

$$\lambda_k(\Omega_\eta) - \lambda_k(\Omega) \le 4k^2 e^{1/4\pi} \lambda_k(\Omega_\eta)\lambda_k(\Omega)^{(d+2)/2}[E(\Omega) - E(\Omega_\eta)]. \tag{2.21}$$

Indeed, if η is small enough, one can prove (see (2.10)-(2.11)) that $\lambda_k(\Omega_\eta) \le 2\lambda_k(\Omega)$ and get

$$\lambda_k(\Omega_\eta) - \lambda_k(\Omega) \le C_{\eta,\lambda_k(\Omega),c,d}(|\Omega|^{\frac{2}{d}} - |\Omega_\eta|^{\frac{2}{d}}). \tag{2.22}$$

The constant $C_{\eta,\lambda_k(\Omega),c,d}$ is smaller when η and $\lambda_k(\Omega)$ are smaller. The dependence of $C_{\eta,\lambda_k(\Omega),c,d}$ on all parameters, including η is explicit. As a consequence, we get

$$\lambda_k(\Omega_\eta) + C_{\eta,\lambda_k(\Omega),c,d}|\Omega_\eta|^{\frac{2}{d}} \le \lambda_k(\Omega) + C_{\eta,\lambda_k(\Omega),c,d}|\Omega|^{\frac{2}{d}}, \tag{2.23}$$

which leads for any value $C_{\eta,\lambda_k(\Omega),c,d} \le \frac{\lambda_k(\Omega)}{c^{\frac{2}{d}}}$ to the inequality

$$\lambda_k(\Omega_\eta)|\Omega_\eta|^{\frac{2}{d}} \le \lambda_k(\Omega)|\Omega|^{\frac{2}{d}}. \tag{2.24}$$

This implies that $\lambda_k(\tilde{\Omega}) \le \lambda_k(\Omega)$.

Clearly, inequalities $\lambda_i(\tilde{\Omega}) \le \lambda_i(\Omega)$ also hold for $i = 1, \dots, k-1$.

Moreover, the set Ω_η being a subsolution for the torsion energy, it is bounded and has finite perimeter, controlled only by Ω, η and d. This holds as well for the set $\tilde{\Omega}$, with rescaling factors coming from the ratio $\frac{|\Omega|}{|\Omega_\eta|}$.

Provided that the constant η is chosen small enough such that this ratio is not larger than 2 and that $C_{\eta,\lambda_k(\Omega),c,d} \le \frac{\lambda_k(\Omega)}{c^{\frac{2}{d}}}$, we conclude the proof.

□

Further remarks

The perimeter constraint. A natural question is to ask if Theorem 2.10 could hold under a further constraint $Per(\Omega) \le c_2$. Of course, this question becomes interesting, as soon as the constant c_2 is smaller than the perimeter of the optimal set in Theorem 2.10. In order to deal with these kind of questions, in [221] a second surgery result is proved, with the purpose of having a finer control of the perimeter. This result asserts, roughly speaking, that one can also decrease the perimeter of a set if its diameter is large in Lemma 2.9. For this purpose, the surgery procedure is performed in a different way.

Lemma 2.14. (surgery of the perimeter) *For every $K, P, c > 0$, there exist $D > 0$ depending only on K, P, c and the dimension d, such that for every quasi-open set $\Omega \subset \mathbb{R}^d$ with $|\Omega| = c$, $Per(\Omega) \le P$, there exists a quasi-open set $\tilde{\Omega}$ of the same measure, with $\operatorname{diam}(\tilde{\Omega}) \le D$, $Per(\tilde{\Omega}) \le Per(\Omega)$ such that if for some $k \in \mathbb{N}$ it holds $\lambda_k(\Omega) \le K$, then $\lambda_i(\tilde{\Omega}) \le \lambda_i(\Omega)$ for all $1 \le i \le k$.*

A consequence of this lemma concerns the following spectral optimization problem

$$\min\{F(\lambda_1(\Omega), \dots, \lambda_k(\Omega)) : \Omega \subseteq \mathbb{R}^d, |\Omega| = c_1, Per(\Omega) \le c_2\}. \tag{2.25}$$

Theorem 2.15. *Provided that $F : \mathbb{R}^k \to \mathbb{R}$ is non-decreasing in each variable and lower semicontinuous, for every $c_1, c_2 > 0$ such that $c_2 \ge \frac{\mathcal{H}^{d-1}(\partial B_1)}{|B_1|^{\frac{d-1}{d}}} c_1^{\frac{d-1}{d}}$, problem (2.25) has a solution in the class of measurable sets.*

We refer to [330] for details on shape optimization problems on measurable sets. Roughly speaking, this means that the minimum is attained on a quasi-open set Ω, for which there exists a measurable set A such that $\Omega \subseteq A$, $|A| = c_1$ and $Per(A) \le c_2$ (see [221] for details).

Optimization in specific classes of sets. An interesting task is to search for the extremal sets of spectral functionals in some specific classes of sets, e.g. the class of convex subsets of $\mathbb{R}^d$ (satisfying, or not, a constraint on measure, perimeter or diameter), the class of simply connected sets open sets of $\mathbb{R}^2$, the class of N-gones of $\mathbb{R}^2$, etc. As a general fact, one can notice that the existence question has a much more direct answer, as soon as those geometric or topological constraints are imposed. For instance, Theorems 2.1, 2.10 can be rephrased in the class of open convex sets (in any dimension of the space) or in the class of open sets in $\mathbb{R}^2$ whose complement have at most l connected components (l is a fixed natural number) (see [207, Sections 4.6, 4.7 and Chapters 5, 6]). In the family of convex sets, very interesting phenomena may occur leading to optimal sets which are locally of polygonal type. We refer the reader to [622, 623] and Section 3.5 of Chapter 3.

Other boundary conditions, higher order operators. In this chapter we discussed the existence questions only for functionals depending on the spectrum of the Laplace operator with Dirichlet boundary conditions. A good question is whether or not similar results hold for the Laplace operator with other boundary conditions.

Working with different boundary conditions requires us to completely change the functional framework. For instance, when optimizing spectral functionals associated with the Neumann Laplacian, similar techniques as in the proof of Theorem 2.1 can hardly be used. In fact, the functional space one has to use for the Neumann Laplacian is $H^1(\Omega)$. For different sets Ω, the spaces $H^1(\Omega)$ are not naturally embedded in a "good" functional space, unless (uniform) geometric requirements are satisfied by the different sets. The existence question for general functionals depending on the spectrum of the Neumann Laplacian is completely open (see [207, Chapter 7]).

For Robin boundary conditions (or for the Steklov problem) one could use the theory of special functions of bounded variations in order to handle existence, at least in some specific situations. The regularity of the boundaries of the optimal sets, relies here on the theory of free discontinuity problems. We refer the reader to the discussion around Theorem 4.24 in Chapter 4 and to [216, 217] for an introduction to the topic.

For the bi-Laplace operator with Dirichlet boundary conditions, a similar result to Theorem 2.1 holds true in a bounded design region, while for other boundary conditions or $D = \mathbb{R}^d$, the question is open.

Asymptotic behavior for large k. An interesting question is to understand the behavior of a sequence of solutions of problem (2.5) when k goes to $+\infty$. Only partial answers are known: in two dimensions of the space, if the measure constraint is replaced by a perimeter constraint, then any sequence of optimal domains converges to the disc, as it was recently proved by Bucur and Freitas. The question is to understand if a similar result continues to be true for the measure constraint, in any dimension of the space. A key problem is to prove that all the optimal sets for problem (2.5) are uniformly bounded, independently on k. Even partial results, asserting that subsequences of solutions have a geometric limit, would be of interest.

Jimmy Lamboley and Michel Pierre

3 Regularity of optimal spectral domains

3.1 Introduction

The main goal of this chapter is to review known results and open problems about the regularity of optimal shapes for the minimization problems

$$\min\left\{\lambda_k(\Omega),\quad \Omega\subset D, |\Omega| = a\right\}, \tag{3.1}$$

where D is a given open subset of $\mathbb{R}^d$, $|\cdot|$ denotes the Lebesgue measure, $a \in (0, |D|)$, $k \in \mathbb{N}^\star$ and $\lambda_k(\Omega)$ is the k-th eigenvalue of the Laplace operator on Ω with homogeneous Dirichlet boundary conditions. We will also consider the regularity question for penalized versions of (3.1) and discuss the possibility of singularities appearing for optimal shapes, either for (3.1) or for related problems involving convexity constraints.

We refer to Chapter 2 for all of the necessary definitions and for the question of existence of optimal shapes. In particular, if D is bounded or if $D = \mathbb{R}^d$, then Problem (3.1) has a solution (say $\Omega^\star$) in the family of *quasi-open* subsets of $\mathbb{R}^d$ (*as explained in Chapter 2, the eigenvalues $\lambda_k(\Omega)$ may be well-defined for all quasi-open sets Ω with finite measure as well as the space $H^1_0(\Omega)$*).

We analyze the question of the regularity of this optimal shape $\Omega^\star$.

This turns out to be a difficult and wide open question. It is even difficult to decide whether $\Omega^\star$ is open or not and this is not completely understood yet.

Is $\Omega^\star$ always open? Is at least one of the optimal $\Omega^\star$ open? What is the regularity of the optimal k-th eigenfunctions $u_{\Omega^\star}$?

As recalled in Chapter 1 and Chapter 2, if $D = \mathbb{R}^d$ or more generally if D is 'large enough', then we know that:

- $\Omega^\star$ is a ball if $k = 1$,
- $\Omega^\star$ is the union of two disjoint identical balls if $k = 2$,

Jimmy Lamboley: CEREMADE, Université Paris-Dauphine, CNRS, PSL Research University, 75775 Paris, France, E-mail: lamboley@ceremade.dauphine.fr
Michel Pierre: IRMAR, ENS Rennes, CNRS, UBL, av Robert Schumann, 35170 - Bruz, France, E-mail: michel.pierre@ens-rennes.fr

with uniqueness in both cases up to translations (and sets of zero-capacity). *Here, D 'large enough' means that, when $k = 1$, it can contain a ball of volume a, and when $k = 2$, it can contain two disjoint identical balls whose total volume is a.* Thus full regularity holds for the optimal shape in these two cases. The question remains however open for 'large' D with $k \geq 3$ and for any k with 'small' D. Then, the regularity analysis of the optimal shapes in (3.1) is very similar to the analysis of the optimal shapes for the Dirichlet energy, namely

$$\min \left\{ G_f(\Omega),\ \Omega \subset D,\ |\Omega| = a \right\}, \tag{3.2}$$

where $f \in L^\infty(D)$ is given and

$$G_f(\Omega) = \int_\Omega \left[\frac{1}{2} |\nabla u_\Omega|^2 - f\, u_\Omega \right], \quad u_\Omega \in H_0^1(\Omega),\ -\Delta u_\Omega = f \text{ in } \Omega. \tag{3.3}$$

(*The solution u_Ω of this Dirichlet problem is classically defined when Ω is an open set with finite measure. As explained in Chapter 2, this definition may be extended to the case when Ω is only a quasi-open set with finite measure.*)

Actually, for these two problems (3.1) and (3.2), the analysis of the regularity follows the same main steps and offers the following main features. They will provide the content of Sections 3.2 and 3.3.

1. **The situation is easier when the state function is nonnegative !** For the Dirichlet energy case (3.2), for instance in dimension two, full regularity of the boundary holds for positive data f, inside D (see [187] and Paragraph 3.2.3.1 below). On the other hand, even in dimension two, it is easily seen that *singularities do necessarily occur* at each point of the boundary of the optimal set $\Omega^\star$ in the neighborhood of which the state function $u_{\Omega^\star}$ (as defined in (3.3)) changes sign. The change of sign of $u_{\Omega^\star}$ does imply that its gradient has to be discontinuous and, therefore, that the boundary cannot be regular near these points. For instance, cusps will then generally occur in dimension two (see e.g. [509]).
For the eigenvalue problem (3.1), state functions are the k-th eigenfunctions on $\Omega^\star$ of the Laplace-Dirichlet operator. Thus the situation (and the analysis) will be quite different if $k = 1$ where the first eigenfunction is nonnegative and if $k \geq 2$ where the eigenfunction changes sign. This partly explains why we devote the specific Section 3.2 to Problem (3.1) with $k = 1$. One more specific feature is that the problem is then equivalent to a minimization problem where the variables are functions rather than domains and we are led to a free boundary formulation (see Paragraph 3.2.1) where one has to understand the regularity of the boundary of $[u_{\Omega^\star} > 0]$. One can essentially obtain as good of regularity results as one did with the Dirichlet energy case and nonnegative data f, see [189]. Here we strongly rely on the seminal paper [25] by Alt-Caffarelli about regularity of free boundaries.
On the other hand, the case $k \geq 2$ is far from being so well understood and we will try to describe what current state of the art is (see Section 3.3).

2. **A first step: regularity of the state function.** For the Dirichlet energy case, the analysis starts by studying the regularity of u_{Ω^*} as defined in (3.3). It is proved (see [188]) that u_{Ω^*} *is locally Lipschitz continuous* on D, for any optimal shape Ω^* and no matter the sign of u_{Ω^*}. This Lipschitz continuity is the optimal regularity we can expect for u_{Ω^*}, as it vanishes on $D \setminus \Omega^*$, and is expected to have a non vanishing gradient on $\partial\Omega^*$ from inside Ω^*. As expected, the proof in the case where u_{Ω^*} changes sign is much more involved and requires for instance the Alt-Caffarelli-Friedman Monotonicity Lemma (proved in [26], [245], see Lemma 3.36 below).
For the optimal eigenvalue problem (3.1) with $k = 1$, it can be proved as well that the corresponding eigenfunction on Ω^* is locally Lipschitz continuous on D (see Theorem 3.16). For $k \geq 2$ and $D = \mathbb{R}^d$, it has been proved in [222] that one of the k-th eigenfunctions is Lipschitz continuous (see Theorem 3.35) (note that the optimal eigenvalue is generally expected to be of multiplicity higher than once). However, in the case where D is bounded and $k \geq 2$, the problem is still not understood. The main difference is that, when $D = \mathbb{R}^d$, Problem (3.1) is equivalent to the penalized version
$$\min\left\{\lambda_k(\Omega) + \mu|\Omega|,\ \ \Omega \subset \mathbb{R}^d\right\}, \tag{3.4}$$
for some convenient $\mu \in (0, \infty)$ (see Proposition 3.33). More regularity information may then be derived on optimal state functions for penalized versions (see below).
3. **Penalized versions.** In order to obtain information on the regularity of Ω^* or u_{Ω^*}, we consider admissible perturbations of Ω^* and use their minimization properties. Obviously, there is more freedom in choosing perturbations on the penalized version (3.4) where the volume constraint $|\Omega| = a$ is relaxed, rather than on the constrained initial version (3.1). The analysis of (3.1) when $k = 1$ starts by showing that (3.1) is *equivalent to the penalized version*
$$\min\left\{\lambda_1(\Omega) + \mu[|\Omega| - a]^+,\ \ \Omega \subset D\right\}, \tag{3.5}$$
for μ large enough (see Proposition 3.7). Analysis of the regularity may then be more easily made on the optimal shapes of (3.5). In Paragraph 3.2.3.2, we make an heuristic analysis of this *"exact penalty"* property for general optimization problems where not only the penalized version converges to the constrained problem as the penalization coefficient $\mu \to \infty$, but more precisely that the two problems are *equivalent* for μ large enough. Optimal such factors μ play the role of Lagrange multipliers. This approach is used again in a local way in Paragraph 3.2.3.3, to prove that the 'pseudo'-Lagrange multiplier does not vanish (see Proposition 3.24). It is also used in Chapter 7 of this book to study the regularity of optimal shapes for similar functionals (see Step 5 in the proof of Theorem 7.13).
4. **How to obtain the regularity of the boundary of Ω^*?** Knowing that the state function is Lipschitz continuous is a first main step in the study of the regularity of the boundary of the optimal set, but obviously not sufficient.

For example when $k = 1$, this boundary can be seen as the boundary of the set $[u_{\Omega^*} > 0]$. If we were in a regular situation (say if u were C^1 on $\overline{\Omega^*}$), then knowing that the gradient of u_{Ω^*} does not vanish at the boundary would imply regularity of this boundary by the implicit function theorem.
Indeed, the next main step is (heuristically) to prove that the gradient of the state function *does not degenerate at the boundary*. This is what is done and then used in Paragraph 3.2.3.3 for the optimal sets of (3.1) when $k = 1$. Full regularity of the boundary is proved in dimension two and regularity of the reduced boundary is proved in any dimension (see Theorem 3.20). Here we strongly rely on the seminal paper [25] by Alt-Caffarelli as explained in details in Section 3.2.3.1. Note that it is also used in Chapter 7 of this book as mentioned at the end of Point 3 above. Nothing like this is known when $k \geq 2$. It is already a substantial piece of information to sometimes know that Ω^* is an open set ! (see Section 3.3).

In Section 3.4, we partially analyze the regularity of Ω^* solution of (3.1) up to the boundary of the box D, when $k = 1$. We notice in particular that it is natural to expect the contact to be tangential (although this is not proved anywhere as far as we know), but we cannot expect in general that the contact will be very smooth; we prove that when D is a strip (too narrow to contain a disc of volume a), the optimal shape is $C^{1,1/2}$ and not $C^{1,1/2+\varepsilon}$ with $\varepsilon > 0$. In order to show that this behavior is not exceptional and is not only due to the presence of a box constraint, we show that a similar property is valid for solutions to the problem

$$\min\left\{\lambda_2(\Omega), \Omega \text{ open and convex}, |\Omega| = a\right\}.$$

This last problem enters the general framework of *convexity constraints*, which is quite challenging from the point of view of calculus of variations. We conclude this chapter with Section 3.5 where we discuss some problems in this framework. They are of the form

$$\min\left\{J(\Omega), \Omega \text{ open and convex}\right\},$$

where J involves λ_1, and possibly other geometrical quantities (such as the volume $|\Omega|$ or the perimeter $P(\Omega)$), and which lead to *singular optimal shapes, such as polygons (in dimension 2)*. Thanks to the convexity constraint, we are allowed to consider the question of maximizing the perimeter and/or the first Dirichlet eigenvalue, and in this direction we discuss a few recent results about the reverse Faber-Krahn inequality.

Remark 3.1. The question of regularity could also be considered for the following optimization problems:

$$\min\left\{\lambda_k(\Omega), \quad \Omega \subset D, P(\Omega) = p\right\}, \qquad \min\left\{P(\Omega) + \lambda_k(\Omega), \quad \Omega \subset D, |\Omega| = a\right\}$$

where P denotes the perimeter (in the sense of geometric measure theory), and D is either a bounded smooth box, or $\mathbb{R}^d$. In these cases, it has been shown in [329, 330] that the regularity of optimal shapes is driven by the presence of the perimeter term. More precisely it can be shown that they exist (which is not trivial if $D = \mathbb{R}^d$) and that they are quasi-minimizers of the perimeter, and therefore smooth outside a singular set of dimension less than $d - 8$.

3.2 Minimization for λ_1

In this section, we focus on the regularity of the optimal shapes of the following problem:

$$\min\left\{\lambda_1(\Omega),\quad \Omega \subset D,\ \Omega \text{ quasi}-\text{open},\ |\Omega| = a\right\}, \tag{3.6}$$

where D is an open set in $\mathbb{R}^d$, $a \in (0, |D|)$ and $k \in \mathbb{N}^*$.

Thanks to the Faber-Krahn inequality, it is well-known that, if D contains a ball of volume a, then this ball is a solution of the problem, and is moreover unique, up to translations (and to sets of zero-capacity). Therefore, the results of this section are relevant only if such a ball does not exist.

3.2.1 Free boundary formulation

We first give an equivalent version of problem (3.6) as a free boundary problem, namely an optimization problem in $H_0^1(D)$ where domains are level sets of functions.

Notation. For $w \in H_0^1(D)$, we will denote $\Omega_w = \{x \in D; w(x) \neq 0\}$.

Recall that for a bounded quasi-open subset Ω of D (see Chapter 2)

$$\lambda_1(\Omega) = \min\left\{\int_\Omega |\nabla v|^2; v \in H_0^1(\Omega),\ \int_\Omega v^2 = 1\right\}. \tag{3.7}$$

Definition 3.2. In this section, we denote by u_Ω any nonnegative minimizer in (3.7), i.e. such that

$$u_\Omega \in H_0^1(\Omega),\quad \int_\Omega |\nabla u_\Omega|^2 = \lambda_1(\Omega),\quad \int_\Omega u_\Omega^2 = 1.$$

Remark 3.3. Choosing in (3.7) $v = v(t) := (u_\Omega + t\varphi)/\|u_\Omega + t\varphi\|_{L^2(\Omega)}$ with $\varphi \in H_0^1(\Omega)$, and using that the derivative at $t = 0$ of $t \mapsto \int_\Omega |\nabla v(t)|^2$ vanishes leads to

$$\forall \varphi \in H_0^1(\Omega),\quad \int_\Omega \nabla u_\Omega \nabla \varphi = \lambda_1(\Omega) \int_\Omega u_\Omega \varphi. \tag{3.8}$$

If Ω is an open set, (3.8) means exactly that $-\Delta u_\Omega = \lambda_1 u_\Omega$ in the sense of distributions in Ω.

Note that if u_Ω is a minimizer in (3.7), so is $|u_\Omega|$. Therefore, with no loss of generality, we can assume that $u_\Omega \geq 0$ and we will always make this assumption in this section on the minimization of $\lambda_1(\Omega)$. If Ω is a connected open set, then $u_\Omega > 0$ on Ω. This is a consequence of the maximum principle applied to $-\Delta u_\Omega = \lambda_1(\Omega)u_\Omega \geq 0$ on Ω. This extends (quasi-everywhere) to the case when Ω is a *quasi-connected* quasi-open set, but the proof requires a little more computation.

Since $\Omega \mapsto \lambda_1(\Omega)$ is nonincreasing with respect to inclusion, any solution of (3.6) is also solution of

$$\min\left\{\lambda_1(\Omega),\ \ \Omega \subset D,\ \Omega \text{ quasi}-\text{open},\ |\Omega| \leq a\right\}. \tag{3.9}$$

The converse is true in most situations, in particular if D is connected, see Remark 3.6, Corollary 3.18 and the discussion in Section 3.2.4.1. Note that it may happen that if D is not connected, then a solution to (3.9) does not satisfy $|\Omega| = a$.

We will first consider Problem (3.9) and this will nevertheless provide a complete understanding of (3.6). We start by proving that (3.9) is equivalent to a free boundary problem.

Proposition 3.4. *1. Let Ω^* be a quasi-open solution of the minimization problem (3.9) and let $u = u_{\Omega^*}$. Then*

$$\int_D |\nabla u|^2 = \min\left\{\int_D |\nabla v|^2; v \in H_0^1(D); \int_D v^2 = 1, |\Omega_v| \leq a\right\}. \tag{3.10}$$

2. Let u be solution of the minimization problem (3.10). Then Ω_u is solution of (3.9).

Proof. For the first point, we choose $v \in H_0^1(D)$ with $|\Omega_v| \leq a$ and we apply (3.9) to $\Omega = \Omega_v$. This gives $\int_D |\nabla u|^2 = \lambda_1(\Omega^*) \leq \lambda_1(\Omega_v)$ and we use the property (3.7) for $\lambda_1(\Omega_v)$ so that

$$\int_D |\nabla u|^2 \leq \min\left\{\int_D |\nabla v|^2; v \in H_0^1(D), \int_D v^2 = 1, |\Omega_v| \leq a\right\}.$$

Equality holds since $u \in H_0^1(\Omega^*) \subset H_0^1(D)$, and $|\Omega_u| = |\Omega^*| \leq a$.

For the second point, let u be a solution of (3.10). Then, $|\Omega_u| \leq a$, $\int_D u^2 = 1$. Let $\Omega \subset D$ quasi-open with $|\Omega| \leq a$ and let u_Ω as in Definition 3.2. Then

$$\lambda_1(\Omega_u) \leq \int_D |\nabla u|^2 \leq \int_D |\nabla u_\Omega|^2 = \lambda_1(\Omega).$$

□

Remark 3.5. We will now work with the functional problem (3.10) rather than (3.9). Note that if D is bounded (or with finite measure), then existence of the minimum u follows easily from the compactness of $H_0^1(D)$ into $L^2(D)$ applied to a minimizing sequence (that we may assume to be weakly convergent in $H_0^1(D)$ and strongly in $L^2(D)$).

Remark 3.6. Two different situations may occur. If D is connected and Ω^* solves (3.9), then $a^* := |[u_{\Omega^*} > 0]| = a$ and $\Omega^* = [u_{\Omega^*} > 0]$. If D is not connected, it may happen that $a^* < a$ and therefore $u_{\Omega^*} > 0$ on some of the connected components of D and identically zero on the others.

Indeed, if $a^* < a$, then for all balls $B \subset D$ with measure less than $a - a^*$ and all $\varphi \in H_0^1(B)$, we may choose $v = v(t) = (u + t\varphi)/\|u + t\varphi\|_{L^2(D)}$ with $u := u_{\Omega^*} \geq 0$ in (3.10). Writing that the derivative at $t = 0$ of $t \mapsto \int_D |\nabla v(t)|^2$ vanishes gives

$$\int_D \nabla u \nabla \varphi = \lambda_a \int_D u\,\varphi \quad \text{with} \quad \lambda_a := \int_D |\nabla u|^2,$$

and this implies : $-\Delta u = \lambda_a u$ in D. The strict maximum principle implies that, in each connected component of D, either $u > 0$ or $u \equiv 0$. If D is connected, we get a contradiction since $a < |D|$. Therefore necessarily $a^* = a$ if D is connected.

We refer to Corollary 3.18 and Proposition 3.29 for a complete description of the regularity when D is not connected.

3.2.2 Existence and Lipschitz regularity of the state function

3.2.2.1 Equivalence with a penalized version

We will first prove that (3.10) is equivalent to a penalized version.

Proposition 3.7. *Assume $|D| < +\infty$. Let u be a solution of (3.10) and $\lambda_a := \int_D |\nabla u|^2$. Then, there exists $\mu > 0$ such that*

$$\int_D |\nabla u|^2 \leq \int_D |\nabla v|^2 + \lambda_a \left[1 - \int_D v^2\right]^+ + \mu\,[|\Omega_v| - a]^+, \quad \forall v \in H_0^1(D). \tag{3.11}$$

Remark 3.8. Given a quasi-open set $\Omega \subset D$, and choosing $v = u_\Omega$ in (3.11), we obtain the penalized 'domain' version of (3.9), where Ω^* is solution of (3.9)

$$\lambda_1(\Omega^*) \leq \lambda_1(\Omega) + \mu[|\Omega] - a]^+, \quad \forall \Omega \subset D, \ \Omega \text{ quasi-open}. \tag{3.12}$$

Proof of Proposition 3.7.. Note first that, by definition of u and of λ_a, for all $v \in H_0^1(D)$ with $|\Omega_v| \leq a$, we have $\int_D |\nabla v|^2 - \lambda_a \int_D v^2 \geq 0$, or

$$\int_D |\nabla u|^2 \leq \int_D |\nabla v|^2 + \lambda_a \left[1 - \int_D v^2\right]. \tag{3.13}$$

Let us now denote by $J_\mu(v)$ the right-hand side of (3.11) and let u_μ be a minimizer of $J_\mu(v)$ for $v \in H_0^1(D)$ (its existence follows by compactness of $H_0^1(D)$ into $L^2(D)$, see also Remark 3.5). Up to replacing u_μ by $|u_\mu|$, we may assume $u_\mu \geq 0$. Using that $J_\mu(u_\mu) \leq J_\mu(u_\mu/\|u_\mu\|_2)$, we also deduce that $\|u_\mu\|_2^2 = \int_D u_\mu^2 \leq 1$.

For the conclusion of the proposition, it is sufficient to prove $|\Omega_{u_\mu}| \leq a$ since then

$$J_\mu(u_\mu) \leq J_\mu(u) = \int_D |\nabla u|^2 \leq J_\mu(u_\mu),$$

where this last inequality comes from (3.13).

In order to obtain a contradiction, assume that $|\Omega_{u_\mu}| > a$ and introduce $u^t := (u_\mu - t)^+$. Then $J_\mu(u_\mu) \le J_\mu(u^t)$. This implies, using $|\Omega_{u^t}| > a$ for t small, that

$$\int_{[0<u_\mu<t]} |\nabla u_\mu|^2 + \mu\,\big|[0 < u_\mu < t]\big| \le \lambda_a \int_{[0<u_\mu<t]} u_\mu^2 + 2t\lambda_a \int_D u_\mu.$$

Using the coarea formula (see e.g. [370], [432]), this may be rewritten for $t \le t_0 \le \sqrt{\mu/\lambda_a}$ as

$$\int_0^t ds \int_{[u_\mu=s]} \left[|\nabla u_\mu| + \frac{\mu - \lambda_a s^2}{|\nabla u_\mu|}\right] d\mathcal{H}^{d-1} \le 2t\lambda_a \int_D u_\mu \le 2t\lambda_a |\Omega_{u_\mu}|^{1/2}.$$

But the function $x \in (0,\infty) \mapsto x + (\mu - \lambda_a s^2)x^{-1} \in [0,\infty)$ is bounded from below by $2\sqrt{\mu - \lambda_a s^2}$ and also by $2\sqrt{\mu - \lambda_a t_0^2}$ as soon as $s^2 \le t^2 \le t_0^2 \le \mu/\lambda_a$. Therefore, it follows that

$$\forall t \in [0, t_0),\quad 2\sqrt{\mu - \lambda_a t_0^2} \int_0^t \int_{[u_\mu=s]} d\mathcal{H}^{d-1} \le 2t\lambda_a |\Omega_{u_\mu}|^{1/2}. \tag{3.14}$$

We now use the isoperimetric inequality: $\int_{[u_\mu=s]} d\mathcal{H}^{d-1} \ge C(d)\,\big|[u_\mu > s]\big|^{\frac{d-1}{d}}$. We divide the inequality by t and we let $t \to 0$, then $t_0 \to 0$, to deduce

$$2\sqrt{\mu}\, C(d) |\Omega_{u_\mu}|^{\frac{d-1}{d}} \le 2\lambda_a |\Omega_\mu|^{1/2}, \quad \text{and finally } 2\sqrt{\mu}\, C(d)\, a^{\frac{d-2}{2d}} \le 2\lambda_a.$$

Thus, if $d \ge 2$, $|\Omega_{u_\mu}| > a$ is impossible if $\mu > \mu^* := \lambda_a^2 C(d)^{-2} a^{(2-d)/d}$. Therefore the conclusion of Proposition 3.7 holds for any $\mu > \mu^*$.

If $d = 1$, we have $\sqrt{\mu} C(1) \le \lambda_a |\Omega_\mu|^{1/2}$. On the other hand, by definition of u_μ we also have $|\Omega_{u_\mu}| \le a + \lambda_1(\Omega_1)/\mu$ for some fixed $\Omega_1 \subset D$ with $|\Omega_1| = a$. We deduce an upper bound for μ as well. □

Remark 3.9. With respect to the heuristic remarks made in Paragraph 3.2.3.2, it is interesting to notice that our problem here is not in a 'differentiable setting'. However, we do perform some kind of differentiation in the direction of the perturbations $t \mapsto (u_\mu - t)^+$. This provides the upper bound μ^* on μ which plays the role of a Lagrange multiplier. This remark is a little more detailed in Paragraph 3.2.3.2. Note that μ^* does not depend on $|D|$. The assumption $|D| < \infty$ was used only to prove existence of the minimizer u_μ.

Remark 3.10 (Sub- and super-solutions). Note that to prove Proposition 3.7, we only use perturbations of the optimal domain Ω_u *from inside*. This means that the same result is valid for *shape subsolutions where (3.10) is assumed only for functions v for which $\Omega_v \subset \Omega_u$*.

Next, we will prove Lipschitz continuity of the functions u solutions of the penalized problem (3.11). Interestingly, Lipschitz continuity will hold for *super-solutions of (3.11) which are defined when the inequality (3.11) is valid only for perturbations* from outside, *i.e. such that $\Omega_u \subset \Omega_v$*.

3.2.2.2 A general sufficient condition for Lipschitz regularity

We now state a general result to prove Lipschitz regularity of functions independently of shape optimization. It applies to signed functions as well and will be used again in the minimization of the k-th eigenvalue.

Proposition 3.11. *Let $U \in H_0^1(D)$, bounded and continuous on D and let $\omega := \{x \in D; U(x) \neq 0\}$. Assume ΔU is a measure such that $\Delta U = g$ on ω with $g \in L^\infty(\omega)$ and*

$$|\Delta|U||(B(x_0, r)) \leq Cr^{d-1} \tag{3.15}$$

for all $x_0 \in D$ with $B(x_0, 2r) \subset D$, $r \leq 1$ and $U(x_0) = 0$. Then U is locally Lipschitz continuous on D. If moreover $D = \mathbb{R}^d$, then U is globally Lipschitz continuous.

Remark 3.12. Note that if U is locally Lipschitz continuous on D with $\Delta U \geq 0$, then for a test function φ with

$$\begin{aligned} &\varphi \in C_0^\infty(B(x_0, 2r)),\ B(x_0, 2r) \subset D,\ 0 \leq \varphi \leq 1, \\ &\varphi \equiv 1 \text{ on } B(x_0, r),\ \|\nabla \varphi\|_{L^\infty(B)} \leq C/r, \end{aligned} \tag{3.16}$$

we have

$$\Delta U(B(x_0, r)) \leq \int_D \varphi d(\Delta U) = -\int_D \nabla \varphi \nabla U \leq$$
$$\|\nabla U\|_{L^\infty} |\Omega_\varphi| \|\nabla \varphi\|_{L^\infty} \leq C \|\nabla U\|_{L^\infty} r^{d-1}.$$

This indicates that the estimate (3.15) is essentially a necessary condition for the Lipschitz continuity of U. This theorem states that the converse holds in some cases which are relevant for our analysis as it will appear in the next paragraph.

Remark 3.13. In the proof of Proposition 3.11, as in [188], we will use the following identity which is useful to estimate the variation of functions:

$$⨍_{\partial B(x_0,r)} U(x) d\sigma(x) - U(x_0) = C(d) \int_0^r s^{1-d} \left[\int_{B(x_0,s)} d(\Delta U)\right] ds. \tag{3.17}$$

This is easily proved for regular functions U by integration in s of

$$\frac{d}{ds} ⨍_{\partial B(0,1)} U(x_0 + s\xi) d\sigma(\xi) - ⨍_{\partial B(0,1)} \nabla U(x_0 + s\xi) \cdot \xi = C(d) s^{1-d} \int_{B(x_0,s)} \Delta U,$$

which implies that for $a.e.$ $0 < r_1 < r_2$,

$$⨍_{\partial B(x_0,r_2)} U(x) d\sigma(x) - ⨍_{\partial B(x_0,r_1)} U(x) d\sigma(x) = C(d) \int_{r_1}^{r_2} \left[s^{1-d} \int_{B(x_0,s)} \Delta U\right] ds.$$

It extends to functions $U \in H^1(D)$ where ΔU is a measure with $\int_0^r s^{1-d} \int_{B(x_0,s)} d(|\Delta U|) ds < \infty$. We may then consider that U is precisely defined at x_0 as:

$$U(x_0) = \lim_{r \to 0^+} ⨍_{\partial B(x_0,r)} U(x) d\sigma(x), \tag{3.18}$$

and (3.17) holds with this precise definition of $U(x_0)$.

Proof of Proposition 3.11. We want to prove that $\nabla U \in L^\infty_{loc}(D)$. We can first claim that $\nabla U = 0$ a.e. on $D \setminus \omega$. On the open set ω, we have $\Delta U = g \in L^\infty(\omega)$ so that at least $U \in C^1(\omega)$.

Let us denote $D_\delta = \{x \in D; d(x, \partial D) > \delta\}$ (we start with the case $D \neq \mathbb{R}^d$). We will bound $\nabla U(x_0)$ for $x_0 \in \omega \cap D_\delta$. The meaning of the constant C will vary but always depend only on δ, $\|U\|_{L^\infty(D)}$, $\|g\|_{L^\infty(D)}$, d and on the constant C in the assumption (3.15).

Let $y_0 \in \partial\omega$ be such that $|x_0 - y_0| = d(x_0, \partial\omega) := r_0$. Then $r_0 > 0$ and $B(x_0, r_0) \subset \omega$. We have $U(y_0) = 0$ since $y_0 \in \partial\omega$ and U is continuous. Let us introduce $s_0 := \min\{r_0, 1\}$, $B_0 := B(x_0, s_0)$ and $V \in H^1_0(B_0)$ such that $\Delta V = g$ on B_0. Since $g \in L^\infty$, by scaling we obtain

$$\|V\|_{L^\infty(B_0)} \le C s_0^2, \ \|\nabla V\|_{L^\infty(B_0)} \le C s_0, \ C = C(\|g\|_{L^\infty}).$$

Since $U - V$ is harmonic on B_0, we also have $|\nabla(U - V)(x_0)| \le \frac{d}{s_0}\|U - V\|_{L^\infty(B_0)}$ so that

$$|\nabla U(x_0)| \le |\nabla V(x_0)| + d s_0^{-1}\|U - V\|_{L^\infty(B_0)} \le C\left[s_0 + s_0^{-1}\|U\|_{L^\infty(B_0)}\right]. \tag{3.19}$$

If $s_0 \ge \delta/16$, we deduce from (3.19): $|\nabla U(x_0)| \le C(\delta, \|U\|_{L^\infty}, \|g\|_{L^\infty})$. We now assume $\delta \le 16$.

If $s_0 < \delta/16$ i.e. $r_0 = s_0 < \delta/16$, since $x_0 \in D_\delta$, $d(y_0, \partial D) \ge d(x_0, \partial D) - d(x_0, y_0) \ge \delta - r_0 \ge 15 r_0$ which implies $B(x_0, r_0) \subset B(y_0, 2r_0) \subset B(y_0, 8r_0) \subset D$. Thanks to assumption (3.15), $U(y_0) = 0$ and to formula (3.17) applied with U replaced by $|U|$, we deduce $⨍_{\partial B(y_0, 4r_0)} |U(z)| d\sigma(z) \le C r_0$. Finally, using the representation $(U - V)(x) = ⨍_{B(y_0, 4r_0)} U(z) P_x(z) d\sigma(z)$ for all $x \in B(y_0, 2r_0)$ where $P_x(\cdot)$ is the Poisson kernel at x, we have

$$\|U - V\|_{L^\infty(B_0)} \le \|U - V\|_{L^\infty(B(y_0, 2r_0))} \le C ⨍_{\partial B(y_0, 4r_0)} |U(z)|\, d\sigma(z) \le C r_0.$$

This together with (3.19) (where $s_0 = r_0$) and $\|V\|_{L^\infty(B_0)} \le C r_0^2$, this implies $|\nabla u(x_0)| \le C$.

Now if $D = \mathbb{R}^d$, either $\omega = \mathbb{R}^d$ and (3.19) gives the estimate ($r_0 = +\infty$, $s_0 = 1$), or $\omega \neq \mathbb{R}^d$: then we argue just as above, replacing $\delta/16$ by 1 in the discussion. □

In Proposition 3.11, the function U is assumed to be continuous on D. For our optimal eigenfunctions, this will be a consequence of the following lemma.

Lemma 3.14. *Let $U \in H^1_0(D)$ such that ΔU is a measure satisfying*

$$|\Delta U|\left(B(x_0, r)\right) \le C r^{d-1}, \tag{3.20}$$

for all $x_0 \in D$ with $B(x_0, 2r) \subset D$, $r \le 1$. Then U is continuous on D.

Proof. Assumption (3.20) implies that $\int_0^r s^{1-d}|\Delta U|(B(x_0, s)) < \infty$ so that (3.18) and (3.17) hold. Let $x_0, y_0 \in D$ and $r > 0$ small enough so that $B(x_0, 2r) \subset D$, $B(y_0, 2r) \subset D$.

We deduce, using (3.20) again and the representation (3.18):

$$|U(x_0) - U(y_0)| \le \left| \fint_{\partial B(x_0,r)} U - \fint_{\partial B(y_0,r)} U \right| + C\,r \le \\ \fint_{\partial B(0,r)} |U(x_0+\xi) - U(y_0+\xi)|\,d\sigma(\xi) + C\,r.$$

But by continuity of the trace operator from $H^1(B(0,r))$ into $L^1(B(0,r))$, this implies

$$|U(x_0) - U(y_0)| \le C(r)\|U(x_0+.) - U(y_0+.)\|_{H^1(B(0,r))} + C\,r.$$

Thus

$$\limsup_{y_0 \to x_0} |U(x_0) - U(y_0)| \le C\,r.$$

Since this is valid for all r sufficiently small, continuity of U at x_0 follows and therefore continuity on D as well. □

Remark 3.15. Looking at the proof, we easily see that the assumptions could be weakened in Lemma 3.14: $U \in W_0^{1,1}(D)$ would be sufficient and r^{d-1} could be replaced in (3.20) by $r^{d-2}\varepsilon(r)$ with $\varepsilon(r)/r$ integrable on $(0,1)$.

3.2.2.3 Lipschitz continuity of the optimal eigenfunction

Theorem 3.16. *Let u be a solution of (3.10). Then u is locally Lipschitz continuous on D.*

Proof. Up to replacing u by $|u|$, we may assume that $u \ge 0$. We will first show that $U = u$ satisfies the assumptions of Lemma 3.14. It will follow that u is continuous on D. Therefore, we will have $-\Delta u = \lambda_a u$ on the open set $\omega = [u > 0]$ (see Remark 3.3). Then we will prove (see also Remark 3.17 below) that

$$-\Delta u \le \lambda_a u \text{ in } D. \tag{3.21}$$

This will imply that Δu is a measure and also, by an easy bootstrap that $u \in L^\infty(D)$. Thus the assumptions of Proposition 3.11 will be satisfied and local Lipschitz continuity on D will follow.

By Proposition 3.7, u is also solution of Problem (3.11). We apply this inequality with $v = u + t\varphi$, $t > 0$, $\varphi \in H_0^1(D)$. Then

$$0 \le \int_D 2\nabla u \nabla \varphi + t|\nabla \varphi|^2 + \lambda_a \left[-2u\varphi - t\varphi^2\right]^+ + \frac{\mu}{t}\left[|\Omega_{u+t\varphi}| - a\right]^+. \tag{3.22}$$

Choosing first $\varphi = -p_n(u)\psi$ where $\psi \in C_0^\infty(D)$, $\psi \ge 0$, and $p_n(r) = \min\{r^+/n, 1\}$, we obtain with $q_n(r) = \int_0^r p_n(s)ds$ and after letting $t \to 0$ (note that $|\Omega_{u+t\varphi}| = |\Omega_u| \le a$)

$$0 \le \int_D -2p_n'(u)|\nabla u|^2\psi - 2\nabla q_n(u)\nabla\psi + 2\lambda_a u p_n(u)\psi.$$

Note that $up_n(u) \to u^+ = u$, $q_n(u) \to u^+ = u$ in a nondecreasing way as n increases to $+\infty$. Using $p'_n(u)|\nabla u|^2 \geq 0$, we obtain at the limit that $\Delta u + \lambda_a u \geq 0$ in the sense of distributions in D, whence (3.21).

Choosing $\varphi \in C_0^\infty(D)^+$ in (3.22) leads to $-2\int_D \nabla u \nabla \varphi \leq \int_D t|\nabla \varphi|^2 + \frac{\mu}{t}|\Omega_\varphi|$ or

$$2\langle \Delta u + \lambda_a u, \varphi\rangle \leq \int_D 2\lambda_a u\varphi + t|\nabla\varphi|^2 \quad + \frac{\mu}{t}|\Omega_\varphi|. \tag{3.23}$$

Minimizing over $t \in (0, \infty)$ gives

$$\langle \Delta u + \lambda_a, \varphi\rangle \leq \int_D \lambda_a u\varphi + \|\nabla\varphi\|_{L^2}\,[\mu|\Omega_\varphi|]^{1/2}. \tag{3.24}$$

Let now $x_0 \in D$ such that $B(x_0, 2r) \subset D$ and let $\varphi \in C_0^\infty(B(x_0, 2r))^+$ as in (3.16). Using also $u \in L^\infty$, we deduce that

$$|\Delta u|\left(B(x_0, r)\right) \leq (\Delta u + \lambda_a u)\left((B(x_0, r)\right) + \lambda_a \int_{B(x_0,r)} u \leq Cr^{d-1},$$

whence the estimate (3.15).

□

Remark 3.17. Here, we use the positivity of u. Actually, u is an eigenfunction for the eigenvalue λ_a on Ω_u. Since Ω_u is open, we know that $\Delta u + \lambda_a u = 0$ on Ω_u (see Remark 3.3). Since $u \geq 0$, one can prove that $\Delta u + \lambda_a u \geq 0$ on D. To prove this, use the test functions $\varphi = -p_n(u)\psi$ which satisfy $\Omega_\varphi \subset \Omega_u$ and therefore belong to $H_0^1(\Omega_u)$. Thus applying (3.8) in Remark 3.3 with this φ is sufficient (and we finish as above).

This positivity of the measure $\Delta u + \lambda_a u$ allows to directly estimate the mass of $|\Delta u|$ on balls only with the information (3.24). This will not be the case when dealing with k-th eigenfunctions when $k \geq 2$ (see the remarks and comments on the use of the Monotonicity Lemma 3.36).

Let us now state a corollary of Proposition 3.16 for the initial actual shape optimization problem (3.6).

Corollary 3.18. *Assume D is open and with finite measure. Then there exists an open set Ω^* which is solution of (3.6). Moreover, for any (quasi-open) solution Ω^* of (3.6), u_{Ω^*} is locally Lipschitz continuous on D. If D is connected, then all solutions Ω^* of (3.6) are open.*

Remark 3.19. If D is not connected, then it may happen that Ω^* is not open: we refer for instance to Example 3.28. However u_{Ω^*} is always locally Lipschitz continuous. Let us mention that the existence of an optimal open set for (3.6) had first been proved in [469]. A different penalization was used and it was proved that the corresponding state function converged to a Lipschitz optimal eigenfunction.

Proof of Corollary 3.18. If D is of finite measure, as already seen, Problem (3.10) has a solution u. By Theorem 3.16, it is locally Lipschitz continuous on D. In particular Ω_u is open. If $|\Omega_u| = a$, then $\Omega^* := \Omega_u$ is an open solution of (3.6). If $|\Omega_u| < a$, then any open set Ω^* satisfying $\Omega_u \subset \Omega^* \subset D$, $|\Omega^*| = a$ is also a solution since then, by monotonicity $\lambda_1(\Omega^*) \le \lambda_1(\Omega_u)$ (and there are such Ω^* like for instance $\Omega^* := \Omega_u \cup B(x_0, r) \cap D$ where $x_0 \in D$ and r is chosen so that $|\Omega^*| = a$).

Now let Ω^* be a solution of (3.6). Then, by Proposition 3.4, u_{Ω^*} is solution of the minimization problem (3.10). By Theorem 3.16, it is locally Lipschitz continuous in D. As proved in Remark 3.6, if D is connected, then $\Omega^* = [u_{\Omega^*} > 0]$. Therefore Ω^* is open. □

3.2.3 Regularity of the boundary

In this section, we go deeper in the analysis of the free boundary problem (3.10), and we explain a strategy for proving the regularity of the boundary of any solution to (3.6). Here is the main result of this section:

Theorem 3.20 (Brianҫon-Lamboley [189]). *Assume D is open, bounded and connected. Then any solution of (3.6) satisfies:*

1. *Ω^* has locally finite perimeter in D and*

$$\mathcal{H}^{d-1}((\partial\Omega^* \setminus \partial^*\Omega^*) \cap D) = 0, \tag{3.25}$$

where $\mathcal{H}^{d-1}$ is the Hausdorff measure of dimension $d-1$, and $\partial^*\Omega^*$ is the reduced boundary (in the sense of sets with finite perimeter, see [370] or [432]).

2. *There exists $\Lambda > 0$ such that*

$$\Delta u_{\Omega^*} + \lambda_1(\Omega^*) u_{\Omega^*} = \sqrt{\Lambda}\mathcal{H}^{d-1} \lfloor \partial\Omega^*,$$

in the sense of distribution in D, where $\mathcal{H}^{d-1}\lfloor\partial\Omega^*$ is the restriction of the $(d-1)$-Hausdorff measure to $\partial\Omega^*$.

3. *$\partial^*\Omega^*$ is an analytic hypersurface in D.*
4. *If $d = 2$, then the whole boundary $\partial\Omega^* \cap D$ is analytic.*

3.2.3.1 Regularity for free boundary problems

In this section, we give a small overview of the literature about regularity of free boundaries, related to the problem (3.6) or (3.9); it is not exhaustive since the literature on this subject is huge.

It has been seen in Section 3.2.1 that problem (3.9) is equivalent to a free boundary problem: this will allow us to use some of the deep and well-known results on the subject. The seminal paper [25] is dealing with the following model problem:

$$\min\left\{\int_D |\nabla u|^2 + \int_D g(x)^2 \mathbb{1}_{\Omega_u},\quad u \in H^1(D),\quad u = u_0 \text{ on } \partial D\right\} \tag{3.26}$$

where $u_0 \in H^1(D)$ is a given positive boundary data, D is an open bounded set, and $\Omega_u = \{u \neq 0\}$. In [25], different results were given about the regularity of the free boundary $\partial\Omega_u$ where u solves (3.26). Let us first notice that it is expected to obtain some regularity from the optimality condition, which is

$$\Delta u = 0 \text{ in } \Omega_u, \qquad |\nabla u| = g \text{ on } \partial\Omega_u \cap D, \tag{3.27}$$

where this has to be understood in a weak sense (see below). In the paper [589], it is shown that, if $\partial\Omega_u$ is assumed to be $C^{1,\alpha}$ (locally inside D) for some $\alpha > 0$ and g is smooth (say analytic), then $\partial\Omega_u$ is locally analytic. As usual though, the most difficult part is to obtain regularity for $\partial\Omega_u$ from scratch, and in particular, to give sense to (3.27). In [442], these results have been adapted to a different situation, namely

$$\min\left\{\int_D |\nabla u|^2 - 2fu + g^2 \mathbb{1}_{\Omega_u},\quad u \in H^1(D), u = u_0 \text{ on } \partial D\right\} \tag{3.28}$$

where f is a given nonnegative bounded function (in [442] they actually have $D = \mathbb{R}^d$ and no boundary condition: nevertheless, their results can easily be adapted to this situation, and for more clarity we prefer to deal with this problem here). In that case, the Euler-Lagrange optimality condition is given by

$$\begin{cases} -\Delta u &= f \quad \text{in} \quad \Omega_u, \\ |\nabla u| &= g \quad \text{on} \quad \partial\Omega_u \cap D. \end{cases} \tag{3.29}$$

These equations, especially the second one that defines the free boundary condition, make sense a priori only if it is known that u and $\partial\Omega_u$ are smooth. Therefore, there are different ways of formulating this boundary condition in a weak sense:

- Shape derivative formulation (see [25, Theorem 2.5]): for any $\xi \in C_0^\infty(D)$,

$$\lim_{\varepsilon\to 0} \int_{\partial\{u>\varepsilon\}} (|\nabla u|^2 - g^2)(\xi \cdot \nu)\, d\mathcal{H}^{d-1} = 0, \tag{3.30}$$

  where ν is the outward normal vector.
- Weak solution ([25, 442]): $u \in H^1(D)$ is called a weak solution of (3.29) if
  1. u is continuous and *nonnegative*,
  2. u satisfies, in the sense of distribution in D:

$$\Delta u + f = g\mathcal{H}^{d-1} \lfloor \partial^* \Omega_u \text{ in } D, \tag{3.31}$$

  3. there exists $0 < c \leq C$ such that for all balls $B_r(x) \subset D$ with $x \in \partial\Omega_u$,

$$c \leq \frac{1}{r} \fint_{\partial B_r(x)} u\, d\mathcal{H}^{d-1} \leq C.$$

As stated in [25, Theorem 5.5], if g is continuous, then local minima for (3.26) are weak solutions in this sense, and in particular $\mathcal{H}^{d-1}(\partial\Omega_u \setminus \partial^*\Omega_u) = 0$, so in (3.31) the term $\mathcal{H}^{d-1}\lfloor\partial^*\Omega_u$ can be replaced by $\mathcal{H}^{d-1}\lfloor\partial\Omega_u$ although this is not true in general (note that in [442] they use a weaker notion of weak solutions, though this one is more suitable for regularity results).

- Viscosity formulation: though we will not use this framework here, let us emphasize the complete regularity theory developed for *viscosity solutions to* (3.29) (see [242–244, 248]). Lots of proofs have been dramatically simplified compared to the original paper [25], see also [331].

For our purpose, we focus on the use of the notion of weak solutions, and here are the two main results about the regularity for these solutions:

Theorem 3.21 (Theorem 8.2 in [25], Theorem 2.17(a) in [442]). *Suppose $f \in L^\infty(D)$, $f \geq 0$, g is Hölder continuous in D and there exists $c > 0$ such that $g \geq c$ in D. Then if u is a weak solution (see the definition above), $\partial^*\Omega_u$ is locally $C^{1,\alpha}$ in D, for some $\alpha > 0$, and moreover*

$$\mathcal{H}^{d-1}((\partial\Omega_u \setminus \partial^*\Omega_u) \cap D) = 0. \tag{3.32}$$

In the two-dimensional case, we can improve this statement.

Theorem 3.22 (Theorem 8.3 in [25]). *Under the same assumption as Theorem 3.21, if moreover $d = 2$ and*

$$\fint_{B_r \cap \Omega_u} (g^2 - |\nabla u|^2)_+ \underset{r\to 0}{\longrightarrow} 0$$

then

$$\partial\Omega_u = \partial^*\Omega_u \quad (in\ D)$$

and therefore $\partial\Omega_u$ is locally $C^{1,\alpha}$ in D.

These two results are the main achievement of [25], see Sections 5 to 8 in this paper. This relies in particular on the proof of the fact that "flatness implies regularity", in other words if the boundary is assumed to be a bit flat (therefore avoiding the singularity described in Paragraph 3.2.4.2), then it is actually smooth.

Different generalizations or simplifications of the proofs of these results did appear in the literature after the paper [25]. A complete regularity theory for viscosity solutions of (3.29) has been developed, and we can find in [331] a simplified proof of a version of Theorem 3.21 for these viscosity solutions. See also Section 3.2.4.2 for further results and in particular an improvement of (3.32).

Once a $C^{1,\alpha}$-regularity is obtained for $\partial^*\Omega_u$ (or $\partial\Omega_u$), then with extra assumptions on the data f and g, more regularity on $\partial^*\Omega_u$ can easily be obtained: namely as proven in [589], if f is $C^{m,\beta}$ and g is $C^{m+1,\beta}$, then $\partial^*\Omega_u$ (or $\partial\Omega_u$ is $d = 2$) is $C^{m+2,\beta}$; moreover, if f and g are analytic, so is $\partial^*\Omega_u$.

3.2.3.2 Some heuristic remarks on "exact penalty" property

As seen in 3.2.2.1 and as it will be used again in 3.2.3.3 and in Section 3.3, regularity results are proved by strongly using the so-called "exact penalty" property. It says that our constrained problems (like (3.9), (3.48)) are *equivalent* to penalized versions (like (3.12), (3.39), (3.40), (3.49) or (3.51))) at least for well chosen or large enough penalty factors: this property goes quite beyond the weaker property that the penalized version 'converges' to the constrained one as the penalty factor tends to ∞), see Propositions 3.7, 3.24, 3.33 and Remark 3.34. The general question is analyzed in the literature of optimization theory: we refer for instance to [528, Section VII] for the basic theory or to [140] for some interesting results in this direction. Our optimization problems do not seem to fit in their demanding framework (in particular for differentiability or convexity assumptions), but they present the same features nevertheless.

Let us describe some main questions and answers, together with 'heuristic proofs' inspired from this literature, for the following abstract optimization problem

$$\min\left\{f(x),\quad g(x) = a\right\} \tag{3.33}$$

where $f, g : X \to [0, \infty)$, $a > 0$ and X is a real Banach space. Let us say we want to prove that any solution x_0 of (3.33) is also solution of the penalized following version:

$$\min\left\{f(x) + \mu\left[g(x) - a\right],\quad g(x) \geq a\right\}, \tag{3.34}$$

(the same problem with the constraint $g(x) \leq a$ instead, is similar). We concentrate on the two following questions:

- If x_0 is a solution of (3.33), is it still a (local) solution of (3.34) for some (finite) value of μ? Or in the terminology introduced in Remark 3.10, is x_0 a super-solution of the penalized version (3.34) for some $\mu > 0$?
- If so, what is the link between the best (smallest) possible μ and the Lagrange multiplier Λ associated with x_0, that is the real number Λ such that

$$f'(x_0) + \Lambda\, g'(x_0) = 0\ ? \tag{3.35}$$

The proof of the regularity result of Theorem 3.20 and in particular its point 2), strongly relies on this comparison and on the main fact that $\Lambda > 0$.

Let us start with the first question. In many situations, the answer to this question is yes. As a heuristic proof of this fact, let us assume that f, g are smooth enough (say C^1), and that there exists a solution x_μ to (3.34) for every $\mu \in (0, \infty)$. First notice that if $g(x_\mu) = a$, then it is admissible for (3.33), and we easily deduce that x_0 is also a solution to (3.34). Assume instead that $g(x_\mu) > a$ for every $\mu > 0$ and let us look for a contradiction. To that end, we write the optimality condition for (3.34), which gives (since the constraint is not saturated)

$$f'(x_\mu) + \mu\, g'(x_\mu) = 0. \tag{3.36}$$

Assuming some compactness on the set $\{x_\mu\}_{\mu>0}$, x_μ converges (up to a subsequence) to some $\widetilde{x}_0$ as $\mu \to +\infty$. It is clear, using $f(x_\mu) + \mu(g(x_\mu) - a) \le f(x_0)$ that $g(\widetilde{x}_0) = a$ (and that $\widetilde{x}_0$ solves (3.33)). But then (3.36) leads to a contradiction as $\mu \to \infty$ at least if we (naturally) asssume that a is not a critical value of g (i.e. $g'(\widetilde{x}_0) \neq 0$).

We now examine the second question, and we first notice that a necessary condition on μ so that x_0 (solution of (3.33)) be also solution of (3.34) is $\mu \ge \Lambda$, where Λ is the Lagrange multiplier associated with x_0 as defined in (3.35). Indeed, the optimality condition for (3.34) is then satisfied at x_0 and classically means (Karush-Kuhn-Tucker conditions) that there exists γ_μ such that

$$f'(x_0) + \mu g'(x_0) + \gamma_\mu g'(x_0) = 0, \qquad \text{and } \gamma_\mu \in (-\infty, 0],$$

the sign of γ_μ coming from the fact that the constraint is "$g \ge a$". Therefore, again if a is not a critical value of g, this implies: $\Lambda = \mu + \gamma_\mu \le \mu$.

It is then natural to hope that for any $\mu > \Lambda$, a solution of (3.33) is also a solution of (3.34). With some strong convexity assumptions, this is actually the case (see e.g. [140]). Since we are far from this kind of 'convex' situation, let us refer heuristically to a weaker and local result which will be quite in the spirit of one of the main steps in the proof of Theorem 3.20. More precisely, let us replace (3.34) by the following 'local' version where $h > 0$ is fixed:

$$\min \left\{ f(x) + \mu\left[g(x) - a\right],\ g(x) \in [a, a+h] \right\}. \tag{3.37}$$

Then the previous remarks made for (3.34) are still valid: if f, g are C^1 and that we have some adequate compactness properties, for μ large enough, x_0 is solution of this penalized problem; moreover, this requires $\mu \ge \Lambda$ where Λ is defined in (3.35).

Let us prove that $\mu > \Lambda$ is indeed (generally) sufficient. For this, we assume that Λ is the only Lagrange multiplier for the constrained problem (3.33) or more generally that it is the upper (finite) bound of the Lagrange multipliers (we indeed have to take into account that there may be several solutions to (3.33), associated to different Lagrange multipliers).

Let $\mu = \Lambda + \varepsilon$, $\varepsilon > 0$ be fixed. Let x_h be the solution of (3.37) and assume $g(x_h) > a$ for every $h > 0$. Equation (3.36) is satisfied as before. Using compactness, we may assume that x_h converges to some $\widetilde{x}_0$ which is a solution of (3.33) since the constraints on x_h lead to $g(\widetilde{x}_0) = a$. Therefore, on one hand, passing to the limit in (3.36), we obtain

$$f'(\widetilde{x}_0) + (\Lambda + \varepsilon) g'(\widetilde{x}_0) = 0,$$

while on the other hand $f'(\widetilde{x}_0) + \lambda g'(\widetilde{x}_0) = 0$ holds for some $\lambda \le \Lambda$, whence a contradiction (assuming as before that $g'(\widetilde{x}_0) \neq 0$).

3.2.3.3 Penalization of the volume constraint in Problem (3.10)

Our Problem (3.10) was proved to be equivalent to the penalized version (3.11) in Proposition 3.7 (see also (3.12) for a penalized "domain" version). We can check that, according to the previous heuristic analysis, if our problem was differentiable, then the best penalized factor μ^* would essentially be the Lagrange multiplier Λ of the constraint problem. Let us formally make the computation. The Euler-Lagrange equation for Problem (3.6) inside D is

$$-|\nabla u_{\Omega^*}|^2 + \Lambda = 0 \text{ on } \partial\Omega^* \text{ or also } -\partial_\nu u_{\Omega^*} = \sqrt{\Lambda} \text{ on } \partial\Omega^*.$$

Integrating $-\Delta u_{\Omega^*} = \lambda_a u_{\Omega^*}$ on Ω^* gives

$$P(\Omega^*)\sqrt{\Lambda} = -\int_{\partial\Omega^*} \partial_\nu u_{\Omega^*} = -\int_{\Omega^*} \Delta u_{\Omega^*} = \lambda_a \int_{\Omega^*} u_{\Omega^*} \le \lambda_a |\Omega^*|^{1/2}.$$

Inserting now the isoperimetric inequality $P(\Omega^*) \ge C(d)|\Omega^*|^{(d-1)/d}$, we obtain exactly the estimate $\Lambda \le \mu^*$ where μ^* is defined at the end of the proof of Proposition 3.7. The proof of this proposition is nothing but a rigourous justification of the computation we just made here.

Now, the proof of the regularity of the boundary stated in Theorem 3.20 will require us to make a "local" penalized version of the type (3.37). But the heuristic tools described in the previous paragraph are difficult to justify rigorously. Let us list the difficulties we face to apply such a strategy to (3.6) and lead to a complete proof of Proposition 3.24:

- Compactness and continuity arguments are used several times, including in order to get existence for problems (3.34) or (3.37). This requires us to obtain H^1 bounds, and weak H^1 continuity of the functionals.
- We use Euler-Lagrange equations, which requires that functionals are differentiable. As the natural choice of space is $X = H_0^1(D)$ it is important to notice here that the functionals (mainly the constraint) are not classically differentiable. Therefore at every step of the proof we write the Euler-Lagrange equation in a very weak way, using shape derivatives: in other words, if $u \in H_0^1(D)$ is a minimizer for our problem, we compare u to $u \circ T_t$ where t is a small parameter, and $T_t : D \to D$ is a family of smooth vector fields, close to the Identity in the C^1 norm. This gives, in our case: for all $\Phi \in C_0^\infty(D, \mathbb{R}^d)$,

$$\int_D 2(D\Phi\nabla u, \nabla u) - \int_D |\nabla u|^2 \nabla\cdot\Phi + \lambda_a \int_D u^2 \nabla\cdot\Phi = \Lambda \int_{\Omega_u} \nabla\cdot\Phi$$

  (which is another way to formulate (3.30)).
- Because of the difficulties stated above, we need to localize the argument. We fix B a ball centered at a point of $\partial\Omega_{u_0}$ and prove that if the radius of B is small enough,

then we can penalize the constraint if we restrict the test-functions to $H_0^1(B) \subset H_0^1(D)$.

- For our purpose, we also need to study the penalization procedure for sets Ω such that $|\Omega| < a$ (see (3.39)). Using the monotonicity of λ_1, it is easy to see that having a vanishing penalization $\mu = 0$ is valid. However, it is important for the rest of the proof to explain that some positive values of the penalization parameter μ are also valid, at least when $|\Omega|$ is close to a. This relies on the fact, as explained above, that this parameter μ can be chosen to be as close to Λ (the Lagrange multiplier for problem (3.6)) as one wants, and on the fact that Λ is positive. This fact is highly non-trivial, see Remark 3.25.

Remark 3.23. Of course, if we were studying the following penalized problem

$$\min\left\{\lambda_1(\Omega) + \mu|\Omega|,\ \Omega \subset D\right\}, \tag{3.38}$$

we would not face the same difficulties. This problem is easier to analyze, and the result of Theorem 3.20 is actually also valid for solutions of (3.38).

When $D = \mathbb{R}^d$ (or more generally when D is star-shaped), using the homogeneity properties of the functionals $\Omega \mapsto \lambda_1(\Omega), |\Omega|$, one can find an explicit μ so that problem (3.6) is equivalent to (3.38). This is proved in Proposition 3.33 (even for the λ_k-problem). Therefore, in a sense, the regularity theory is easier in that case (though as we noticed before, this case is not relevant for $D = \mathbb{R}^d$ since we already know that Ω^* is a ball).

We now state the following theorem, which is a main step in the proof of Theorem 3.20: let u be a solution of (3.10) and B_R be a ball included in D and centered on $\partial\Omega_u \cap D$. We define

$$\mathcal{F} = \{v \in H_0^1(D), u - v \in H_0^1(B_R)\}.$$

We denote $J(w) = \int_D |\nabla w|^2 - \lambda_a \int_D w^2$ for $w \in H_0^1(D)$. For $h > 0$, we denote by $\mu_-(h)$ the biggest $\mu_- \geq 0$ such that,

$$\forall\, v \in \mathcal{F} \text{ such that } a - h \leq |\Omega_v| \leq a,\ J(u) + \mu_-|\Omega_u| \leq J(v) + \mu_-|\Omega_v|. \tag{3.39}$$

We also define $\mu_+(h)$ as the smallest $\mu_+ \geq 0$ such that,

$$\forall\, v \in \mathcal{F} \text{ such that } a \leq |\Omega_v| \leq a + h,\ J(u) + \mu_+|\Omega_u| \leq J(v) + \mu_+|\Omega_v|. \tag{3.40}$$

Proposition 3.24. *Let u, B_R and $\mathcal{F}$ as above. Then for R small enough (depending only on u, a and D), there exist $\Lambda > 0$ and $h_0 > 0$ such that,*

$$\forall\, h \in (0, h_0),\ 0 < \mu_-(h) \leq \Lambda \leq \mu_+(h) < +\infty,$$

and, moreover,

$$\lim_{h\to 0} \mu_+(h) = \lim_{h\to 0} \mu_-(h) = \Lambda. \tag{3.41}$$

The proof is detailed in [189], and follows the heuristic proof from 3.2.3.2 with the main steps and difficulties described just above.

Remark 3.25. The most difficult statement here (and the most needed one later) is the fact that $\Lambda > 0$ which implies that $\mu_-(h) > 0$ for h small enough. The proof of this fact can be found in [187, Propositions 6.1 and 6.2] or [189, Proposition 2.6]: the main idea is to prove that if $\Lambda = 0$, then $-\Delta u = \lambda u$ in the whole box D, as in a weak sense, $|\nabla u| = \Lambda = 0$ on $\partial\Omega_u$, which implies that the measure Δu has no singularity across $\partial\Omega_u$. Once this fact is proven, we easily get a contradiction.

Remark 3.26. Note that a related but different approach has been followed in [13] where they deal with the constrained version of (3.26), namely

$$\min\left\{\int_D |\nabla u|^2, \quad u \in H^1(D), \quad |\Omega_u| = a, \quad u = u_0 \text{ on } \partial D\right\}. \tag{3.42}$$

They introduce the penalized version

$$\min\left\{\int_D |\nabla u|^2 + f_\varepsilon(|\Omega_u|), \quad u \in H^1(D), \quad u = u_0 \text{ on } \partial D\right\}, \tag{3.43}$$

where $f_\varepsilon(x) = \frac{1}{\varepsilon}(x - a)$ if $x \geq a$ and $f_\varepsilon(x) = \varepsilon(x - a)$ if $x \leq a$. They prove that the regularity theory can be applied to u_ε, and that for ε small enough, u_ε is such that $|\Omega_{u_\varepsilon}| = a$ and therefore u_ε actually solves (3.26). The results look very similar and are indeed based on similar observations, namely that the main point is to prove the estimate $0 < c \leq \mu_\varepsilon \leq C$ where $|\nabla u_\varepsilon|^2 = \mu_\varepsilon$ is the optimality condition for (3.43). But, we notice two main differences with the approach described above:

- First, the strategy in [13] leads to the weaker result that there exists smooth solutions to the constrained problem (3.42) (namely u_ε for ε small enough), while we get here that any solution is smooth,
- Next, the authors use in [13] the regularity theory in order to assert that solutions of (3.43) are actually also solutions to (3.42) for ε small enough. For this step, they apply the regularity theory from [25] to u_ε, in order to prove the estimate $0 < c \leq \mu_\varepsilon \leq C$ mentioned before. The approach described here relies on weaker properties and can therefore be applied to a wider class of examples. Several ideas can for instance be applied to the analysis of (3.2) where f is not a priori assumed to be nonnegative (see [187]). A complete regularity theory is however still to be done for this signed situation as we discuss in the next Section.

It is of course interesting to apply the ideas of [13] to (3.6). This has been done in the literature in [866]. See also [469] where a similar penalization is studied to prove that there exists an open set which solves (3.6), and Chapter 7 of the present book where the same penalization is used as well.

3.2.3.4 Conclusion

Let u be a solution of (3.10) and $B_R, \mathcal{F}$ as in 3.2.3.3. Before giving a sketch of the proof of Theorem 3.20, we need the following Lemma, which gives a rigorous meaning to the fact that, at a point $x \in \partial\Omega_u$, the gradient of u is bounded from above and from below.

Lemma 3.27. *There exist $C_1, C_2, r_0 > 0$ such that, for $B(x_0, r) \subset B$ with $r \leq r_0$,*

$$\begin{aligned} &\text{if } \frac{1}{r} \fint_{\partial B(x_0,r)} u \geq C_1 \quad \text{then} \quad u > 0 \text{ on } B(x_0, r),\\ &\text{if } \frac{1}{r} \fint_{\partial B(x_0,r)} u \leq C_2 \quad \text{then} \quad u \equiv 0 \text{ on } B(x_0, r/2). \end{aligned} \tag{3.44}$$

The detailed proof is given in [189], and here we describe the main ingredients. The first part was already proven earlier as we have seen that u is Lipschitz, but one can now give a slightly different proof of this fact, following the arguments from [25]. The idea is to use (3.40) for a suitable test function, namely v such that $-\Delta v = \lambda_a v$ in $B(x_0, r)$ and $v = u$ on $B(x_0, r)^c$. Again, it is interesting to notice that we only use perturbation from outside, namely v such that $\Omega_u \subset \Omega_v$.

The second part says that the gradient does not degenerate on the free boundary. As we noticed before, this relies on the fact that $\Lambda > 0$ and that we can penalize the volume constraint for some positive $\mu_-(h)$. As in the previous case, we now use (3.39) for v defined such that $v = 0$ in $B(x_0, r/2)$, $-\Delta v = \lambda_a u$ on $B(x_0, r) \setminus B(x_0, r/2)$ and $v = u$ on $B(x_0, r)^c$.

With this result, we are in position to give the main steps for the proof of Theorem 3.20.

1. The proof is now, using (3.44) in Lemma 3.27, the same as in [442] or in [25]: we first show a volume density estimate, namely that that there exists C_1, C_2 and r_0 such that, for every $B(x_0, r) \subset B$ with $r \leq r_0$,
$$0 < C_1 \leq \frac{|B(x_0, r) \cap \Omega_u|}{|B(x_0, r)|} \leq C_2 < 1,$$
and also an estimate for the measure $(\Delta u + \lambda_a u)$
$$C_1 r^{d-1} \leq (\Delta u + \lambda_a u)(B(x_0, r)) \leq C_2 r^{d-1}.$$
The proof is the same as in [442] with $\lambda_a u$ instead of f. It gives directly (using classical Geometric Measure Theory arguments, see section 5.8 in [370]) the first point of Theorem 3.20, namely that Ω^* has local finite perimeter and $\mathcal{H}^{d-1}((\partial\Omega_u \setminus \partial^*\Omega_u) \cap D) = 0$.
2. For the second point, we see that $\Delta u + \lambda_a u$ is absolutely continous with respect to $\mathcal{H}^{d-1} \lfloor \partial\Omega_u$ which is a Radon measure (using the first point), so we can use Radon-Nikodym's Theorem. To compute the Radon's derivative, we argue as in Theorem

2.13 in [442] or (4.7,5.5) in [25]. The main difference is that here, we have to use (3.41) in Proposition 3.24 to show that, if u_0 denotes a blow-up limit of $u(x_0+rx)/r$ (when r goes to 0), then u_0 is such that,

$$\int_{B(0,1)} |\nabla u_0|^2 + \Lambda|\{u_0\neq 0\} \cap B(0,1)| \leq \int_{B(0,1)} |\nabla v|^2 + \Lambda|\{v\neq 0\} \cap B(0,1)|,$$

for every v with $v = u_0$ outside $B(0,1)$. To show this, in [25] or in [442] the authors use only perturbations in $B(x_0, r)$ with r going to 0, so using (3.41), we get the same result. We can compute the Radon's derivative and get (in B)

$$\Delta u + \lambda_a u = \sqrt{\Lambda}\mathcal{H}^{d-1}\lfloor \partial\Omega_u,$$

which means we have proven the second point in Theorem 3.20.

3. Now, u is a weak-solution in the sense recalled in 3.2.3.1, and with Theorem 3.21, we directly get the regularity of $\partial^*\Omega_u$.
4. If $d = 2$, in order to have the regularity of the whole boundary, we have to show that Theorem 6.6 and Corollary 6.7 in [25] (which are stated for solutions and not weak solutions) are still true for our problem. The corollary comes directly from the theorem. So we need to show that, if $d = 2$ and $x_0 \in \partial\Omega_u$, then

$$\lim_{r\to 0} \fint_{B(x_0,r)} \max\{\Lambda - |\nabla u|^2, 0\} = 0. \tag{3.45}$$

As in [25, Theorem 6.6], the idea is to use (3.39) with $v = \max\{u - \varepsilon\zeta, 0\}$ and $\zeta \in C_0^\infty(B)^+$, and $h = |0 < u \leq \varepsilon\zeta| \leq |\{\zeta\neq 0\}|$, and we get

$$\int_{\{0<u<\varepsilon\zeta\}} (\Lambda - |\nabla u|^2) \leq \int_{\{u\geq\varepsilon\zeta\}} \varepsilon^2|\nabla\zeta|^2 + (\Lambda - \mu_-(h))h.$$

The only difference now with [25] is the last term. Using Proposition 3.24 again, we see that $(\Lambda - \mu_-(h))h = o(h)$, so we can choose the same kind of ζ and ε as in [25] to get (3.45) (see also Theorem 5.7 in [187] for more details). □

3.2.4 Remarks and perspectives

3.2.4.1 About the connectedness assumption

In this paragraph, we discuss the hypothesis "D is connected" in Theorem 3.20. We begin with the following example, taken from [188] which proves that the optimal set Ω^* may be irregular if D is not connected, although u_{Ω^*} is very regular.

Example 3.28. *(from [188])* We choose $D = D_1 \cup D_2$, where D_1, D_2 are disjoint disks in $\mathbb{R}^2$ of radius R_1, R_2 with $R_1 > R_2$. If $a = \pi R_1^2 + \varepsilon$, then the solution u of (3.10) coincides with the first eigenfunction of D_1 and is identically 0 on D_2, and thus $\Omega_u = D_1$ and $|\Omega_u| < a$.

In this case, we can choose any open subset ω of D_2 with $|\omega| = \varepsilon$. Then $\Omega^* := D_1 \cup \omega$ is a solution of (3.9). Since ω may be chosen to be as irregular as one wants, this proves that optimal domains are not regular in general.

We are however able to prove the following proposition.

Proposition 3.29 (The non-connected case). *If we suppose that D is not connected and with finite measure, the problem (3.10) still has a solution u which is locally Lipschitz continuous in D. If ω is any open connected component of D, we have three cases:*
1. *either $u > 0$ on ω,*
2. *or $u \equiv 0$ on ω,*
3. *or $0 < |\Omega_u \cap \omega| < |\omega|$: in that case $\Omega_u \subset \omega$ and $\partial\Omega_u$ has the same regularity as stated in Theorem 3.20.*

If $|\Omega_u| < a$, then only the first two cases can appear.

Remark 3.30. It follows from Proposition 3.29 that we obtain the same regularity as in the connected case. Indeed, in the first two cases, $\partial\Omega_u \cap \omega = \emptyset$.

Remark 3.31. To summarize, in all cases, there exists a solution Ω^* to (3.9) which is regular in the sense of Theorem 3.20, but there may be some other non regular optimal shape. And if D is connected, any optimal shape is regular.

Proof of Proposition 3.29. The existence and the Lipschitz regularity are stated in Theorem 3.16. In particular Ω_u is open. If $|\Omega_u| < a$, then by Remark 3.6, we are in one the two first situations. Let us now assume that $|\Omega_u| = a$. Then, as proved below:

A) either $|\Omega_u \cap \omega| = |\omega|$: then we are again in the first situation,
B) or $|\Omega_u \cap \omega| < |\omega|$: then $\Omega_u \subset \omega$ and u is solution of (3.10) with D replaced by ω. In particular, the regularity result of Theorem 3.20 applies (whence Point 3.).

To obtain A) and B), recall first that $u_{|\omega} \in H_0^1(\omega)$: indeed, u is the limit in $H_0^1(D)$ of $u_n \in C_0^\infty(D)$. Then $(u_n)_{|\omega} \in C_0^\infty(\omega)$ and converges in $H^1(\omega)$ to $u_{|\omega}$. Since also $u \in H_0^1(\Omega_u)$, it follows that $u_{|\Omega_u \cap \omega} \in H_0^1(\Omega_u \cap \omega)$. On the other hand, $-\Delta u = \lambda_a u$ on Ω_u and therefore on $\Omega_u \cap \omega$ so that $\lambda_1(\Omega_u \cap \omega) = \lambda_a$.

In the case A), we remark that, by minimality in (3.10) (since $|\omega| \le a$), and by monotonicity of $\lambda_1(\cdot)$

$$\lambda_a = \int_D |\nabla u|^2 \le \int_D |\nabla u_\omega|^2 = \lambda_1(\omega) \le \lambda_1(\Omega_u \cap \omega) = \lambda_a.$$

We also have $\lambda_a = \int_{\Omega_u \cap \omega} |\nabla u|^2 / \int_{\Omega_u \cap \omega} u^2 = \int_\omega |\nabla u|^2 / \int_\omega u^2$. Since ω is connected, $\lambda_1(\omega)$ is simple and therefore $u_\omega = u_{|\omega} / \|u_{|\omega}\|_{L^2(\omega)}$ and $u = u_\omega > 0$ on ω (see Remark 3.6).

In the case B), if we had $|\Omega_u \cap \omega| < a$, then we could find an open set $\hat{\omega}$ such that

$$\Omega_u \cap \omega \subset \hat{\omega} \subset \omega,\ \lambda_1(\hat{\omega}) < \lambda_1(\Omega_u \cap \omega) = \lambda_a,\ |\hat{\omega}| \leq a,$$

and this would be a contradiction with the minimality of λ_a. Thus $|\Omega_u \cap \omega| = a \geq |\Omega_u|$, that is $|\Omega_u \cap \omega| = |\Omega_u|$ which implies that $u = 0$ a.e. (and therefore eveywhere) on the complement of ω. Whence $\Omega_u \subset \omega$. □

3.2.4.2 Full regularity and improvement of the estimate of the singular set

It is natural to ask whether the regularity stated in Theorem 3.20 can be improved. As it mainly relies on the use of the theory of free boundary regularity for problem (3.29), we state here the different improvements that have been made in the literature since the original paper [25]:

- When concerned with regularity for weak solutions of (3.29), it is important to notice that regularity does not occur in general in dimension 3 or higher. Indeed, we recall here the construction of [25, Example 2.7]: using the spherical coordinates (r, ϕ, θ) so that $x = r(\cos\phi \sin\theta, \sin\phi \sin\theta, \cos\theta)$, we search for a function u of the form $rh(\theta)$, harmonic in $B(0,1) \subset \mathbb{R}^3$:

  $$\begin{aligned}\Delta u &= \tfrac{1}{r^2}\partial_r(r^2\partial_r u) + \tfrac{1}{r^2 \sin\theta}\partial_\theta(\sin\theta\partial_\theta u) + \tfrac{1}{r^2\sin^2\theta}\partial^2_\phi u = \\ &= \tfrac{1}{r\sin\theta}\left[2\sin\theta.h(\theta) + (\sin(\theta).h')'(\theta)\right].\end{aligned}$$

  Adding the condition $h'(\frac{\pi}{2}) = 0$ we obtain the explicit solution

  $$h(\theta) = 2 + \cos(\theta)\log\frac{1-\cos\theta}{1+\cos\theta}.$$

  Then, defining $u(x) = rh_+(\theta)$ and θ_0 the unique zero of h in $(0, \frac{\pi}{2})$, we have that $\Omega_u = \{(r,\theta,\phi), \theta_0 < \theta < \pi - \theta_0\}$ is singular. Moreover, it is easy to notice that in the basis (e_r, e_θ, e_ϕ) we have

  $$\nabla u = \partial_r u.e_r + \frac{1}{r}\partial_\theta u.e_\theta + \frac{1}{r\sin\theta}\partial_\phi u.e_\varphi = h(\theta)e_r + h'(\theta)e_\theta \text{ in } \Omega_u,$$

  so that

  $$|\nabla u| = h'(\theta_0) \text{ on } \partial\Omega_u \setminus \{0\}.$$

  It is easy to see that this function satisfies the shape derivative formulation of the free boundary (3.30), and is also a weak solution defined in Section 3.2.3.1. This proves that there exist critical sets for problem (3.26) in $\mathbb{R}^3$ that are singular. It can be seen that this example is not a minimizer; see also [533].
  Nevertheless, it is proven in [874] by G. Weiss that, for weak solutions (as defined in Section 3.2.3.1), in the case $f = 0$ and g is a positive constant:

  $$\dim_{\mathcal{H}}(\partial\Omega_u \setminus \partial^*\Omega_u) \leq d - 3, \tag{3.46}$$

(and in the case $d = 3$ it is known that singularities are isolated points) which recovers Theorem 3.22 and improves the estimate (3.32). This result is optimal since the previous critical set in $\mathbb{R}^3$ had a 0-dimensional singularity.
- In another work of G. Weiss [875], a better estimate is obtained for the singular set of minimizers for (3.26), and he also gives a strategy to obtain an optimal estimate: indeed he proves that there exists $k^* \in \mathbb{N} \cup \{+\infty\}$ such that if u solves (3.26), then

$$\dim_{\mathcal{H}}(\partial\Omega_u \setminus \partial^*\Omega_u) \leq d - k^* \tag{3.47}$$

(where we understand $\partial\Omega_u \setminus \partial^*\Omega_u = \emptyset$ if $d - k^* < 0$) and he gives a characterization of k^* as the minimal dimension so that there exists a singular homogeneous minimizer. This result relies on a monotonicity formula to prove that blow-ups are homogeneous, and a dimension reduction argument, similar to Federer's strategy in the regularity theory for perimeter minimizers. It is known from [25] that there is no singularity in dimension 2 (see Theorem 3.22) for weak solutions (and so for minimizers). It is also known that $k^* \geq 3$ and therefore Weiss recovers the estimate (3.46) of weak solutions, and gives a strategy to improve it.
- After Weiss' results, three improvements have been given about the number k^*:
  1. in [246] it is proven there is no singular cone in $\mathbb{R}^3$, and therefore $k^* \geq 4$,
  2. in [332] it is proven that there exists a singular cone in $\mathbb{R}^7$, and therefore $k^* \leq 7$.
  3. in [563] it is proven that there is no singular cone in $\mathbb{R}^4$, and therefore $k^* \geq 5$.

  Let us conclude by mentioning that a conjecture is $k^* = 7$.

It is natural to expect that the regularity for the free boundary problem (3.10) is very similar, therefore we propose:

Open problem 3.32. *Are solutions Ω^* to problem* (3.6) *such that*

$$\partial\Omega^* = \partial^*\Omega^* \quad \text{if } d < 7, \quad \text{and} \quad \dim_{\mathcal{H}}(\partial\Omega^* \setminus \partial^*\Omega^*) \leq d - 7 \quad \text{if } d \geq 7\,?$$

Of course, this open problem can be decomposed into two different open questions: first show that Weiss' results can be applied to the solutions of (3.6), and that an estimate like (3.47) can be obtained with a critical exponent $k^\sharp$ (which may actually be the same as k^*), then identify $k^\sharp$.

3.3 Minimization for λ_k

Here we consider the general minimization problems with $k \geq 1$, $\mu \in (0, \infty)$, $0 < a < |D|$:

$$\min\{\lambda_k(\Omega);\ \Omega \subset D,\ \Omega \text{ quasi-open } |\Omega| = a\}, \tag{3.48}$$

$$\min\{\lambda_k(\Omega) + \mu|\Omega|;\ \Omega \subset D,\ \Omega \text{ quasi-open}\}. \tag{3.49}$$

Applying the existence results of Chapter 2, we know that (3.48) and (3.49) have solutions when $D = \mathbb{R}^d$ (and this is a highly nontrivial result), or if D is bounded. Moreover, they are bounded and with finite perimeter. Actually, the existence proof requires one to prove a priori regularity properties for the expected optima (see Chapter 2)).

3.3.1 Penalized is equivalent to constrained in $\mathbb{R}^d$

We remark that when $D = \mathbb{R}^d$, the two problems (3.48) and (3.49) are equivalent for a good choice of μ. Obviously, any solution $\Omega^\star$ of (3.49) is a solution of (3.48) with $a = |\Omega^\star|$, and this for all $\mu \in (0, \infty)$. Conversely

Proposition 3.33. *Let $D = \mathbb{R}^d$. Then a solution of (3.48) is solution of (3.49) with $\mu = 2\lambda_k(\Omega^\star)/ad$.*

Proof. Let $\Omega^\star$ be a solution of (3.48) and let $\mu := 2\lambda_k(\Omega^\star)/ad$. Then for all quasi-open set Ω with $|\Omega| < +\infty$, we have

$$\lambda_k(\Omega^\star) \le \lambda_k\left((a|\Omega|^{-1})^{1/d}\Omega\right) = (a^{-1}|\Omega|)^{2/d}\lambda_k(\Omega), \tag{3.50}$$

which implies that for all $t > 0$

$$g(t) := a^{2/d}t^{-2}\lambda_k(\Omega^\star) + a^{1+2/d}\mu\, t^d \le t^{-2}|\Omega|^{2/d}\lambda_k(\Omega) + a^{1+2/d}\mu\, t^d.$$

The minimum of $t \in (0, \infty) \mapsto g(t)$ is reached at $t = 1$ so that $g(1) \le t^{-2}|\Omega|^{2/d}\lambda_k(\Omega) + a^{1+2/d}\mu\, t^d$ for all $t \in (0, \infty)$. Choosing $t = (a^{-1}|\Omega|)^{1/d}$ leads to

$$a^{-2/d}g(1) = \lambda_k(\Omega^\star) + \mu|\Omega^\star| \le \lambda_k(\Omega) + \mu|\Omega|.$$ □

Remark 3.34. It is interesting to notice that, if D is star-shaped (say around the origin), that is $[\Omega \subset D \Rightarrow t\Omega \subset D,\ \forall t \in [0, 1]]$, then *a solution $\Omega^\star$ of (3.48) is also a super-solution of (3.49)* with the same μ as in Proposition 3.33 that is

$$\lambda_k(\Omega^\star) + \mu a \le \lambda_k(\Omega) + \mu|\Omega| \text{ for all } \Omega^\star \subset \Omega \subset D. \tag{3.51}$$

Indeed, we do the same proof as above, using that if $\Omega^\star \subset \Omega \subset D$, then $a|\Omega|^{-1} \le 1$ so that $(a|\Omega|^{-1})^{1/d}\Omega \subset D$ and (3.50) holds. The rest of the proof remains unchanged.

3.3.2 A Lipschitz regularity result for optimal eigenfunctions

The following result is proved in [222], [864].

Theorem 3.35. *Let $D = \mathbb{R}^d$ and let $\Omega^\star$ be a solution of the minimization problem (3.48) or (3.49). Then there exists an eigenfunction associated with $\lambda_k(\Omega^\star)$ which is Lipschitz continuous.*

The proof of this result is much more involved than for the case $k = 1$. According to Proposition 3.11, it would be sufficient to prove that one of the eigenfunctions $u = u_{\Omega^*}$ satisfies

$$|\Delta|u||(B(x_0, r)) \le Cr^{d-1}, \tag{3.52}$$

around each x_0 where $u(x_0) = 0$. This property can actually be proved under the extra (strong) assumption that $\lambda_k(\Omega^*) > \lambda_{k-1}(\Omega^*)$. This is very restrictive, but it is nevertheless a starting point for the proof.

If this strict inequality does not hold, the strategy of [222] is to consider a series of auxiliary approximate shape optimization problems involving the lower i-th eigenvalues for $i \le k$. Using extensively the fact that (3.52) holds even for *super-solutions*, they prove that the approximate state functions are uniformly Lipschitz continuous and converge to one of the k-th eigenfunctions. The result follows.

Let us describe a little more the main steps of this proof.

1. A first step in reaching (3.52) is to try to prove that, for all functions $\varphi \in C_0^\infty(B(x_0, r))$, we have an estimate like we proved in the case $k = 1$ (see (3.24)), namely
$$|\langle \Delta u + \lambda_k u, \varphi\rangle| \le C\,|B(x_0, r)|^{1/2}\|\nabla\varphi\|_{L^2}. \tag{3.53}$$
This will actually be proved only under the extra assumption that $\lambda_k(\Omega^*) > \lambda_{k-1}(\Omega^*)$, see below. We emphasize that this estimate will hold for any *super-solution* of the penalized problem (3.49).
As we saw in the proof of Theorem 3.16, (3.53) directly implies (3.52) when $u \ge 0$ (see Remark 3.17 for more comments on this point). In the general case, (3.53) only provides an estimate of $|\Delta u^+|(B(x_0, r))$ in terms of $|\Delta u^-|(B(x_0, r))$ and conversely. Therefore we need one more piece of the puzzle to bound both of them. This is given by the Monotonicity Lemma 3.36 below.
2. At this point, it is at least possible to prove the continuity of u. The proof is rather direct at points x_0 where $u(x_0)\neq 0$ (since we formally have $\Delta u + \lambda_a u = 0$ around this x_0). Continuity at points x_0 of the free boundary where $u(x_0) = 0$ is more involved and uses the estimate (3.53).
3. The extra information required to estimate $|\Delta(u^+ + u^-)|(B(x_0, r))$ may be done around points x_0 where $u(x_0) = 0$ thanks to the following celebrated *Monotonicity Lemma* by Caffarelli-Jerison-Kenig [245]:

Lemma 3.36. *Let $U \in H^1(B(0, R))$ and*

$$\Delta U^+ \ge -a, \quad \Delta U^- \ge -a \quad \text{on } B(0, R) \text{ for some } a \ge 0.$$

Set

$$\Phi(r) := \left(\frac{1}{r^2}\int_{B(0,r)} \frac{|\nabla U^+|^2}{|x|^{d-2}}\right)\left(\frac{1}{r^2}\int_{B(0,r)} \frac{|\nabla U^-|^2}{|x|^{d-2}}\right).$$

Then,

- *if $a = 0$, $r \in (0, R) \mapsto \Phi(r)$ is nondecreasing.*
- *in all cases, there exists C such that*

$$\forall r \in (0, R/2), \quad \Phi(r) \le C \left[1 + \int_{B(0,R)} U^2 \right]. \tag{3.54}$$

The case $a = 0$ was first proved in the seminal paper [26] by Alt-Caffarelli-Friedman and in this case, we do have an actual monotonicity property. The estimate (3.54) implied by this monotonicity lemma when $a = 0$ was extended to the non-homogeneous case in [245]. This was proved under a continuity assumption for U which was later dropped by B. Velichkov [864], [863].
This lemma essentially says that for such functions U, both ∇U^+ and ∇U^- cannot be bad at the same time around x_0. Thus, if there exists some control on one by the other (and this is given by (3.53)), then we control both of them. The complete proof of this may be found in [188] (see also in [222], [864]).
4. Now, the question is: how does one prove (3.53)? The starting point is the 'super-solution property'

$$\lambda_k(\Omega^\star) \le \lambda_k(\Omega^\star \cup B(x_0, r)) + \mu\, |B(x_0, r)| \tag{3.55}$$

implied by (3.49). In the case of $k = 1$, we apply the optimal property to the optimal eigenfunction u and to the perturbation $u + t\varphi$, $\varphi \in C_0^\infty(B(x_0, r)$ to obtain

$$\int_D |\nabla u|^2 \le \int_D |\nabla(u + t\varphi)|^2 + \mu\, |B(x_0, r)|,$$

from which we easily deduce (3.53) (see the proof of Theorem 3.16 for such an argument). But, for higher eigenfunctions $k \ge 2$, one cannot work so easily with test functions due to the more complex variational functional characterization. Then, two main ideas are used in [222].
5. **Case** $\lambda_k(\Omega^\star) > \lambda_{k-1}(\Omega^\star)$: it can be proved (see [222],[864]) that for all $v \in H_0^1(B(x_0, r))$ with $\int |\nabla v|^2 \le 1$ and for $r \in (0, r_0)$,

$$\lambda_k(\Omega^\star \cup B(x_0, r)) \le \frac{\int |\nabla(u + v)|^2 + (\lambda_{k-1}(\Omega^\star) + 1) \int |\nabla v|^2}{\int (u + v)^2 - \int |\nabla v|^2/2}.$$

Plugging this information into (3.55) gives after a simple computation

$$|\langle \Delta u + \lambda_k(\Omega^\star)u, v\rangle| \le C|B(x_0, r)| + C_k \int |\nabla v|^2, \quad C_k = 1 + \lambda_{k-1}(\Omega^\star)/2 + \lambda_k(\Omega^\star)/4.$$

The choice of $v := |B(x_0, r)|^{1/2}\varphi/\|\nabla\varphi\|_{L^2}$ (r small enough) leads to the estimate (3.53) and the expected (3.52) follows as well as Lipschitz continuity.
6. **Case** $\lambda_k(\Omega^\star) = \lambda_{k-1}(\Omega^\star)$: this is the most difficult case, and also the most frequent since optimal eigenvalues are often repeated. Then the idea of [222] is to consider the problem with $\varepsilon > 0$ small:

$$\min\{(1 - \varepsilon)\lambda_k(\Omega) + \varepsilon\lambda_{k-1}(\Omega) + 2\mu\, |\Omega|; \Omega \supset \Omega^\star\}, \tag{3.56}$$

which does have a quasi-open solution (see Chapter 2). Then, there are two cases (A) and (B):

(A) Suppose there exists a sequence of optimizers Ω_{ε_n} with $\lim_{n\to\infty} \varepsilon_n = 0$ and such that $\lambda_k(\Omega_{\varepsilon_n}) > \lambda_{k-1}(\Omega_{\varepsilon_n})$. From (3.56), using also that $\lambda_{k-1}(\cdot)$ is nonincreasing for the inclusion, we have

$$\lambda_k(\Omega_{\varepsilon_n}) + \frac{2\mu}{1-\varepsilon}|\Omega_{\varepsilon_n}| \le \lambda_k(\Omega) + \frac{2\mu}{1-\varepsilon}|\Omega|,$$

for all $\Omega \supset \Omega_{\varepsilon_n}$. As in the points 4) and 5) above (where we used only the supersolution property of the optimizer), we deduce that $u_n := u_{\Omega_{\varepsilon_n}}$ is Lipschitz continuous and it can be checked that the Lipschitz constant does not depend on ε. At this step, convergence arguments must be used to prove that Ω_ε converges in some weak sense, but strong enough so that its limit is solution of the limit problem

$$\min\{\lambda_k(\Omega) + 2\mu|\Omega|; \Omega \supset \Omega^*\}. \tag{3.57}$$

It is easily seen that a solution of this problem necessarily coincides with Ω^*. Finally, it can be checked that the limit of the uniformly Lipschitz sequence u_n converges to one of the k-th eigenvalue of Ω^*. The statement of Theorem 3.35 follows.

(B) Suppose there exists $\varepsilon_0 \in (0,1)$ such that Ω_{ε_0} is a solution of (3.56) and $\lambda_k(\Omega_{\varepsilon_0}) = \lambda_{k-1}(\Omega_{\varepsilon_0})$. But Ω_{ε_0} is then also solution of (3.57) and therefore coincides with Ω^*. Using that $\lambda_k(\cdot)$ is nonincreasing for the inclusion, we deduce from (3.56) and $\Omega_{\varepsilon_0} = \Omega^*$ that

$$\lambda_{k-1}(\Omega^*) + 2\mu\varepsilon_0^{-1}|\Omega^*| \le \lambda_{k-1}(\Omega) + 2\mu\varepsilon_0^{-1}|\Omega|,$$

for all $\Omega^* \subset \Omega$. Thus Ω^* is also a *super-solution* for $\lambda_{k-1}(\cdot)+2\mu\varepsilon_0^{-1}|\cdot|$. Therefore, one can start again the discussion:

(B1) $\lambda_{k-1}(\Omega^*) > \lambda_{k-2}(\Omega^*)$,

(B2) $\lambda_{k-1}(\Omega^*) = \lambda_{k-2}(\Omega^*)$,

and we repeat the same analysis with adequate auxiliary problems, in the same spirit, and at most a finite number of times. Theorem 3.35 follows.

Remark 3.37. It is not known whether all k-th eigenfunctions are Lipschitz continuous. As proved in the same paper [222], it is the case when minimizing functions of the eigenvalues which involve all of them like

$$\min\left\{\sum_{j=1}^{p} \lambda_j(\Omega) + \mu|\Omega|;\ \Omega \subset \mathbb{R}^d,\ \Omega \text{ quasi-open}\right\}. \tag{3.58}$$

In order to understand this fact (see Corollary 3.40 below), let us mention that the strategy for Theorem 3.35 can be generalized to deal with problems like

$$\min\left\{F(\lambda_{k_1}(\Omega)), \dots, F(\lambda_{k_p}(\Omega)) + \mu|\Omega|;\ \Omega \subset \mathbb{R}^d,\ \Omega \text{ quasi-open}\right\}, \tag{3.59}$$

where $0 < k_1 < k_2 < \ldots < k_p$ and $F : \mathbb{R}^p \to [0, \infty)$ is locally bi-Lipschitz function, increasing in each variable. Indeed, for such functionals, the following result holds.

Theorem 3.38. *(see [222]) Let Ω^* be a bounded optimal shape of (3.59) (or even only a bounded* super-solution*). Then there exists a sequence of orthonormal eigenfunctions $u_{k_1}, \ldots, u_{k_p}$ corresponding to each of the eigenvalues $\lambda_{k_1}, \ldots, \lambda_{k_p}$ which are Lipschitz continuous. Moreover*

- *if $\lambda_{k_j}(\Omega^*) > \lambda_{k_j-1}(\Omega^*)$ for some j, then the full eigenspace corresponding to $\lambda_{k_j}(\Omega^*)$ consists of Lipschitz continuous functions;*
- *if $\lambda_{k_j}(\Omega^*) = \lambda_{k_{j-1}}(\Omega^*)$ for some j, then there exists at least $k_j - k_{j-1} + 1$ orthonormal eigenfunctions corresponding to $\lambda_{k_j}(\Omega^*)$ which are Lipschitz continuous.*

Remark 3.39. Note the difference between λ_{k_j-1} and $\lambda_{k_{j-1}}$ in the above theorem.

As a consequence of this theorem, the following holds for any optimal solution of Problem (3.58).

Corollary 3.40. *Let Ω^* be a solution of (3.58). Then, all eigenfunctions corresponding to the eigenvalues $\lambda_j(\Omega^*)$, $j = 1, \ldots, p$ are Lipschitz continuous on $\mathbb{R}^d$ and Ω^* is equal a.e. to an open set.*

Proof. By the previous theorem, all eigenfunctions corresponding to the eigenvalues $\lambda_j(\Omega^*)$ are Lipschitz continuous. Now let $\Omega^{**} := \cup_{j=1}^p [u_j \neq 0]$ where $u_1, \ldots, u_p$ is an orthonormal set of normalized eigenfunctions corresponding respectively to the $\lambda_j(\Omega^*)$, $j = 1, \ldots, p$. This set Ω^{**} is open and $\Omega^{**} \subset \Omega^*$. Moreover, $u_j \in H_0^1([u_j \neq 0]) \subset H_0^1(\Omega^{**})$ and satisfies $\int_{\Omega^*} \nabla u_j \nabla \varphi = \lambda_j(\Omega^*) \int_{\Omega^*} u_j \varphi$ for all $\varphi \in H_0^1(\Omega^*)$ and therefore for all $\varphi \in H_0^1(\Omega^{**})$. Thus, all the $\lambda_j(\Omega^*)$ are also eigenvalues on Ω^{**}, with at least the same multiplicity. Due to the monotonicity for the inclusion, we actually have $\lambda_j(\Omega^{**}) = \lambda_j(\Omega^*)$ for all $j = 1, \ldots, p$. Now, using also the optimality of Ω^*, we may write

$$\sum_{j=1}^{p} \lambda_j(\Omega^*) + \mu|\Omega^*| \leq \sum_{j=1}^{p} \lambda_j(\Omega^{**}) + \mu|\Omega^{**}| = \sum_{j=1}^{p} \lambda_j(\Omega^*) + \mu|\Omega^{**}|.$$

Since $|\Omega^{**}| \leq |\Omega^*|$, this implies that equality holds and thus proves that Ω^* is open up to a set of zero Lebesgue measure. □

Remark 3.41. As proved in [222], this corollary may be extended in two directions:

- first $\sum_j \lambda_j(\cdot)$ may be replaced by $F\big(\lambda_1(\cdot), \ldots, \lambda_p(\cdot)\big)$ where $F : \mathbb{R}^p \to [0, \infty)$ is locally bi-Lipschitz and increasing with respect to each variable;
- then, to the pure constrained problem, namely

$$\min\{F(\lambda_1(\Omega), \ldots, \lambda_p(\Omega)); \Omega \text{ quasi-open}, |\Omega| = 1\}. \tag{3.60}$$

Let us explain why these extensions hold.

- Extension to F (with the penalized term $\mu|\cdot|$) is done as in Corollary 3.40 by using Theorem 3.38 above.
- Extension to the constrained problem is done by proving that an optimal solution of (3.60) is a super-solution of the penalized version for some $\mu > 0$ (in the spirit of Proposition 3.33 and Remark 3.34). Indeed, if Ω^* is an optimal set of (3.60), and Ω a quasi-open set with finite measure such that $\Omega^* \subset \Omega$, then letting $t := [|\Omega|/|\Omega^*|]^{1/d} > 1$, we have

$$\begin{cases} F(\dots,\lambda_j(\Omega^*),\dots) \le F(\dots,\lambda_j(\Omega/t),\dots) = F(\dots,t^2\lambda_j(\Omega),\dots) \\ \le F(\dots,\lambda_j(\Omega),\dots) + \|F\|_{Lip}(t^2-1)\sum_j \lambda_j(\Omega) \\ \le F(\dots,\lambda_j(\Omega),\dots) + \|F\|_{Lip}(t^d-1)\sum_j \lambda_j(\Omega^*) \\ \le F(\dots,\lambda_j(\Omega),\dots) + \mu[|\Omega| - |\Omega^*|], \end{cases}$$

with $\mu = \|F\|_{Lip}|\Omega^*|^{-1}\sum_j \lambda_j(\Omega^*)$.

Open problem 3.42. *Concerning the minimization of $\lambda_k(\Omega)$, $k \ge 2$ as in Problems (3.48) or (3.49):*
- *Does there exist an optimal solution which is open?*
- *Are all k-th eigenvalues Lipschitz continuous?*

Partial answers are given next when $k = 2$.

3.3.3 More about $k = 2$

3.3.3.1 An example with singular solutions

We go back to problems (3.48)-(3.49) with $k = 2$. First, let us give an example showing that, as for $k = 1$, if the box D is not connected, then a quite different qualitative behavior may occur. We saw (see Corollary 3.18) that the optimal first eigenfunction is nevertheless (locally) Lipschitz continuous and consequently, there is an optimal set which is open, but optimal sets are not all open. For $k = 2$, the situation is even worse since the second eigenfunctions which are optimal for (3.48) may not be regular. This is seen in the following example.

Example 3.43. Let $D := D_1 \cup D_2 \subset \mathbb{R}^2$ where D_1, D_2 are *disjoint open disks* of radius respectively $R_1 > 0$ and $R_2 = R_1(1+2\varepsilon)$, $\varepsilon > 0$. Let $a := \pi R_1^2[1+(1+\varepsilon)^2]$. Let Ω^* be an optimal quasi-open solution of (3.48) with $k = 2$, namely

$$\lambda_2(\Omega^*) = \min\{\lambda_2(\Omega);\ \Omega \subset D,\ \Omega \text{ quasi-open},\ |\Omega| = a\},\quad \Omega^* \subset D,\ |\Omega^*| = a. \tag{3.61}$$

By monotonicity, $\lambda_2(\Omega^*) \ge \lambda_2(D) = \lambda_1(D_1)$. Actually, equality holds since, by minimality, $\lambda_2(\Omega^*) \le \lambda_2(D_1 \cup D_3)$ where D_3 is the disk of radius $R_1(1+\varepsilon)$ with the same center as D_2 (note that $|D_1 \cup D_3| = a$). And $\lambda_2(D_1 \cup D_3) = \lambda_1(D_1)$. Thus, $D_1 \cup D_3$ is also optimal.

Now we may perturb D_3 (for instance near its boundary) into an open set $D' \subset D_2$ so that:

a) $|D'| = |D_3|$, which means $|D_1 \cup D'| = a$,
b) $\lambda_1(D_3) \le \lambda_1(D') < \lambda_1(D_1)$ and therefore $\lambda_2(D_1 \cup D') \le \lambda_1(D_1) = \lambda_2(\Omega^*)$ ($\le \lambda_2(D_1 \cup D')$ by optimality);
c) the boundary of D' is irregular.

Then, since $\lambda_2(D_1 \cup D') = \lambda_1(D_1)(= \lambda_2(\Omega^*))$, $|D_1 \cup D'| = a$, $D_1 \cup D'$ is also a solution of the above problem, but it is not regular (one could even choose D' so that it be only quasi-open and not a.e. equal to an open set).
Now, we can perturb D' into D'' so that $|D''| = |D_3|$, and $\lambda_1(D'') = \lambda_1(D_1)$ (for instance by taking off larger and larger circles from D'). In this case, $D_1 \cup D''$ is still optimal, but one of its second eigenfunctions (namely, the first eigenfunction of D'') is not regular.

Note that, in this situation, it may happen that a solution of (3.48) is not a solution of (3.49), regardless of the value of $\mu > 0$. For instance, there does not exist any $\mu > 0$ such that $D_1 \cup D_3$ is a solution of (3.49) although it is solution of (3.48) as we just saw. Indeed, let D_4 be the unit disk with the same center as D_2. Then

$$\begin{cases} \lambda_2(D_1 \cup D_4) + \mu|D_1 \cup D_4| = \lambda_1(D_1) + \mu[|D_1| + |D_4|] \\ < \lambda_1(D_1) + \mu[|D_1| + |D_3|] = \lambda_2(D_1 \cup D_3) + \mu|D_1 \cup D_3|. \end{cases}$$

The same remark is valid for $D_1 \cup D'$ or $D_1 \cup D''$.

On the other hand, as indicated in Remark 3.34, if D is star-shaped, then a solution of (3.48) is also a super-solution of (3.49) for some adequate $\mu > 0$. It is very likely that the regularity analysis used to prove Theorem 3.38 would extend from $D = \mathbb{R}^d$ to "good" boxes D, locally inside D.

3.3.3.2 There are open optimal sets

As a partial answer to the open problems indicated at the end of Section 3.3.2, let us mention the following result proved in [224, 864] that uses the regularity result of Theorem 3.16.

Theorem 3.44. *Let $D \subset \mathbb{R}^d$ be open, connected and with finite measure. Let Ω^* be a solution of*

$$\min\{\lambda_2(\Omega) + \mu|\Omega|,\ \Omega \subset D,\ \Omega \text{ quasi-open}\}. \tag{3.62}$$

Then, Ω^* is a.e. equal to an open set.

Proof. Let us indicate the main steps of the proof. We denote by u_1, u_2 a first and a second eigenfunction on Ω^* with $\int_{\Omega^*} u_1 u_2 = 0$, $u_1 \ge 0$. We denote $\Omega_1 = [u_1 > 0]$, $\Omega^+ = [u_2 > 0]$, $\Omega^- = [u_2 < 0]$.

Let us first prove that we may choose u_2 so that both Ω^+, Ω^- are not empty (*this uses the fact that D is connected and may not hold otherwise as seen in Example 3.43*). Assume that $u_2 \geq 0$ on D (that is $\Omega^- = \emptyset$). Then, $\lambda_2(\Omega^*) = \lambda_1(\Omega^+)$ and the relation $\int_{\Omega^*} u_1 u_2 = 0$ implies $u_1 \equiv 0$ on Ω^+. We have $\lambda_1(\Omega_1) = \lambda_1(\Omega^*)$. Let us prove that

$$\lambda_1(\Omega^*) = \lambda_2(\Omega^*). \tag{3.63}$$

It will follow that we may replace u_2 by $u_2 - u_1$ so that the 'new' Ω^+ and Ω^- are not empty as expected.

Assume by contradiction that $\lambda_1(\Omega^*) < \lambda_2(\Omega^*)$, that is $\lambda_1(\Omega_1) < \lambda_1(\Omega^+)$. Since D is connected and open, we may find $x_0 \in \partial\Omega^+$ and $r > 0$ such that, if $\Omega^r := \Omega^+ \cup B(x_0, r)$, then,

$$\Omega^r \subset D, \quad |\Omega^r| > |\Omega^+|, \quad \lambda_1(\Omega^r) \in \left(\lambda_1(\Omega_1^r), \lambda_1(\Omega^+)\right),$$

where Ω_1^r is chosen so that $\Omega_1^r \subset \Omega_1 \setminus \Omega^r$ and $|\Omega_1^r \cup \Omega^r| = |\Omega_1 \cup \Omega^+|$. Note that we use the continuity at 0 of $r \in [0, r_0) \mapsto (\lambda_1(\Omega^r), \lambda_1(\Omega_1^r))$. We then have

$$\lambda_2(\Omega_1^r \cup \Omega^r) + \mu|\Omega_1^r \cup \Omega^r| = \lambda_1(\Omega^r) + \mu|\Omega_1^r \cup \Omega^r| < \lambda_2(\Omega^*) + \mu|\Omega^*|,$$

which is a contradiction with the minimality of Ω^*. Whence (3.63).

Thus we have Ω^+, Ω^- not empty and

$$\lambda_2(\Omega^*) = \lambda_1(\Omega^+) = \lambda_1(\Omega^-) = \lambda_2(\Omega^+ \cup \Omega^-), \quad |\Omega^+ \cup \Omega^-| = |\Omega^*|,$$

the last identity coming from the minimality: $\lambda_2(\Omega^*) + \mu|\Omega^*| \leq \lambda_2(\Omega^+ \cup \Omega^-) + \mu|\Omega^+ \cup \Omega^-|$.

Now we remark that Ω^+ (resp. Ω^-), are *subsolutions* of (3.62). Indeed, for $\omega \subset \Omega^+$, we may write

$$\lambda_2(\Omega^*) + \mu|\Omega^*| \leq \lambda_2(\omega \cup \Omega^-) + \mu|\omega \cup \Omega^-|.$$

This is the same as

$$\lambda_1(\Omega^+) + \mu[|\Omega^+| + |\Omega^-|] \leq \lambda_1(\omega) + \mu[|\omega| + |\Omega^-|],$$

or

$$\lambda_1(\Omega^+) + \mu[\Omega^+| \leq \lambda_1(\omega) + \mu|\omega|,$$

whence the subsolution property. The same holds for Ω^-. It follows from a (nontrivial) result in [224, 864] that there exist two *open* sets $D^+, D^- \subset D$ such that

$$\Omega^+ \subset D, \ \Omega^- \subset D, \ D^+ \cap \Omega^- = \emptyset, \ D^- \cap \Omega^+ = \emptyset.$$

This result relies on the fact that subsolutions of (3.62) are also subsolutions for the torsion energy as explained in Chapter 2 (see (2.7) and Paragraph 2.5). Let us show that Ω^+ is solution of the following problem

$$\min\{\lambda_1(\Omega); \Omega \subset D^+, \ \Omega \text{ quasi}-\text{open}, \ |\Omega| = |\Omega^+|\}. \tag{3.64}$$

Assume by contradiction that there exists $\Omega \subset D^+$ such that $\lambda_1(\Omega) < \lambda_1(\Omega^+)[= \lambda_1(\Omega^-) = \lambda_2(\Omega^*)]$, $|\Omega| = |\Omega^+|$. Then we argue as above by introducing $\Omega_r^- := \Omega^- \cup B(x_0, r)$ for some $x_0 \in \partial\Omega^-$, $r > 0$ such that

$$\Omega_r^- \subset D, \ |\Omega_r^-| > |\Omega^-|, \ \lambda_1(\Omega_r^-) \in \left(\lambda_1(\Omega^r), \lambda_1(\Omega^-)\right),$$

where Ω^r is chosen so that $\Omega^r \subset \Omega \setminus \Omega_r^-$ and $|\Omega_r^- \cup \Omega^r| = |\Omega \cup \Omega^-|$. We then have

$$\lambda_2(\Omega_r^- \cup \Omega^r) + \mu|\Omega_r^- \cup \Omega^r| = \lambda_1(\Omega_r^-) + \mu|\Omega_r^- \cup \Omega^r| < \lambda_2(\Omega^*) + \mu|\Omega^*|,$$

which is a contradiction with the minimality of Ω^*. Thus Ω^+ is solution of (3.64).

Since D^+ is open, we may apply Theorem 3.16 which says that u_2 is locally Lipschitz continuous on D^+. It follows that $\Omega^+ = [u_2 > 0]$ is open. Similarly, Ω^- is open so that $\Omega^+ \cup \Omega^-$ is an open optimal set with the same measure as Ω^*. □

The same question can be asked for the constrained problem:

Open problem 3.45. *Are the minimal shapes of*

$$\min\{\lambda_2(\Omega); \Omega \subset D, \ \Omega \text{ quasi-open}, \ |\Omega| = a\}$$

open subsets of D ?

3.4 Singularities due to the box or the convexity constraint

In this section, we study the regularity up to the boundary of the box D for the problem

$$\min\left\{\lambda_1(\Omega), \quad \Omega \subset D, \Omega \text{ quasi-open}, |\Omega| = a\right\}, \tag{3.65}$$

where D is a *smooth* open set of $\mathbb{R}^2$. If Ω^* solves (3.65), we expect the contact between $\partial\Omega^*$ and ∂D to be a bit smooth (see below), but as we will see, the smoothness is in general limited.

In order to insist on the fact that this appearance of a (mild) singularity is not only due to the box, we also show that a similar behavior applies to the solutions of the following problem:

$$\min\left\{\lambda_2(\Omega), \Omega \text{ open and convex}, |\Omega| = a\right\}. \tag{3.66}$$

This problem is peculiar because of the convexity constraint: If we drop this constraint, it is well-known that the solution of this problem is any disjoint union of two balls of volume $a/2$ (see Figure 3.1), which is clearly not convex, therefore Problem (3.66) is interesting on its own. It has been studied in [508] where it is proven, in particular, that the optimal shape is not a stadium (convex hull of two tangent balls). They also obtain a rough description of the optimal shape, under the a priori assumption

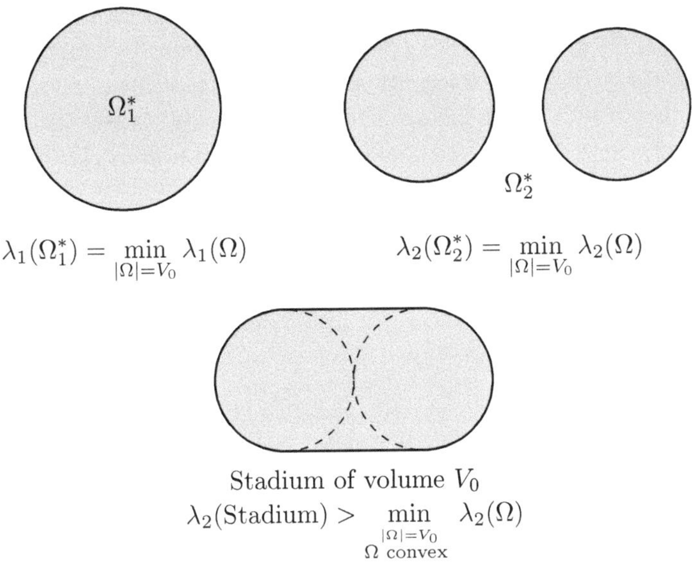

Fig. 3.1. Minimization of the first two eigenvalues under volume constraint

that the shape is $C^{1,1}$, and has a simple geometry. But as shown in this Section, the a priori $C^{1,1}$-regularity assumption is too optimistic (see also Remark 3.48 for a related comment).

The previous sections were concerned with the regularity of the pieces of the free boundary $\partial\Omega^*$ where it is expected to give sense to the optimality condition $|\nabla u|^2 = \Lambda$ where $\Lambda \in (0, \infty)$ is a Lagrange multiplier and u is the eigenfunction associated to the eigenvalue under study. Here we focus on the global regularity of the boundary $\partial\Omega^*$. We divide this section in 3 paragraphs. First we describe the situation and the possible regularity that we can expect from the optimality condition, seen as a partially overdetermined problem. Then we focus on a particular case of (3.65) where D is a strip. Finally we give some partial answers for (3.66). Since the conclusions are only partial, we describe open problems that would be interesting to fully investigate.

3.4.1 Regularity for partially overdetermined problem

For both of these problems, (3.65) or (3.66), if Ω^* is an optimal shape, the boundary $\partial\Omega^*$ can be decomposed into two subsets, namely the free boundary $\Gamma_1 \subset \partial\Omega^*$ where one can write an optimal condition for optimality, and the set $\Gamma_2 \subset \partial\Omega^* \setminus \Gamma_1$ of saturation of the constraint. More precisely,

- For problem (3.65), $\Gamma_1 = \partial\Omega^* \cap D$; we have seen that this boundary is locally smooth, and that $|\nabla u|^2 = \Lambda$ on Γ_1, where u is the normalized first eigenfunction of Ω^* and $\Lambda \in (0, \infty)$ is a Lagrange multiplier for the volume constraint. The set Γ_2 is equal to $\partial\Omega^* \cap \partial D$. If this set is empty (or reduced to one point), then by Serrin's result on overdetermined boundary problems (see [805]), the set Ω^* must be a ball (which is the unconstrained minimizer, so the box is irrelevant). When this set is not empty (which is the case when D does not contain any ball of volume a), it means that the constraint $\Omega \subset D$ is active.
- For problem (3.66), one can define

$$\Gamma_1 := \{x \in \partial\Omega^* \ / \ \exists r > 0 \text{ such that } B_r(x) \cap \Omega^* \text{ is strictly convex}\} \tag{3.67}$$

(where we understand an open set ω to be 'strictly convex' if $\forall(x, y) \in \overline{\omega}$ with $x \neq y, \forall t \in (0, 1), tx + (1 - t)y \in \omega$). The set Γ_1 is a relatively open subset of $\partial\Omega^*$. It will improperly be called the strictly convex part of the boundary. This set can be understood as the part of $\partial\Omega^*$ where the curvature is positive, though one has to be careful since the curvature is defined in a weak sense, and is a priori only a measure, see Section 3.4.3 for more details. We can define Γ_2 as the union of all closed nontrivial segments which are included in $\partial\Omega^*$. This set represents the part of $\partial\Omega^*$ where the convexity constraint is saturated (vanishing curvature). It is a priori difficult to understand the structure of the sets Γ_1 and Γ_2: in particular, with the previous definitions, it is not true in general that $\Gamma_1 \cup \Gamma_2 = \partial\Omega^*$. We notice that it is possible to construct a $C^{1,1}$ convex domain such that $\Gamma_1 = \emptyset$ and Γ_2 is strictly contained (and dense) in $\partial\Omega^*$. Such singular convex set can be obtained by taking the epigraph of $f : [0, 1] \to \mathbb{R}$ such that $f'' = \mathbb{1}_K$ (where f'' is understood in the sense of distributions) and K is a compact set with positive measure and empty interior. Nevertheless, if one assumes that Γ_2 is made of a finite number of segments, then $\partial\Omega^* = \Gamma_1 \cup \Gamma_2$.

Let us focus here on the regularity of a point which is at the intersection of $\overline{\Gamma_1}$ and Γ_2. In both cases, the situation is the following:

- On the side Γ_1, one has the overdetermined equation $|\nabla u(x)|^2 = \Lambda$, where either u is the first eigenfunction, or the second eigenfunction of Ω^* (depending on whether Ω^* solves (3.65) or (3.66)). This fact is not easy to prove in the case (3.66), as a smooth deformation of a strictly convex set does not necessarily remain convex, see Section 3.4.3.
- On the side Γ_2, we have an information about the geometry, *up to the intersection point*: either this part is flat (for solutions of (3.66)) or it is smooth (for solutions of (3.65), assuming the box D is smooth).

Notice first that it is likely to expect the contact to be C^1: indeed, assume that the boundary is piecewise smooth around x_0 and that there is a (convex and non-flat) corner at x_0. Then it is classical that $\nabla u(x)$ goes to 0 when $x \to x_0$ in $\overline{\Omega}$, but this would contradict the fact that $|\nabla u(x)|^2 = \Lambda$ on Γ_1. This proof is not completely valid as we do not know, even applying the results of the previous sections, that $\partial\Omega^*$ is piecewise smooth. However, it implies that we expect the optimal shape to be at least C^1. For problem (3.66), we will give a proper proof of this fact in Section 3.4.3; for problem (3.65), it seems that this is not proved anywhere yet.

In the following result, knowing that the contact is C^1, we prove higher regularity, and analyze the possible singularity near such a point.

Proposition 3.46 ([621]). *Let Ω be an open bounded subset of $\mathbb{R}^2$, $x_0 \in \partial\Omega$, $\gamma_1 \subset \partial\Omega$ and $\gamma_2 \subset \partial\Omega$ two relatively open connected sets, such that*
- $\overline{\gamma_1} \cap \overline{\gamma_2} = \{x_0\}$
- $\overline{\gamma_1 \cup \gamma_2}$ *is C^1, and $\overline{\gamma_2}$ is C^∞.*

We assume there exists $u \in C^2(\Omega) \cap C^1(\Omega \cup \gamma_1) \cap L^\infty(\Omega)$ satisfying

$$\begin{cases} -\Delta u &= f(u) & \text{in } \Omega \\ u &= 0 & \text{on } \partial\Omega \\ |\nabla u|^2 &= \Lambda > 0 & \text{on } \gamma_1, \end{cases} \tag{3.68}$$

where $f : \mathbb{R} \to \mathbb{R}$ is a C^∞ function, and $f(u) \geq 0$ in a neighborhood of x_0. Then,
- *either $\overline{\gamma_1 \cup \gamma_2}$ is C^∞,*
- *or there exists $k \in \mathbb{N}^*$ such that $\overline{\gamma_1 \cup \gamma_2}$ is $C^{k,\frac{1}{2}}$ and $\forall \varepsilon > 0$, $\overline{\gamma_1 \cup \gamma_2}$ is not $C^{k,\frac{1}{2}+\varepsilon}$.*

Sketch of proof: thanks to the fact that we are in a two-dimensional framework, we use the conformal mapping of the set Ω (or only a neighborhood of x_0 in $\overline{\Omega}$). There is a biholomorphic map $\phi : \mathbb{H} \to \Omega$ where $\mathbb{H} = \{z \in \mathbb{C}, Im(z) < 0\}$, which is such that $\phi(0) = x_0$, $\phi^{-1}(\gamma_1) \subset \mathbb{R}_-$, $\phi^{-1}(\gamma_2) \subset \mathbb{R}_+$. The regularity of $\partial\Omega$ and γ_2 can be read on the regularity of the trace of $Arg(\phi')$, respectively on $\partial\mathbb{H}$ or $\mathbb{R}_+$ (as it is a parametrization of the angle of the tangent vector to $\partial\Omega$, see for example [773]), and the regularity of $|\nabla u|$ on γ_1 can be read on the regularity of the trace of $\log(|\phi'|)$ on $\mathbb{R}_-$ (it is seen by transporting (3.68) on $\mathbb{H}$ and studying the regularity of $u \circ \phi$). These two functions $\log(|\phi'|)$ and $Arg(\phi')$ are harmonic and conjugated to each other. This harmonicity provides a mixed boundary value problem for each of them. The analysis of singularities for such a problem leads to the fact that either $Arg(\phi')$ is smooth on $\overline{\mathbb{H}}$, or its behavior near 0 is of the form $r^{k+1/2}\cos(k\varphi/2)$ for some $k \in \mathbb{N}^*$, where (r, φ) are radial coordinates in $\mathbb{H}$, see [439]. This leads to the expected result, as the regularity of $Arg(\phi')$ on $\overline{\mathbb{H}}$ near 0 implies the regularity of $\partial\Omega$ in a neighborhood of x_0. □

3.4.2 Minimization of λ_1 in a strip

We focus here on the particular case where $D = \mathbb{R} \times (-M, M)$ is a strip in $\mathbb{R}^2$. Thanks to the symmetries of this box, we can give a complete description of the regularity of optimal shapes for (3.65). It is expected that a similar behavior happens for more general boxes, see the end of this Section for open problems.

Proposition 3.47. *Let $a > 0$ and $D = \mathbb{R} \times (-M, M)$ for some $M > 0$. Let $\Omega^* \subset \mathbb{R}^2$ be a solution of* (3.65).
We assume that the contact between $\partial\Omega^*$ and ∂D is tangential. Then
- *either Ω^* is a disk,*
- *or $\left[\ \partial\Omega^* \in C^{1,\frac{1}{2}},\ \text{and}\ \forall\varepsilon > 0,\ \partial\Omega^* \notin C^{1,\frac{1}{2}+\varepsilon}\ \right]$.*

Remark 3.48. This result is a bit surprising, considering the behavior of solutions to the constrained isoperimetric problem

$$\min\left\{P(\Omega),\quad \Omega \subset D,\quad |\Omega| = a\right\}.$$

In this case, the contact between Ω^* optimal and the boundary of the box is expected to be $C^{1,1}$. In dimension 2 for example, this fact is easy to understand since the optimal shape is made of arcs of circles, touching the boundary tangentially (it is easy to see, for example writing optimality conditions near the contact point, that having a non-flat angle of contact is not optimal). For a more general result, see [832].

Sketch of proof: (see also [620, 621])
1) It is well known that the solution of (3.65) is the ball of volume a, if this one is admissible (included in D). If such a ball does not exist, we already saw that any optimal shape Ω^* should touch the boundary of the box on a nontrivial set. Since the cylindrical box $D = \mathbb{R} \times (-M, M)$ has two orthogonal symmetry axes, one can prove using two Steiner symmetrizations, that Ω^* also has two axes of symmetry and is vertically and horizontally convex (see for example [390] for more details), and therefore the free boundary $\Gamma_1 = \partial\Omega^* \cap D$ necessarily has exactly two connected components, and the remaining boundary $\Gamma_2 = \partial\Omega^* \cap \partial D$ is the union of two segments.

2) If we assume that the contact is tangential, we know that the full boundary is C^1, and applying Proposition 3.46 around one "corner" (a point of $\overline{\Gamma_1} \cap \Gamma_2$), we get that $\partial\Omega^*$ is $C^{1,\frac{1}{2}}$ or at least $C^{2,\frac{1}{2}}$ (by symmetry the regularity at each corner is the same).

3) We then assume the contact is C^2 and seek a contradiction: to that end we adapt an argument from [508]. The idea is that from the regularity of Ω^*, we know that u is C^2 on $\overline{\Omega^*}$, and from this information, we will obtain a contradiction by studying the nodal sets of $\partial_x u$, showing that one of them (denoted ω) is such that $\partial_x u$ is a first eigenfunction for the Dirichlet-Laplacian on ω, which is indeed a contradiction with the strict monotonicity of the eigenvalues, since $\lambda_1(\Omega^*) < \lambda_2(\Omega^*) = \lambda_1(\omega)$ while $\omega \subset \Omega^*$. As we know that $-\Delta\partial_x u = \lambda_2(\Omega^*)\partial_x u$ on Ω^*, one only needs to check that such ω can be chosen such that $\partial_x u = 0$ on $\partial\omega$.

To find ω, we show that, thanks to the C^2-regularity of u, differentiating tangentially $|\nabla u|^2$ on $\Gamma_1 = \partial\Omega^* \cap D$ gives $\partial_{xy}u(x_0) = 0$ where x_0 is (say) the upper left "corner" of the optimal shape. We also know $\partial_x u(x_0) = 0$ (as $u = 0$ on the upper segment), so from the strong maximum principle, x_0 belongs to the closure of both the sets $[\partial_x u > 0]$ and $[\partial_x u < 0]$. Let us denote by ω one connected component of $[\partial_x u < 0]$ that has points in a neighborhood of x_0. From the symmetries of Ω^*, u is even, so ω is on the left of the vertical axis of symmetry of Ω^*. We can easily check that $\partial_x u \geq 0$ on the left part of the free boundary and that $\partial_x u = 0$ on $\partial D \cap \partial\Omega^*$. We then conclude that $\partial_x u = 0$ on $\partial\omega$, which completes the contradiction and the proof. □

We conclude this section with the following open problem:

Open problem 3.49. *Concerning Problem* (3.65) *where D is a smooth open set in* $\mathbb{R}^2$, *can one prove that any optimal shape* Ω^* *is globally* $C^{1,\frac{1}{2}}$*?*

Of course, a similar question in higher dimension can be asked, but the regularity is already limited by the possible singularities of the free boundary, so it makes more sense to obtain a more satisfying regularity theory of the free boundary first, as it is done in $\mathbb{R}^2$ so far, see Section 3.2.4.2.

3.4.3 Minimization of λ_2 with convexity constraint

We address here the question of the regularity of an optimal shape Ω^* for problem (3.66).

Theorem 3.50. *Let* $a > 0$ *and let* $\Omega^* \subset \mathbb{R}^2$ *be a solution of the minimization problem* (3.66), *that is to say an optimal convex set of given area for the second Dirichlet-Laplacian eigenvalue.*
We assume:

$$\Omega^* \text{ contains at most a finite number of segments in its boundary.} \tag{3.69}$$

Then

$$\Omega^* \text{ is } C^{1,\frac{1}{2}}, \text{ and } \forall\, \varepsilon > 0,\ \Omega^* \text{ is not } C^{1,\frac{1}{2}+\varepsilon}. \tag{3.70}$$

Remark 3.51. About assumption (3.69)**:** as we noticed before (see (3.67)) the boundary of a convex shape contains two specific subsets, Γ_1 the strictly convex part, and Γ_2 the flat parts, but in general, even if the set Ω^* is only slightly regular, these pieces of the boundary can have a highly non-trivial structure. We know that Γ_2 is not empty and contains at least two segments (see the proof below). We notice that it is announced in [507, 508] that Γ_2 is made exactly of two segments (and then that these segments are parallel), but it seems that the proof is not complete. This explains our geometric assumption, which does not appear in [507, 508], but is implicitly used in these papers. It is even not known whether the strictly convex part Γ_1 as defined in (3.67), is not empty, even if Ω^* is assumed to be smooth (say C^1, since we have a proof of this fact, see below). With assumption (3.69), everything becomes more simple, and one can focus on the singularities at junction points between a flat part and a strictly convex part.

Remark 3.52. As noted below, the first step in the proof of this result is to prove that $\Omega^\star$ is C^1. This fact is actually very general, namely the result in [205] states that any optimal shape for

$$\min\left\{F(\lambda_1(\Omega),\dots,\lambda_k(\Omega))+\mu|\Omega|,\quad \Omega\subset D,\quad \Omega \text{ open and convex}\right\}$$

(where $F:\mathbb{R}^k\to\mathbb{R}$ is Lipschitz continuous, $\mu\in(0,\infty)$ and D is open) is C^1. Compared with the results in Section 3.3, this one is much easier since the convexity guarantees some a priori regularity for any optimal shape $\Omega^\star$. The difficulty arises when dealing with the convexity constraint and establishing that it is C^1, see also Section 3.5.

Sketch of proof: 1) As we noticed in the beginning of Section 3.4.1, we want to apply Proposition 3.46, and to that end, one needs first to prove an a priori C^1-regularity of $\Omega^\star$. This can be done with an argument taken from [205] (see Remark 3.52), that we briefly reproduce here (in $\mathbb{R}^2$ for simplicity, but this argument is valid in any dimension): we notice in particular that this argument does not use (3.69). Because of the a priori convexity, proving C^1 regularity is equivalent to proving that $\partial\Omega^\star$ has no corners. By contradiction, if $\Omega^\star$ had such a corner, cutting this corner at a size ε into a set $\Omega^\star_\varepsilon$ would lead to $\lambda_2(\Omega^\star_\varepsilon)=\lambda_2(\Omega^\star)+o(\varepsilon^2)$ (this relies on the fact that, in a weak sense ∇u goes to 0 at a convex corner, see [205] for more details) while $|\Omega^\star|-|\Omega^\star_\varepsilon|\geq c\varepsilon^2$ for some c (depending on the angle of the corner). By a classical scaling argument (see Section 3.3.1), $\Omega^\star$ minimizes (among convex domains) $\lambda_2(\Omega)+\alpha|\Omega|$ for a suitable $\alpha>0$, while $\Omega^\star_\varepsilon$ has a lower energy than $\Omega^\star$, because of the previous estimates, which is a contradiction.

2) The next step to prepare the application of Proposition 3.46, is to write the optimality condition. One can prove that the second eigenvalue of $\Omega^\star$ is simple so that it is shape differentiable. Then, on the strictly convex part, one can write $|\nabla u|^2=\Lambda$, by focusing on smooth deformations supported on Γ_1, and taking the convex hull, see [508] for more details.

We are also interested in the existence of a segment in the boundary: this fact is easy since u has a nodal line that hits the boundary of $\Omega^\star$ at exactly two points (from a result of Melas-Alessandrini, see [20, 705]), where ∇u must vanish, which is incompatible with the optimality condition just proven before.

3) We then apply Proposition 3.46 and conclude that either the shape is $C^{1,1/2}$ and not $C^{1,1/2+\varepsilon}$, or at least $C^{2,1/2}$.

4) Then assuming that $\Omega^\star$ is $C^{2,1/2}$, a similar argument (but slightly more involved) as in the previous section, leads us to a contradiction. This argument is written in [508, Theorem 10] with the idea that the optimal shape contains only two parallel segments in its boundary. But this is unfortunately an open problem (see Remark 3.51). Nev-

ertheless, we notice that their proof is still valid up to minor modifications and we reproduce roughly the arguments here. One knows that there exists one segment on the boundary, touching the nodal line of u: we denote it by Σ and we fix it at the x-axis. Then we focus on $\partial_x u$ which solves $-\Delta\partial_x u = \lambda_2(\Omega^\star)\partial_x u$ and vanishes on Σ. We next seek for connected components of $\{\partial_x u \neq 0\}$ so that $\partial_x u$ vanishes on their boundary. To that end, we write an optimality condition on $\Sigma = [a, b]$: this is not classical since, for most deformations supported in Σ, the convexity is not preserved. But with the convex hull method, as explained in [508, Theorem 7], one can write $|\nabla u|^2(x, 0) = \Lambda + w''(x)$, where $w \geq 0$ has triple roots at a and b, and $(x, 0)$ parametrizes Σ. This implies that $\partial_{xy} u$ vanishes at least three times inside Σ. But since $|\nabla u|^2 = \Lambda$ on the strictly convex part, we also know that $\partial_{xy} u$ vanishes at a and b, *thanks to the* C^2*-regularity of* u *on* $\overline{\Omega^\star}$. From there we know that, in a neighborhood of Σ, there are at least 6 connected components of $\{\partial_x u \neq 0\}$ (be careful that some of these "locally" connected components may be part of the same "globally" connected components of $\{\partial_x u \neq 0\}$). We study where the nodal lines can end. This is restricted to $\Sigma \cup \{N\} \cup \Sigma_{/\!/}$ where:
- N is the point of intersection of the nodal line of u which is not in Σ (we recall that the nodal line of u touches the boundary exactly at two points).
- $\Sigma_{/\!/}$ is the parallel segment of Σ in $\partial\Omega^\star$, which may be restricted to one point.
We deduce that $\partial_x u$ has at least two nodal domains (whose union is denoted ω) so that it vanishes on their boundary. But then, $\lambda_2(\omega) = \lambda_2(\Omega^\star)$, which contradicts the strict monotonicity of λ_2. □

The set $\Omega^\star$ is poorly understood and a better understanding of its geometry would help to analyze its regularity. Therefore, we propose the following open questions, which are supported by numerical evidence:

Open problem 3.53. *Prove that* $\Omega^\star$*, solution of* (3.66)*, satisfies:*
- $\partial\Omega^\star$ *contains a finite number of segments.*
- $\Omega^\star$ *has two orthogonal axes of symmetry.*

3.5 Polygons as optimal shapes

In the previous section, we saw that mild singularities can happen in shape optimization involving eigenvalues. In this section, we will see that we can obtain much stronger singularities. We will provide examples where, in dimension 2, the optimal shapes are actually polygons.

The results we are going to describe here can be found in the papers [622–624], see also [215, 451]. The general framework is to study optimization problems under a convexity constraint, in the spirit of (3.66):

$$\min\left\{J(\Omega),\quad \Omega \subset \mathbb{R}^d,\ \text{ open, convex}\right\}.$$

It is much easier to obtain existence for such problems, compared to similar optimization problems without convexity constraints. Indeed, the convexity provides much stronger compactness properties, and allows us to investigate unusual optimizations, like maximizing the perimeter, or maximizing eigenvalues. The remaining difficulty is usually to avoid minimizing sequences with diameter going to ∞ or collapsing (converging to something flat). To avoid these behaviors and enforce existence, we will focus on the following constraints, though our strategy can be applied to more general situations:

$$\min\left\{J(\Omega),\quad \Omega\subset\mathbb{R}^d,\ \text{open, convex},\ B(0,a)\subset\Omega\subset B(0,b)\right\},\tag{3.71}$$

where $0<a<b<\infty$.

3.5.1 General result about the minimization of a weakly concave functional

A first general result, stated in dimension 2, asserts that for a wide class of shape functionals, the optimal sets, under convexity constraints, happen to be polygons. In order to describe this result, we recall the following classical parametrization of 2-dimensional convex domains with polar coordinates $(r,\theta)\in[0,\infty)\times\mathbb{T}$, where $\mathbb{T}=\mathbb{R}/2\pi\mathbb{Z}$:

$$\Omega_u := \left\{(r,\theta)\in[0,\infty)\times\mathbb{R}\ ;\ r<\frac{1}{u(\theta)}\right\},\tag{3.72}$$

where u is a positive and 2π-periodic function and is called the gauge function of Ω_u. A simple computation shows that the curvature of Ω_u is

$$\kappa_{\partial\Omega_u}=\frac{u''+u}{\left(1+\left(\frac{u'}{u}\right)^2\right)^{3/2}}.\tag{3.73}$$

This implies that Ω_u is convex if and only if $u''+u\geq 0$, which has to be understood in the sense of $H^{-1}(\mathbb{T})$ if u is not C^2. More precisely, if $u\in H^1(\mathbb{T})$ then $u''+u\geq 0$ if and only if

$$\forall\, v\in H^1(\mathbb{T})\ \text{ with } v\geq 0,\quad \int_{\mathbb{T}}(uv-u'v')\,d\theta\geq 0.$$

Throughout this section, any function defined on $\mathbb{T}$ is considered as the restriction to $\mathbb{T}$ of a 2π-periodic function on $\mathbb{R}$, with the same regularity. Moreover, it is clear that Ω_u and u share the same regularity.

With this parametrization, considering $j(u)=J(\Omega_u)$, Problem (3.71) is equivalent to

$$\min\left\{j(u),\ u''+u\geq 0,\ u\in\mathcal{U}_{ad}\right\},\quad \text{where } \mathcal{U}_{ad}=\{u\in W^{1,\infty}(\mathbb{T}),\ 1/u\in[a,b]\}.\tag{3.74}$$

Then we have the following result proven in [623, Theorem 3].

Theorem 3.54. *Let $u_0 > 0$ be a solution for (3.74) and* $\mathbb{T}_{in} := \left\{\theta \in \mathbb{T},\, a < \frac{1}{u_0(\theta)} < b\right\}$. *Assume $j : W^{1,\infty}(\mathbb{T}) \to \mathbb{R}$ is C^2 and that there exist $s \in [0, 1)$, $\alpha > 0$, $\beta \in \mathbb{R}$ such that, for any $v \in W^{1,\infty}(\mathbb{T})$, we have*

$$j''(u_0)(v, v) \leq -\alpha \|v'\|^2_{L^2(\mathbb{T})} + \beta \|v\|^2_{H^s(\mathbb{T})}. \tag{3.75}$$

Then

$$u_0'' + u_0 \text{ is a finite sum of Dirac masses in } \mathbb{T}_{in}.$$

In this statement, the assumption made on j can be seen as a weak concavity property. It implies that, if v has a small support around some point x_0 (like $v(x) = v_0(x_0 + \sigma x)$ with σ large), then $j''(u_0)(v, v) < 0$. With this remark, the conclusion of the statement appears natural, as it says that minimizers are locally a sum of Dirac masses, while Dirac masses can be seen as extremal points among nonnegative measures, and it is a general fact that minimizers of concave functionals are expected to be extremal.

Geometrically, this result is a tool to extract sufficient conditions on the functional J so that solutions of (3.71) are polygons: indeed, formulae (3.73) implies that Ω_u is polygonal if and only if $u'' + u$ is a sum of Dirac masses.

Since it is not directly related to eigenvalues, we do not describe the proof of this result. Let us notice though that it is highly inspired by the paper [616] which deals with the Newton's problem of minimal resistance, one of the oldest shape optimization problems with a convexity constraint.

We conclude noticing that a similar result can be obtained if one adds to (3.74) a constraint of the form $m(u) = 0$ where $m : W^{1,\infty}(\mathbb{T}) \to \mathbb{R}^d$ is C^2 and such that $m'(u_0)$ is onto and $\|m''(u_0)(v, v)\| \leq \beta' \|v\|^2_{H^s(\mathbb{T})}$, for some $\beta' \in \mathbb{R}$ and $s \in [0, 1)$, see [623, Theorem 4]. We will use this fact for volume constrained problems.

3.5.2 Examples

In order to introduce the list of examples we are interested in, let us recall the reverse isoperimetric inequality which, in the framework of convex geometry is due to Ball (see [93]). It can be stated as the fact that the optimization problem

$$\max_{\Omega \in \mathcal{O}} \left\{ \min_{T \in GL_d(\mathbb{R})} \left\{ \frac{P(T(\Omega))}{|T(\Omega)|^{\frac{d-1}{d}}} \right\} \right\}, \tag{3.76}$$

where $\mathcal{O} = \{\Omega \subset \mathbb{R}^d \text{ open, bounded, convex, centrally symmetric}\}$, is solved by the unit cube. This can be understood as the maximization of the isoperimetric ratio among centrally symmetric convex bodies, where shapes are considered up to linear inversible transformations. In [622], some shape optimization problems with a similar behavior were introduced, like:

$$\min \left\{ \mu|\Omega| - P(\Omega),\ \ \Omega \text{ open, convex, } B(0, a) \subset \Omega \subset B(0, b) \right\}, \tag{3.77}$$

whose solutions are trivial for $\mu = 0$ or $\mu = +\infty$, but for which it is expected to obtain interesting optimal shapes for $\mu \in (0, \infty)$. Theorem 3.54 applies for such problems, and if Ω^* solves (3.77), then it is polygonal inside $B(0, b) \setminus \overline{B(0, a)}$. To go further, a full description for any parameters (a, b, μ) has been achieved in [148], where they show in particular that, for $\mu \in (\frac{1}{2b}, \frac{2}{a})$, the optimal shape is actually a full polygon (which means that the contact between Ω^* and the boundary of the annulus is a finite set of points).

In this section, we will study similar problems, involving the first eigenvalue of the Dirichlet-Laplacian, which can be listed in two sets of examples:

- The shape functional involves again a maximization of the perimeter: in that case, the regularity/singularity of the optimal shape is driven by the perimeter which is the leading term, see Examples 3.55 and 3.56,
- The shape functional does not contain a perimeter term, but involves a maximization of the first eigenvalue. In that case, we only have partial results, though there are a few indications that the behavior should be the same as in the previous cases, see Example 3.57.

Example 3.55 (Negative perimeter penalization). One can study

$$\min\{F(|\Omega|, \lambda_1(\Omega)) - P(\Omega)\ ;\ \Omega \text{ convex}, B(0, a) \subset \Omega \subset B(0, b)\} \tag{3.78}$$

where $F : (0, +\infty) \times (0, +\infty) \to \mathbb{R}$ is C^2. Then any optimal shape Ω^* is such that each connected component of the free boundary $\partial\Omega^* \setminus (\partial B(0, a) \cup \partial B(0, b))$ is polygonal.

This relies on Theorem 3.54, and the following properties of the second order derivatives of the perimeter, the volume, and the eigenvalue: denoting $p(u) = P(\Omega_u)$, $a(u) = |\Omega_u|$, $\ell(u) = \lambda_1(\Omega_u)$, we have

$$|a''(u)(v, v)| \le \beta_1 \|v\|^2_{L^2(\mathbb{T})}, \tag{3.79}$$

$$p''(u)(v, v) \ge \alpha |v'|^2_{L^2(\mathbb{T})} - \beta_2 \|v\|^2_{L^2(\mathbb{T})}, \tag{3.80}$$

$$|\ell''(u)(v, v)| \le \beta_1 \|v\|^2_{H^{1/2}}(\mathbb{T}). \tag{3.81}$$

The first two estimates are easily obtained by direct computations, while the third one is much more involved as one has to prove it for irregular domains, namely only convex domains. In a smooth setting though, it is easy to get such estimates, using the classical formula for the second order shape derivative of λ_1, see (3.92). In [623], we deal with general convex domains in dimension 2, and obtain a weaker version of (3.81) (sufficient for our purpose) where $1/2$ is replaced by $1/2 + \varepsilon$ where $\varepsilon > 0$. A new approach is introduced in [624] which leads to $\varepsilon = 0$ (it is written there for energy functionals, but the same method can be adapted to the eigenvalue case), and the approach in [624] works in any dimension.

Example 3.56 (Volume constraint and negative perimeter penalization). We can also consider a similar problem with a volume constraint:

$$\min\{J(\Omega) := F(\lambda_1(\Omega)) - P(\Omega)\ ;\ \Omega \text{ convex in } \mathbb{R}^2,\ |\Omega| = a\} \tag{3.82}$$

where $a \in (0, +\infty)$. Again, any optimal shape of (3.82) is a polygon, using a volume constraint version of Theorem 3.54 (see the remarks following its statement).

Studying minimizing sequences that converge to a segment, one may prove that for a large class of functionals F, there exists an optimal shape. For example, making good use of the estimate $\lambda_1(\Omega) \geq \frac{\pi^2}{4\mathrm{Diam}(\Omega)^2}$ if $F(x) \geq cx^{1/2}$ at $x \to \infty$ for some c large enough, one can prove that there exists a solution to (3.82). Thus considering, for $\mu \in (0, \infty)$ the problem

$$\min\{J(\Omega) := \mu\lambda_1(\Omega) - P(\Omega)\ ;\ \Omega \text{ convex in } \mathbb{R}^2,\ |\Omega| = a\}$$

where there is a competition between minimizing the eigenvalue and maximizing the perimeter, our previous statement asserts that the regularity of the solutions is driven by the perimeter term, which means that solutions are polygons.

Example 3.57 (Reverse Faber-Krahn inequality). In [214], motived by the question of adapting the classical Mahler inequality, replacing the area by the first Dirichlet eigenvalue, the authors were naturally led to question the reverse Faber-Krahn inequality in the same spirit as in (3.76): is the cube solution of

$$\max_{\Omega\in\mathcal{O}}\left\{\min_{T\in GL_d(\mathbb{R})}\left\{\lambda_1(T(\Omega))|T(\Omega)|^{\frac{2}{d}}\right\}\right\}, \tag{3.83}$$

where $\mathcal{O} = \{\Omega \subset \mathbb{R}^d$, open, bounded, convex, centrally symmetric$\}$?

This question, even if $d = 2$, is certainly very difficult. As in (3.77), it motivates the following optimization problems:

$$\max\left\{\mu|\Omega| + \lambda_1(\Omega),\quad \Omega \text{ open, convex, } B(0,a) \subset \Omega \subset B(0,b)\right\}, \tag{3.84}$$

or

$$\max\left\{\lambda_1(\Omega),\quad \Omega \text{ open, convex, } \Omega \subset B(0,b),\ |\Omega| = a\right\}, \tag{3.85}$$

where again $\mu \in \mathbb{R}_+$.

Contrary to the previous examples, the leading term for the geometry of optimal shapes is no longer the perimeter, but the first Dirichlet eigenvalue. Thus, in order to apply Theorem 3.54, we would need a convexity property of the functional λ_1 in the spirit of (3.80). More precisely, we wonder whether

$$\ell''(u)(v,v) \geq \alpha|v|^2_{H^{1/2}(\mathbb{T})} - \beta_2\|v\|^2_{L^2(\mathbb{T})}, \tag{3.86}$$

where $\ell(u) = \lambda_1(\Omega_u)$, Ω_u is the set whose gauge function is u and is only assumed to be convex and $|\cdot|_{H^{1/2}(\mathbb{T})}$ is the $H^{1/2}$-semi-norm. Such a result would imply that any optimal shape for the previous problems (3.84), (3.85) is locally polygonal inside

$B(0,b) \setminus \overline{B(0,a)}$ or $B(0,b)$ respectively. Unfortunately, even if (3.81) was obtained in full generality for convex sets, we are only able to obtain (3.86), when assuming that the deformation v is supported on a set where u is smooth enough. Therefore, one obtains as a weaker result that if Ω^* is an optimal shape and $\gamma \subset \partial\Omega^*$, then γ cannot, at the same time, be smooth and have a strictly positive curvature. We generalize this result in higher dimension in the next paragraph.

Another interesting result from [214] is that, in the class of convex axisymmetric octagons having vertices at the points $(\pm l, 0)$ and $(0, \pm l)$, the square is a solution of

$$\max\{\lambda_1(\Omega)|\Omega|\}.$$

These results suggest the following open problems:

Open problem 3.58.
- *Solutions to* (3.84) *and* (3.85) *are polygonal inside the box constraints.*
- *The square is solution of* (3.83).

3.5.3 Remarks on the higher dimensional case

In the multi-dimensional case, the convexity constraint in shape optimization is much less understood, though there are some results in this direction, see [215, 451] and the work of T. Lachand-Robert, see for example [251, 615, 616]. We describe in this section some results from [624], which can be applied to Examples from Section 3.5.2.

We can use again the parametrization of convex bodies with their gauge function and obtain a result of the type of Theorem 3.54, but whose conclusion will not allow us to prove that optimal shapes are polyhedra.

If $d \geq 2$, and $u : \mathbb{S}^{d-1} \to (0, \infty)$ is given, $\mathbb{S}^{d-1} = \{x \in \mathbb{R}^d,\ |x| = 1\}$, we can consider

$$\Omega_u := \left\{(r,\theta) \in [0,\infty) \times \mathbb{S}^{d-1},\quad r < \frac{1}{u(\theta)}\right\}. \tag{3.87}$$

The function u is again called the gauge function of Ω_u. The set Ω_u is convex if and only if the 1-homogeneous extension of u, denoted by the same letter and given by $u(x) = |x|u(x/|x|)$, is convex in $\mathbb{R}^d$ (in this section, we will refer to this property by saying that $u : \mathbb{S}^{d-1} \to \mathbb{R}$ is convex), see [801, Section 1.7] for example. In this way, we describe every bounded convex open set containing the origin. Throughout this section, the regularity of any function defined on $\mathbb{S}^{d-1}$ is seen as the regularity on $\mathbb{R}^d \setminus \{0\}$ of its 1-homogeneous extension, and it is classical that it is equivalent to the regularity of the set Ω_u itself.

With this parametrization, considering $j(u) = J(\Omega_u)$, problem (3.71) is equivalent to

$$\min\left\{j(u),\ u : \mathbb{S}^{d-1} \to (0,\infty) \text{ convex},\ a \leq 1/u \leq b\right\}. \tag{3.88}$$

Then in the same spirit as Theorem 3.54, we can prove the following where we denote $|v|^2_{H^1(\mathbb{S}^{d-1})} = \int_{\mathbb{S}^{d-1}} |\nabla_\tau v|^2 d\theta$, ∇_τ = tangential gradient on $\mathbb{S}^{d-1}$.

Theorem 3.59. *Let $u_0 > 0$ be a solution for (3.88). Assume $j : W^{1,\infty}(\mathbb{S}^{d-1}) \to \mathbb{R}$ is C^2 and that there exist $s \in [0, 1)$, $\alpha > 0$, $\beta, \gamma \in \mathbb{R}$ such that, for any $v \in W^{1,\infty}(\mathbb{S}^{d-1})$, we have*

$$j''(u_0)(v, v) \leq -\alpha|v|^2_{H^1(\mathbb{S}^{d-1})} + \beta\|v\|^2_{H^s(\mathbb{S}^{d-1})}. \tag{3.89}$$

Then the set

$$T_{u_0} = \{v \in W^{1,\infty}(\mathbb{S}^{d-1})/\exists\varepsilon > 0, \forall|t| < \varepsilon,\ u_0 + tv \text{ convex}, a \leq u_0 + tv \leq b\}, \tag{3.90}$$

is a linear vector space of finite dimension.

This theorem is a generalization of Theorem 3.54. Nevertheless, in dimension 3 or higher, it is not true that $[dim\,(T_{u_0}) < +\infty]$ implies that Ω_{u_0} is a polyhedra.

Example 3.60 (Negative perimeter penalization). This result applies to

$$\min\{F(|\Omega|, \lambda_1(\Omega)) - P(\Omega)\ ;\ \Omega \text{ convex}, B(0, a) \subset \Omega \subset B(0, b)\}$$

as explained in [624] (adapting the computation done for the Dirichlet energy to the case of the first eigenvalue). Using the fact that the set defined in (3.90) is of finite dimension, we easily deduce that if ω is a C^2 relatively open subset of $\partial\Omega^* \cap \{x, a < |x| < b\}$, then the Gauss curvature of Ω^* vanishes on ω (otherwise $C_0^\infty(\omega) \subset T_{u_0}$ which is a contradiction with the finite dimension property).

Like when $d = 2$, we cannot obtain as good results for Reverse Faber-Krahn type problems. However, in the spirit of [215, Theorem 4.5], we can prove the following;

Proposition 3.61. *Let Ω^* be a solution of (3.84) (resp. (3.85)) in $\mathbb{R}^d$. If $(\partial\Omega^* \cap B(0, b)) \setminus \overline{B(0, a)}$ (resp. $\partial\Omega^* \cap B(0, b)$) contains a relatively open set ω of class C^2, then its Gauss curvature vanishes on ω.*

Though this result is new, the computations, and the observation that the second shape derivative of λ_1 always has a sign, which is the main ingredient in the proof below, can be found in [809].

Sketch of proof: We focus first on the case where Ω^* solves (3.84). Let us assume that the Gauss curvature of ω is positive at one point x_0, and is therefore (by C^2 assumption) greater than some $\alpha > 0$ in a neighborhood $\hat{\omega} \subset \omega$ of x_0. Then if $\varphi \in C_c^\infty(\hat{\omega})$, the set $\Omega_t = (Id + t\varphi v)(\Omega^*)$ (where v is the normal vector to $\partial\Omega^*$, well defined on the support of φ) is admissible in the sense that it is still convex and satisfies the box constraint. Therefore the optimality conditions are

$$\frac{d}{dt}\lambda_1(\Omega_t)_{|t=0} = -\mu\frac{d}{dt}\text{Vol}(\Omega_t)_{|t=0}, \quad \frac{d^2}{dt^2}\left[\lambda_1 + \mu\text{Vol}\right](\Omega_t)_{|t=0} \leq 0. \tag{3.91}$$

On the other hand, from classical formula for first and second order derivative (see for example [510]), denoting by u the first eigenfunction of Ω^*, we have

$$\frac{d}{dt}\mathrm{Vol}(\Omega_t)_{|t=0} = \int_\omega \varphi, \quad \frac{d}{dt}\lambda_1(\Omega_t)_{|t=0} = -\int_\omega (\partial_\nu u)^2\varphi,$$

$$\frac{d^2}{dt^2}\left[\lambda_1 + \mu\mathrm{Vol}\right](\Omega_t)_{|t=0} = \int_\omega \left[2V_\varphi \partial_\nu V_\varphi + H\left((\partial_\nu u)^2 + \mu\right)\varphi^2\right], \tag{3.92}$$

where V_φ solves

$$\begin{cases} -\Delta V_\varphi &= \lambda_1(\Omega^*)V_\varphi - u\int_\omega (\partial_\nu u)^2\varphi \text{ in } \Omega^*, \\ V_\varphi &= -\varphi\partial_\nu u \text{ on } \partial\Omega^* \quad \text{and } \int_{\Omega^*} uV_\varphi = 0. \end{cases}$$

By the first relation of (3.91) and the above formula, we have $\mu = (\partial_\nu u)^2$ on $\hat{\omega}$. Since $H \geq 0$, it follows

$$\begin{cases} \frac{d^2}{dt^2}\left[\lambda_1 + \mu\mathrm{Vol}\right](\Omega_t)_{|t=0} &\geq 2\int_\omega V_\varphi\partial_\nu V_\varphi = 2\int_{\Omega^*}\left(|\nabla V_\varphi|^2 - \lambda_1(\Omega^*)V_\varphi^2\right), \\ &\geq \alpha|\varphi|^2_{H^{1/2}(\omega)} - \beta\|\varphi\|^2_{L^2(\omega)}, \end{cases} \tag{3.93}$$

for some $\alpha, \beta > 0$, where the last inequality is obtained as follows: first, we use (recall that $V_\varphi = \sqrt{\mu}\,\varphi$ on $\hat{\omega}$)

$$\mu\|\varphi\|^2_{H^{1/2}(\partial\hat{\omega})} = \|V_\varphi\|^2_{H^{1/2}(\partial\Omega^*)} \leq C\|V_\varphi\|^2_{H^1(\Omega^*)} = C\left[\int_{\Omega^*}|\nabla V_\varphi|^2 + V_\varphi^2\right]. \tag{3.94}$$

Then the L^2-norm of V_φ may be estimated from above by introducing the solution ψ of

$$\psi \in H_0^1(\Omega^*), \quad -\Delta\psi - \lambda_1(\Omega^*)\psi = V_\varphi \text{ in } \Omega^*, \quad \int_{\Omega^*}\psi\, u = 0.$$

This solution exists since $\int_{\Omega^*} u\,V = 0$. Moreover,

$$[\lambda_2(\Omega^*) - \lambda_1(\Omega^*)]\int_{\Omega^*}\psi^2 \leq \int_{\Omega^*}|\nabla\psi|^2 - \lambda_1(\Omega^*)\psi^2 = \int_{\Omega^*}V_\varphi\psi,$$

so that $\|\psi\|_{L^2(\omega)} \leq C\|V_\varphi\|_{L^2(\omega)}$ for some $C > 0$ (we use $\lambda_2(\Omega^*) - \lambda_1(\Omega^*) > 0$ since Ω^* is convex). Multiplying the equation in ψ by V_φ gives

$$\int_{\Omega^*}V_\varphi^2 = -\int_{\partial\Omega^*}V_\varphi\partial_\nu\psi = -\sqrt{\mu}\int_\omega \varphi\partial_\nu\psi \leq \sqrt{\mu}\,\|\varphi\|_{L^2(\hat{\omega})}\|\partial_\nu\psi\|_{L^2(\hat{\omega})}. \tag{3.95}$$

Now, the equation in ψ implies that

$$\|\Delta\psi\|_{L^2(\Omega^*)} \leq \lambda_1(\Omega^*)\|\psi\|_{L^2(\Omega^*)} + \|V_\varphi\|_{L^2(\Omega^*)} \leq C\|V_\varphi\|_{L^2(\Omega^*)}.$$

And near ω, the H^2-norm of ψ is controlled by the L^2-norms of $\Delta\psi$ and ψ so that $\|\partial_\nu\psi\|_{L^2(\hat{\omega})} \leq C\|V_\varphi\|_{L^2(\Omega^*)}$. Finally, going back to (3.95) leads to $\|V_\varphi\|_{L^2(\Omega^*)} \leq C\|\varphi\|_{L^2(\hat{\omega})}$.

This, together with (3.94), proves the estimate (3.93). Since it is valid for any $\varphi \in C_c^\infty(\hat{\omega})$, this contradicts the second part of (3.91) and ends the proof in the case of (3.84).

For the case where Ω^* solves (3.85), a similar proof is valid, μ is the Lagrange multiplier for the volume constraint, and one only has to restrict to deformations φ such that $\int_\omega \varphi = 0$ (to preserve the volume constraint at the first order), but the same computations still leads to a contradiction. □

Dorin Bucur, Pedro Freitas, and James Kennedy

4 The Robin problem

4.1 Introduction

Given a domain Ω in $\mathbb{R}^d$ we consider the problem

$$\begin{cases} -\Delta u = \lambda u & \text{in } \Omega\,, \\ \dfrac{\partial u}{\partial \nu} + \alpha u = 0 & \text{on } \partial\Omega\,, \end{cases} \tag{4.1}$$

where ν is the outer unit normal to Ω and the boundary parameter α is a real constant. For $\alpha \neq 0$ this boundary condition is usually called a Robin condition or, more rarely, a boundary condition of the third kind (the first being Dirichlet and the second being Neumann) or even a Fourier boundary condition. However, when viewed properly, the eigenvalue problem (4.1) contains as special cases the Neumann ($\alpha = 0$), Steklov ($\lambda = 0$) and Dirichlet problems (in the limiting case as α goes to infinity). These connections will be illustrated below, where we discuss the behaviour of the eigenvalues as the boundary parameter varies from $-\infty$ to $+\infty$. In this sense, this is the most general eigenvalue problem for the Laplace operator and, accordingly, not only was it the last problem for which isoperimetric results were obtained, but it is also the only one for which the isoperimetric problem for the first eigenvalue has not yet been completely solved.

Since the eigenvalues of the Neumann and Dirichlet problems are bounded from above and below, respectively, it is to be expected that this will influence the way Robin isoperimetric inequalities behave. Indeed, for positive values of α, the Robin eigenvalues are also bounded from below and we should thus expect an extension of the Faber–Krahn result in this case. For negative α, the situation is reversed and we should expect a Neumann type result. These differences become clear from the corresponding Rayleigh quotient for the first eigenvalue which now includes a boundary term, namely,

$$\lambda_1(\Omega, \alpha) = \inf_{0 \neq u \in H^1(\Omega)} \frac{\int_\Omega |\nabla u|^2\, dx + \alpha \int_{\partial\Omega} u^2\, d\sigma}{\int_\Omega u^2\, dx}. \tag{4.2}$$

Dorin Bucur: Institut Universitaire de France, Laboratoire de Mathématiques, CNRS UMR 5127, Université de Savoie, Campus Scientifique, 73376 Le-Bourget-Du-Lac, France, E-mail: dorin.bucur@univ-savoie.fr

Pedro Freitas: Department of Mathematics, Faculty of Human Kinetics & Group of Mathematical Physics, Faculty of Sciences, Universidade de Lisboa, Campo Grande, Edifício C6, 1749-016 Lisboa, Portugal, E-mail: psfreitas@fc.ul.pt

James Kennedy: Institute of Analysis, University of Ulm, Helmholtzstr. 18, D-89069 Ulm, Germany & Institute of Analysis, Dynamics and Modelling, University of Stuttgart, Pfaffenwaldring 57, D-70569 Stuttgart, Germany, E-mail: james.kennedy@mathematik.uni-stuttgart.de

For positive α it is clear that eigenvalues remain non-negative, although it is not obvious at first sight that the infimum of $\lambda_1(\Omega, \alpha)$ among domains with a given fixed measure and fixed α is not zero. On the other hand, for negative α, using a constant test function in the Rayleigh quotient above, it is seen that the first eigenvalue must be strictly negative and smaller than α times the measure of the boundary divided by the measure of the domain, cf. (4.22). Furthermore, and as we discuss in the relevant sections below, it is expected that approaching the Neumann problem has some more subtle consequences concerning optimizers of higher eigenvalues: although for large values of the boundary parameter (always assuming a given fixed measure of the domain), the optimizers should become close to those of the Dirichlet problem, they should behave differently as α approaches zero. A glimpse of what is at play here may be gathered from the fact that a domain with precisely k disjoint components will have k zero Neumann eigenvalues; thus the optimal k^{th} eigenvalue for such a domain will approach zero as α goes to zero and, indeed, the picture obtained numerically in [46] points to the existence of disconnected optimizers in this region, see also Figure 4.5 below.

The first result concerning isoperimetric inequalities for problem (4.1) for positive α was obtained by Bossel in 1986. In [169] she proved that in two dimensions the Faber–Krahn inequality extends from the Dirichlet case to all such values of α and thus the disk is the minimizer among domains with the same area. Although as we saw above this was expected to be the case, the proof is not only non-trivial but, as is often the case in such problems, uses a different technique from those for the Dirichlet and Neumann problems. Furthermore, the extension to the d-dimensional case was not immediate and it was twenty years before the full solution in any dimension was given by Daners [319]. In Section 4.5.1 we give precise statements of these results and present the main ideas of their proofs in some detail.

This central inequality, which has become known as the Bossel–Daners inequality, may be viewed as a Poincaré inequality with a trace term; in this context we also discuss related inequalities such as the Saint-Venant inequality and extensions to arbitrary open sets, as well as an application of the method of proof to inequalities for the principal eigenvalue involving the Cheeger constant of the domain.

A result similar to the Krahn–Szegő inequality, namely that as a consequence of the result for the first eigenvalue the second eigenvalue is necessarily minimized by two (disjoint) equal balls was proven by the third author in [586]. As mentioned above, the situation for higher eigenvalues becomes more complex, also in the sense that now optimizers are expected to depend on the boundary parameter. So far the evidence for this is mostly numerical and in two dimensions [46].

The transitions which take place as α goes through zero and becomes negative are even more striking. On the one hand, and as we saw above, we should now be looking for maximizers. The ball, which due to the geometric isoperimetric inequality is the domain for which the derivative of the first eigenvalue as a function of α is smallest at $\alpha = 0$ (cf. (4.16)), is thus the natural candidate to maximize the first eigenvalue.

This was conjectured in 1977 by Bareket [110] and further local evidence was provided by Ferone, Nitsch and Trombetti in [380], who showed that the ball is a local maximizer. However, it turns out that globally this will not be the case, as asymptotically the ball loses out to spherical shells. This was shown by the second author and Krejčiřík in [400], and it is the only situation known so far for the Laplace operator where the first eigenvalue is not optimized by a ball. On the other hand, it was also shown in the same paper that in two dimensions the disk is still the maximizer for small negative values of α. It remains an open problem whether or not this also holds in higher dimensions and whether the role of the maximizers is indeed taken by spherical shells.

The remaining part of this chapter is divided into five sections, which are structured as follows. After a short section summarizing the most important definitions and properties of the Robin Laplacian, Section 4.3 provides an overview of the Robin eigencurves, i.e., the curves of eigenvalues as functions of the boundary parameter $\alpha \in \mathbb{R}$ on a fixed domain, and discusses their relation to the Dirichlet, Neumann and Steklov problems. Section 4.4 examines the asymptotic behaviour of these curves as $\alpha \to \pm\infty$. We then address the optimization of the first eigenvalue for positive and negative values of α in Sections 4.5.1 and 4.5.2, respectively. The focus is mostly on inequalities of Faber–Krahn type, but for positive α we also briefly review other known bounds for eigenvalues, including the above-mentioned Saint-Venant and Cheeger-type inequalities. The higher eigenvalues are treated in Section 4.6.

4.2 Basic properties of the Robin Laplacian

From an operator-theoretic point of view, the Robin Laplacian has all the basic properties of its Dirichlet and Neumann counterparts, which can also be proved by the same means, such as the theory of sesquilinear (or, since we are only interested in real-valued functions, bilinear) forms and self-adjoint operators in a Hilbert space. In the following paragraphs, we will summarize briefly the properties we shall need in the sequel, omitting the routine proofs, and also point out a few differences from the Dirichlet case. A more complete treatment of the Robin Laplacian can be found in many places (albeit usually only for positive α, which however is often allowed to be a function), for example in [51, 97, 318] for $\alpha > 0$ and in [419, 614] for negative or indefinite α; although a single comprehensive, elementary treatment seems to be lacking in the standard literature.

Suppose that $\Omega \subset \mathbb{R}^d$ is a bounded Lipschitz domain. Formally, we will consider in this chapter the (negative) Robin Laplacian as the operator on $L^2(\Omega)$ associated with the symmetric bilinear form $a : H^1(\Omega) \times H^1(\Omega) \to \mathbb{R}$,

$$a(u, v) = \int_\Omega \nabla u \cdot \nabla v \, dx + \int_{\partial\Omega} \alpha u v \, d\sigma, \qquad (4.3)$$

where α may in general be a function defined on $\partial\Omega$, but will always be taken here to be a real constant, $\alpha \in \mathbb{R}$; if $u = v$, we will also write $a(u)$ for $a(u, u)$. Due to the trace inequality (see, e.g., [704, Theorem 3.37]), for any $\alpha \in \mathbb{R}$, this form, also called a quadratic form, is bounded from above and semi-bounded from below (i.e. $L^2(\Omega)$-elliptic): there exist $c_1, c_2 > 0$ depending only on Ω and α such that

$$a(u) + c_1\|u\|_2^2 \geq c_2\|u\|_{H^1(\Omega)}^2$$

for all $u \in H^1(\Omega)$. It follows that the operator on $L^2(\Omega)$ associated with a, which is easily seen to be given by

$$\begin{aligned} D(-\Delta_\alpha) = \{u \in L^2(\Omega) : \Delta u \in L^2(\Omega), \frac{\partial u}{\partial \nu} \text{ exists} \\ \text{in the sense of distributions and } = -\alpha u \text{ in } L^2(\partial\Omega)\}, \\ -\Delta_\alpha u = -\Delta u, \end{aligned}$$

is self-adjoint and bounded from below, has compact resolvent due to the compactness of the embedding $H^1(\Omega) \hookrightarrow L^2(\Omega)$ and hence, by the spectral theorem, has a sequence of eigenvalues

$$\lambda_1 = \lambda_1(\Omega, \alpha) \leq \lambda_2 = \lambda_2(\Omega, \alpha) \leq \ldots \to +\infty,$$

where as usual we repeat each eigenvalue according to its (finite) multiplicity, with $\lambda_1 < \lambda_2$ if Ω is connected. The theory of self-adjoint operators immediately yields the variational (min-max and max-min) characterization of the eigenvalues: as was noted in the introduction, for λ_1 we have

$$\lambda_1(\Omega, \alpha) = \inf_{0 \neq u \in H^1(\Omega)} \frac{a(u)}{\|u\|_2^2} = \inf_{0 \neq u \in H^1(\Omega)} \frac{\int_\Omega |\nabla u|^2 \, dx + \int_{\partial\Omega} \alpha u^2 \, d\sigma}{\int_\Omega u^2 \, dx}, \tag{4.4}$$

where the infimum is attained (only) at eigenfunctions associated with $\lambda_1(\Omega, \alpha)$, and where as usual the quotient on the right-hand side of (4.4) will be called the Rayleigh quotient; for λ_n we have

$$\begin{aligned} \lambda_n(\Omega, \alpha) &= \inf_{\substack{M \subset H^1(\Omega) \\ \dim M = n}} \sup_{0 \neq u \in M} \frac{\int_\Omega |\nabla u|^2 \, dx + \int_{\partial\Omega} \alpha u^2 \, d\sigma}{\int_\Omega u^2 \, dx} \\ &= \sup_{\substack{N \subset H^1(\Omega) \\ \operatorname{codim} N = n-1}} \inf_{0 \neq u \in N} \frac{\int_\Omega |\nabla u|^2 \, dx + \int_{\partial\Omega} \alpha u^2 \, d\sigma}{\int_\Omega u^2 \, dx}, \end{aligned} \tag{4.5}$$

where M and N are subspaces of $H^1(\Omega)$ of dimension n and codimension $n-1$, respectively; infimum and supremum are always achieved, again (only) at the corresponding eigenfunctions. An immediate consequence of (4.5) is that $\lambda_n(\Omega, \alpha)$ is an increasing function of $\alpha \in \mathbb{R}$ for fixed $n \geq 1$ and $\Omega \subset \mathbb{R}^d$, since this is true of the Rayleigh quotient for each fixed u. (Actually, one can show that $\lambda_n(\Omega, \alpha)$ is strictly increasing in α. Moreover, a corresponding statement holds when α is a function on $\partial\Omega$; for example, if $\alpha_1, \alpha_2 \in C(\partial\Omega)$, say, with $\alpha_1 \le \alpha_2$ on $\partial\Omega$ and $\alpha_1 < \alpha_2$ somewhere, then $\lambda_n(\Omega, \alpha_1) < \lambda_n(\Omega, \alpha_2)$ for all $n \ge 1$. See [791].)

In particular, the Robin eigenvalues are sandwiched between the Neumann and Dirichlet ones when $\alpha > 0$, since for $\alpha = 0$ we have the Neumann problem, and for the Dirichlet eigenvalues we replace $H^1(\Omega)$ by the smaller space $H_0^1(\Omega)$ in (4.5):

$$\mu_n(\Omega) = \lambda_n(\Omega, 0) \le \lambda_n(\Omega, \alpha) \le \lambda_n(\Omega) \qquad \text{for all } \alpha \in (0, +\infty). \tag{4.6}$$

It also follows that $\lambda_1(\Omega, \alpha)$ is a concave function of α on a fixed domain $\Omega \subset \mathbb{R}^d$, since it is the infimum of linear functions of α. In fact, $\lambda_n(\Omega, \alpha)$ is a locally concave function of $\alpha > 0$ for every $n \ge 1$ (i.e. except at an at most countable set of α's; for $\alpha < 0$ the situation is more complicated), cf. Figure 4.2 and the discussion after (4.12). In fact, one can show that each curve $\lambda_n(\Omega, \alpha)$ is piecewise analytic as a function of $\alpha \in \mathbb{R}$; this follows, for example, since the family of operators $(-\Delta_\alpha)_{\alpha\in\mathbb{R}}$ is holomorphic of type (B) in the sense of Kato (see [576, Section VII.4.2 and Remark VII.4.22]), the corresponding family of forms (4.3) obviously being holomorphic in α for each fixed $u \in H^1(\Omega)$. The dependence of the eigenvalues on α will be examined in more detail in Sections 4.3 and 4.4; see in particular Section 4.3.2.

The eigenfunctions, when normalized, form an orthonormal basis of $L^2(\Omega)$. We shall denote the eigenfunction associated with $\lambda_n(\Omega, \alpha)$ by $\psi_n = \psi_n(\Omega, \alpha)$; if λ_n is a multiple eigenvalue, say $\lambda_n = \lambda_{n+1} = \dots = \lambda_{n+k}$, then we choose $\psi_n, \dots, \psi_{n+k}$ as an arbitrary basis of the corresponding eigenspace, unless we have reason to make a more specific choice. As usual, if Ω is connected, the Krein–Rutman theorem (or a direct argument involving the characterization (4.4)) guarantees that not only is the first eigenvalue $\lambda_1(\Omega, \alpha)$ simple for each fixed α, but the corresponding eigenfunction ψ_1 may be chosen to be strictly positive in Ω. Moreover, all the eigenfunctions enjoy good regularity properties.

Proposition 4.1. *Let $\Omega \subset \mathbb{R}^d$ be a bounded, Lipschitz domain. Any eigenfunction ψ_n of the Robin Laplacian $-\Delta_\alpha$ on Ω is analytic in Ω and satisfies in addition $\psi_n \in H^1(\Omega) \cap C(\overline{\Omega})$.*

The analyticity is immediate for solutions of the Helmholtz equation $-\Delta u = \lambda u$ in an open set; continuity up to the boundary when $\alpha > 0$ follows from combining [318, Corollary 5.5] with [869, Corollary 2.9], or alternatively from [213, Lemma 2.1]; when $\alpha < 0$, see [320, Corollary 4.2]. Obviously, if Ω is more regular, then we have more regularity of ψ_n up to the boundary; for example, if $\alpha > 0$ and Ω is of class C^2, then it can be shown using classical regularity estimates that $\psi_n \in W^{2,p}(\Omega)$ for all $p \in (1, +\infty)$ (use [9, Theorem 4.2]), so in particular $\psi_n \in C^1(\overline{\Omega})$ as well.

4.2.1 Domain monotonicity and rescaling

Let us end this section with a few remarks on *domain monotonicity* and homothety of the eigenvalues: we recall that in the Dirichlet case, if $\Omega \subset \Omega'$, then

$$\lambda_n(\Omega) \geq \lambda_n(\Omega') \qquad \text{for all } n \geq 1,$$

which follows from the variational characterization and the fact that $H_0^1(\Omega)$ may be identified with a subset of $H_0^1(\Omega')$. This does *not* hold for the Robin eigenvalues in general, as was already observed in [424] for positive α; see also [748] for counterexamples for both positive and negative α when $n = 1$. One possibility for $\alpha > 0$ is to take Ω' with extremely "rough" (rapidly oscillating) boundary, which in general causes the eigenvalues to increase, cf. [317]. However, for some special choices of Ω' and $n = 1$ domain monotonicity is known to hold, most notably

$$\lambda_1(\Omega, \alpha) \geq \lambda_1(\widetilde{B}, \alpha) \qquad \text{for all } \alpha > 0, \tag{4.7}$$

where $\widetilde{B}$ is the ball circumscribing the arbitrary (sufficiently smooth, bounded) domain Ω. This was proved in [748] in two and three dimensions, but is easily generalized to higher dimensions; see also [644] for another proof using the maximum principle (but note that this is weaker than the Bossel–Daners inequality, Theorem 4.23, as follows from the rescaling property (4.10)). A different proof is also given in [424], where the corresponding, reversed inequality is obtained for $\alpha < 0$:

$$\lambda_1(\Omega, \alpha) \leq \lambda_1(\widetilde{B}, \alpha) \qquad \text{for all } \alpha < 0,$$

where $\widetilde{B}$ is again the smallest (or indeed any) ball containing Ω.

Various other results, such as for convex polyhedra, can also be found in [424]. Nothing seems to be known for the higher eigenvalues, and indeed this is a much harder problem than it seems at first glance, as the following simple observations demonstrate.

Proposition 4.2. *Fix $n \geq 2$. If $\Omega \subset \mathbb{R}^d$ is a bounded, Lipschitz domain with at least n connected components, then for* any *ball $\widetilde{B} \subset \mathbb{R}^d$ there exists $\alpha > 0$ such that*

$$\lambda_n(\Omega, \alpha) < \lambda_n(\widetilde{B}, \alpha).$$

Proof. Since Ω has at least n connected components, for its nth Neumann eigenvalue we have $\mu_n(\Omega) = \lambda_n(\Omega, 0) = 0$, while for any $\widetilde{B}$ we obviously have $\mu_n(\widetilde{B}) = \lambda_n(\widetilde{B}, 0) > 0$, since $n \geq 2$. Since on any given domain, each eigenvalue is a continuous function of α (cf. (4.12)), it follows that for $\alpha > 0$ small enough we must have $\lambda_n(\widetilde{B}, \alpha) > \lambda_n(\Omega, \alpha)$. □

Remark 4.3. *There exist examples of convex domains $\Omega \subset \Omega'$ for which $\lambda_2(\Omega, 0) = \mu_2(\Omega) < \mu_2(\Omega') = \lambda_2(\Omega', 0)$ (for example, if Ω is a long, thin rectangle along the diagonal* *of a square Ω'; see [505, Section 1.3.2]). Since $\lambda_n(\Omega, \alpha)$ is a smooth function of $\alpha \in \mathbb{R}$, it follows that*

$$\lambda_2(\Omega, \alpha) < \lambda_2(\Omega', \alpha) \tag{4.8}$$

for all $|\alpha|$ small enough. On the other hand, one can easily find $\Omega \subset \Omega'$ for which $\mu_2(\Omega) > \mu_2(\Omega')$, for example, concentric balls, for which in particular the reversed inequality to (4.8) holds. Thus no general inequality between convex sets is possible. We also see immediately that apart from λ_1 we cannot expect the direction of any inequality to change upon going from positive to negative α.

Open problem 4.4. *Do there exist classes of domains $\Omega \subset \Omega'$ for which domain monotonicity holds between $\lambda_n(\Omega, \alpha)$ and $\lambda_n(\Omega', \alpha)$ for general n, possibly under restrictions on $\alpha \neq 0$?*

If we dilate a given domain Ω by a factor of t, i.e., if we consider $t\Omega := \{tx : x \in \Omega\}$ for some $t > 0$, then both the Dirichlet and Neumann eigenvalues scale accordingly:

$$\lambda_n(t\Omega) = \frac{1}{t^2}\lambda_n(\Omega), \qquad \mu_n(t\Omega) = \frac{1}{t^2}\mu_n(\Omega) \qquad \text{for all } n \geq 1.$$

For the Robin problem, if we perform the same rescaling, we obtain

$$\lambda_n\left(t\Omega, \frac{\alpha}{t}\right) = \frac{1}{t^2}\lambda_n(\Omega, \alpha). \tag{4.9}$$

In particular, the boundary parameter in the rescaled problem changes depending on the rescaling, in accordance with the fact the equation in which α appears is not adimensional. This will have major implications for the optimization of the higher eigenvalues among all domains of given volume; see Section 4.6.2. Formula (4.9) also implies a monotonicity property with respect to dilations. Indeed, for every $t \geq 1, \alpha > 0$ we have

$$\lambda_1(t\Omega, \alpha) = \frac{1}{t^2}\lambda_1(\Omega, t\alpha) \leq \frac{1}{t}\lambda_1(\Omega, \alpha) \leq \lambda_1(\Omega, \alpha), \tag{4.10}$$

where the first inequality is a direct consequence of the monotonicity of the Rayleigh quotient.

4.3 A picture of Robin eigencurves: connecting the Dirichlet, Neumann and Steklov problems

When the boundary parameter α vanishes, problem (4.1) reduces to the Neumann eigenvalue problem. Formally, it might be expected that we then approach the Dirichlet problem as α goes to infinity. As we will see, this is indeed the case, but not in full generality.

4.3.1 Robin eigencurves in one dimension

To understand what happens we shall begin by considering the one-dimensional case, say on the interval (0, 1). In this instance, problem (4.1) simplifies to

$$\begin{cases} -u'' = \lambda u & \text{on } (0,1) \\ -u'(0) + \alpha u(0) = 0 \\ u'(1) + \alpha u(1) = 0 \end{cases}$$

and solutions of the above equation are of the form $u(x) = c_1 \cos(\sqrt{\lambda}x) + c_2 \sin(\sqrt{\lambda}x)$, yielding the following equation for the eigenvalues

$$\alpha^2 + 2\alpha\sqrt{\lambda}\cot(\sqrt{\lambda}) - \lambda = 0. \tag{4.11}$$

Although to determine the eigenvalues we would need to solve for λ as a function of α, it is in fact sufficient for our present purposes (and much simpler) to proceed the other way round and solve the quadratic equation in α to obtain

$$\alpha = -\sqrt{\lambda}\cot(\sqrt{\lambda}) \pm \sqrt{\lambda \csc^2(\sqrt{\lambda})}.$$

From this we see immediately that for each value of λ there are two (and only two) corresponding values of α, except at eigenvalues of the Dirichlet problem, that is, when $\lambda = \lambda_k = k^2\pi^2$ ($k \in \mathbb{N}$). For each k, and depending on whether one approaches $k^2\pi^2$ from above or below, one of the two branches provides a finite value, corresponding to the Neumann spectrum at α equal to zero, which in this case, except for the zero eigenvalue, coincides with the Dirichlet spectrum. The other branch provides the solution with α approaching either plus or minus infinity and λ approaching $k^2\pi^2$. For the latter solutions we have

$$\alpha = -\frac{4\left(\pi^2k^2\right)}{\lambda - \pi^2k^2} - 3 + \left(\frac{1}{12} + \frac{1}{4\pi^2k^2}\right)\left(\lambda - \pi^2k^2\right) + o\left(\lambda - \pi^2k^2\right), \quad \text{as } \lambda \to k^2\pi^2,$$

from which we may derive

$$\lambda = \pi^2k^2 - \frac{4\pi^2k^2}{\alpha} + \frac{12\pi^2k^2}{\alpha^2} + o(1), \quad \text{as } \alpha \to \pm\infty.$$

For negative values of λ the above functions simplify to

$$f_+(t) = -\sqrt{t}\tanh\left(\frac{\sqrt{t}}{2}\right) \text{ and } f_-(t) = -\sqrt{t}\coth\left(\frac{\sqrt{t}}{2}\right),$$

where $t = -\lambda \geq 0$, yielding the two negative branches of the $\lambda - \alpha$ eigencurves which go to $-\infty$ as α goes to $-\infty$. It is also clear that on these two branches

$$\alpha = -\sqrt{t} + \textit{exponentially small terms}$$

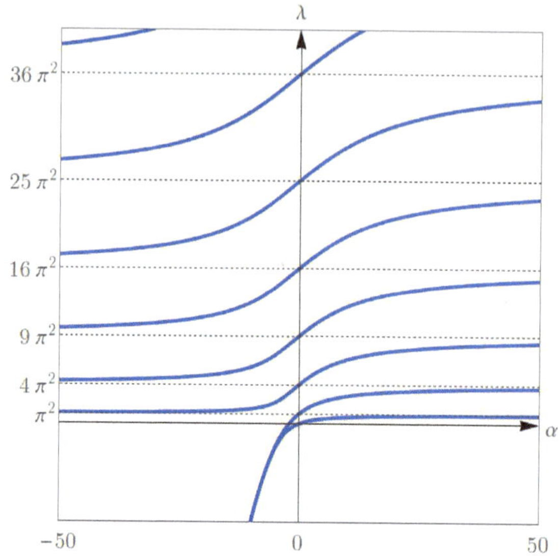

Fig. 4.1. Eigenvalues of the one-dimensional problem (4.11) as a function of the boundary parameter α.

as t goes to $+\infty$, in agreement with the first term of the known asymptotics for the Robin problem with large negative boundary parameter – see Section 4.4.2 below. These are the only branches which are unbounded in λ.

A graph of the resulting eigencurves, together with the corresponding asymptotes at the Dirichlet eigenvalues, is shown in Figure 4.1.

We see that although all eigenvalues approach those of the Dirichlet problem as α goes to $+\infty$, this is not necessarily the case as α goes to $-\infty$. In fact, in the latter situation eigenvalues are divided into two types, those that converge to $-\infty$ and those that again converge to Dirichlet eigenvalues, now from above.

4.3.2 Robin eigencurves in higher dimensions

In the general case of a bounded domain Ω in $\mathbb{R}^d$ corresponding to equation (4.1), each of the eigencurves $\lambda_n(\Omega, \alpha)$ is a piecewise smooth (analytic) function of the real parameter α (possibly even across $\alpha = \pm\infty$). In fact, the existence of points where there might not be analyticity will depend on the labelling of the eigenvalues and occur when two eigencurves cross each other, meaning that it is possible to choose eigenvalue branches such that they are indeed analytic for all real values of α – we refer again to [576, Chapter VII].

From the variational formulation of the nth eigenvalue (4.5), it is clear that each λ_n is an increasing function of α. Furthermore, if $\lambda_n(\Omega, \alpha)$ is a simple eigenvalue for some $\alpha \in \mathbb{R}$, with corresponding eigenfunction ψ_n, then

$$\frac{d}{d\alpha}\lambda_n(\Omega, \alpha) = \frac{\int_{\partial\Omega} \psi_n^2 \, d\sigma}{\int_{\Omega} \psi_n^2 \, dx}. \tag{4.12}$$

This formula for the derivative of Robin eigenvalues is well known and a derivation may, for instance, be found in [46] (for $\alpha > 0$; the proof is however the same for general $\alpha \in \mathbb{R}$). It also implies that each $\lambda_n(\Omega, \alpha)$ is a locally concave function of $\alpha > 0$, except possibly at points where eigencurves cross, since explicitly calculating the corresponding tangent to the eigencurve at the point $(\alpha, \lambda_n(\Omega, \alpha))$ and using the characterization of λ_n as an infimum shows that the tangent must lie above the eigencurve passing through $(\alpha, \lambda_n(\Omega, \alpha))$.

To get a full qualitative picture of the eigencurves of the Robin problem we shall now relate it to another eigenvalue problem, namely the Steklov problem, defined by

$$\begin{cases} -\Delta u = 0 & \text{in } \Omega \\ \dfrac{\partial u}{\partial \nu} = \sigma u & \text{on } \partial\Omega \end{cases}. \tag{4.13}$$

It is clear that an eigenvalue σ of (4.13) corresponds to a zero eigenvalue of (4.1) with $\alpha = -\sigma$. Since for dimensions higher than one the Steklov problem has an infinite number of eigenvalues (plus the zero eigenvalue; see Chapter 5), we see that there must be an infinite number of crossings of the horizontal axis by the Robin eigencurves. From the monotonicity, these eigenvalues must remain negative for larger (in absolute value) values of α.

We again choose a specific example for which it is possible to compute the eigenvalues explicitly to illustrate this behaviour, in this instance the two-dimensional disk – it is possible to perform a similar analysis for d-dimensional balls in exactly the same way; see [400] for a slightly different but essentially equivalent approach where such an analysis is carried out for the first eigenvalue of balls and spherical shells. By using separation of variables we are led to the following transcendental relation between α and λ

$$\alpha J_k\left(\sqrt{\lambda}\right) + \frac{1}{2}\sqrt{\lambda}\left[J_{k-1}\left(\sqrt{\lambda}\right) - J_{k+1}\left(\sqrt{\lambda}\right)\right] = 0, \tag{4.14}$$

where J_k denotes the Bessel functions of the first kind.

As before, solving for α allows us to obtain the form of the eigencurves in a simple fashion and to derive the asymptotic expansion for α at the Dirichlet eigenvalues, namely,

$$\alpha = -\frac{2j_{k,l}^2}{\lambda - j_{k,l}^2} - 1 + \frac{1}{6}\left(\frac{1-k^2}{j_{k,l}^2} + 1\right)\left(\lambda - j_{k,l}^2\right) + \mathrm{o}\left(\lambda - j_{k,l}^2\right), \text{ as } \lambda \to j_{k,l}^2,$$

where now $j_{k,l}$ denotes the lth zero of J_k. From this it follows that there are branches satisfying the following asymptotics

$$\lambda = j_{k,l}^2 - \frac{2j_{k,l}^2}{\alpha} + \mathrm{o}(\alpha^{-1}), \text{ as } \alpha \to \pm\infty.$$

For negative values of λ the relation between J_k and the modified Bessel functions of the first kind I_k

$$J_k(it) = e^{k\pi i/2} I_k(t) \quad (t > 0)$$

allows us to write the above as

$$\alpha = -\frac{\sqrt{-\lambda}\left[I_{k-1}\left(\sqrt{-\lambda}\right) + I_{k+1}\left(\sqrt{-\lambda}\right)\right]}{2I_k\left(\sqrt{-\lambda}\right)},$$

yielding the following asymptotic expansion

$$\alpha = -\sqrt{-\lambda} + \frac{1}{2} + \frac{1-4k^2}{8\sqrt{-\lambda}} - \frac{(2k+1)\left(4k^2-8k-13\right)(1-2k)^2}{256\lambda} + o(\lambda^{-1}), \tag{4.15}$$

as $\lambda \to -\infty$.

The corresponding eigencurves are plotted in Figure 4.2 for the first eigenvalues of each of the first five families of solutions ($k = 0, \ldots, 4$).

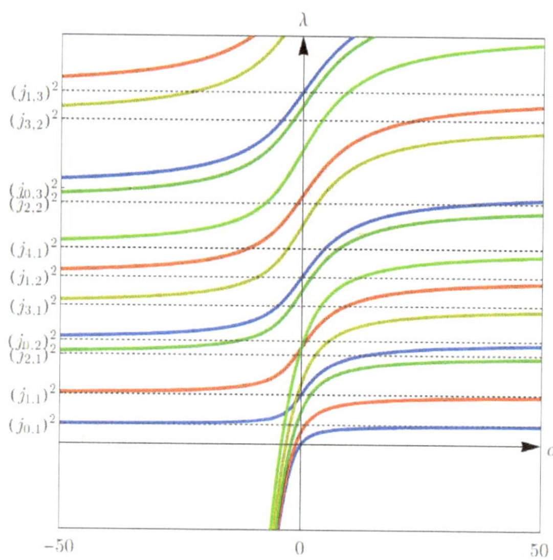

Fig. 4.2. Eigenvalues of the disk as a function of the boundary parameter α; the low eigenvalues of the first five families of solutions of equation (4.14) are shown.

The above analysis illustrates how the Robin problem relates to the three other classical spectral problems considered in this book, namely, the Dirichlet, Neumann and Steklov problems, in the particular case of balls. In the next section we will present results characterizing the asymptotic behaviour of eigenvalues for general domains Ω in any dimension as α goes to plus and minus infinity, and see that the behaviour described above holds in the more general case.

In the special case when $n = 1$ and $\alpha = 0$, i.e., where $\lambda_1(\Omega, 0) = \mu_1(\Omega) = 0$ is the first Neumann eigenvalue of Ω, the eigenfunction is constant, and (4.12) reduces to the well-known formula

$$\left.\frac{d}{d\alpha}\right|_{\alpha=0} \lambda_1(\Omega, \alpha) = \frac{|\partial\Omega|}{|\Omega|}, \tag{4.16}$$

which has appeared in [424, 426, 614], for example.

4.4 Asymptotic behaviour of the eigenvalues

4.4.1 Large positive values of the boundary parameter

As we saw above, on a line or a disk, as $\alpha \to +\infty$, each eigenvalue $\lambda_n(\Omega, \alpha)$ converges to a Dirichlet eigenvalue. In fact the assertion that this is true on a general bounded domain goes back at least as far as the classical book of Courant–Hilbert[4.1].

Proposition 4.5. *Let $\Omega \subset \mathbb{R}^d$ be a bounded Lipchitz domain. Then for each $n \geq 1$,*

$$\lambda_n(\Omega) = \lim_{\alpha \to +\infty} \lambda_n(\Omega, \alpha) = \sup\{\lambda_n(\Omega, \alpha) : \alpha \in \mathbb{R}\}. \tag{4.17}$$

We recall that $\lambda_n(\Omega)$ is the nth Dirichlet eigenvalue of Ω (counted according to multiplicities). This statement is routine to prove: firstly, as we have seen, $\lambda_n(\Omega, \alpha)$ is an increasing function of α which is bounded from above by $\lambda_n(\Omega)$. Hence the second equality in (4.17), including the existence of the limit, is trivial. The rest is a standard argument using, for example, the fact that any corresponding sequence of eigenfunctions is bounded in $H^1(\Omega)$ as $\alpha \to +\infty$.

In this section we will focus on the asymptotic behaviour of the curves $\lambda_n(\Omega, \alpha)$, together with complementary estimates in terms of easily computable functions of α. General bounds on λ_1 in terms of α and/or geometric and other accessible properties of the domain Ω, of which inequalities of Faber–Krahn type are the most important, will be considered in detail in Section 4.5.

In fact very little seems to be known in the general case. Estimates on λ_n are difficult to obtain, and there are virtually no bounds available on the higher eigenvalues (mirroring the fact that the variational characterization becomes far more involved). This also provides additional reasons to investigate the asymptotic behaviour of $\lambda_n(\Omega, \alpha)$ as $\alpha \to +\infty$, but in fact this does not seem to have been done. Possibly the

4.1 More precisely, they claim this in Ch. VI §2.1 on p. 411 of Volume I [292] and write that the assertion "may best be proved by investigating the nature of the eigenfunctions more closely. Since this will not be done until later, we refrain from carrying out the proof at this point (cf. Volume II)." However, no such proof seems to have been included in Volume II.

only general bound, which at least yields an estimate on the rate of convergence to the Dirichlet spectrum which is probably optimal, can be found in [383, 384]:

$$\lambda_n(\Omega) - C\lambda_n(\Omega)^2\alpha^{-1} \le \lambda_n(\Omega, \alpha) \ (\le \lambda_n(\Omega)) \tag{4.18}$$

for any $n \ge 1$, on a C^2-domain $\Omega \subset \mathbb{R}^d$, is obtained, where the constant $C = C(\Omega) > 0$ does not depend on $n \ge 1$. The approach used to prove (4.18) is essentially an estimate on the norm of the difference of the resolvents of (shifted) Dirichlet and Robin Laplacians, since this in turn implies an estimate on the differences of the corresponding eigenvalues.

At least for the first eigenvalue, the next term in the expansion of λ_1 is known, namely

$$\lambda_1(\Omega, \alpha) = \lambda_1(\Omega) - \frac{\int_{\partial\Omega} \left(\frac{\partial \psi_1^D}{\partial \nu}\right)^2 d\sigma}{\int_\Omega (\psi_1^D)^2\, dx}\alpha^{-1} + o(\alpha^{-1}) \qquad \text{as } \alpha \to +\infty, \tag{4.19}$$

where ψ_1^D is the first eigenfunction of the Dirichlet Laplacian, which was first announced in [385] and proved in [382]. Regarding the higher eigenvalues, (4.18) implies that

$$\lambda_n(\Omega) - \lambda_n(\Omega, \alpha) = O(\alpha^{-1}) \qquad \text{as } \alpha \to +\infty,$$

for each fixed $n \ge 1$. It is perhaps surprising that not even the first term of the asymptotic expansion in the general case $n \ge 2$ seems to appear in the literature, although one would expect an analogous formula to (4.19) to hold with ψ_1^D replaced by ψ_n^D, at least in the case where $\lambda_n(\Omega)$ is a simple eigenvalue.

Open problem 4.6. *Find more terms in the asymptotic expansion of $\lambda_1(\Omega) - \lambda_1(\Omega, \alpha)$ as $\alpha \to +\infty$.*

Open problem 4.7. *Obtain a corresponding expansion of $\lambda_n(\Omega) - \lambda_n(\Omega, \alpha)$ in powers of α as $\alpha \to +\infty$. (How) do the powers of α and constants in the expansion depend on smoothness and curvature properties of $\partial\Omega$?*

4.4.2 Large negative values of the boundary parameter

The general situation for $\alpha < 0$ is, as we saw in the case of the disk, more involved: on any Lipschitz domain $\Omega \subset \mathbb{R}^d$, there exists a sequence of eigencurves tending to $-\infty$.

Proposition 4.8. *Let $\Omega \subset \mathbb{R}^d$ $d \ge 2$ be bounded and Lipschitz. Then for any* fixed $n \ge 1$, *we have $\lambda_n(\Omega, \alpha) \to -\infty$ as $\alpha \to -\infty$.*

As Figure 4.2 already suggests, in general $\lambda_n(\Omega, \alpha)$ will not be described by a smooth eigencurve, but rather, a positive eigencurve may be “overtaken” as $\alpha \to -\infty$ by another shooting off to $-\infty$. In general, there are eigencurves which do in fact stay positive and converge to an element of the Dirichlet spectrum as $\alpha \to -\infty$ – here and in the sequel, when speaking of eigencurves, we shall always mean the analytic branches rather than the piecewise smooth curves corresponding to λ_n for some fixed n; we will denote an arbitrary, fixed analytic branch by $\lambda_\alpha = \lambda_\alpha(\Omega)$. Sometimes one also speaks of accumulation points of the set $\{\lambda_n(\Omega,\alpha) : n \in \mathbb{N}$ and $\alpha \in \mathbb{R}\}$ with respect to α as $\alpha \to -\infty$. In light of Proposition 4.8 we see that the positive eigencurves (or, equivalently, points of accumulation) cannot be analyzed by variational means.

Proposition 4.9. *Let $\Omega \subset \mathbb{R}^d$ be bounded and Lipschitz.*
(a) Suppose that $\lambda_\alpha(\Omega)$ is an analytic curve of eigenvalues for which the corresponding eigenfunctions ψ_α are normalized to have $L^2(\Omega)$-norm 1. Then $\lambda := \lim_{\alpha\to-\infty} \lambda_\alpha(\Omega) = \inf\{\lambda_\alpha(\Omega) : \alpha \in \mathbb{R}\}$ exists and is an eigenvalue of the Dirichlet Laplacian if and only if ψ_α remains bounded in $H^1(\Omega)$ as $\alpha \to -\infty$.
(b) Every Dirichlet eigenvalue of Ω is an accumulation point of $\{\lambda_n(\Omega,\alpha) : n \in \mathbb{N}$ and $\alpha \in \mathbb{R}\}$, i.e., there is an analytic branch converging to it.

The first paper to study this problem was probably [614], where it was noted that Propositions 4.8 and 4.9 both hold on balls and rectangles. Proposition 4.8 is considered standard and follows from a variational argument: for any fixed $n \geq 1$, take for example the first n Neumann eigenfunctions $\psi_1^N, \dots, \psi_n^N$ of Ω as test functions; their span is an n-dimensional subspace of $L^2(\Omega)$. Since none of them vanish identically on $\partial\Omega$ (as no function is simultaneously a Neumann and a Dirichlet eigenfunction), we have $\int_{\partial\Omega} (\psi_k^N)^2 \, d\sigma > 0$ for all $k = 1, \dots, n$. It follows that the Rayleigh quotient of each ψ_k^N tends to $-\infty$ as $\alpha \to -\infty$; the min-max characterization (4.5) yields immediately $\lambda_n(\Omega,\alpha) \to -\infty$ as $\alpha \to -\infty$.

Proposition 4.9(a) follows from a similar argument to the proof of Proposition 4.5, or alternatively appears in an equivalent form in [250, Section 2]; namely, if one chooses the normalization $\|\psi_\alpha\|_{H^1(\Omega)} = 1$, then $(\lambda_\alpha)_{\alpha<0}$ accumulates at a Dirichlet eigenvalue if and only if $\|\psi_\alpha\|_{L^2(\Omega')}$ remains bounded away from zero on every compact subset $\Omega' \subset \Omega$. A proof of (b) can also be found in [250, Section 3], where a procedure for deriving the entire asymptotic expansion of $|\lambda_\alpha - \lambda_0|$ in powers of α^{-1} is given (λ_0 being the given Dirichlet eigenvalue). Although the proof is only given for simple Dirichlet eigenvalues, and is not phrased in terms of eigencurves, it is easy to see that at least Proposition 4.9 (b) holds in general, since the method of proof involves controlling the difference of the eigenvalues in terms of the operator norm of a suitably chosen difference of operators. This method also yields an estimate on the rate of convergence to the Dirichlet eigenvalues. This was indeed done in [250], in fact before the corresponding, independent, analysis for $\alpha \to +\infty$ in [382, 384, 385]. The following theorem summarizes the results of [250, Section 3] (taking $N = 0$ there).

Theorem 4.10 (Cakoni–Chaulet–Haddar). *Suppose that $\lambda_\alpha(\Omega)$ is an analytic branch of eigenvalues of* (4.1) *such that $\lambda_\alpha(\Omega) \to \lambda_n(\Omega)$ for some $n \geq 1$. Then*

$$\lambda_\alpha(\Omega) = \lambda_n(\Omega) + O(\alpha^{-1}) \qquad \text{as } \alpha \to -\infty.$$

There is, however, still a small gap in our understanding of the possible types of behaviour of branches λ_α: namely, it remains to be shown that divergence to $-\infty$ and convergence to the Dirichlet spectrum are the *only* possibilities.

Open problem 4.11. *Show that every analytic branch of eigenvalues of the set $\{\lambda_n(\Omega, \alpha) : n \in \mathbb{N} \text{ and } \alpha \in \mathbb{R}\}$ either tends to $-\infty$ or converges to a Dirichlet eigenvalue of Ω as $\alpha \to -\infty$. To do this, it is sufficient to show that any analytic branch whose eigenfunctions accumulate at the boundary eventually tends to $-\infty$. Here "accumulation at the boundary" means if $\|\psi_\alpha\|_{H^1(\Omega)} = 1$ for all $\alpha < 0$, then for any compact set $\omega \subset \Omega$, we have $\|\psi_\alpha\|_{L^2(\omega)} \to 0$ as $\alpha \to -\infty$.*

4.4.2.1 The leading term in the asymptotic expansion of λ_1 as $\alpha \to -\infty$

The question of the asymptotic behaviour of $\lambda_1(\Omega, \alpha)$ and, to a lesser extent, $\lambda_n(\Omega, \alpha)$ as $\alpha \to -\infty$—that is, the eigenvalue curves which diverge to $-\infty$, which we will refer to as the unbounded curves—, is one which has received considerable attention in the last few years, beginning with [614]. Before discussing this in detail, let us first give the following instructive variational estimate on $\lambda_1(\Omega, \alpha)$, which was first observed in [425, Theorem 2.3].

Proposition 4.12. *Let $\Omega \subset \mathbb{R}^d$ be a bounded, Lipschitz domain in dimension $d \geq 2$. Then for any $\alpha < 0$, we have*

$$\lambda_1(\Omega, \alpha) < -\alpha^2. \tag{4.20}$$

Proof. Writing $x = (x_1, \dots, x_d) \in \mathbb{R}^d$, we use the test function $\varphi(x) = e^{-\alpha x_d}$ (in fact, we could take any function of the form $\varphi(x) = e^{-\alpha x \cdot \nu}$, where ν is an arbitrary unit vector in $\mathbb{R}^d$). Then $\nabla\varphi = (0, \dots, 0, -\alpha e^{-\alpha x_d})$ and

$$a(\varphi) = \alpha^2 \int_\Omega e^{-2\alpha x_d}\, dx + \alpha \int_{\partial\Omega} e^{-2\alpha x_d}\, d\sigma.$$

Denoting the outer unit normal to Ω by $\nu = (\nu_1, \dots, \nu_d)$, applying the divergence theorem to $V := (0, \dots, 0, e^{-2\alpha x_d})$, we obtain

$$\int_{\partial\Omega} e^{-2\alpha x_d}\, d\sigma \geq \int_{\partial\Omega} e^{-2\alpha x_d} \nu_d\, d\sigma = \int_{\partial\Omega} V \cdot \nu\, d\sigma = \int_\Omega \operatorname{div} V\, dx = -2\alpha \int_\Omega e^{-2\alpha x_d}\, dx.$$

It follows that

$$a(\varphi) \leq \alpha^2 \int_\Omega e^{-2\alpha x_d}\, dx - 2\alpha^2 \int_\Omega e^{-2\alpha x_d}\, dx = -\alpha^2 \int_\Omega \varphi^2\, dx.$$

Noting that φ cannot be an eigenfunction corresponding to $\lambda_1(\Omega, \alpha)$, the variational characterization (4.4) yields the assertion. □

See also [653, Example 2.4]: in one dimension, on the half-line $(0, +\infty)$, the function $\varphi(x) = e^{-\alpha x}$ satisfies $\varphi''(x) = \alpha^2\varphi(x)$, and indeed, $-\alpha^2$ is the bottom of the spectrum of (4.1) on $\Omega := (0, +\infty)$. A corresponding statement is true on the d-dimensional half-space $\mathbb{R}^d_+$.

Remark 4.13. *Strictness of the inequality* (4.20) *alternatively follows by noting that the inequality*

$$\int_{\partial\Omega} e^{-2\alpha x_d}\, d\sigma \geq \int_{\partial\Omega} e^{-2\alpha x_d} \nu_d \, d\sigma$$

is itself strict unless $\nu_d \equiv 1$, *that is,* Ω *is itself a half-space. In fact, by noting that*

$$\nu_d \leq 1 - \frac{1}{2}(1 - \nu_d)^2,$$

we can improve (4.20) *using the above argument to*

$$\lambda_1(\Omega, \alpha) \leq -\alpha^2 + \frac{\alpha}{2} \frac{\int_{\partial\Omega} e^{-2\alpha x_d}(1 - \nu_d)^2\, d\sigma}{\int_\Omega e^{-2\alpha x_d}\, dx}. \tag{4.21}$$

The second term in (4.21) *can be thought of as measuring a certain weighted distance of* Ω *to the half-space* $\mathbb{R}^d_-$. *Obviously, a corresponding inequality holds for any other half-space. (We thank Lorenzo Brasco for this observation.)*

We also remark in passing that other than (4.20) and (4.21) there seem to be no other known general estimates on $\lambda_1(\Omega, \alpha)$ (or indeed $\lambda_n(\Omega, \alpha)$) for $\alpha < 0$ other than the reversed Faber–Krahn-type bound discussed in Section 4.5.2, the few results on domain monotonicity mentioned in Section 4.2.1, and trivial bounds such as

$$\lambda_1(\Omega, \alpha) < \alpha \frac{|\partial\Omega|}{|\Omega|} \tag{4.22}$$

(obtainable by using the constant function 1 as a test function).

As observed above for the ball, the inequality in Proposition 4.12 becomes an equality in the asymptotic limit, independent of both the space dimension and the volume of the domain, if the domain is of class C^1. This independence is to be expected, since the eigenfunction will necessarily become concentrated near the boundary as $\alpha \to -\infty$.

Theorem 4.14 (Lacey–Ockendon–Sabina; Lou–Zhu). *Let* $\Omega \subset \mathbb{R}^d$ *be bounded and* C^1. *Then*

$$\lambda_1(\Omega, \alpha) = -\alpha^2 + o(\alpha^2) \qquad \text{as } \alpha \to -\infty.$$

This was first proved in [614] for smooth (C^2) domains Ω whose boundary can be mapped smoothly and invertibly onto the sphere. For these domains, the annular region near $\partial\Omega$ can thus be mapped onto an annulus, for which Theorem 4.14 can be shown explicitly. Studying the transformation (i.e. the change of variables) more carefully allows one to prove the statement for Ω. For general C^1 domains, Theorem 4.14 was proved in [679] by showing convergence as $\alpha \to -\infty$ of the eigenfunctions associated with $\lambda_1(\Omega, \alpha)$ under a suitable renormalization (a type of blow up) at the points where they reach their maxima to the solution of a problem on a half-space, which can again be analyzed explicitly. In fact, the statement from [679] is more general: if we replace $\alpha \in \mathbb{R}$ by a function, i.e., our boundary condition becomes $\frac{\partial u}{\partial \nu} + \alpha b(x)u = 0$ for a function $b \in C(\partial\Omega)$ which is strictly positive somewhere, then the asymptotic expression becomes

$$\lambda_1(\Omega, \alpha) = -\alpha^2(\max_{x\in\partial\Omega} b(x))^2 + o(\alpha^2).$$

The situation changes if Ω is less regular, for example, if it has corners. It was already noted in [614] that for a *planar* domain $\Omega \subset \mathbb{R}^2$ with m corners having half-angles $\theta_1, \dots, \theta_m$, using test functions of the form $\varphi(x) = e^{\alpha \sin^{-1}(\theta_j)\nu\cdot x}$ (for a suitable direction ν and translation of Ω), one obtains the asymptotic inequality

$$\lambda_1(\Omega, \alpha) \lesssim -\left[\max_{j=1,\dots,m} \sin^{-1}(\theta_j)\right]^2 \alpha^2 \qquad \text{as } \alpha \to -\infty, \tag{4.23}$$

which was shown in [614] to be an asymptotic equality on triangles. There, it was also conjectured that asymptotic equality should hold far more generally, at least if the angles are all outward-pointing (i.e. the half-angles $\theta_j \le \pi/2$ for all j). This problem was subsequently taken up and studied in far more detail in [653]. Of importance for the analysis are "model cones". More precisely, we suppose $\Omega \subset \mathbb{R}^d$ is piecewise smooth with corners in the following sense: Ω satisfies the interior cone condition and, for each point $z \in \partial\Omega$, there exists an infinite model cone K_z with vertex at 0 and a smooth diffeomorphism f_z with $f_z(0) = z$ and $f_z'(0) = \mathrm{Id}$, mapping K_z locally (near 0) onto Ω. If $\partial\Omega$ is locally C^1 at z, then K_z is a half-space. Since K_z has infinite volume, the spectrum of $-\Delta_\alpha(K_z)$ is no longer discrete, and we will now use the notation $\lambda_1(K_z, \alpha)$ to denote the corresponding spectral bound:

$$\lambda_1(K_z, \alpha) = \inf\{|\lambda| : \lambda \in \sigma(-\Delta_\alpha(K_z))\}. \tag{4.24}$$

The following theorem summarizes the key results, namely Theorems 3.2, 3.5 and 3.6, of [653].

Theorem 4.15 (Levitin–Parnovski). *Suppose $\Omega \subset \mathbb{R}^d$ satisfies the condition described above. Then*

$$\lambda_1(\Omega, \alpha) = -\left[\sup_{z\in\partial\Omega} \lambda_1(K_z, 1)\right]^2 \alpha^2 + o(\alpha^2) \qquad \text{as } \alpha \to -\infty.$$

Moreover, $\lambda_1(K_z, 1) = 1$ if K_z is, or contains, a half-space. If $d = 2$ and K_z has half-angle $\theta \le \pi/2$, then $\lambda_1(K_z, 1) = \sin^{-2}(\theta)$ with corresponding eigenfunction $\psi(x) = e^{-\alpha \sin^{-1}(\theta)x_1}$ (after a rotation).

This proves in particular the conjecture made in [614] mentioned above. In higher dimensions, a two-sided estimate on $\lambda_1(K_z, 1)$ was given (in terms of the angle θ), but the actual value does not seem to be known.

Open problem 4.16. *Determine the correct value of $C > 0$ in the asymptotic expansion*

$$\lambda_1(\Omega, \alpha) = -C\alpha^2 + o(\alpha^2) \qquad \text{as } \alpha \to -\infty$$

for piecewise smooth domains $\Omega \subset \mathbb{R}^d$, $d \geq 3$, satisfying a uniform cone condition, by determining the value of $\lambda_1(K_0, 1)$, where K_0 is a d-dimensional cone with vertex at the origin, in terms of the angle of opening of K_0.

For a generalization of both Theorem 4.14 and Theorem 4.15 to elliptic problems with weights and with a Dirichlet condition on part of the boundary, see [284].

If Ω is allowed to have outward-pointing cusps, then Theorem 4.15 no longer holds, and indeed, the asymptotic behaviour of λ_1 no longer has to be of the form $-C\alpha^2$. For example, if one takes the model domain

$$\Omega_p := \{(x_1, x_2) \in \mathbb{R}^2 : x_1 > 0,\ |x_2| < x_1^p\}, \qquad p > 1,$$

then, as noted in [653, Example 3.6], choosing test functions of the form $\varphi(x) = e^{\alpha x_1^{q(p)}}$ with $q(p) = 2 - p$ for $p < 2$ and $q(p) = 2$ for $p \geq 2$ leads to

$$\lambda_1(\Omega_p, \alpha) \leq -C \begin{cases} |\alpha|^{2/(2-p)} & \text{if } 1 < p < 2, \\ |\alpha|^n & \text{for any } n > 0, \text{ if } p \geq 2. \end{cases}$$

Obviously Ω_p is unbounded, but imposing the additional constraint $|x_1| < 1$ will not affect the asymptotics. Note however that for a sufficiently sharp cusp, the trace mapping of $H^1(\Omega)$ into $L^2(\partial\Omega)$ ceases to be compact (here for $p = 2$) and, eventually, ceases to exist altogether (for $p > 2$). In this case $\lambda_1(\Omega, \alpha)$ should be understood as the spectral bound, cf. (4.24). In fact, compactness of the trace is equivalent to $\lambda_1(\Omega, \alpha) > -\infty$, see [49, Proposition 8.1(b)]. There is a fairly extensive literature on general spectral properties of the Robin problem with $\alpha < 0$ and the corresponding Steklov problem on domains with model cusps; see for example [721–723] and the references therein.

There is still a certain gap between the two sides: for a Lipschitz domain we certainly have $\lambda_1(\Omega, \alpha) > -\infty$, but only a fairly special class of Lipschitz domains was considered in [653].

Open problem 4.17. *Suppose $\Omega \subset \mathbb{R}^d$ is an* arbitrary *bounded, Lipschitz domain. Does there exists a constant $C = C(\Omega) > 0$ such that $\lambda_1(\Omega, \alpha) = -C\alpha^2 + o(\alpha^2)$ as $\alpha \to -\infty$?*

4.4.2.2 The leading term in the asymptotic expansion of λ_n as $\alpha \to -\infty$

For the higher eigenvalues, the picture is much the same as for λ_1.

Theorem 4.18 (Daners–Kennedy). *Let $\Omega \subset \mathbb{R}^d$ be a bounded domain of class C^1. Then for each* fixed $n \geq 1$, *we have*

$$\lambda_n(\Omega, \alpha) = -\alpha^2 + o(\alpha^2) \qquad \text{as } \alpha \to -\infty. \tag{4.25}$$

This result first appeared in [322] and was proved using a variational argument: roughly speaking, if for fixed $n \geq 1$, we choose n test functions of the form $\varphi_{k,\alpha}(x) = e^{\alpha v_k \cdot x}$ for n different unit vectors $v_1, \ldots, v_n \in \mathbb{R}^d$ (cf. the proof of Proposition 4.12), then for $\alpha < 0$ large enough, the functions $\varphi_{k,\alpha}$ are "almost" orthogonal to each other in $L^2(\Omega)$ (since each has mass concentrated at a different part of the boundary) and all have Rayleigh quotient $-\alpha^2$. Hence $\lambda_n(\Omega, \alpha) \lesssim -\alpha^2$ as $\alpha \to -\infty$; since $\lambda_1(\Omega, \alpha) \sim -\alpha^2$, this forces $\lambda_n(\Omega, \alpha) \sim -\alpha^2$ for all $n \geq 1$. However, very little seems to be known about the higher eigenvalues on domains with corners. A special class of domains having two (or more) equal corners was considered in [499]. As observed in [614], separation of variables on a rectange yields *four* eigenvalues exhibiting the behaviour $-2\alpha^2$, while all other negative ones behave like $-\alpha^2$. This, plus the test function principles used extensively above, raise the following question.

Open problem 4.19. *Suppose that $\Omega \subset \mathbb{R}^2$ is a bounded, piecewise smooth domain having $m \geq 1$ corners with half-angles $\theta_1 \leq \ldots \leq \theta_m < \pi/2$. Is it true that the first m eigenvalues have the asymptotic behaviour*

$$\lambda_n(\Omega, \alpha) \sim -\sin^{-2}(\theta_n)\,\alpha^2 \qquad \text{as } \alpha \to -\infty,$$

for $n = 1, \ldots, m$? How does $\lambda_n(\Omega, \alpha)$ behave for fixed *$n \geq m$? Investigate the corresponding situation in higher dimensions and for more general Ω.*

We emphasize that since we are considering n to be fixed, this only concerns those eigencurves tending to $-\infty$. On that subject, let us also note the following partial refinement of Open Problem 4.11, another small gap in the literature which, although fairly clear, does not yet seem to have been addressed explicitly:

Open problem 4.20. *Prove that all eigencurves λ_α tending to $-\infty$ (say, on a sufficiently smooth domain $\Omega \subset \mathbb{R}^d$) exhibit the behaviour*

$$\lambda_\alpha \lesssim -\alpha^2 + o(\alpha^2) \qquad \text{as } \alpha \to -\infty,$$

i.e., prove that there are no negative eigencurves which behave asymptotically like a lower power of α.

This is supported by explicit calculations on the disk (cf. (4.15)) and the rectangle. In the general case it should be amenable to an analysis similar to the one used below, for example in [374, 740], to obtain higher order terms in the asymptotic expansion, based on Dirichlet-Neumann bracketing and a transformation of the boundary region, since this technique allows one to estimate all negative eigenvalues of $-\Delta_\alpha$ in terms of those of operators living near the boundary which can be described fairly explicitly.

4.4.2.3 Two- and three-term asymptotics for the negative eigencurves

Having observed that $\lambda_n(\Omega, \alpha)$ displays the same (leading) asymptotics independent of $n \geq 1$ (fixed), the dimension and the volume of the domain, the next natural question is to determine more terms in the asymptotic expansion of $\lambda_1(\Omega, \alpha)$ or more generally $\lambda_n(\Omega, \alpha)$, since at some point properties of the domain, in particular the curvature of its boundary, should make themselves felt. At the time of writing, this is a very active field of research: there is a number of papers on, or related to, this topic which have appeared only within the last couple of years [373, 374, 400, 499, 739, 740]; we also mention the preprints [197, 198, 496, 497, 601]. The state-of-the art on general smooth domains consists of two-term asymptotics, which may be summarized as follows.

Theorem 4.21 (Pankrashkin–Popoff; Exner–Minakov–Parnovski). *Let $\Omega \subset \mathbb{R}^d$ be a bounded, connected, C^3 domain, $d \geq 2$. Denote by γ_{max} the maximum mean curvature of the boundary $\partial\Omega$. Then for any fixed $n \geq 1$, we have*

$$\lambda_n(\Omega, \alpha) = -\alpha^2 + (d-1)\gamma_{max}\alpha + O(\alpha^{2/3}) \qquad \text{as } \alpha \to -\infty. \tag{4.26}$$

If, additionally, Ω is of class C^4, then the error estimate may be improved to $O(\alpha^{1/2})$.

This was proved in full generality by Pankrashkin and Popoff in [740], the two-dimensional case having been proved slightly earlier by Exner, Minakov and Parnovski in [374] (with slightly stronger assumptions on Ω and a weaker error estimate); the preprint [601] gives a generalisation for the first eigenvalue, $n = 1$, to the p-Laplacian and requires less regularity on Ω, namely $C^{1,1}$, albeit with a worse error term. The asymptotic expansion (4.26) in two dimensions had previously been obtained for the principal eigenvalue $\lambda_1(\Omega, \alpha)$, with similar techniques, in [739].

The paper [374] and the preprint [496] also investigate the higher terms in the asymptotic expansion on smooth planar domains, where the eigenvalues of a one-dimensional Schrödinger operator related to the curvature of the boundary appear. Assume that $\partial\Omega$ is a smooth closed Jordan curve of length L, parametrized by a variable $t \in [0, L]$, and denote by $\gamma : [0, L] \to \mathbb{R}$ the (signed) curvature of $\partial\Omega$. Denote by $\mu_1, \mu_2, \mu_3, \ldots$ the eigenvalues, ordered by increasing size and repeated according to their multiplicities, of the Schrödinger operator

$$S := -\frac{d^2}{dt^2} - \frac{1}{4}\gamma^2(t) \qquad \text{in } L^2(0, 1). \tag{4.27}$$

Then the following asymptotic inequality was obtained in [374].

$$\lambda_n(\Omega, \alpha) \geq -\alpha^2 + \gamma_{\max}\alpha - \frac{1}{4}\gamma_{\max} + \mu_n + O\left(\frac{\ln|\alpha|}{\alpha}\right) \qquad \text{as } \alpha \to -\infty. \tag{4.28}$$

The full asymptotic expansion for smooth planar domains for which the curvature achieves a unique, non-degenerate maximum, is given in [496].

Open problem 4.22. *Obtain the zero*th *order (i.e. third) term in the asymptotic expansion of* $\lambda_n(\Omega, \alpha)$ *on sufficiently smooth, general domains* $\Omega \subset \mathbb{R}^d$.

Any analysis of this nature has as its starting point the technique of Dirichlet–Neumann bracketing (see [789, Sec. XIII.15]) in order to obtain tight two-sided estimates on the eigenvalue(s) in terms of operators living in a thin tubular neighbourhood of the boundary $\partial\Omega$, by imposing additional Dirichlet or Neumann conditions at the boundary of the neighbourhood inside Ω. The neighbourhood itself may be thought of as being locally like a rectangular prism, $\mathbb{R}^{d-1} \times (0, \varepsilon)$; after a suitable change of variables, one sees that the bounding operators in the tubular region can be controlled in the limit $\alpha \to -\infty$ by a direct sum of operators, one acting on the boundary (cf. (4.27)) and the other in the normal direction. A careful analysis of these operators yields the asymptotics or, alternatively, bounds of the form (4.28). This analysis is reminiscent of, and seems to have been in large part inspired by, a similar asymptotic analysis of Schrödinger operators on $\mathbb{R}^d$ with an attractive δ-interaction supported on a surface (i.e. $\partial\Omega$), as the strength of the interaction tends to infinity; see, e.g., [371, 375, 376].

In [373] the results of [374] (i.e. Theorem 4.21 in two dimensions) are generalized to a class of smooth, unbounded, planar domains, the lower bound corresponding to (4.28) now depending on whether the eigenvalue in question is below or above the bottom of the (possibly non-empty) essential spectrum.

The paper [499] is possibly the only work to date establishing two-term asymptotics when Ω has corners. More precisely, the authors investigate the special case where Ω belongs to a model class of piecewise smooth planar domains formed by the intersection of two infinite sectors having the same angle of opening: in this case, the first two eigenvalues have the same first term in their expansions and differ only in the second and above. The authors interpret these corners as geometric wells; in this language, it is natural that the deepest well (i.e., the sharpest angle at the boundary) should determine the principal term of the spectral asymptotics. Their analysis here is inspired by methods developed for the semiclassical analysis of Schrödinger operators with multiple wells, as in [480].

We also mention [198], which deals with higher dimensions, among other things.

4.4.2.4 Consequences for Faber–Krahn-type inequalities when $\alpha < 0$

The formula (4.26) has implications for certain eigenvalue comparisons of (reversed) Faber–Krahn or domain monotonicity type, at least in the asymptotic regime. For example, it was observed in [740] that if Ω is a star-shaped domain other than a ball and B is a ball of the same volume, then since $\gamma_{\max}(B) < \gamma_{\max}(\Omega)$, for any $n \geq 1$ there exists $\alpha_n^* < 0$ such that

$$\lambda_n(\Omega, \alpha) < \lambda_n(B, \alpha) \qquad \text{for all } \alpha < \alpha_n^*.$$

This holds in the same way for any pair of domains Ω_1, Ω_2 for which $\gamma_{\max}(\Omega_1) < \gamma_{\max}(\Omega_2)$. This will be examined in more detail in Section 4.5.2 in the context of a reversed Faber–Krahn inequality for $\lambda_1(\Omega, \alpha)$ with $\alpha < 0$ (see in particular Theorem 4.31), and again briefly in Section 4.6.3 for the higher eigenvalues.

The same principle also implies that among *general* domains of given volume, the ball cannot maximize the first eigenvalue (or indeed any other) for large enough $\alpha < 0$, since a large, thin annulus will have a smaller maximum mean curvature. This was first observed and used in [400], where (4.26) was derived for λ_1 on balls and annuli (with a slightly weaker error term, but independently of, and with different methods from, the other works [374, 740] etc.), by using asymptotic expansions for the Bessel functions corresponding to the eigenfunctions. The conclusion is that $\lambda_1(B, \alpha)$ cannot be maximal among all domains of given volume for all $\alpha < 0$, refuting a Faber–Krahn-type conjecture. This will be discussed in more detail in Section 4.5.2.

4.5 Isoperimetric inequalities and other eigenvalue estimates

4.5.1 Positive parameter: Faber–Krahn and other inequalities

We now consider the shape optimization problem for *fixed* $\alpha > 0$. As in the Dirichlet case, we have an inequality of Faber–Krahn type: the ball is the unique minimizer of the first eigenvalue of the Robin Laplacian among all domains of given volume, for any fixed $\alpha > 0$.

Theorem 4.23 (Bossel–Daners). *Let $\Omega \subset \mathbb{R}^d$ be a bounded Lipschitz domain and let B be a ball with $|B| = |\Omega|$. Then for any $\alpha > 0$*

$$\lambda_1(\Omega, \alpha) \geq \lambda_1(B, \alpha), \tag{4.29}$$

with equality if and only if Ω is a ball.

It is said that Krahn himself attempted to prove Theorem 4.23,[4.2] although the authors could not find a reliable first-hand source for this assertion. The first positive result in this direction goes back to 1957, when Payne and Weinberger proved via a direct variational argument that if Ω is a smooth domain, then $\lambda_1(\Omega, \alpha)$ is at least as large as the corresponding first eigenvalue of any ball circumscribing Ω (for the same $\alpha > 0$), cf. (4.7). This is weaker than Theorem 4.23, since $\lambda_1(\Omega, \alpha)$ is at least monotonic with respect to *homothetic* rescalings of Ω (for fixed $\alpha > 0$), cf. (4.10). Further upper and lower bounds for $\lambda_1(\Omega, \alpha)$ predating Theorem 4.23 can be found in [514, 750, 757, 829].

4.2 For example, Bossel refers to this in the introduction of [170] as Krahn's conjecture, and Catherine Bandle informs us (private communication) that Payne also used this appellation for the conjecture.

For smooth domains, inequality (4.29) was first established in two dimensions in [169] (see also [168, 170]) and in higher dimensions in [319], which also filled in a number of technical details not covered in [169]. The case of equality among sufficiently smooth domains was treated in [321] and a generalization to the p-Laplacian was given in [311]. The proof for the p-Laplacian follows the same scheme as the linear case, leading to the suggestion that – as in the Dirichlet case – the proof does not rely on the linear structure of the operator, but is essentially of variational nature. As we shall see at the end of the section, the key feature allowing us to reproduce the proof of the linear case is the equal balance between the homogeneities of the Dirichlet energy, the boundary energy and the zero order term which, for instance, is not valid anymore for the Saint-Venant inequality. In [335], the result of Theorem 4.23 is extended to nonlinear (p-Laplacian type) anisotropic operators, where the Euclidean norm is replaced by a generic norm.

A subsequent refinement of the proof in [213], which reduced a number of auxiliary results to variational arguments, established Theorem 4.23 for a larger class of domains, including the statement on the case of equality. The main ideas of the proof of Theorem 4.23 will be described in Section 4.5.1.3 below.

Other interesting results which are not of Faber–Krahn type will be considered in Section 4.5.1.5 (see also Sections 4.2.1 and 4.4.1).

4.5.1.1 Extension of the Bossel–Daners inequality to arbitrary open sets

In Theorem 4.23 a crucial requirement is that the set Ω is Lipschitz. For Dirichlet boundary conditions, the Faber–Krahn inequality does not require any smoothness of the domain and implies the following Poincaré inequality in $\mathbb{R}^d$

$$\forall u \in H^1(\mathbb{R}^d),\ |\{u \neq 0\}| = c,\ \int_{\mathbb{R}^d} |\nabla u|^2 dx \geq \lambda_1^D(B) \int_{\mathbb{R}^d} u^2 dx,$$

where B is the ball of measure c and $\lambda_1^D(B)$ is the first Dirichlet eigenvalue of the Laplacian on the ball. When viewed in this way, the set Ω from the Faber–Krahn inequality is in fact the (quasi-open) set $\{u \neq 0\}$. The passage from open sets to *quasi-open* sets is simply a consequence of the density of smooth functions (which are nonzero on an open set) in the space $H^1(\mathbb{R}^d)$.

Similar questions can be asked for the Robin Laplacian. Does the Bossel–Daners inequality extend to arbitrary open sets? Does this imply a general Poincaré type inequality?

Assume Ω is a bounded Lipschitz open set and $u \in H^1(\Omega)$. Then Theorem 4.23 implies

$$\int_\Omega |\nabla u|^2 dx + \alpha \int_{\partial\Omega} u^2 d\sigma \geq \lambda_1(B,\alpha) \int_\Omega u^2 dx, \qquad (4.30)$$

which is a Poincaré inequality with trace term. If Ω is arbitrary, the inequality above needs clarification, since the trace of u on $\partial\Omega$ is not defined in the usual sense. For instance, one can ask if (4.30) holds for all functions which belong to $H^1(\mathbb{R}^d)$, in which case the trace exists on $\partial\Omega$ pointwise $\mathcal{H}^{d-1}$-a.e. A different approach to the trace was followed in [51, 318]. The authors used the so-called Maz'ja space, consisting of the completion of $H^1(\Omega) \cap C(\overline{\Omega})$ for the norm $\|\cdot\|_{H^1(\Omega)} + \|\cdot\|_{L^2(\partial\Omega,\mathcal{H}^{d-1})}$. This space, which in general is smaller than the usual Sobolev space, allows one to define properly the Robin Laplacian and its spectrum. It is a natural question to ask if the Bossel–Daners inequality, in particular in the form (4.30), still holds. Another situation that one can imagine is that of "cracked" domains, e.g. a bounded Lipschitz set Ω from which one removes a smooth $(d-1)$-dimensional surface, denoted Γ. The Robin Laplacian is well defined on the set $\Omega \setminus \Gamma$, with the observation that a function $u \in H^1(\Omega \setminus \Gamma)$ may have two different traces on both sides of Γ (denoted by u^+ and u^-, in the sequel). In order to understand the Bossel–Daners inequality and inequality (4.30) on $\Omega \setminus \Gamma$ one could imagine that the crack is approximated by tubular neighbourhoods with vanishing widths. In the limit, one can naturally expect that on Γ both traces u^+ and u^- have to be counted.

The complete answer to the questions above was given in [216] (Theorem 4.24 below). The main idea is to transform the Bossel–Daners inequality on arbitrary sets into a functional inequality of Poincaré type, in which the best constant is identified to be the first eigenvalue on the ball. From this perspective, an element of $H^1(\Omega)$ is imagined to be extended by zero on $\mathbb{R}^d \setminus \Omega$, so that the boundary of Ω is a region where the function u jumps from its inner trace (say u^+) to zero (the external trace, say u^-). This approach is consistent with the example above involving the inner set Γ, which again can be seen as the region where the functions jumps from u^- to u^+.

A convenient definition of a space of functions having jumps, which covers all the examples above, involves the total variation of the gradient and relies on the space of functions of bounded variation. A function $v : \mathbb{R}^d \to \mathbb{R}$ is a *special* function of bounded variation if $v \in L^1(\mathbb{R}^d)$ and its distributional gradient Dv is a Radon measure which can be decomposed as the sum of a volume and surface part

$$Dv = \nabla v\, dx + (v^+ - v^-)\nu_v \mathcal{H}^{d-1} \llcorner_{J_v},$$

where $\nabla v\, dx$ is the absolutely continuous part with respect to the Lebesgue measure, J_v is the jump set (where the approximate upper and lower limits v^+ and v^- either do not exist or are different) and ν_v is a normal vector field on J_v. The space of special functions of bounded variations is denoted by $SBV(\mathbb{R}^d)$.

The result below extends the Bossel–Daners inequality to arbitrary sets and can also be seen as a general Poincaré inequality with trace terms.

Theorem 4.24 (Bucur–Giacomini). *Let $c > 0$ be given and let $u : \mathbb{R}^d \to \mathbb{R}$, $u \in SBV(\mathbb{R}^d) \cap L^2(\mathbb{R}^d)$. Assume that $|\{u \neq 0\}|$ equals c. The following inequality holds:*

$$\int_{\mathbb{R}^d} |\nabla u|^2 dx + \alpha \int_{J_u} (u^+)^2 + (u^-)^2 d\mathcal{H}^{d-1} \geq \lambda_1(B, \alpha) \int_{\mathbb{R}^d} u^2 dx,$$

where B is the ball of volume c. Equality holds if and only if u is the first Robin eigenfunction of the ball B extended by zero on $\mathbb{R}^d \setminus B$ (up to a translation in space).

The hypotheses of this theorem do not require that $\nabla u \in L^2(\mathbb{R}^d)$, $u^+, u^- \in L^2(J_u, \mathcal{H}^{d-1})$. Nevertheless, if this is not the case, the inequality becomes trivial.

4.5.1.2 The Saint-Venant inequality and best constant for general Poincaré inequalities with trace terms in $\mathbb{R}^d$

Bandle and Wagner raised in [99] the question of the validity of the Saint-Venant inequality related to the maximization of the torsional rigidity in the context of Robin boundary conditions.

For a bounded Lipschitz $\Omega \subset \mathbb{R}^d$ one solves

$$\begin{cases} -\Delta u_\Omega = 1 & \text{in } \Omega, \\ \frac{\partial u_\Omega}{\partial n} + \alpha u_\Omega = 0 & \text{on } \partial\Omega \end{cases}$$

and defines the "elastic" torsional rigidity of Ω as $T(\Omega) = \int_\Omega u_\Omega dx$. The question raised in [99] was to prove that $T(\Omega, \alpha) \leq T(B, \alpha)$, where B is the ball of the same measure as Ω. An easy computation leads to

$$\frac{1}{T(\Omega, \alpha)} = \min \left\{ \frac{\int_\Omega |\nabla u|^2 dx + \alpha \int_{\partial\Omega} u^2 d\sigma}{\left(\int_\Omega |u| dx\right)^2} : u \in H^1(\Omega) \setminus \{0\} \right\}, \tag{4.31}$$

and thus relates this question to a Faber–Krahn-type inequality similar to the Bossel–Daners one.

Bandle and Wagner proved that the ball is a local maximizer for the elastic torsional rigidity, in the sense that $T(\Omega, \alpha) \leq T(B, \alpha)$ provided Ω is a small perturbation of the ball.

The proof of the Saint-Venant inequality was given in [217, 218].

Theorem 4.25 (Bucur–Giacomini). *Let $\Omega \subset \mathbb{R}^d$ be a bounded Lipschitz domain and let B be a ball with $|B| = |\Omega|$. Then for any $\alpha > 0$*

$$T(\Omega, \alpha) \leq T(B, \alpha),$$

with equality if and only if Ω is a ball.

The proof of Theorem 4.25 is based on the analysis of free discontinuity problems; a proof of the Saint-Venant inequality via the use of a suitable "H-function" following the ideas of Theorem 4.23 (cf. (4.33)) has not yet been found, and indeed, this seems to be a challenge.

As seen in Theorem 4.24 and in the reinterpretation of the torsional rigidity as a non-linear eigenvalue problem in (4.31), the inequalities of Bossel–Daners type can be seen as a particular type of Poincaré inequality with trace term in $\mathbb{R}^d$, where the best constant is identified as the eigenvalue of the ball. Poincaré inequalites may hold for various integral norms, so it is natural to ask the following question: given $1 \le p, q, r < +\infty$, is it true that for every open Lipschitz set Ω of volume c, for every $u \in W^{1,p}(\Omega)$ the following holds:

$$\int_\Omega |\nabla u|^p dx + \alpha\Big(\int_{\partial\Omega} |u|^q d\sigma\Big)^{\frac{p}{q}} \ge C_c\Big(\int_\Omega |u|^r dx\Big)^{\frac{p}{r}} \ ? \quad \textbf{(p,q,r)-inequality} \tag{4.32}$$

The constant C_c is expected to depend only on the measure of Ω and on α and to correspond to the case of equality achieved on the ball of measure c. Let us identify this Poincaré inequality by the triple (p, q, r).

The Bossel–Daners theorem refers to the case $(2, 2, 2)$, while the results from [213, 311] concern the case (p, p, p), for $p \in (1, +\infty)$. The case $(2, 2, 1)$ corresponds to the Saint–Venant inequality.

The limit case $(p, p^\sharp, p^*)$, with $p \in [1, d)$, $p^\sharp = \frac{(d-1)p}{d-p}$ and $p^* = \frac{pd}{d-p}$ is of particular interest, being an extension of the Sobolev inequality to functions with "jumps". It was first discussed in a slightly different setting by Brézis and Lieb in [186]. Its main particularity is its scale invariance and it was solved by Maggi and Villani in [683, 684] using mass transportation theory.

A quite general result in the spirit of Theorem 4.24, which covers the situations $(2, 2, r)$ with $r \in [1, 2]$, was proved in [217], where the best constant C_c was identified to correspond to the ball. Its proof is based on the analysis of free discontinuity problems and is likely to extend to more general situations (p, q, r) with $p \in (1, +\infty)$, $q > p$, $1 < r < \frac{pd}{d-1}$.

It is also important to mention that there are known situations in which the best constant in inequality (p, q, r) is not attained for the ball. In [201], it was proved that for the case $(2, 1, 2)$, there are values of c and $\alpha > 0$ such that the ball is not optimal.

4.5.1.3 Proof of Theorem 4.23

In general, the Robin problem is considered to be significantly harder than the corresponding Dirichlet problem, and the former lacks many of the useful properties of the latter, such as monotonicity with respect to domain inclusion (see Section 4.2.1).

The additional difficulty here is due principally to the appearance of the boundary integral in the Rayleigh quotient (4.2), and indeed, one can convince oneself quite easily that the usual symmetrization techniques no longer work, since in general symmetrization will not decrease both the Dirichlet integral and the boundary integral.

Here, we will give a brief description of the method of proof of Theorem 4.23, which relies on a functional of the level sets of the first eigenfunction, inspired from a conformal invariant known as *extremal length*. This idea also applies to the proof of Theorem 4.24. The presentation below is inspired by [213].

Let $\Omega \subset \mathbb{R}^d$ be a bounded Lipschitz domain. For the rest of this subsection, $\alpha > 0$ will be fixed, and so we will denote the first eigenvalue simply by $\lambda_1(\Omega)$, suppressing the dependence on α, and the corresponding first eigenfunction by ψ_1. As noted in Section 4.2 and in particular Prop 4.1, $\psi_1 \in H^1(\Omega) \cap C^1(\Omega) \cap C(\overline{\Omega})$ is simple and can be chosen strictly positive in Ω; let us normalize ψ_1 so that $\|\psi_1\|_\infty = 1$.

We denote by ψ_1^B the first eigenfunction of the ball, positive and normalized in L^∞ as above. It can be easily seen that ψ_1^B is radially symmetric, decreasing, and that $|\nabla \ln(\psi_1^B)|$ is maximal on the boundary of the ball, where it is equal to α.

As usual for inequalities such as Theorem 4.23, we need to consider the level sets of ψ_1, which we will denote by $U_t := \{x \in \Omega : \psi_1(x) > t\}$. Note that, in opposition to the Dirichlet case, Ω is not in general a level set itself, and so for any given $t > 0$ the boundary ∂U_t of U_t may consist of two parts: an "internal part" $S_t := \{x \in \Omega : \psi_1(x) = t\}$ and an "external part" $\Gamma_t := \partial\Omega \cap \partial U_t$.

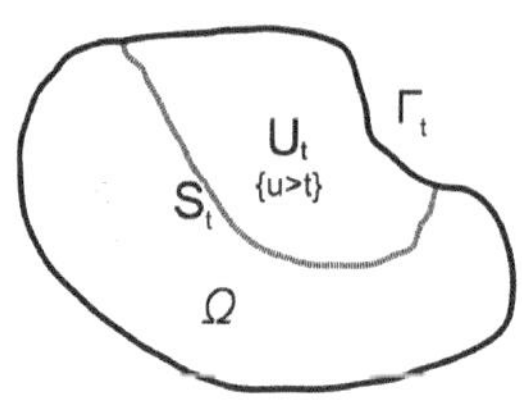

Fig. 4.3. Depiction of the internal and external boundaries of the level set of a function u.

The key tool used to prove Theorem 4.23 is a functional of the level sets U_t: for an arbitrary positive, measurable function $\varphi : \Omega \to [0, +\infty)$, we set

$$H_\Omega(U_t, \varphi) := \frac{1}{|U_t|}\left(\int_{S_t} \varphi\, d\sigma + \int_{\Gamma_t} \alpha\, d\sigma - \int_{U_t} \varphi^2\, dx\right). \tag{4.33}$$

The idea to consider this functional goes back at least to Bossel, who was inspired by a conformal invariant called extremal length due to Ahlfors and Beurling. Its close relation to λ_1 and ψ_1 is given by:

Lemma 4.26. *For the choice $\varphi := |\nabla\psi_1|/\psi_1 = |\nabla \ln(\psi_1)|$, we have*

$$\lambda_1(\Omega) = H_\Omega(U_t, |\nabla\psi_1|/\psi_1) \tag{4.34}$$

for almost all $t \in (0,1)$.

Sketch of proof. This identity is obtained formally by multiplying the equation $-\Delta\psi_1 = \lambda_1\psi_1$ with the function $\frac{1}{\psi_1}$, integrating over U_t and using integration by parts. If Ω is smooth (C^2), in which case also $\inf_\Omega \psi_1 > 0$, this computation is quite direct. In the case of a Lipschitz set Ω, the computation is slightly more involved, since the equation cannot be used in the strong form. One has to use the weak form of the equation with a test function obtained by a cut off around a level of $\frac{1}{\psi_1}$, suitably extended in Ω. A passage to the limit relying on the co-area formula leads to (4.34). □

This lemma will lead to the central observation that, for any non-negative function φ in $L^2(\Omega)$ essentially distinct from $|\nabla\psi_1|/\psi_1$, $\lambda_1(\Omega)$ is a strict upper bound on $\operatorname{ess\,inf}_t H_\Omega(U_t,\varphi)$:

Lemma 4.27. *Suppose φ is a non-negative function $L^2(\Omega)$ satisfying $\varphi \neq |\nabla\psi_1|/\psi_1$ on a set of positive measure in Ω. Then there exists a set $S = S(\varphi) \subset (0,1)$ of positive measure such that*

$$\lambda_1(\Omega) > H_\Omega(U_t,\varphi)$$

for all $t \in S$.

Sketch of proof. Since we shall suppress all the technical details, we shall assume that all the computations below can be done for a.e. $t \in (0,1)$, which is for instance the case if Ω is of class C^2. In the general case, one has to justify their validity. The idea is to prove that the inequality

$$H_\Omega(U_t,\varphi) \le \lambda_1(\Omega) - \frac{1}{t|U_t|}\frac{d}{dt}\left(t\int_{U_t}\left(\varphi - \frac{|\nabla\psi_1|}{\psi_1}\right)\frac{|\nabla\psi_1|}{\psi_1}\,dx\right)$$

holds a.e. $t \in (0,1)$. In a second step, by contradiction one can assume that

$$\lambda_1(\Omega) \le H_\Omega(U_t,\varphi), \text{ a.e. } t \in (0,1).$$

This implies that the mapping

$$(0,1) \ni t \mapsto t\int_{U_t}\left(\varphi - \frac{|\nabla\psi_1|}{\psi_1}\right)\frac{|\nabla\psi_1|}{\psi_1}\,dx$$

is non-increasing. Since it vanishes for $t = 1$ and has limit zero at 0^+, it must in fact be constant, which leads to $\varphi = \frac{|\nabla\psi_1|}{\psi_1}$ a.e. on Ω. □

Proof of Theorem 4.23. We are now in a position to prove Theorem 4.23. This follows from combining Lemma 4.27 with a rearrangement argument involving the first eigenfunction on B. We consider the L^∞ normalized, positive, first eigenfunction ψ_1^B on the ball and denote by φ the rearrangement of the function $\frac{|\nabla\psi_1^B|}{\psi_1^B}$ on Ω, following the level sets U_t, i.e., we take φ as the function on Ω such that

$$\varphi|_{\partial S_t} := \frac{|\nabla \psi_1^B|}{\psi_1^B}\Big|_{\partial B^{r(t)}}$$

for all $t \in (0,1)$, where $B^{r(t)}$ is the ball centred at the origin having the same volume as U_t. This is possible since the level sets of $\frac{|\nabla \psi_1^B|}{\psi_1^B}$ are given by concentric balls, just as the function ψ_1^B itself, so that $\varphi|_{\partial S_t}$ is constant for each t. At this point, we apply Lemma 4.27 to obtain the existence of some $t \in (0,1)$ such that

$$\lambda_1(\Omega) \geq H_\Omega(U_t, \varphi).$$

By classical properties of the rearrangement and the (geometric) isoperimetric inequality, we have

$$H_\Omega(U_t, \varphi) \geq H_B(B^t, \frac{|\nabla \psi_1^B|}{\psi_1^B}) = \lambda_1(B).$$

The last equality is a consequence of Lemma 4.26.

Consequently, the inequality is proved. The case of equality follows by a simple analysis of the equality $H_\Omega(U_t, \varphi) = H_B(B^t, \frac{|\nabla \psi_1^B|}{\psi_1^B})$. □

4.5.1.4 Inequalities of Cheeger type

The method of proof of the Bossel–Daners inequality also implies an inequality for λ_1 in terms of the *Cheeger constant* of a domain $\Omega \subset \mathbb{R}^d$, which we will consider in the form

$$h(\Omega) = \inf \left\{ \frac{|\partial \omega|}{|\omega|} : \omega \text{ compactly contained in } \Omega \right\}.$$

This measure of the geometry of Ω, originally introduced by Cheeger on general Riemannian manifolds, is known to be closely related to the first Dirichlet eigenvalue of the domain (or manifold) via

$$\lambda_1(\Omega) \geq \frac{1}{4} h(\Omega)^2;$$

see, e.g., [577] for more details. Obviously such an inequality cannot hold in general for $\lambda_1(\Omega, \alpha)$ for all $\alpha > 0$, but it was already observed by Bossel in [169] that the H-function (4.33) can be easily used to obtain a variant.

Theorem 4.28 (Bossel). *Let $\Omega \subset \mathbb{R}^d$ be a bounded, Lipschitz domain. Then*

$$\lambda_1(\Omega, \alpha) \geq h(\Omega)\alpha - \alpha^2 \qquad \text{for all } \alpha > 0. \tag{4.35}$$

If, in addition, $\alpha \geq \frac{1}{2} h(\Omega)$, then

$$\lambda_1(\Omega, \alpha) \geq \frac{1}{4} h(\Omega)^2. \tag{4.36}$$

Note that the two bounds are equal when $\alpha = \frac{1}{2}h(\Omega)^2$. Bossel only gave the proof of this theorem for smooth domains in two dimensions, but exactly the same idea works in the general case: namely, for (4.35), we simply choose the constant test function $\varphi := \alpha$; then Lemma 4.27 implies the existence of a level $t \in (0, 1)$ such that

$$\lambda_1(\Omega, \alpha) \geq H_\Omega(U_t, \alpha) = \frac{1}{|U_t|}\left(\int_{S_t} \alpha \, d\sigma + \int_{\Gamma_t} \alpha \, d\sigma - \int_{U_t} \alpha^2 \, dx\right)$$
$$= \frac{|\partial U_t|}{|U_t|}\alpha - \frac{1}{|U_t|}\int_{U_t} \alpha^2 \, dx.$$

Since by definition $|\partial U_t|/|U_t| \geq h(\Omega)$, we immediately obtain (4.35). An analogous calculation with $\varphi = h(\Omega)/2$ under the assumption $\alpha \geq h(\Omega)/2$ immediately yields (4.36).

In fact, inequalities (4.35) and (4.36) are strict. For instance, concerning (4.36), this is a consequence of the monotonicity $\lambda_1(\Omega, \alpha) \geq \lambda_1(\Omega, \frac{1}{2}h(\Omega))$ and of the strict inequality $\lambda_1(\Omega, \frac{1}{2}h(\Omega)) > \frac{1}{4}h(\Omega)^2$. The last inequality is indeed strict, since a necessary condition for equality to occur is that for some level t, the set U_t must be a ball. This implies that the set Ω is a ball, in which case the inequality turns out to be strict.

This result was briefly mentioned by Daners in [319], but appears to have been largely forgotten since then; in particular, no further progress has been made. The bound in (4.35) is of the correct form for small $\alpha > 0$, in that its leading term is linear, but it is not clear if (4.35) is generally optimal: by the formula for the derivative of λ_1 at zero (4.16) and the fact that λ_1 is analytic we have

$$\lambda_1(\Omega, \alpha) = \frac{|\partial\Omega|}{|\Omega|}\alpha + O(\alpha^2) \qquad \text{as } \alpha \to 0,$$

and in general $|\partial\Omega|/|\Omega| > h(\Omega)$. Thus we ask the following open-ended question.

Open problem 4.29. *Is it possible to use the same method to obtain a tighter lower bound on $\lambda_1(\Omega, \alpha)$ than (4.35) and (4.36), at least for some $\alpha > 0$, possibly in terms of a different quantity than $h(\Omega)$?*

4.5.1.5 Other inequalities for the principal eigenvalue

Here we will collate and briefly discuss various other bounds for $\lambda_1(\Omega, \alpha)$ which depend on $\alpha > 0$ together with geometric or other, generally easy to compute, properties of Ω.

The paper [829] was possibly the first to obtain both upper and lower bounds on $\lambda_1(\Omega, \alpha)$ of this nature; the simplest of these, valid on a planar domain Ω, have the form

$$\lambda_1(\Omega)\left(1 + \frac{\lambda_1(\Omega)}{\alpha q_1(\Omega)}\right)^{-1} \leq \lambda_1(\Omega, \alpha) \leq \lambda_1(\Omega)\left(1 + \frac{4\pi}{\alpha|\partial\Omega|}\right)^{-1},$$

where $q_1(\Omega)$ is the smallest eigenvalue of the Steklov-type fourth-order problem

$$\Delta^2 u = 0 \quad \text{in } \Omega, \qquad u = \Delta u - q\frac{\partial u}{\partial \nu} = 0 \quad \text{on } \partial\Omega.$$

The techniques used are essentially variational. The lower bound is obtained via a characterization of $1/\lambda_1(\Omega, \alpha)$ together with a decomposition of the space of test functions: each $u \in H^1(\Omega)$ can be written as the sum of a weakly harmonic function having the same boundary value and a weak solution of the Dirichlet boundary value problem.

The recent paper [600] addresses the question of optimizing $\lambda_1(\Omega, \alpha)$ for fixed Ω in terms of the function $\alpha \in L^1(\partial\Omega)$ with fixed integral mean and support in a prescribed section of the boundary (this problem admits a maximum but no minimum, nor infinimum). Two-sided bounds on $\lambda_1(\Omega, \alpha)$ (for constant $\alpha > 0$) are also obtained for convex Ω in terms of α and the inradius r_Ω, mirroring similar known results for the Dirichlet problem, the lower bound being based on a functional inequality of Hardy type for the Robin problem:

$$\frac{1}{4}\frac{\alpha}{r_\Omega(1 + \alpha r_\Omega)} \le \lambda_1(\Omega, \alpha) \le 2\lambda_1(B_1)\frac{\alpha}{r_\Omega(1 + \alpha r_\Omega)},$$

where $\lambda_1(B_1)$ is the first eigenvalue of the Dirichlet Laplacian on the unit ball in $\mathbb{R}^d$. Further bounds, in particular for convex domains, can be found in [830].

Further interesting results may be found in [304, 830].

4.5.2 Negative parameter

4.5.2.1 Bareket's conjecture

When α is negative, using a constant test function in the Rayleigh quotient (4.2) shows that the first eigenvalue will be negative for any bounded, Lipschitz domain Ω; see (4.22), or (4.20). Furthermore, by choosing a sequence of domains Ω_n of fixed volume with $|\partial\Omega_n| \to +\infty$, by (4.22) we must have $\lambda_1(\Omega_n, \alpha) \to -\infty$, making it clear that we can no longer have a minimizer for $\lambda_1(\Omega, \alpha)$ among domains of given volume. In fact, if Ω is unbounded, or has rough enough boundary, then for $\alpha < 0$ the associated bilinear form (4.3) will no longer be semi-bounded from below, so formally $\lambda_1(\Omega, \alpha) = -\infty$.

On the other hand, the fact that $\lambda_1(\Omega, \alpha)$ is now bounded from above suggests that we should we seek a maximizer instead. In the two-dimensional case it was conjectured by Bareket [110] in 1977, and again more recently in any dimension [193], that this maximizer should once again be the ball:

Conjecture 4.30 (Baraket). *Let $\Omega \subset \mathbb{R}^d$ be a bounded, sufficiently smooth domain, and denote by B a ball with $|B| = |\Omega|$. Then $\lambda_1(\Omega, \alpha) \le \lambda_1(B, \alpha)$ for all negative values of α.*

Supporting evidence is provided by the formula (4.16) for the derivative of λ_1 with respect to α at zero, on a given domain Ω. By the classical geometric isoperimetric inequality this is smallest when Ω is a ball, yielding in particular that for any bounded Lipschitz domain $\Omega \subset \mathbb{R}^d$ there exists $\alpha_0(\Omega) < 0$ such that

$$\lambda_1(\Omega, \alpha) \le \lambda_1(B, \alpha) \qquad \text{for all } \alpha \in [\alpha_0(\Omega), 0]; \tag{4.37}$$

of course, the constant α_0 depends *a priori* on Ω and thus whether or not the conjecture holds on an interval $(\alpha_1, 0)$ for some negative value α_1 requires showing that this dependence may be lifted, making the bound uniform.

Bareket provided further evidence in [110]: a proof of the conjecture for all planar domains for which $|\partial\Omega|$ is large relative to α, more precisely, those for which

$$|\partial\Omega| \ge 2\sqrt{\pi|\Omega|} - |\Omega|\alpha.$$

She also showed that the disk is a local minimizer with respect to a certain class of area-preserving perturbations. Recently, this was extended by Ferone, Nitsch and Trombetti in [380], who showed that the ball is a local maximizer in a strong sense, namely with respect to the L^∞ topology, in any space dimension.

However, while for $\alpha > 0$ the Faber–Krahn inequality for the *Dirichlet* Laplacian may also be viewed as "supporting evidence" for the validity of the Bossel–Daners inequality, the Szegő–Weinberger inequality for the first Neumann eigenvalue does not play the same role for $\alpha < 0$, since that corresponds to $\lambda_2(\Omega, 0)$. Nevertheless, the problem of maximizing $\lambda_1(\Omega, \alpha)$ for $\alpha < 0$ does seem to bear a certain resemblance to the Neumann problem.

It was thus a surprise when it was shown in 2015 that Conjecture 4.30 cannot hold in general, even in two dimensions [400] – although the independent development of the two-term asymptotics for $\lambda_1(\Omega, \alpha)$ as $\alpha \to -\infty$ at about the same time turns out to support the negation of the conjecture, cf. Section 4.4.2.4. More precisely, the first eigenvalue of a ball must be smaller than that of spherical shells with the same volume, for sufficiently large negative α (depending on the particular shell).

Theorem 4.31 (Freitas–Krejčiřík). *Given a ball B_r of radius r, there exist positive numbers r_1 and r_2 with $r_1 < r_2$ such that the spherical shell defined by*

$$S_{r_1,r_2} = \left\{x \in \mathbb{R}^d : r_1 < |x| < r_2\right\}$$

with the same volume as B_r satisfies $\lambda_1(B_r, \alpha) < \lambda_1(S_{r_1,r_2}, \alpha)$ for all sufficiently large negative values of α.

Proof. The proof consists in determining the asymptotic expansions for both B_r and S_{r_1,r_2} as α goes to $-\infty$. Since both domains are rotationally symmetric the solution may again be obtained by separation of variables. Since the first eigenvalue is simple, it follows that the first eigenfunction (which may be chosen to be positive) is also rotationally symmetric. In the case of spherical shells, the solution is thus obtained by solving the boundary value problem

$$\begin{cases} -r^{-(d-1)}[r^{d-1}\psi'(r)]' = \lambda\psi(r) \\ -\psi'(r_1) + \alpha\psi(r_1) = 0\,, \\ \psi'(r_2) + \alpha\psi(r_2) = 0\,. \end{cases} \tag{4.38}$$

for $r \in [r_1, r_2]$ – the solution for balls is similar and somewhat simpler, so we will focus on the case of shells alone. Solutions of equation (4.38) are given in terms of the modified Bessel functions I_m and K_m with $m = (d-2)/2$ and λ being the solution of the transcendental equation resulting from the system of boundary conditions in (4.38).

Using the known asymptotic expansions for $I_m(x)$ and $K_m(x)$ as x approaches infinity [4], it is possible to derive the asymptotic expansion of $\lambda_1(\alpha)$ as α goes to $-\infty$. A detailed analysis may be found in [400], leading to

$$\lambda_1(S_{r_1,r_2}, \alpha) = -\alpha^2 + \frac{d-1}{r_2}\alpha + o(\alpha) \quad \text{(spherical shells)}\,,$$
$$\lambda_1(B_r, \alpha) = -\alpha^2 + \frac{d-1}{r}\alpha + o(\alpha) \quad \text{(ball)}\,.$$

Since r_2 may be picked arbitrarily as long as it is larger than or equal to r, if r_1 is then chosen so that the two volumes are equal, we have

$$\lambda_1(S_{r_1,r_2}, \alpha) - \lambda_1(B_r, \alpha) = (d-1)\left(\frac{1}{r_2} - \frac{1}{r}\right)\alpha + o(\alpha).$$

Noting that α is negative we obtain that the difference above will become positive for sufficiently large α. □

In Figure 4.4 we provide a two-dimensional example of two eigencurves for $\lambda_1(S_{r_1,r_2}, \alpha)$ and $\lambda_1(B_r, \alpha)$ in two dimensions, where it is clear that while the first eigenvalue for the unit disk starts above that of the shell with radii 1 and $\sqrt{2}$, the situation is reversed for $\alpha \le \alpha_1 \approx -8.7$.

4.5.2.2 A reversed Faber–Krahn inequality

The above result naturally begs the question as to whether the situation depicted in Figure 4.4 may be extended up to $\alpha = 0$, either using shells or another type of domains, or whether there is a limiting negative value $\alpha_\star$, depending only on the volume of Ω, such that the ball remains a maximizer for α in $[\alpha_\star, 0]$. A partial answer to this question was provided in [400] for the planar case, where it was shown that there is indeed a limiting value of α such that the disk remains a maximizer for negative values of the boundary parameter larger than this.

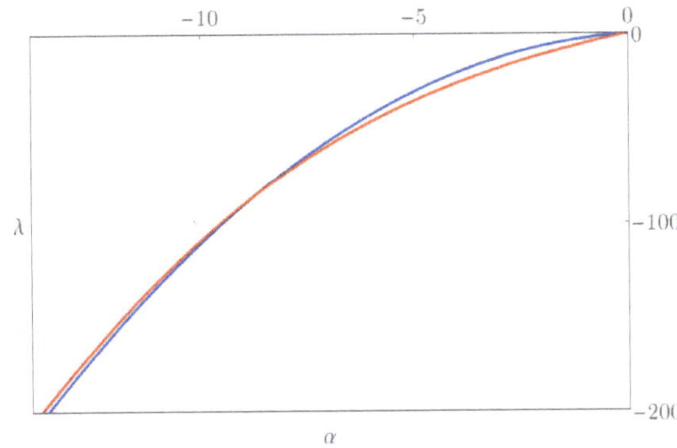

Fig. 4.4. Eigenvalues $\lambda_1(S_{r_1,r_2}, \alpha)$ (red) and $\lambda_1(B_r, \alpha)$ (blue) as a function of the boundary parameter α with $d = 2$, $r = r_1 = 1$ and $r_2 = \sqrt{2}$.

Theorem 4.32 (Freitas–Krejčiřík). *Let $\Omega \subset \mathbb{R}^2$ be a bounded domain of class C^2 and denote by B a ball with $|B| = |\Omega|$. Then there exists a constant $\alpha_\star < 0$, depending only on $|\Omega|$, such that*

$$\lambda_1(\Omega, \alpha) \leq \lambda_1(B, \alpha)$$

for all $\alpha \in [\alpha_\star, 0]$.

Sketch of proof. As was mentioned above, from the derivative given by equation (4.16) it follows that for any given domain Ω satisfying the conditions in the theorem there exists a negative value of $\alpha_1 = \alpha_1(\Omega)$ such that $\lambda_1(\Omega, \alpha) < \lambda_1(B, \alpha)$ for α in $(\alpha_1, 0)$. The main point in proving the above result is thus to ensure that the choice of the values of α_1 may be done uniformly in Ω without it approaching zero.

The setting under which this was done in [400] uses a coordinate system to parametrize planar domains which was used by Payne and Weinberger in [750]. This is based on parallel coordinates which, as in [400], are based on the outer boundary only. This allows us to pass from the Rayleigh quotient (4.2) to the Rayleigh quotient corresponding to a two-dimensional spherical shell (annulus) with the same area as Ω, but where the boundary condition at the inner circle now is modified to Neumann boundary conditions as a result of the parallel coordinates used – the Robin boundary conditions at the outer boundary are preserved, with the same boundary parameter. We may thus bound the original Rayleigh quotient by a variational problem in one dimension. More precisely, and denoting by $\mu_1(S_{r_1,r_2}, \alpha)$ the first eigenvalue of the shell S_{r_1,r_2} with Robin and Neuman boundary conditions at the outer and inner circles, respectively, we obtain

$$\lambda_1(\Omega, \alpha) \leq \inf_{\psi \neq 0} \frac{\displaystyle\int_{r_1}^{r_2} \psi'(r)^2\, r\, dr + \alpha\, r_2\, \psi(r_2)^2\, dr}{\displaystyle\int_{r_1}^{r_2} \psi(r)^2\, r\, dr} =: \mu_1(S_{r_1,r_2}, \alpha),$$

In order for the result in the theorem to be implied by the above inequality, one would need to have

$$\mu_1(S_{r_1,r_2}, \alpha) \leq \lambda_1(B, \alpha). \tag{4.39}$$

However, in view of Theorem 4.31 we already know that this cannot hold in general. In fact, the first two terms in the asymptotics for $\mu_1(S_{r_1,r_2}, \alpha)$ as α goes to $-\infty$ are the same as those for $\lambda_1(S_{r_1,r_2}, \alpha)$ given in the proof of Theorem 4.31. There remains the possibility that (4.39) holds uniformly for small α. It turns out that this is indeed the case, with the proof relying on a detailed study of the intersection between the eigencurves associated with the two eigenvalues $\mu_1(S_{r_1,r_2}, \alpha)$ and $\lambda_1(B, \alpha)$ for negative α. These intersections are given by solutions of the following equations in λ and α

$$\begin{aligned} \mu_1(S_{r_1,r_2}, \alpha):\quad & K_1\left(\sqrt{-\lambda}r_1\right)\left[\sqrt{-\lambda}I_1\left(\sqrt{-\lambda}r_2\right) + \alpha I_0\left(\sqrt{-\lambda}r_2\right)\right] \\ & -I_1\left(\sqrt{-\lambda}r_1\right)\left[\sqrt{-\lambda}K_1\left(\sqrt{-\lambda}r_2\right) - \alpha K_0\left(\sqrt{-\lambda}r_2\right)\right] = 0 \\ \lambda_1(B, \alpha):\quad & \sqrt{-\lambda}I_1\left(\sqrt{-\lambda}r\right) + \alpha I_0\left(\sqrt{-\lambda}\right) = 0 \end{aligned} \tag{4.40}$$

where r is the radius of B, the ball with the same area as S_{r_1,r_2}, and any pair (α, λ) which is a solution of the above system corresponds to an intersection of the two eigencurves.

For each fixed value of r_1 any intersection of the two eigencurves for a negative value of α must take place below a negative value depending on r_1, say $\alpha_1(r_1)$. Thus for r_1 in a compact set not containing zero we have that $\alpha_1(r_1)$ stays bounded away from zero and it remains to consider what happens as r_1 approaches zero and infinity. In the latter case, we have from the derivative at zero and the fact that the eigencurve is a concave function of α, which is a consequence of the Rayleigh quotient (4.2), that any intersections must lie under the tangent at the eigencurve at the origin. Hence they are bounded away from zero.

The case where r_1 approaches zero is more involved and relies on solving with respect to α in the second equation in (4.40) to obtain

$$\alpha = -\sqrt{-\lambda}\frac{I_1(\sqrt{-\lambda}r)}{I_0(\sqrt{-\lambda}r)}.$$

For fixed r the above function vanishes when λ is zero and is strictly decreasing for negative λ. Replacing this in the first equation yields an equation of the form $F(r_1, r, \lambda) = 0$, for which it is then necessary to prove that any negative solution in λ remains bounded away from zero – for the details of this last part, see [400]. This, together with the monotonicity property just mentioned concludes the proof of the theorem. □

Remark 4.33. *The above inequality may be extended to Lipschitz domains in a similar way to what was done by Daners in [319] for Theorem 4.23. The main point is that every bounded Lipschitz set Ω can be approximated by a sequence Ω_n of C^2 domains satisfying a uniform cone condition such that the Lebesgue measures are kept constant and the* *surface area converges, yielding the corresponding convergence of the Robin eigenvalues of the Laplacian. Although the argument used by Daners in [319] was for the positive boundary parameter case, it works in a similar fashion for negative values of the parameter. The key property of this approximation which ensures convergence is the following continuity result: let Ω be a Lipschitz domain and Ω_n a sequence of bounded C^2 domains of $\mathbb{R}^d$ satisfying a uniform cone condition such that*

$$1_{\Omega_n} \to 1_\Omega$$

in $L^1(\mathbb{R}^d)$. Then, for every sequence u_n such that $u_n \in H^1(\Omega_n)$ and $(1_{\Omega_n} u_n, 1_{\Omega_n} \nabla u_n) \to (1_\Omega u, 1_\Omega \nabla u)$ weakly in $L^2(\mathbb{R}^d, \mathbb{R}^{d+1})$, for some $u \in H^1(\Omega)$, we have

$$\int_{\partial\Omega_n} u_n^2 d\sigma \to \int_{\partial\Omega} u^2 d\sigma.$$

The natural conjecture to make at this point is that there exists a value of α where the maximizer stops being the disk and becomes an annulus. From the result in [380] mentioned above, namely, the fact that the disk remains a local maximizer, it is also to be expected that the bifurcation does not take place from the disk but that simply at this value of α the first eigenvalue of an annulus with a positive inner radius r_1 becomes as high as that of the disk with the same area and there are thus two distinct maximizers at this point. As α becomes more negative, the optimal domain now becomes an annulus, with the inner radius depending on the value of α.

Open problem 4.34. *Prove that there exists a maximizer of $\lambda_1(\Omega, \alpha)$ for each fixed negative value of α under a volume restriction, which has spherical symmetry. Furthermore, show that there exists a strictly negative number α^* such that for α in $(\alpha^*, 0)$ this maximizer is a ball, while for α smaller than α^* it becomes a spherical shell with radii depending on α.*

Recent numerical work carried out in [38] supports this conjecture in two and three dimensions. In that paper it was also shown that if we replace the area restriction by a perimeter restriction in two dimensions, then the disk remains a maximizer for all negative values of α. The proof uses similar techniques to those in the proof of Theorem 4.32. The corresponding result in higher dimensions remains open.

Open problem 4.35. *Prove that in dimensions three and higher the ball is a maximizer of* (4.1) *under a surface area restriction for all negative values of α.*

4.6 The higher eigenvalues

For the higher eigenvalues, essentially nothing is known when $\alpha < 0$, so we shall restrict ourselves to considering the case $\alpha > 0$ in the next two subsections and briefly return to the former at the end. However, even when $\alpha > 0$, we now start to see phenomena appearing which are unknown, and even impossible, in the Dirichlet case – at least, beginning with the third eigenvalue.

4.6.1 The second eigenvalue

For the second eigenvalue and $\alpha > 0$, the minimizer is, as in the Dirichlet case, the disjoint union of two equal balls.

Theorem 4.36 (Kennedy). *Suppose $\Omega \subset \mathbb{R}^d$ is a bounded Lipschitz domain. Denote by B_2 any fixed domain consisting of the disjoint union of two equal balls, each of volume $|\Omega|/2$. Then, for any given $\alpha > 0$, we have*

$$\lambda_2(\Omega, \alpha) \geq \lambda_2(B_2, \alpha),$$

with equality if and only if Ω is itself a disjoint union of two equal balls.

The inequality first appeared in [586]; the paper [583] covered the case of equality and generalized the statement to the p-Laplacian. The principle is the same as in the Dirichlet case: for $\alpha > 0$ fixed, we suppose Ω to be connected, take any eigenfunction ψ_2 associated with λ_2, consider the nodal domains

$$\Omega^+ := \{x \in \Omega : \psi_2(x) > 0\} \qquad \text{and} \qquad \Omega^- := \{x \in \Omega : \psi_2(x) < 0\}.$$

and denote by B^+ and B^- balls that have the same volume as Ω^+ and Ω^-, respectively. Then ψ_2 is the equal to the first eigenfunction of a mixed Dirichlet-Robin problem on $\Omega^\pm$ (i.e., with a Dirichlet condition on the interface $\partial\Omega^\pm \cap \Omega$), which by the variational characterization of the latter and the Bossel–Daners inequality must be larger than the first Robin eigenvalues $\lambda_1(B^+, \alpha)$ and $\lambda_1(B^-, \alpha)$. More precisely, we may apply Theorem 4.24 to the functions $\psi_2|_{B^+}$ and $\psi_2|_{B^-}$ (extended by zero and treated as special functions of bounded variation on $\mathbb{R}^d$) to obtain

$$\lambda_2(\Omega, \alpha) \geq \max\{\lambda_1(B^+, \alpha), \lambda_1(B^-, \alpha)\} \geq \lambda_2(B_2, \alpha),$$

with equality if and only if Ω coincides with B_2.

The original proof in [583, 586], which predates Theorem 4.24 (the Bossel–Daners inequality on arbitrary open sets), used a domain approximation argument to deal with the possibility that $\Omega^\pm$ might not be Lipschitz.

4.6.2 Higher eigenvalues (positive boundary parameter)

For the third eigenvalue and above, the situation becomes far more interesting. As we have seen, for $\alpha > 0$ "large", the eigenvalues are close to their Dirichlet counterparts, cf. (4.18); but for $\alpha > 0$ "small", we are close to the Neumann problem. Therefore, domains consisting of n connected components will have a very small n^{th} eigenvalue. In particular, for $\alpha > 0$ small enough, the union of n equal disjoint balls is asymptotically optimal for λ_n. In general there is also no minimizer of λ_n among all domains of fixed volume which is independent of $\alpha > 0$, as the following proposition, which first appeared in [583], shows.

Proposition 4.37 (Kennedy). *Let $\Omega \subset \mathbb{R}^d$ be a bounded, not necessarily connected Lipschitz domain, and denote by B_n any disjoint union of n equal balls, such that $|B_n| = |\Omega|$. Then there exists $\alpha^*(\Omega)$ such that*

$$\lambda_n(\Omega, \alpha) \geq \lambda_n(B_n, \alpha)$$

for all $\alpha \in (0, \alpha^*(\Omega))$, with equality if and only if Ω is itself a disjoint union of n equal balls.

Sketch of proof. If Ω has fewer than n connected components, then its nth Neumann eigenvalue $\mu_n(\Omega) = \lambda_n(\Omega, 0)$ must be strictly positive. Since all eigenvalues on Ω depend continuously on $\alpha \in \mathbb{R}$, it follows that $\lambda_n(\Omega, \alpha) > 0$ for $\alpha > 0$ small enough. The same argument works if fewer than n components contribute to the eigenvalues $\lambda_1(\Omega, 0), \ldots, \lambda_n(\Omega, 0)$.

If Ω has more than n connected components, then by removing superfluous ones and inflating the best n components we can decrease λ_n. If Ω has exactly n components, then we can apply Theorem 4.23 to each of them; hence we may assume without loss of generality that Ω is a union of n disjoint balls. We now take $\alpha_\Omega > 0$ small enough that for all $\alpha \in (0, \alpha_\Omega)$ the first n eigenvalues of Ω are exactly the n first eigenvalues of these balls. Since, for any fixed $\alpha > 0$, the first eigenvalue of a ball decreases strictly monotonically as its radius increases, it follows immediately that $\lambda_n(\Omega, \alpha) \geq \lambda_n(B_n, \alpha)$. ☐

Since for any fixed domain Ω, we have that $\lambda_n(\Omega, \alpha)$ converges to the nth Dirichlet eigenvalue of Ω as $\alpha \to +\infty$, and since in general B_n is not optimal for the nth Dirichlet eigenvalue, it follows in particular that in such cases there cannot be an optimizer of λ_n independent of $\alpha > 0$. Note however that Proposition 4.37 is analogous to (4.37): since $\alpha^*(\Omega)$ depends on Ω, it is an open problem to show that B_n is genuinely a minimizer of λ_n for $\alpha > 0$ small enough.

Open problem 4.38. *Prove that, for any fixed $n \geq 3$, there exists $\alpha_n^* > 0$, depending only on n, the volume $c > 0$ and the dimension $d \geq 2$, such that*

$$\lambda_n(\Omega, \alpha) \geq \lambda_n(B_n, \alpha),$$

for all sufficiently smooth domains $\Omega \subset \mathbb{R}^d$ with $|\Omega| = c$ and all $\alpha \in (0, \alpha_n^*)$. Here B_n is the union of n equal, disjoint balls of volume c/n.

Table 4.1. Conjectured value of α_n^* for dimension 2 and volume $c = 1$.

n	3	4	5	6	7	8	9	10
α_n^*	14.512	16.757	18.735	20.524	22.168	23.699	25.137	26.496

The two-dimensional optimization problem was studied numerically in [46], which found evidence supporting the existence of such an $\alpha_n^* > 0$; in fact, the numerics together with properties of zeros of Bessel functions suggest that in fact $\alpha_n^* \to +\infty$ as $n \to \infty$, for fixed dimension d and volume c; see Figure 4.5 and Table 4.1. In Figure 4.5, the numerically conjectured optimizers for the first few eigenvalues are displayed as a function of α. In each case, for sufficiently large $\alpha > 0$ the optimizer is similar to the Dirichlet minimizer, deforming gradually with α; for example, for λ_4 the ratio of the two balls is adjusted with α. However, just as was conjectured to be the case for $\lambda_1(\Omega, \alpha)$ with $\alpha < 0$, it appears that there are threshold values of α, where a new type or class of domain overtakes ("leapfrogs") the previous minimizer; for small enough $\alpha > 0$ the union of n balls, B_n, wins out. For the higher eigenvalues, this behaviour becomes more involved, as there are more combinations of minimizers of lower eigenvalues which become optimal for at least a small range of α.

A study of the eigenvalues of unions of balls allows one to prove a number of statements about the minimal attainable value of $\lambda_n(\Omega, \alpha)$, which, for fixed dimension d and given volume $|\Omega| = c > 0$, we will denote by $\lambda_n^*(c, \alpha)$. More precisely, by using the fact that $\lambda_1(B, \alpha)$ does *not* scale with the dimensionally normalized volume $|B|^{2/d}$ for fixed $\alpha > 0$, cf. (4.9), we can prove the following result. Here ω_d denotes the volume of the ball of unit radius in d dimensions.

Theorem 4.39 (Antunes–Freitas–Kennedy). *For any given $c > 0$, $n \geq 1$ and $\alpha > 0$, we have*

$$\lambda_n^*(c, \alpha) \leq \lambda_n(B_n, \alpha) \leq d\alpha \left(\frac{\omega_d}{c}\right)^{\frac{1}{d}} \cdot n^{\frac{1}{d}}.$$

Moreover, the dimensionally normalized optimal gap satisfies

$$[\lambda_{n+1}^*(c, \alpha)]^{\frac{d}{2}} - [\lambda_n^*(c, \alpha)]^{\frac{d}{2}} \to 0$$

as $n \to \infty$.

The optimal gap can be analyzed by adapting an argument of Colbois and El Soufi [278]. Of particular interest here is the divergence from the Weyl asymptotics,[4.3]

4.3 Note that the first term of the Weyl asymptotics obviously has to be the same as for the Dirichlet and Neumann Laplacians; for $\alpha > 0$, we can simply note that the Robin eigenvalues lie between their Dirichlet and Neumann counterparts.

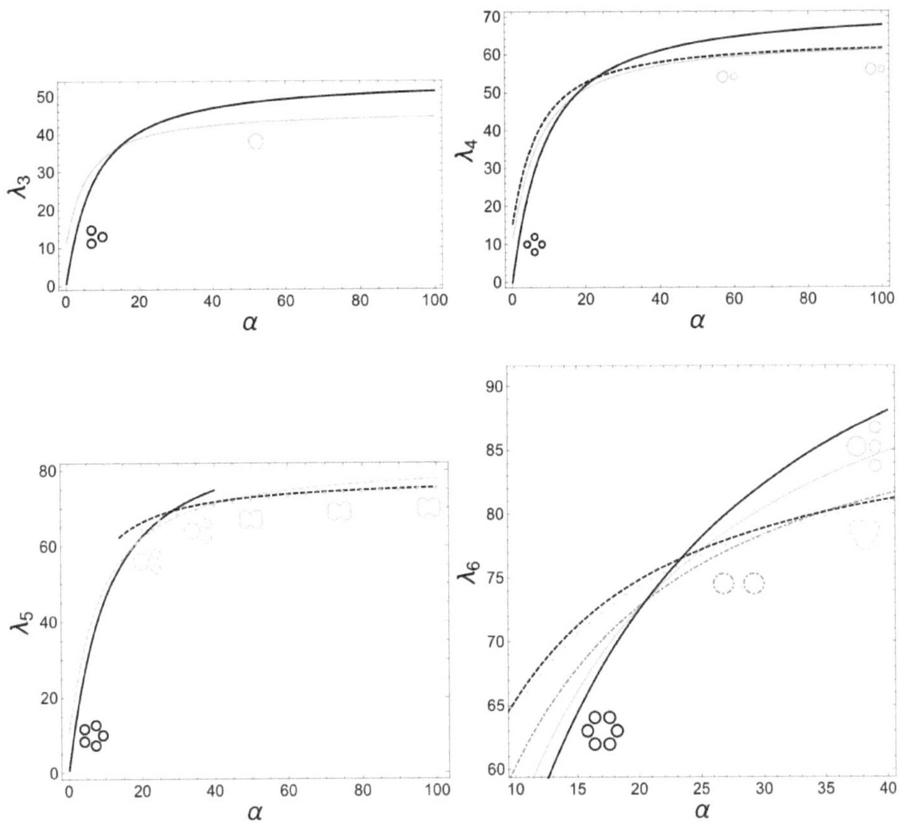

Fig. 4.5. The conjectured Robin optimizers for λ_3 through λ_6 in dimension $d = 2$ with volume $c = 1$, as a function of α. For λ_6 only the region $\alpha \in [10, 40]$ is shown. The solid black curve shows the value of $\lambda_n(B_n, \alpha)$; the black dashed curve gives the value of $\lambda_n(D, \alpha)$ for the domain $D = D(n)$ conjectured to be the Dirichlet minimizer of λ_n. The grey curves show "classes" of domains which are optimal for intermediate values of α; the optimal domain within each class deforms gradually with α.

that on any fixed domain Ω and for fixed $\alpha > 0$ we have $\lambda_n(\Omega, \alpha) \sim C(d, |\Omega|)n^{\frac{2}{d}}$: here $\lambda_n^\star(c, \alpha) = O(n^{\frac{1}{d}})$. This is however purely a consequence of the above-mentioned scaling properties of the Robin problem. Indeed, a similar phenomenon for the Dirichlet eigenvalues is immediately ruled out by classical inequalities such as those of Li and Yau [660]

$$\lambda_n(\Omega) \geq \left(C_d|\Omega|^{-\frac{2}{d}}\right) n^{\frac{2}{d}}$$

for a dimensional constant $C_d > 0$.

Finally, let us note that the actual existence of a minimizer of the nth eigenvalue ($n \geq 3$) under a volume constraint is itself completely open in all dimensions. This is likely to be an extremely difficult problem, as we have essentially none of the tools used in the (already difficult) Dirichlet case.

Open problem 4.40. *Prove that, for any fixed $n \geq 3$ and $\alpha > 0$, a minimizer of $\lambda_n(\Omega, \alpha)$ exists over all bounded domains $\Omega \subset \mathbb{R}^d$ of given volume $|\Omega| = c > 0$.*

4.6.3 Higher eigenvalues (negative boundary parameter)

When $\alpha < 0$, nothing is known regarding the maximizer of $\lambda_n(\Omega, \alpha)$, not even when $n = 2$, a case which is usually accessible. However, the Szegő–Weinberger inequality together with a continuity argument at $\alpha = 0$ suggests that the ball should be maximal for $\alpha < 0$ small enough:

Open problem 4.41. *Prove that there exists $\alpha^* < 0$ such that for any bounded Lipschitz domain $\Omega \subset \mathbb{R}^d$*

$$\lambda_2(\Omega, \alpha) \leq \lambda_2(B, \alpha)$$

for all $\alpha \in (\alpha^*, 0)$, where B is a ball with $|\Omega| = |B|$.

Similarly, by a result of Girouard, Nadirashvili and Polterovich [428], $\mu_3(\Omega)$ is maximized under a volume constraint, at least among simply connected planar domains Ω, by the disjoint union of two equal disks (in the degenerate limit). The usual continuity argument suggests the same should hold for $\lambda_3(\Omega, \alpha)$ when $\alpha < 0$ is small enough.

On the other hand, in the asymptotic limit $\alpha \to -\infty$, the ball cannot be optimal for any n, and indeed for the same reason as for λ_1 outlined in Section 4.5.2: Theorem 4.21 implies that appropriately chosen spherical shells S_{r_1,r_2} eventually have a more negative nth eigenvalue than B, since they have larger $\gamma_{\max}$.

Proposition 4.42. *Let $n \geq 1$. There exists a spherical shell $S_{r_1,r_2} = \{x \in \mathbb{R}^d : r_1 < |x| < r_2\}$ with the same volume as B and $\alpha_*(n) < 0$ such that $\lambda_n(B, \alpha) < \lambda_n(S_{r_1,r_2}, \alpha)$ for all $\alpha < \alpha_*(n)$.*

It thus seems reasonable that, as in the case $n = 1$, for the higher eigenvalues we can expect the maximizer to be an appropriate spherical shell, or at least to converge to one in an appropriate sense, for $\alpha < 0$ large negative. For small $\alpha < 0$, there is however no reason to expect spherical symmetry in general – or any other particular properties, apart from the fact that we expect "convergence" to the (conjectured) Neumann maximizers as $\alpha \to 0$.

Open problem 4.43. *Prove that for each $n \geq 1$ and $\alpha < 0$ there actually exists a maximizer of $\lambda_n(\Omega, \alpha)$ among all domains Ω of given volume. Prove that for any $n \geq 1$ there exist $\alpha_*(n) < 0$ such that the maximizer of $\lambda_n(\Omega, \alpha)$ is a spherical shell for all $\alpha < \alpha_*(n)$. Does there exist $\alpha^*(n) \in (\alpha_*(n), 0)$, $n \geq 4$, such that for all $\alpha \in (\alpha^*(n), 0)$ the maximizer is connected (at least in two dimensions, as appears to be the case for the Neumann maximizers, cf. Figure 11.2)?*

Acknowledgements: We would like to thank Pedro Antunes for providing us with Figure 4.5, Catherine Bandle for sharing her recollections in connection with the history of the Bossel–Daners inequality, and Lorenzo Brasco and Alexandre Girouard for a careful reading of the chapter and for making a number of observations and suggestions which have contributed to its improvement. The work of P.F. was partially supported by FCT (Portugal) through project PTDC/MAT-CAL/4334/2014.

Alexandre Girouard and Iosif Polterovich

5 Spectral geometry of the Steklov problem

Note from the authors

This chapter is a reprint of the article *Spectral geometry of the Steklov problem* by the same authors, published in the Journal of Spectral Theory. We would like to thank the Journal of Spectral Theory and the European Mathematical Society for granting permission to reproduce the paper in this book. Spectral geometry of the Steklov problem is a rapidly developing subject, and there have been a number of important advances since the original version of this article has appeared. In the present text, we have added references to some of these new results in the footnotes. In order to make this chapter coherent with the rest of the book, the dimension is denoted by d, and the trivial Steklov eigenvalue is now denoted by $\sigma_1 = 0$, as opposed to $\sigma_0 = 0$ in the journal version of this article. The numeration has been also changed.

5.1 Introduction

5.1.1 The Steklov problem

Let Ω be a compact Riemannian manifold of dimension $d \geq 2$ with (possibly non-smooth) boundary $M = \partial\Omega$. The *Steklov problem* on Ω is

$$\begin{cases} \Delta u = 0 & \text{in } \Omega, \\ \frac{\partial u}{\partial n} = \sigma u & \text{on } M. \end{cases}$$

where Δ is the Laplace-Beltrami operator acting on functions on Ω, and $\frac{\partial u}{\partial n}$ is the outward normal derivative along the boundary M. This problem was introduced by the Russian mathematician V.A. Steklov at the turn of the 20th century (see [613] for a historical discussion). It is well known that the spectrum of the Steklov problem is discrete as long as the trace operator $H^1(\Omega) \to L^2(\partial\Omega)$ is compact (see [50]). In this case, the eigenvalues form a sequence $0 = \sigma_1 \leq \sigma_2 \leq \sigma_3 \leq \cdots \nearrow \infty$. This is true under some mild regularity assumptions, for instance if Ω has Lipschitz boundary (see [725, Theorem 6.2]).

The present paper focuses on the geometric properties of Steklov eigenvalues and eigenfunctions. A lot of progress in this area has been made in the last few years, and

Alexandre Girouard: Département de mathématiques et de statistique, Pavillon Alexandre-Vachon, Université Laval, Québec, QC, G1V 0A6, Canada, E-mail: alexandre.girouard@mat.ulaval.ca
Iosif Polterovich: Département de mathématiques et de statistique, Université de Montréal, C. P. 6128, Succ. Centre-ville, Montréal, QC, H3C 3J7, Canada, E-mail: iossif@dms.umontreal.ca

some fascinating open problems have emerged. We will start by explaining the motivation for studying the Steklov spectrum. In particular, we will emphasize the differences between this eigenvalue problem and its Dirichlet and Neumann counterparts.

5.1.2 Motivation

The Steklov eigenvalues can be interpreted as the eigenvalues of the *Dirichlet-to-Neumann operator* $\mathcal{D} : H^{1/2}(M) \to H^{-1/2}(M)$ which maps a function $f \in H^{1/2}(M)$ to $\mathcal{D}f = \frac{\partial Hf}{\partial n}$, where Hf is the harmonic extension of f to Ω. The study of the Dirichlet-to-Neumann operator (also known as the voltage-to-current map) is essential for applications to electrical impedance tomography, which is used in medical and geophysical imaging (see [858] for a recent survey). The Steklov spectrum also plays a fundamental role in the mathematical analysis of photonic crystals (see [606] for a survey).

A rather striking feature of the asymptotic distribution of Steklov eigenvalues is its unusually (compared to the Dirichlet and Neumann cases) high sensitivity to the regularity of the boundary. On one hand, if the boundary of a domain is smooth, the corresponding Dirichlet-to-Neumann operator is pseudodifferential and elliptic of order one (see [850, pp. 37-38]). As a result, one can show, for instance, that a surprisingly sharp asymptotic formula for Steklov eigenvalues (5.3) holds for smooth surfaces. However, this estimate already fails for polygons (see section 5.3). It is in fact likely that for domains which are not C^∞-smooth but only of class C^k for some $k \geq 1$ that the rate of decay of the remainder in eigenvalue asymptotics depends on k. To our knowledge, for domains with Lipschitz boundaries, even one-term spectral asymptotics have not yet been proved. A summary of the available results is presented in [11] (see also [12]).

One of the oldest topics in spectral geometry is shape optimization. Here again, the Steklov spectrum holds some surprises. For instance, the classical result of Faber–Krahn for the first Dirichlet eigenvalue $\lambda_1(\Omega)$ states that among Euclidean domains with fixed measure, λ_1 is minimized by a ball. Similarly, the Szegő–Weinberger inequality states that the first nonzero Neumann eigenvalue $\mu_2(\Omega)$ is maximized by a ball. In both cases, no topological assumptions are made. The analogous result for Steklov eigenvalues is Weinstock's inequality, which states that among planar domains with fixed perimeter, σ_2 is maximized by a disk provided that Ω is simply–connected. In contrast with the Dirichlet and Neumann case, this assumption cannot be removed. Indeed the result fails for appropriate annuli (see section 5.4.2). Moreover, maximization of the first nonzero Steklov eigenvalue among all planar domains of given perimeter is an open problem. At the same time, it is known that for simply–connected planar domains, the k-th normalized Steklov eigenvalue is maximized in the limit by a disjoint union of $k-1$ identical disks for any $k \geq 2$ [892]. Once again, for the Dirichlet and Neumann eigenvalues the situation is quite different: the extremal domains for $k \geq 3$ are known only at the level of experimental numerics, and, with a few exceptions, are expected to have rather complicated geometries, see the pictures in Chapter 11.

Probably the most well–known question in spectral geometry is "Can one hear the shape of a drum?", or whether there exist domains or manifolds that are isospectral but not isometric. Apart from some easy examples discussed in section 5.5, no examples of Steklov isospectral non-isometric manifolds are presently known. Their construction appears to be even trickier than for the Dirichlet or Neumann problems. In particular, it is not known whether there exist Steklov isospectral Euclidean domains which are not isometric. Note that the standard transplantation techniques (see [127, 232, 233]) are not applicable for the Steklov problem, as it is not clear how to reflect Steklov eigenfunctions across the boundary.

New challenges also arise in the study of the nodal domains and the nodal sets of Steklov eigenfunctions. One of the problems is to understand whether the nodal lines of Steklov eigenfunctions are dense at the "wave-length scale", which is a basic property of the zeros of Laplace eigenfunctions. Another interesting question is the nodal count for the Dirichlet-to-Neumann eigenfunctions. We touch upon these topics in section 5.6.

Let us conclude this discussion by mentioning that the Steklov problem is often considered in the more general form

$$\frac{\partial u}{\partial n} = \sigma \rho u, \tag{5.1}$$

where $\rho \in L^\infty(\partial\Omega)$ is a non-negative weight function on the boundary. If Ω is two-dimensional, the Steklov eigenvalues can be thought of as the squares of the natural frequencies of a vibrating free membrane with its mass concentrated along its boundary with density ρ (see [619]). A special case of the Steklov problem with the boundary condition (5.1) is the sloshing problem, which describes the oscillations of fluid in a container. In this case, $\rho \equiv 1$ on the free surface of the fluid and $\rho \equiv 0$ on the walls of the container. There is an extensive literature on the properties of sloshing eigenvalues and eigenfunctions, see [105, 388, 602] and references therein.

Since the present survey is directed towards geometric questions, in order to simplify the analysis and presentation we focus on the pure Steklov problem with $\rho \equiv 1$.

5.1.3 Computational examples

The Steklov spectrum can be explicitly computed in a few cases. Below we discuss the Steklov eigenvalues and eigenfunctions of cylinders and balls using separation of variables.

Example 5.1. *The Steklov eigenvalues of a unit disk are*

$$0, 1, 1, 2, 2, \ldots, k, k, \ldots .$$

The corresponding eigenfunctions in polar coordinates (r, ϕ) *are given by*

$$1, r\sin\phi, r\cos\phi, \dots, r^k \sin k\phi, r^k \cos k\phi, \dots.$$

Example 5.2. *The Steklov eigenspaces on the ball* $B(0, R) \subset \mathbb{R}^d$ *are the restrictions of the spaces* H_k^d *of homogeneous harmonic polynomials of degree* $k \in \mathbb{N}$ *on* $\mathbb{R}^d$. *The corresponding eigenvalue is* $\sigma = k/R$ *with multiplicity*

$$\dim H_k^d = \binom{d+k-1}{d-1} - \binom{d+k-3}{d-1}.$$

This is of course a generalization of the previous example.

Example 5.3. *This example is taken from [279]. Let* Σ *be a compact Riemannian manifold without boundary. Let*

$$0 = \lambda_1 < \lambda_2 \le \lambda_3 \cdots \nearrow \infty$$

be the spectrum of the Laplace-Beltrami operator Δ_Σ *on* Σ, *and let* (u_k) *be an orthonormal basis of* $L^2(\Sigma)$ *such that*

$$\Delta_\Sigma u_k = \lambda_k u_k.$$

Given any $L > 0$, *consider the cylinder* $\Omega = [-L, L] \times \Sigma \subset \mathbb{R} \times \Sigma$. *Its Steklov spectrum is given by*

$$0, 1/L, \ \sqrt{\lambda_k}\tanh(\sqrt{\lambda_k}L), \ \sqrt{\lambda_k}\coth(\sqrt{\lambda_k}L).$$

and the corresponding eigenfunctions are

$$1, \ t, \ \cosh(\sqrt{\lambda_k}t)u_k(x), \ \sinh(\sqrt{\lambda_k}t)u_k(x).$$

In sections 5.3.1 and 5.4.2 we will discuss two more computational examples: the Steklov eigenvalues of a square and of annuli.

5.1.4 Plan of the chapter

The chapter is organized as follows. In section 5.2 we survey results on the asymptotics and invariants of the Steklov spectrum on smooth Riemannian manifolds. In section 5.3 we discuss asymptotics of Steklov eigenvalues on polygons, which turns out to be quite different from the case of smooth planar domains. Section 5.4 is concerned with geometric inequalities. In section 5.5 we discuss Steklov isospectrality and spectral rigidity. Finally, section 5.6 deals with the nodal geometry of Steklov eigenfunctions and the multiplicity bounds for Steklov eigenvalues.

5.2 Asymptotics and invariants of the Steklov spectrum

5.2.1 Eigenvalue asymptotics

As above, let $d \geq 2$ be the dimension of the manifold Ω, so that the dimension of the boundary $M = \partial\Omega$ is $d-1$. As was mentioned in the introduction, the Steklov eigenvalues of a compact manifold Ω with boundary $M = \partial\Omega$ are the eigenvalues of the Dirichlet-to-Neumann map. It is a first order elliptic pseudodifferential operator which has the same principal symbol as the square root of the Laplace-Beltrami operator on M. Therefore, applying the results of Hörmander [540, 541][5.1] we obtain the following Weyl's law for Steklov eigenvalues:

$$\#(\sigma_j < \sigma) = \frac{\operatorname{Vol}(\mathbb{B}^{d-1})\operatorname{Vol}(M)}{(2\pi)^{d-1}}\sigma^{d-1} + \mathcal{O}(\sigma^{d-2}),$$

where $\mathbb{B}^{d-1}$ is a unit ball in $\mathbb{R}^{d-1}$. This formula can be rewritten

$$\sigma_j = 2\pi\left(\frac{j}{\operatorname{Vol}(\mathbb{B}^{d-1})\operatorname{Vol}(M)}\right)^{\frac{1}{d-1}} + \mathcal{O}(1). \tag{5.2}$$

In two dimensions, a much more precise asymptotic formula was proved in [429]. Given a finite sequence $C = \{\alpha_1, \cdots, \alpha_k\}$ of positive numbers, consider the following union of multisets (i.e. sets with multiplicities): $\{0, \ldots\ldots, 0\} \cup \alpha_1\mathbb{N} \cup \alpha_1\mathbb{N} \cup \alpha_2\mathbb{N} \cup \alpha_2\mathbb{N} \cup \cdots \cup \alpha_k\mathbb{N} \cup \alpha_k\mathbb{N}$, where the first multiset contains k zeros and $\alpha\mathbb{N} = \{\alpha, 2\alpha, 3\alpha, \ldots, n\alpha, \ldots\}$. We rearrange the elements of this multiset into a monotone increasing sequence[5.2] $S(C)$. For example, $S(\{1\}) = \{0, 1, 1, 2, 2, 3, 3, \cdots\}$ and $S(\{1, \pi\}) = \{0, 0, 1, 1, 2, 2, 3, 3, \pi, \pi, 4, 4, 5, 5, 6, 6, 2\pi, 2\pi, 7, 7, \cdots\}$. The following sharp spectral estimate was proved in [429].

Theorem 5.4. *Let Ω be a smooth compact Riemannian surface with boundary M. Let $M_1, \cdots, M_k$ be the connected components of the boundary $M = \partial\Omega$, with lengths $\ell(M_i)$, $1 \leq i \leq k$. Set $R = \left\{\frac{2\pi}{\ell(M_1)}, \cdots, \frac{2\pi}{\ell(M_k)}\right\}$. Then*

$$\sigma_j = S(R)_j + \mathcal{O}(j^{-\infty}), \tag{5.3}$$

where $\mathcal{O}(j^{-\infty})$ means that the error term decays faster than any negative power of j.

In particular, for simply–connected surfaces we recover the following result proved earlier by Rozenblyum and Guillemin–Melrose (see [354, 794]):

$$\sigma_{2j} = \sigma_{2j+1} + \mathcal{O}(j^{-\infty}) = \frac{2\pi}{\ell(M)}j + \mathcal{O}(j^{-\infty}). \tag{5.4}$$

5.1 The authors thank Y. Kannai for providing them a copy of L. Hörmander's unpublished manuscript [540].

5.2 In this chapter, the sequence starts with $S(C)_1 = 0$, as opposed to $S(C)_0 = 0$ in the original paper.

The idea of the proof of Theorem 5.4 is as follows. For each boundary component M_i, $i = 1, \dots, k$, we cut off a "collar" neighbourhood of the boundary and smoothly glue a cap onto it. In this way, one obtains k simply–connected surfaces, whose boundaries are precisely $M_1, \dots, M_k$, and the Riemannian metric in the neighbourhood of each M_i, $i = 1, \dots k$, coincides with the metric on Ω. Denote by Ω^* the union of these simply–connected surfaces. Using an explicit formula for the full symbol of the Dirichlet-to-Neumann operator [646] we notice that the Dirichlet-to-Neumann operators associated with Ω and Ω^* differ by a smoothing operator, that is, by a pseudodifferential operator with a smooth integral kernel; such operators are bounded as maps between any two Sobolev spaces $H^s(M)$ and $H^t(M)$, $s, t \in \mathbb{R}$. Applying pseudodifferential techniques, we deduce that the corresponding Steklov eigenvalues of $\sigma_j(\Omega)$ and $\sigma_j(\Omega^*)$ differ by $\mathcal{O}(j^{-\infty})$. Note that a similar idea was used in [529]. Now, in order to study the asymptotics of the Steklov spectrum of Ω^*, we map each of its connected components to a disk by a conformal transformation and apply the approach of Rozenblyum-Guillemin-Melrose which is also based on pseudodifferential calculus.

5.2.2 Spectral invariants

The following result is an immediate corollary of Weyl's law (5.2).

Corollary 5.5. *The Steklov spectrum determines the dimension of the manifold and the volume of its boundary.*

More refined information can be extracted from the Steklov spectrum of surfaces.

Theorem 5.6. *The Steklov spectrum determines the number k and the lengths $\ell_1 \geq \ell_2 \geq \dots \geq \ell_k$ of the boundary components of a smooth compact Riemannian surface. Moreover, if $\{\sigma_j\}$ is the monotone increasing sequence of Steklov eigenvalues, then:*

$$\ell_1 = \frac{2\pi}{\limsup_{j\to\infty}(\sigma_{j+1} - \sigma_j)}.$$

This result is proved in [429] by a combination of Theorem 5.4 and certain number-theoretic arguments involving the Dirichlet theorem on simultaneous approximation of irrational numbers.

As was shown in [429], a direct generalization of Theorem 5.6 to higher dimensions is false. Indeed, consider four flat rectangular tori: $T_{1,1} = \mathbb{R}^2/\mathbb{Z}^2$, $T_{2,1} = \mathbb{R}/2\mathbb{Z} \times \mathbb{R}/\mathbb{Z}$, $T_{2,2} = \mathbb{R}^2/(2\mathbb{Z})^2$ and $T_{\sqrt{2},\sqrt{2}} = \mathbb{R}^2/(\sqrt{2}\mathbb{Z})^2$. It was shown in [344, 741] that the disjoint union $\mathcal{T} = T_{1,1} \sqcup T_{1,1} \sqcup T_{2,2}$ is Laplace–Beltrami isospectral to the disjoint union $\mathcal{T}' = T_{2,1} \sqcup T_{2,1} \sqcup T_{\sqrt{2},\sqrt{2}}$. It follows from Example 5.3 that for any $L > 0$, the two disjoint unions of cylinders $\Omega_1 = [0, L] \times \mathcal{T}$ and $\Omega_2 = [0, L] \times \mathcal{T}'$ are Steklov isospectral. At the same time, Ω_1 has four boundary components of area 1 and two boundary components of area 4, while Ω_2 has six boundary components of area 2. Therefore, the collection of areas of boundary components cannot be determined from the Steklov spectrum. Still, the following question can be asked:

Open problem 5.7. *Is the number of boundary components of a manifold of dimension ≥ 3 a Steklov spectral invariant?*

Further spectral invariants can be deduced using the heat trace of the Dirichlet-to-Neumann operator $\mathcal{D}$. By the results of [10, 348, 440], the heat trace admits an asymptotic expansion

$$\sum_{i=1}^{\infty} e^{-t\sigma_i} = \operatorname{Tr} e^{-t\mathcal{D}} = \int_M e^{-t\mathcal{D}}(x,x)\,dx \sim \sum_{k=0}^{\infty} a_k t^{-d+1+k} + \sum_{l=1}^{\infty} b_l t^l \log t. \tag{5.5}$$

The coefficients a_k and b_l are called the *Steklov heat invariants*, and it follows from (5.5) that they are determined by the Steklov spectrum. The invariants $a_0, \ldots, a_{d-1}$, as well as b_l for all l, are local, in the sense that they are integrals over M of corresponding functions $a_k(x)$ and $b_l(x)$ which may be computed directly from the symbol of the Dirichlet-to-Neumann operator $\mathcal{D}$. The coefficients a_k are not local for $k \geq d$ [422, 423] and hence are significantly more difficult to study.

In [764], explicit expressions for the Steklov heat invariants a_0, a_1 and a_2 for manifolds of dimensions three or higher were given in terms of the scalar curvatures of M and Ω, as well as the mean curvature and the second order mean curvature of M (for further results in this direction, see [675]). For example, the formula for a_1 yields the following corollary:

Corollary 5.8. *Let* $\dim \Omega \geq 3$. *Then the integral of the mean curvature over* $\partial\Omega = M$ *(i.e. the total mean curvature of* M*) is an invariant of the Steklov spectrum.*

The Steklov heat invariants will be revisited in section 5.5.

Remark 5.9. *Other spectral invariants have also been studied. For smooth simply connected planar domains it was shown in [353] that the regularized determinant* $\det(\mathcal{D})$ *of the Dirichlet–to–Neumann map is equal to the perimeter of the domain. In fact, on an arbitrary smooth compact Riemannian surface with boundary,* $\det(\mathcal{D})/L(\partial\Omega)$ *is a conformal invariant. This was proved in [441], where an explicit formula for the determinant was given in terms of particular values of Selberg and Ruelle zeta functions and of the Euler characteristic of* Ω.

One should also mention the recent paper [690] where special values of the zeta function are computed for smooth simply connected planar domains, providing a seemingly large number of new spectral invariants which are expressed in terms of the Fourier coefficients of a bihilomorphism from the disk (see also [352]).

Table 5.1. Eigenfunctions obtained by separation of variables on the square $(-1,1) \times (-1,1)$.

Eigenspace basis	Conditions on α	Eigenvalues	Asymptotic behaviour
$\cos(\alpha x)\cosh(\alpha y)$ $\cos(\alpha y)\cosh(\alpha x)$	$\tan(\alpha) = -\tanh(\alpha)$	$\alpha\tanh(\alpha)$	$\frac{3\pi}{4} + \pi j + \mathcal{O}(j^{-\infty})$
$\sin(\alpha x)\cosh(\alpha y)$ $\sin(\alpha y)\cosh(\alpha x)$	$\tan(\alpha) = \coth(\alpha)$	$\alpha\tanh(\alpha)$	$\frac{\pi}{4} + \pi j + \mathcal{O}(j^{-\infty})$
$\cos(\alpha x)\sinh(\alpha y)$ $\cos(\alpha y)\sinh(\alpha x)$	$\tan(\alpha) = -\coth(\alpha)$	$\alpha\coth(\alpha)$	$\frac{3\pi}{4} + \pi j + \mathcal{O}(j^{-\infty})$
$\sin(\alpha x)\sinh(\alpha y)$ $\sin(\alpha y)\sinh(\alpha x)$	$\tan(\alpha) = \tanh(\alpha)$	$\alpha\coth(\alpha)$	$\frac{\pi}{4} + \pi j + \mathcal{O}(j^{-\infty})$
xy		1	

5.3 Spectral asymptotics on polygons

The spectral asymptotics given by formula (5.2) and Theorem 5.4 are obtained using pseudodifferential techniques which are valid only for manifolds with smooth boundaries. In the presence of singularities, the study of the asymptotic distribution of Steklov eigenvalues is more difficult, and the known remainder estimates are significantly weaker (see [11] and references therein). Moreover, Theorem 5.4 fails even for planar domains with corners. This can be seen from the explicit computation of the spectrum for the simplest nonsmooth domain: the square.

5.3.1 Spectral asymptotics on the square

The Steklov spectrum of the square $\Omega = (-1,1) \times (-1,1)$ is described as follows. For each positive root α of the following equations:

$$\tan(\alpha) + \tanh(\alpha) = 0, \quad \tan(\alpha) - \coth(\alpha) = 0,$$
$$\tan(\alpha) + \coth(\alpha) = 0, \quad \tan(\alpha) - \tanh(\alpha) = 0$$

the number $\alpha\tanh(\alpha)$ or $\alpha\coth(\alpha)$ is a Steklov eigenvalue of multiplicity two (see Table 5.1 and Figure 5.1). The function $f(x,y) = xy$ is also an eigenfunction, with a simple eigenvalue $\sigma_4 = 1$. Starting from σ_5, the normalized eigenvalues are clustered in groups of 4 around the odd multiples of 2π:

$$\sigma_{4j+l}L = (2j+1)2\pi + \mathcal{O}(j^{-\infty}), \qquad \text{for } l = 1,2,3,4.$$

This is compatible with Weyl's law since for $k = 4j + l$ it follows that

$$\sigma_k L = \left(\frac{k-l}{2} + 1\right) 2\pi + \mathcal{O}(j^{-\infty}) = \pi k + \mathcal{O}(1).$$

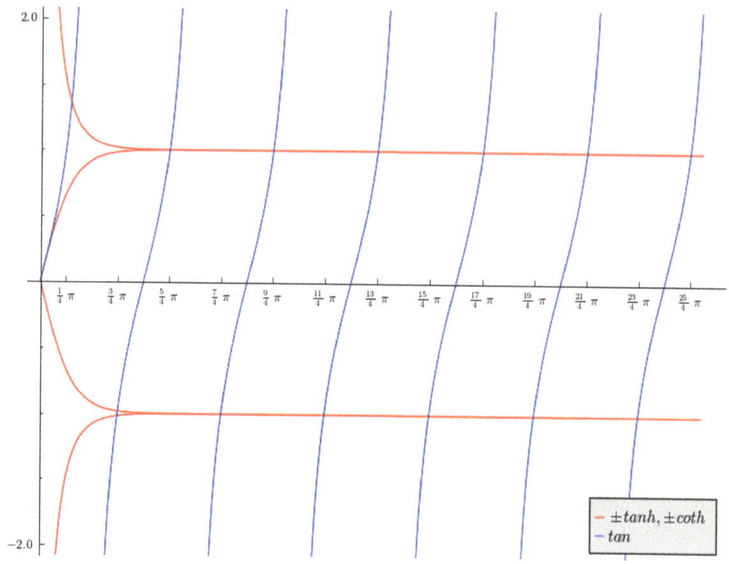

Fig. 5.1. The Steklov eigenvalues of a square. Each intersection corresponds to a double eigenvalue.

Nevertheless, the refined asymptotics (5.4) does not hold.

Let us discuss the spectrum of a square in more detail. Separation of variables quickly leads to the 8 families of Steklov eigenfunctions presented in Table 5.1 plus an “exceptional” eigenfunction $f(x, y) = xy$. One now needs to prove the completeness of this system of orthogonal functions in $L^2(\partial\Omega)$. Using the diagonal symmetries of the square (see Figure 5.2), we obtain symmetrized functions spanning the same eigenspaces. Splitting the eigenfunctions into odd and even parts with respect to the diagonal symmetries, we represent the spectrum as the union of the spectra of four mixed Steklov problems on a right isosceles triangle. In each of these problems the Steklov condition is imposed on the hypotenuse, and on each of the sides the condition is either Dirichlet or Neumann, depending on whether the corresponding eigenfunctions are odd or even when reflected across this side. In order to prove the completeness of this system of Steklov eigenfunctions, it is sufficient to show that the corresponding symmetrized eigenfunctions form a complete set of solutions for each of the four mixed problems.

Let us show this property for the problem corresponding to even symmetries across the diagonal. In this way, one gets a sloshing (mixed Steklov–Neumann) problem on a right isosceles triangle. Solutions of this problem were known since 1840s (see [618]). The restrictions of the solutions to the hypotenuse (i.e. to the side of the

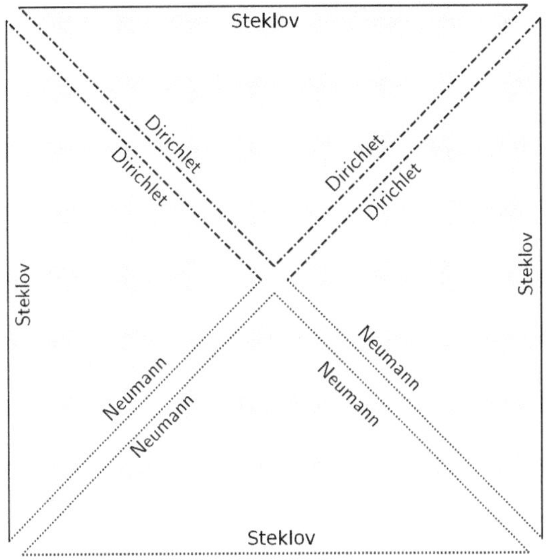

Fig. 5.2. Decomposition of the Steklov problem on a square into four mixed problems on a triangle.

original square) turn out to be the eigenfunctions of the *free beam equation*:

$$\frac{d^4}{dx^4} f = \omega^4 f \qquad \text{on } (-1, 1)$$

$$\frac{d^3}{dx^3} f = \frac{d^2}{dx^2} f = 0 \qquad \text{at } x = -1, 1.$$

This is a fourth order self-adjoint Sturm-Liouvillle equation. It is known that its solutions form a complete set of functions on the interval $(-1, 1)$.

The remaining three mixed problems are dealt with similarly: one reduces the problem to the study of solutions of the vibrating beam equation with either the Dirichlet condition on both ends, or the Dirichlet condition on one end and the Neumann on the other.

Remark 5.10. *The idea to replace the Dirichlet–to–Neumann map on the boundary of a non-smooth domain by a higher order differential problem has been also used in the mathematical analysis of photonic crystals (see [606, section 7.5.3]).*

5.3.2 Numerical experiments

Understanding fine spectral asymptotics for the Steklov problem on arbitrary polygonal domains is a difficult question. We have used software from the FEniCS Project (see http://fenicsproject.org/ and [677]) to investigate the behaviour of the Steklov eigenvalues for some specific examples. This was done using an implementation due to B.

Siudeja [814] which was already applied in [613]. For the sake of completeness, we discuss two of these experiments here.

Example 5.11. *(Equilateral triangle) We have computed the first 60 normalized eigenvalues $\sigma_j L$ of an equilateral triangle. The results lead to a conjecture that*

$$\sigma_{2j}L = \sigma_{2j-1}L + o(1) = \pi(2j-1) + o(1).$$

Example 5.12. *(Right isosceles triangle) For the right isosceles triangle with sides of lengths $1, 1, \sqrt{2}$, we have also computed the first 60 normalized eigenvalues. The numerics indicate that the spectrum is composed of two sequences of eigenvalues, one of which is behaving as a sequence of double eigenvalues*

$$\pi j + o(1)$$

and the other one as a sequence of simple eigenvalues

$$\frac{\pi}{\sqrt{2}}(j + 1/2) + o(1).$$

In the context of the sloshing problem, some related conjectures have been proposed in [388].

5.4 Geometric inequalities for Steklov eigenvalues

5.4.1 Preliminaries

Let us start with the following simple observation. If a Euclidean domain $\Omega \subset \mathbb{R}^d$ is scaled by a factor $c > 0$, then

$$\sigma_k(c\,\Omega) = c^{-1}\sigma_k(\Omega). \tag{5.6}$$

Because of this scaling property, maximizing $\sigma_k(\Omega)$ among domains with fixed perimeter is equivalent to maximizing the normalized eigenvalues $\sigma_k(\Omega)|\partial\Omega|^{1/(d-1)}$ on arbitrary domains. Here and further on we use the notation $|\cdot|$ to denote the volume of a manifold.

All the results concerning geometric bounds are proved using a variational characterization of the eigenvalues. Let $\mathcal{E}(k)$ be the set of all k dimensional subspaces of the Sobolev space $H^1(\Omega)$ which are orthogonal to constants on the boundary $\partial\Omega$, then for each $k \geq 1$,

$$\sigma_{k+1}(\Omega) = \min_{E\in\mathcal{E}(k)} \sup_{0\neq u\in E} R(u), \tag{5.7}$$

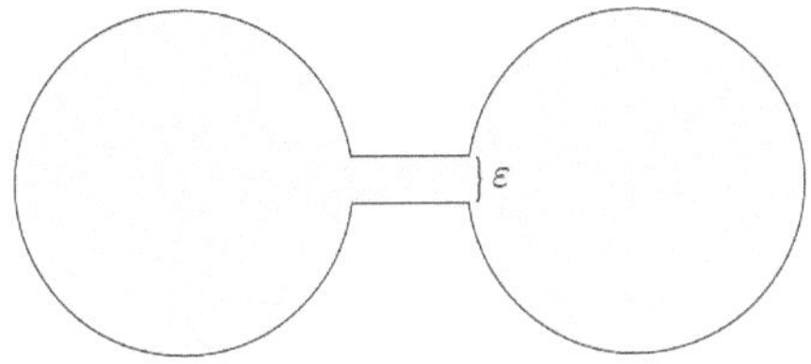

Fig. 5.3. A domain with a thin passage.

where the *Rayleigh quotient* is

$$R(u) = \frac{\int_\Omega |\nabla u|^2 \, dA}{\int_{\partial\Omega} u^2 \, dS}.$$

In particular, the first nonzero eigenvalue is given by

$$\sigma_2(\Omega) = \min\left\{ R(u) \; : \; u \in H^1(\Omega), \; \int_{\partial\Omega} u \, dS = 0 \right\}.$$

These variational characterizations are similar to those of Neumann eigenvalues on Ω, where the integral in the denominator of $R(u)$ would be on the domain Ω rather than on its boundary.

One last observation is in order before we discuss isoperimetric bounds. Let $\Omega_\epsilon := (-1, 1) \times (-\epsilon, \epsilon)$ be a thin rectangle $(0 < \epsilon << 1)$. It is easy to see using using (5.7) that

$$\lim_{\epsilon\to 0} \sigma_k(\Omega_\epsilon) = 0, \qquad \text{for each } k \in \mathbb{N}. \tag{5.8}$$

In fact, it suffices for a family Ω_ϵ of domains to have a *thin collapsing passage* (see Figure 5.3) to guarantee that $\sigma_k(\Omega_\epsilon)$ becomes arbitrarily small as $\epsilon \searrow 0$ (see [892, section 2.2].) This follows from the variational characterization: the idea is to construct a sequence of k pairwise orthogonal test functions that oscillate inside the thin passage and vanish outside. Then the Dirichlet energy of such functions will be very small, while the denominator in the Rayleigh quotient remains bounded away from zero, due to the integration over the side of the passage. Hence, the Rayleigh quotient will tend to zero, yielding (5.8). When considering an isoperimetric constraint, it is therefore more interesting to maximize eigenvalues.

5.4.2 Isoperimetric upper bounds for Steklov eigenvalues on surfaces

On a compact surface with boundary, the following theorem gives a general upper bound in terms of the genus and the number of boundary components.

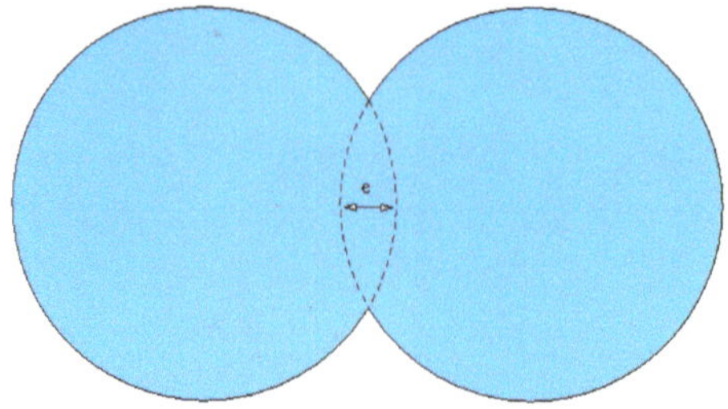

Fig. 5.4. A family of domains Ω_ϵ maximizing $\sigma_3 L$ in the limit as $\epsilon \to 0$.

Theorem 5.13 ([431]). [5.3] *Let Ω be a smooth orientable compact Riemannian surface with boundary $M = \partial\Omega$ of length L. Let γ be the genus of Ω and let l be the number of its boundary components. Then the following holds:*

$$\sigma_p \sigma_q L^2 \leq \begin{cases} \pi^2(\gamma + l)^2(p + q - 2)^2 & \text{if } p + q \text{ is even,} \\ \pi^2(\gamma + l)^2(p + q - 3)^2 & \text{if } p + q \text{ is odd,} \end{cases} \tag{5.9}$$

for any pair of integers $p, q \geq 2$. In particular by setting $p = q = k$ one obtains the following bound:

$$\sigma_k(\Omega)L(M) \leq 2\pi(\gamma + l)(k - 1). \tag{5.10}$$

The proof of Theorem 5.13 is based on the existence of a proper holomorphic covering map $\phi : \Omega \to \mathbb{D}$ of degree $\gamma + l$ (the Ahlfors map), which was proved in [414], and on an ingenious complex analytic argument due to J. Hersch, L. Payne and M. Schiffer [513], who used it to prove inequality (5.9) for planar domains. In this particular case, inequality (5.10) is known to be sharp. Indeed, it was proved in [892] that equality is attained in the limit by a family Ω_ϵ of domains degenerating to a disjoint union of $k - 1$ identical disks (see Figure 5.4). For $k = 2$, inequality (5.10) was proved in [392].

The earliest isoperimetric inequality for Steklov eigenvalues is that of R. Weinstock [873]. For simply–connected planar domains ($\gamma = 0$, $l = 1$), he proved that

$$\sigma_2(\Omega)L(\partial\Omega) \leq 2\pi \tag{5.11}$$

with equality if and only if Ω is a disk. Weinstock used an argument similar to that of G. Szegő [838], who obtained an isoperimetric inequality for the first nonzero Neumann eigenvalue of a simply–connected domain Ω normalized by the measure $|\Omega|$ rather than its perimeter. In fact, Weinstock's proof is the simplest application of the center of mass renormalization (also known as Hersch's lemma, see [428, 430, 519, 802]).

5.3 This result has been recently improved by M. Karpukhin [573]. Karpukhin's bound involves an explicit linear comibination of the genus, the number of boundary components and the index of the eigenvalue. It should also be compared to Theorem 5.18 below, where the constants are implicit, but the estimate is independent on the number of boundary components.

While Szegő's inequality can be generalized to an arbitrary Euclidean domain (see [871]), this is not true for Weinstock's inequality. In particular, as follows from the example below, Weinstock's inequality fails for non-simply–connected planar domains.

Example 5.14. *The Steklov eigenvalues and eigenfunctions of an annulus have been computed in [338]. On the annulus* $A_\epsilon = \mathbb{D}\backslash B(0,\epsilon)$, *there is a radially symmetric Steklov eigenfunction*

$$f(r) = -\left(\frac{1+\epsilon}{\epsilon\log(\epsilon)}\right)\log(r) + 1,$$

with the corresponding eigenvalue $\sigma = \frac{1+\epsilon}{\epsilon\log(1/\epsilon)}$. *All other eigenfunctions are of the form*

$$f_k(r,\theta) = (A_k r^k + A_{-k} r^{-k})T(k\theta) \qquad (\textit{with } k \in \mathbb{N})$$

where $T(k\theta) = \cos(k\theta)$ *or* $T(k\theta) = \sin(k\theta)$. *In order for* $f_k(r,\theta)$ *to be a Steklov eigenfunction it is required that*

$$\frac{\partial}{\partial_r} f_k(1,\theta) = \sigma f_k(1,\theta) \qquad \textit{and} \qquad -\frac{\partial}{\partial_r} f_k(\epsilon,\theta) = \sigma f_k(\epsilon,\theta),$$

which leads to the following system:

$$\begin{pmatrix} \sigma\epsilon^k + k\epsilon^{k-1} & \sigma\epsilon^{-k} - k\epsilon^{-k-1} \\ \sigma - k & \sigma + k \end{pmatrix} \begin{pmatrix} A_k \\ A_{-k} \end{pmatrix} = \begin{pmatrix} 0 \\ 0 \end{pmatrix}.$$

This system has a nonzero solution if and only if its determinant vanishes. After some simplifications, the Steklov eigenvalues of the annulus $A_\epsilon = \mathbb{D} \setminus B(0,\epsilon)$ *are seen to be the roots of the quadratic polynomials*

$$p_k(\sigma) = \sigma^2 - \sigma k\left(\frac{\epsilon+1}{\epsilon}\right)\left(\frac{1+\epsilon^{2k}}{1-\epsilon^{2k}}\right) + \frac{1}{\epsilon}k^2 \qquad (k \in \mathbb{N}).$$

Each of these roots contributes double eigenvalues, corresponding to the choice of a cos *or* sin *function for the angular part* $T(k\theta)$ *of the corresponding eigenfunction. For* $\epsilon > 0$ *small enough, this leads in particular to*

$$\sigma_2(A_\epsilon) = \frac{1}{2\epsilon}\frac{1+\epsilon^2}{1-\epsilon}\left(1 - \sqrt{1 - 4\epsilon\left(\frac{1-\epsilon}{1+\epsilon^2}\right)^2}\right). \tag{5.12}$$

It follows from formula (5.12) that for the annulus $A_\epsilon = B(0,1) \setminus B(0,\epsilon)$ one has

$$\sigma_2(A_\epsilon)L(\partial A_\epsilon) = 2\pi\sigma_2(\mathbb{D}) + 2\pi\epsilon + o(\epsilon) \quad \text{as } \epsilon \searrow 0. \tag{5.13}$$

Therefore, $\sigma_2(A_\epsilon)L(\partial A_\epsilon) > 2\pi\sigma_2(\mathbb{D})$ for $\epsilon > 0$ small enough (see Figure 5.5), and hence Weinstock's inequality (5.11) fails.

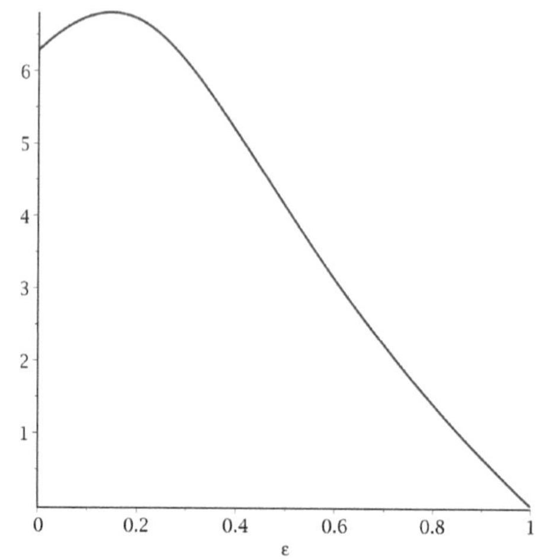

Fig. 5.5. The normalized eigenvalue $\sigma_2(A_\epsilon)L(\partial A_\epsilon)$

Remark 5.15. *One can also compute the Steklov eigenvalues of the spherical shell $\Omega_\epsilon := B(0,1) \setminus B(0,\epsilon) \subset \mathbb{R}^d$ for $d \geq 3$. The eigenvalues are the roots of certain quadratic polynomials which can be computed explicitly. Here again, it is true that for $\epsilon > 0$ small enough, $\sigma_2(\Omega_\epsilon)|\partial\Omega_\epsilon|^{\frac{1}{d-1}} > \sigma_2(\mathbb{B})|\partial\mathbb{B}|^{\frac{1}{d-1}}$. This computation was part of an unpublished undergraduate research project of E. Martel at Université Laval.*

Given that Weinstock's inequality is no longer true for non-simply–connected planar domains, one may ask whether the supremum of $\sigma_2 L$ among all planar domains of fixed perimeter is finite. This is indeed the case, as follows from the following theorem for $k = 2$ and $\gamma = 0$.

Theorem 5.16 ([279]). *There exists a universal constant $C > 0$ such that*

$$\sigma_k(\Omega)L(\partial\Omega) \leq C(\gamma + 1)k. \tag{5.14}$$

Theorem 5.16 leads to the following question:

Open problem 5.17. *What is the maximal value of $\sigma_2(\Omega)$ among Euclidean domains $\Omega \subset \mathbb{R}^d$ of fixed perimeter? On which domain (or in the limit of which sequence of domains) is it realized?*

Some related results will be discussed in subsection 5.4.3. In particular, in view of Theorem 5.26 [392], it is tempting to suggest that the maximum is realized in the limit by a sequence of domains with the number of boundary components tending to infinity.

The proof of Theorem 5.16 is based on N. Korevaar's metric geometry approach [598] as described in [438]. For $k = 2$, inequality (5.14) holds with $C = 4\pi$

(see [594]). For $k = 2$ and $\gamma = 0$, it holds with $C = 2\pi$ [392] (see Theorem 5.26 below). It is also possible to "decouple" the genus γ and the index k. The following theorem was proved by A. Hassannezhad [467], using a generalization of the Korevaar method in combination with concentration results from [281].

Theorem 5.18. *There exists two constants $A, B > 0$ such that*

$$\sigma_k(\Omega)L(\partial\Omega) \le A\gamma + Bk.$$

At this point, we have considered maximization of the Steklov eigenvalues under the constraint of fixed perimeter. This is natural, since they are the eigenvalues of to the Dirichlet-to-Neumann operator, which acts on the boundary. Nevertheless, it is also possible to normalize the eigenvalues by fixing the measure of Ω. The following theorem was proved by F. Brock [192].

Theorem 5.19. *Let $\Omega \subset \mathbb{R}^d$ be a bounded Lipschitz domain. Then*

$$\sigma_2(\Omega)|\Omega|^{1/d} \le \omega_d^{1/d}, \tag{5.15}$$

with equality if and only if Ω is a ball. Here ω_d is the volume of the unit ball $\mathbb{B}^d \subset \mathbb{R}^d$.

Observe that no connectedness assumption is required this time.[5.4] The proof of Theorem 5.19 is based on a weighted isoperimetric inequality for moments of inertia of the boundary $\partial\Omega$. A quantitative improvement of Brock's theorem was obtained in [180] in terms of the *Fraenkel asymmetry* of a bounded domain $\Omega \subset \mathbb{R}^d$,:

$$\mathcal{A}(\Omega) := \inf\left\{\frac{\|1_\Omega - 1_B\|_{L^1}}{|\Omega|} \;:\; B \text{ is a ball with } |B| = |\Omega|\right\}.$$

Theorem 5.20. *Let $\Omega \subset \mathbb{R}^d$ be a bounded Lipschitz domain. Then*

$$\sigma_2(\Omega)|\Omega|^{1/d} \le \omega_d^{1/d}(1 - \alpha_d\mathcal{A}(\Omega)^2), \tag{5.16}$$

where $\alpha_d > 0$ depends only on the dimension.

The proof of Theorem 5.20 is based on a quantitative refinement of the isoperimetric inequality, see also [178] for related results on stability of the Dirichlet and Neumann eigenvalues and Chapter 7 for a complete overview on that topic. It would be interesting to prove a similar stability result for Weinstock's inequality:

Open problem 5.21. *Let Ω be a planar simply–connected domain such that the difference $2\pi - \sigma_2(\Omega)L(\partial\Omega)$ is small. Show that Ω must be close to a disk (in the sense of Fraenkel asymmetry or some other measure of proximity).*

5.4 Recent numerical results [16, 153] show that in two dimensions, the maximizer of σ_k under the area constraint is connected for each $k \ge 2$. Moreover, the maximizing domains appear to be smooth and have exactly k axes of symmetry.

5.4.3 Existence of maximizers and free boundary minimal surfaces

A *free boundary submanifold* is a proper minimal submanifold of some unit ball $\mathbb{B}^d$ with its boundary meeting the sphere $\mathbb{S}^{d-1}$ orthogonally. These are characterized by their Steklov eigenfunctions.

Lemma 5.22 ([393]). *A properly immersed submanifold Ω of the ball $\mathbb{B}^d$ is a free boundary submanifold if and only if the restriction to Ω of the coordinate functions $x_1, \cdots, x_d$ satisfy*

$$\begin{cases} \Delta x_i = 0 & \text{in } \Omega, \\ \frac{\partial x_i}{\partial n} = x_i & \text{on } \partial\Omega. \end{cases}$$

This link was exploited by A. Fraser and R. Schoen who developed the theory of extremal metrics for Steklov eigenvalues. See [392, 393] and especially [394] where an overview is presented.

Let $\sigma^*(\gamma, l)$ be the supremum of $\sigma_2 L$ taken over all Riemannian metrics on a compact surface of genus γ with l boundary components. In [392], a geometric characterization of maximizers was proved.

Proposition 5.23. *Let Ω be a compact surface of genus γ with l boundary components and let g_0 be a smooth metric on Ω such that*

$$\sigma_2(\Omega, g_0)L(\partial\Omega, g_0) = \sigma^*(\gamma, l).$$

Then there exist eigenfunctions $u_1, \cdots, u_d$ corresponding to $\sigma_2(\Omega)$ such that the map

$$u = (u_1, \cdots, u_d) : \Omega \to \mathbb{B}^d$$

is a conformal minimal immersion such that $u(\Omega) \subset \mathbb{B}^d$ is a free boundary submanifold, and u is an isometry on $\partial\Omega$ up to a rescaling by a constant factor.

This result was extended to higher eigenvalues σ_k in [394]. This characterization is similar to that of extremizers of the eigenvalues of the Laplace operator on surfaces (see [361, 362, 720]).

For surfaces of genus zero, Fraser and Schoen could also obtain an existence and regularity result for maximizers, which is the main result of their paper [392].

Theorem 5.24. *For each $l > 0$, there exists a smooth metric g on the surface of genus zero with l boundary components such that*

$$\sigma_2(\Omega, g)L_g(\partial\Omega) = \sigma^*(0, l).$$

Similar existence results have been proved for the first nonzero eigenvalue of the Laplace–Beltrami operator in a fixed conformal class of a closed surface of arbitrary genus, in which case conical singularities have to be allowed (see [556, 756]).

Proposition 5.23 and Theorem 5.24 can be used to study optimal upper bounds for σ_2 on surfaces of genus zero. Observe that inequality (5.10) can be restated as

$$\sigma^*(\gamma, l) \le 2\pi(\gamma + l).$$

This bound is not sharp in general. For instance, Fraser and Schoen [392] proved that on annuli ($\gamma = 0, l = 2$), the maximal value of $\sigma_2(\Omega)L(\partial\Omega)$ is attained by the *critical catenoid* ($\sigma_2 L \sim 4\pi/1.2$), which is the minimal surface $\Omega \subset B^3$ parametrized by

$$\phi(t, \theta) = c(\cosh(t)\cos(\theta), \cosh(t)\sin(\theta), t),$$

where the scaling factor $c > 0$ is chosen so that the boundary of the surface Ω meets the sphere $\mathbb{S}^2$ orthogonally.

Theorem 5.25 ([392]). *The supremum of $\sigma_2(\Omega)L(\partial\Omega)$ among surfaces of genus 0 with two boundary components is attained by the* critical catenoid. *The maximizer is unique up to conformal changes of the metric which are constant on the boundary.*

The uniqueness can be proved using Proposition 5.23 by showing that the critical catenoid is the unique free boundary annulus in a Euclidean ball. The maximization of $\sigma_2 L$ for the Möbius bands has also been considered in [392].

For surfaces of genus zero with arbitrary number of boundary components, the maximizers are not known explicitly, but the asymptotic behaviour for large number of boundary components is understood [392].

Theorem 5.26. *The sequence $\sigma^*(0, l)$ is strictly increasing and converges to 4π. For each $l \in \mathbb{N}$ a maximizing metric is achieved by a free boundary minimal surface Ω_l of area less than 2π. The limit of these minimal surfaces as $l \nearrow +\infty$ is a double disk.*

The results discussed above lead to the following question:

Open problem 5.27. [5.5] *Let Ω be a surface of genus γ with l boundary components. Does there exist a smooth Riemannian metric g_0 such that*

$$\sigma_2(\Omega, g_0)L(\partial\Omega, g_0) \ge \sigma_2(\Omega, g)L(\partial\Omega, g)$$

for each Riemannian metric g?

Free boundary minimal surfaces were used as a tool in the study of maximizers for σ_2, but this interplay can be reversed and used to obtain interesting geometric results.

5.5 Under certain assumptions, the existence of a regular maximizer of σ_k, $k \ge 2$, on an arbitrary Riemannian surface has recently been established in [755]. However, this result does not provide an answer to the open problem.

Corollary 5.28. *For each $l \geq 1$, there exists an embedded minimal surface of genus zero in $\mathbb{B}^3$ with l boundary components satisfying the free boundary condition.*

5.4.4 Geometric bounds in higher dimensions

In dimensions $d = \dim(\Omega) \geq 3$, isoperimetric inequalities for Steklov eigenvalues are more complicated, as they involve other geometric quantities, such as the isoperimetric ratio:

$$I(\Omega) = \frac{|M|}{|\Omega|^{\frac{d-1}{d}}}.$$

For the first nonzero eigenvalue σ_2, it is possible to obtain upper bounds for general compact manifolds with boundary in terms of $I(\Omega)$ and of the relative conformal volume, which is defined below. Let Ω be a compact manifold of dimension d with smooth boundary M. Let $m \in \mathbb{N}$ be a positive integer. The *relative m-conformal volume* of Ω is

$$V_{rc}(\Omega, m) = \inf_{\phi:\Omega \hookrightarrow B^m} \sup_{\gamma \in M(m)} \mathrm{Vol}(\gamma \circ \phi(\Omega)),$$

where the infimum is over all conformal immersions $\phi : \Omega \hookrightarrow \mathbb{B}^m$ such that $\phi(M) \subset \partial\mathbb{B}^m$, and $M(m)$ is the group of conformal diffeomorphisms of the ball. This conformal invariant was introduced in [393]. It is similar to the celebrated conformal volume of P. Li and S.-T. Yau [659].

Theorem 5.29. *[393] Let Ω be a compact Riemannian manifold of dimension d with smooth boundary M. For each positive integer m, the following holds:*

$$\sigma_2(\Omega)|M|^{\frac{1}{d-1}} \leq \frac{d V_{rc}(\Omega, m)^{2/d}}{I(\Omega)^{\frac{d-2}{d-1}}}. \tag{5.17}$$

In case of equality, there exists a conformal harmonic map $\phi : \Omega \to \mathbb{B}^m$ which is a homothety on $M = \partial\Omega$ and such that $\phi(\Omega)$ meets ∂B^m orthogonally. If $d \geq 3$, then ϕ is an isometric minimal immersion of Ω and it is given by a subspace of the first eigenspace.

The proof uses coordinate functions as test functions and is based on the Hersch center of mass renormalization procedure. It is similar to the proof of the Li-Yau inequality [659].

For higher eigenvalues, the following upper bound for bounded domains was proved by B. Colbois, A. El Soufi and the first author in [279].

Theorem 5.30. *Let N be a Riemannian manifold of dimension d. If N is conformally equivalent to a complete Riemannian manifold with non-negative Ricci curvature, then for each domain $\Omega \subset N$, the following holds for each $k \geq 1$,*

$$\sigma_k(\Omega)|M|^{\frac{1}{d-1}} \leq \frac{\alpha(d)}{I(\Omega)^{\frac{d-2}{d-1}}} k^{2/d}. \tag{5.18}$$

where $\alpha(d)$ is a constant depending only on d.

The proof of Theorem 5.30 is based on the methods of metric geometry initiated in [598] and further developed in [438]. In combination with the classical isoperimetric inequality, Theorem 5.30 leads to the following corollary.

Corollary 5.31. *There exists a constant C_d such that for any Euclidean domain $\Omega \subset \mathbb{R}^d$*

$$\sigma_k(\Omega)|\partial\Omega|^{\frac{1}{d-1}} \leq C_d k^{2/d}.$$

Similar results also hold for domains in the hyperbolic space $\mathbb{H}^d$ and in the hemisphere of $\mathbb{S}^d$. An interesting question raised in [279] is whether or not one can replace the exponent $2/d$ in Corollary 5.31 by $1/(d-1)$, which should be optimal in view of Weyl's law (5.2):

Open problem 5.32. *Does there exist a constant C_d such that any bounded Euclidean domain $\Omega \subset \mathbb{R}^d$ satisfies*

$$\sigma_k(\Omega)|\partial\Omega|^{\frac{1}{d-1}} \leq C_d k^{\frac{1}{d-1}}?$$

While it might be tempting to think that inequality (5.18) should also hold with the exponent $1/(d-1)$, this is necessarily false since it would imply a universal upper bound on the isoperimetric ratio $I(\Omega)$ for Euclidean domains.

5.4.5 Lower bounds

In [365], J. Escobar proved the following lower bound.

Theorem 5.33. *Let Ω be a smooth compact Riemannian manifold of dimension ≥ 3 with boundary $M = \partial\Omega$. Suppose that the Ricci curvature of Ω is non-negative and that the second fundamental form of M is bounded below by $k_0 > 0$, then $\sigma_2 > k_0/2$.*

The proof is a simple application of Reilly's formula. In [366], Escobar conjectured the stronger bound $\sigma_2 \geq k_0$, which he proved for surfaces. For convex planar domains, this had already been proved by Payne [745]. Earlier lower bounds for convex and star-shaped planar domains are due to Kuttler and Sigillito [608, 609].

In more general situations (e.g. no convexity assumption), it is still possible to bound the first eigenvalue from below, in a way similar to the classical Cheeger inequality. The classical Cheeger constant associated to a compact Riemannian manifold Ω with boundary $M = \partial\Omega$ is defined by

$$h_c(\Omega) := \inf_{|A| \leq \frac{|\Omega|}{2}} \frac{|\partial A \cap \operatorname{int} \Omega|}{|A|}.$$

where the infimum is over all Borel subsets of Ω such that $|A| \le |\Omega|/2$. In [559] P. Jammes introduced the following Cheeger type constant for the Steklov problem:

$$h_j(\Omega) := \inf_{|A| \le \frac{|\Omega|}{2}} \frac{|\partial A \cap \operatorname{int} \Omega|}{|A \cap \partial\Omega|}.$$

He proved the following lower bound.

Theorem 5.34. *Let Ω be a smooth compact Riemannian manifold with boundary $M = \partial\Omega$. Then*

$$\sigma_2(\Omega) \ge \frac{1}{4} h_c(\Omega) h_j(\Omega) \tag{5.19}$$

The proof of this theorem uses the coarea formula and follows the proof of the classical Cheeger inequality quite closely. Previous lower bounds were also obtained in [365] in terms of a related Cheeger type constant and of the first eigenvalue of a Robin problem on Ω.

5.4.6 Surfaces with large Steklov eigenvalues

The previous discussion immediately raises the question of whether or not there exist surfaces with an arbitrarily large normalized first Steklov eigenvalue. The question was settled by the first author and B. Colbois in [280].

Theorem 5.35. *There exists a sequence $\{\Omega_N\}_{N \in \mathbb{N}}$ of compact surfaces with boundary and a constant $C > 0$ such that for each $N \in \mathbb{N}$, genus$(\Omega_N) = 1 + N$, and*

$$\sigma_2(\Omega_N) L(\partial\Omega_N) \ge CN.$$

The proof is based on the construction of surfaces which are modelled on a family of expander graphs.

Remark 5.36. *The literature on geometric bounds for Steklov eigenvalues is expanding rather fast. There is some interest in considering the maximization of various functions of the Steklov eigenvalues. See [338, 355, 427, 503]. In the framework of comparison geometry, σ_2 was studied is [367] and more recently in [149]. For submanifolds of $\mathbb{R}^d$, upper bounds involving the mean curvatures of $M = \partial\Omega$ have been obtained in [553]. Higher eigenvalues on annuli have been studied in [378]. Isoperimetric bounds for the first nonzero eigenvalue of the Dirichlet-to-Neumann operator on forms have been recently obtained in [785, 786].*

5.5 Isospectrality and spectral rigidity

5.5.1 Isospectrality and the Steklov problem

Adapting the celebrated question of M. Kac "Can one hear the shape of a drum?" to the Steklov problem, one may ask:

Open problem 5.37. *Do there exist planar domains which are not isometric and have the same Steklov spectrum?*

We believe the answer to this question is negative. Moreover, the problem can be viewed as a special case of a conjecture put forward in [564]: two surfaces have the same Steklov spectrum if and only if there exists a conformal mapping between them such that the conformal factor on the boundary is identically equal to one. Note that the "if" part immediately follows from the variational principle (5.7). Indeed, the numerator of the Rayleigh quotient for Steklov eigenvalues is the usual Dirichlet energy, which is invariant under conformal transformations in two dimensions. The denominator also stays the same if the conformal factor is equal to one on the boundary. Therefore, the Steklov spectra of such conformally equivalent surfaces coincide. For simply connected domains, a closely related question is to find out whether a smooth positive function $a \in C^\infty(S^1)$ is determined by the spectrum of $a\mathcal{D}_{\mathbb{D}}$, up to conformal automorphisms of the disk. A positive answer to this question would imply that smooth simply connected domains are spectrally determined (see [564]). In [352], calculations of the zeta function were used to prove a weaker statement – namely, that a family of smooth simply connected planar domains is pre-compact in the topology of a certain Sobolev space.

In higher dimensions, the Dirichlet energy is not conformally invariant, and therefore the approach described above does not work. However, one can construct Steklov isospectral manifolds of dimension $d \geq 3$ with the help of Example 5.3. Indeed, given two compact manifolds M_1 and M_2 which are Laplace-Beltrami isospectral (there are many known examples of such pairs, see, for instance, [232, 436, 834]), consider two cylinders $\Omega_1 = M_1 \times [0, L]$ and $\Omega_2 = M_2 \times [0, L]$, $L > 0$. It follows from Example 5.3 that Ω_1 and Ω_2 have the same Steklov spectra. Recently, examples of higher-dimesional Steklov isospectral manifolds with connected boundaries were announced in [435].

In all known constructions of Steklov isospectral manifolds, their boundaries are Laplace isospectral. The following question was asked in [429]:

Open problem 5.38. *Do there exist Steklov isospectral manifolds such that their boundaries are not Laplace isospectral?*

5.5.2 Rigidity of the Steklov spectrum: the case of a ball

It is an interesting and challenging question to find examples of manifolds with boundary that are uniquely determined by their Steklov spectrum. In this subsection we discuss the seemingly simple example of Euclidean balls.

Proposition 5.39. *A disk is uniquely determined by its Steklov spectrum among all smooth Euclidean domains.*

Proof. Let Ω be an Euclidean domain which has the same Steklov spectrum as the disk of radius r. Then, by Corollary 5.5 one immediately deduces that Ω is a planar domain of perimeter $2\pi r$. Moreover, it follows from Theorem 5.6 that Ω is simply–connected. Therefore, since the equality in Weinstock's inequality (5.11) is achieved for Ω, the domain Ω is a disk of radius r. □

Remark 5.40. *The smoothness hypothesis in the proposition above seems to be purely technical. We have to make this assumption since we make use of Theorem 5.6.*

The above result motivates the following open problem:

Open problem 5.41. *Let $\Omega \subset \mathbb{R}^d$ be a domain which is isospectral to a ball of radius r. Show that it is a ball of radius r.*

Note that Theorem 5.19 does not yield a solution to this problem because the volume $|\Omega|$ is not a Steklov spectrum invariant. Using the heat invariants of the Dirichlet-to-Neumann operator (see subsection 5.2.2), one can prove the following statement in dimension three.

Proposition 5.42. *Let $\Omega \subset \mathbb{R}^3$ be a domain with connected and smooth boundary M. Suppose its Steklov spectrum is equal to that of a ball of radius r. Then Ω is a ball of radius r.*

This result was obtained in [764], and we sketch its proof below. First, let us show that M is simply–connected. We use an adaptation of a theorem of Zelditch on multiplicities [891] proved using microlocal analysis. Namely, since Ω is Steklov isospectral to a ball, the multiplicities of its Steklov eigenvalues grow as $m_k = Ck + O(1)$, where $C > 0$ is some constant and m_k is the multiplicity of the k-th *distinct* eigenvalue (cf. Example 5.2). Then one deduces that M is a Zoll surface (that is, all geodesics on M are periodic with a common period), and hence it is simply–connected [141].

Therefore, the following formula holds for the coefficient a_2 in the Steklov heat trace asymptotics (5.5) on Ω:

$$a_2 = \frac{1}{16\pi}\int_M H_1^2 + \frac{1}{12}.$$

Here $H_1(x)$ denotes the mean curvature of M at the point x, and the term $\frac{1}{12}$ is obtained from the Gauss–Bonnet theorem using the fact that M is simply–connected. We have then: $\int_M H_1^2 = \int_{S_r} H_1^2$, where $S_r = \partial B_r$.

On the other hand, it follows from (5.2) and Corollary 5.8 that $\mathrm{Vol}(M)$ and $\int_M H_1$ are Steklov spectral invariants. Therefore,

$$\mathrm{Area}(M) = \mathrm{Area}(S_r), \quad \int_M H_1 = \int_{S_r} H_1.$$

Hence

$$\sqrt{\mathrm{Area}(M)}\left(\int_M H_1^2\right)^{1/2} - \left|\int_M H_1\right| = \sqrt{\mathrm{Area}(S_r)}\left(\int_{S_r} H_1^2\right)^{1/2} - \left|\int_{S_r} H_1\right| = 0.$$

Since the Cauchy-Schwarz inequality becomes an equality only for constant functions, one gets that H_1 must be constant on M. By a theorem of Alexandrov [18], the only compact surfaces of constant mean curvature embedded in $\mathbb{R}^3$ are round spheres. We conclude that M is itself a sphere of radius r and therefore Ω is isometric to B_r. This completes the proof of the proposition.

5.6 Nodal geometry and multiplicity bounds

5.6.1 Nodal domain count

The study of nodal domains and nodal sets of eigenfunctions is probably the oldest topic in geometric spectral theory, going back to the experiments of E. Chladni with vibrating plates. The fundamental result in the subject is Courant's nodal domain theorem which states that the k-th eigenfunction of the Dirichlet boundary value problem has at most k nodal domains. The proof of this statement uses essentially two ingredients: the variational principle (see section 5.4.1) and the unique continuation for solutions of second order elliptic equations. It can therefore be extended essentially verbatim to Steklov eigenfunctions (see [574, 609]).

Theorem 5.43. *Let Ω be a compact Riemannian manifold with boundary and u_k be an eigenfunction corresponding to the Steklov eigenvalue σ_k. Then u_k has at most k nodal domains.*

Apart from the “interior” nodal domains and nodal sets of Steklov eigenfunctions, a natural problem is to study the boundary nodal domains and nodal sets, that is, the nodal domains and nodal sets of the eigenfunctions of the Dirichlet-to-Neumann operator.

The proof of Courant's theorem cannot be generalized to the Dirichlet-to-Neumann operator because it is nonlocal. The following problem therefore arises:

Fig. 5.6. A surface inside a ball creating only two connected components in the interior and a large number of connected components on the boundary sphere.

Open problem 5.44. *Let Ω be a Riemannian manifold with boundary M. Find an upper bound for the number of nodal domains of the k-th eigenfunction of the Dirichlet-to-Neumann operator on M.*

For surfaces, a simple topological argument shows that the bound on the number of interior nodal domains implies an estimate on the number of boundary nodal domains of a Steklov eigenfunction. In particular, the k-th nontrivial Dirichlet-to-Neumann eigenfunction on the boundary of a simply–connected planar domain has at most $2k$ nodal domains [19, Lemma 3.4].

In higher dimensions, the number of interior nodal domains does not control the number of boundary nodal domains (see Figure 5.6), and therefore new ideas are needed to tackle Open Problem 5.44. However, there are indications that a Courant-type (i.e. $\mathcal{O}(k)$) bound should hold in this case as well. For instance, this is the case for cylinders and Euclidean balls (see Examples 5.2 and 5.3).

5.6.2 Geometry of the nodal sets

The nodal sets of Steklov eigenfunctions, both interior and boundary, remain largely unexplored. The basic property of the nodal sets of Laplace–Beltrami eigenfunctions is their density on the scale of $1/\sqrt{\lambda}$, where λ is the eigenvalue (cf. [889], see also Figure 5.7). This means that for any manifold Ω, there exists a constant C such that for any eigenvalue λ large enough, the corresponding eigenfunction ϕ_λ has a zero in any geodesic ball of radius $C/\sqrt{\lambda}$. This motivates the following questions (see also Figure 5.7):

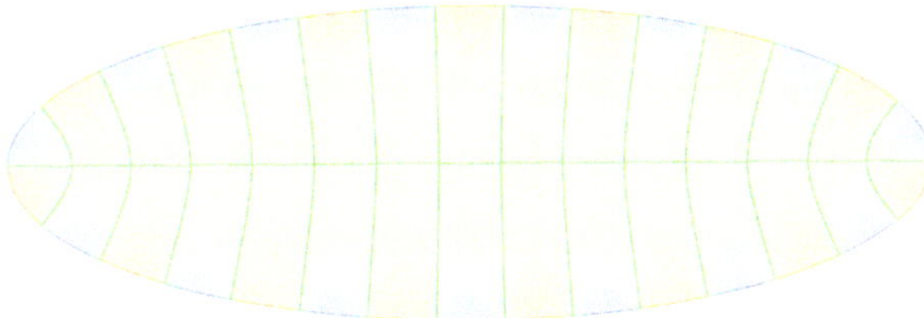

Fig. 5.7. The nodal lines of the 30th eigenfunction on an ellipse.

Open problem 5.45. *(i) Are the nodal sets of Steklov eigenfunctions on a Riemannian manifold Ω dense on the scale $1/\sigma$ in Ω? (ii) Are the nodal sets of the Dirichlet-to-Neumann eigenfunctions dense on the scale $1/\sigma$ in $M = \partial\Omega$?*

For smooth simply–connected planar domains, a positive answer to question (ii) follows from the work of Shamma [807] on the asymptotic behaviour of Steklov eigenfunctions. On the other hand, the explicit representation of eigenfunctions on rectangles implies that there exist eigenfunctions of arbitrary high order which have zeros only on one pair of parallel sides. Therefore, a positive answer to (ii) may hold only under some regularity assumptions on the boundary.

Another fundamental problem in nodal geometry is to estimate the size of the nodal set. It was conjectured by S.-T. Yau that for any Riemannian manifold of dimension d,

$$C_1\sqrt{\lambda} \le \mathcal{H}_{d-1}(\mathcal{N}(\phi_\lambda)) \le C_2\sqrt{\lambda},$$

where $\mathcal{H}_{d-1}(\mathcal{N}(\phi_\lambda))$ denotes the $d-1$-dimensional Hausdorff measure of the nodal set $\mathcal{N}(\phi_\lambda)$ of a Laplace-Beltrami eigenfunction ϕ_λ, and the constants C_1, C_2 depend only on the geometry of the manifold. Similar questions can be asked in the Steklov setting:

Open problem 5.46. *Let Ω be an d-dimensional Riemannian manifold with boundary M. Let u_σ be an eigenfunction of the Steklov problem on Ω corresponding to the eigenvalue σ and let $\phi_\sigma = u_\sigma|_M$ be the corresponding eigenfunction of the Dirichlet-to-Neumann operator on M. Show that*

(i) $C_1\sigma \le \mathcal{H}_{d-1}(\mathcal{N}(u_\sigma)) \le C_2\sigma$,

(ii) $C_1'\sigma \le \mathcal{H}_{d-2}(\mathcal{N}(\phi_\sigma)) \le C_2'\sigma$,

where the constants C_1, C_2, C_1', C_2' depend only on the manifold.

Some partial results on this problem are known. In particular, the upper bound in (ii) was conjectured by [119] and proved in [889] for real analytic manifolds with real analytic boundary. A lower bound on the size of the nodal set $\mathcal{N}(\phi_\sigma)$ for smooth Riemannian manifolds (though weaker than the one conjectured in (ii) in dimensions ≥ 3)

was recently obtained in [868] using an adaptation of the approach of [822] to nonlocal operators.

The upper bound in (i) is related to the question of estimating the size of the zero set of a harmonic function in terms of its frequency (see [445]). In [765], this approach is combined with the methods of potential theory and complex analysis in order to obtain both upper and lower bounds in (i) for real analytic Riemannian surfaces.[5.6] Let us also note that the Steklov eigenfunctions decay rapidly away from the boundary [529], and therefore the problem of understanding the properties of the nodal set in the interior is somewhat analogous to the study of the zero sets of Schrödinger eigenfunctions in the "forbidden regions" (see [446]).

5.6.3 Multiplicity bounds for Steklov eigenvalues

In two dimensions, the estimate on the number of nodal domains allows us to control the eigenvalue multiplicities (see [142, 268]). The argument roughly goes as follows: if the multiplicity of an eigenvalue is high, one can construct a corresponding eigenfunction with a high enough vanishing order at a certain point of a surface. In the neighbourhood of this point the eigenfunction looks like a harmonic polynomial, and therefore the vanishing order together with the topology of a surface yield a lower bound on the number of nodal domains. To avoid a contradiction with Courant's theorem, one deduces a bound on the vanishing order, and hence on the multiplicity.

This general scheme was originally applied to Laplace-Beltrami eigenvalues, but it can be also adapted to prove multiplicity bounds for Steklov eigenvalues. For simply connected surfaces, this idea was used in [19]. For general Riemannian surfaces, interestingly enough, one can obtain estimates of two kinds. Recall that the Euler characteristic χ of an orientable surface of genus γ with l boundary components equals $2 - 2\gamma - l$, and of a non-orientable one is equal to $2 - \gamma - l$. Putting together the results of [392, 558, 560, 574] we get the following bounds:

Theorem 5.47. *Let Σ be a compact surface of Euler characteristic χ with l boundary components. Then the multiplicity $m_k(\Sigma)$ for any $k \geq 2$ satisfies the following inequalities:*

$$m_k(\Sigma) \leq 2k - 2\chi - 2l + 3, \tag{5.20}$$

$$m_k(\Sigma) \leq k - 2\chi + 2. \tag{5.21}$$

Note that the right-hand side of (5.20) depends only on the index of the eigenvalue k and on the genus γ of the surface, while the right-hand side of (5.21) depends also

5.6 See also recent results of J. Zhu [893, 894]. In particular, in [893] the upper bound in (i) was proved for real-analytic Riemannian manifolds of arbitrary dimension.

on the number of boundary components. Inequality (5.21) in this form was proved in [558]. In particular, it is sharp for the first eigenvalue of simply connected surfaces ($\chi = 1$, the maximal multiplicity is two, see also [19]) and for surfaces homeomorphic to a Möbius band ($\chi = 0$, the maximal multiplicity is four). Inequality 5.20 is sharp for surfaces homeomorphic to an annulus ($\chi = 0$, $l = 2$, the maximal multiplicity is three and attained by the critical catenoid, see Theorem 5.25).

While these bounds are sharp in some cases, they are far from optimal for large k. In fact, the following result is an immediate corollary of Theorem 5.4.

Corollary 5.48. *[429] For any smooth compact Riemannian surface Ω with l boundary components, there is a constant N depending on the metric on Ω such that for $j > N$, the multiplicity of σ_j is at most $2l$.*

Remark 5.49. *The multiplicity of the first nonzero eigenvalue σ_2 has been linked to the relative chromatic number of the corresponding surface with boundary in [558].*

Remark 5.50. *It is well-known that the spectrum of the Laplace-Beltrami operator is generically simple [17, 857]. It is likely that the same is true for the Steklov spectrum, however, to our knowledge, such a result has not been established yet.*

For manifolds of dimension $d \geq 3$, no general multiplicity bounds for Steklov eigenvalues are available. Moreover, given a Riemannian manifold Ω of dimension $d \geq 3$ and any non-decreasing sequence of N positive numbers, one can find a Riemannian metric g in a given conformal class, such that this sequence coincides with the first N nonzero Steklov eigenvalues of (M, g) [559].

Theorem 5.51. *Let Ω be a compact manifold of dimension $d \geq 3$ with boundary. Let m be a positive integer and let $0 = s_0 < s_1 \leq \cdots \leq s_m$ be a finite sequence. Then there exists a Riemannian metric g on Ω such that $\sigma_j = s_j$ for $j = 0, \cdots, m$.*

For Laplace-Beltrami eigenvalues, a similar result was obtained in [282]. It is plausible that multiplicity bounds for Steklov eigenvalues in higher dimensions could be obtained under certain geometric assumptions, such as curvature constraints.

Acknowledgment: The authors would like to thank Brian Davies for inviting them to write this survey. The project started in 2012 at the conference on *Geometric Aspects of Spectral Theory* at the Mathematical Research Institute in Oberwolfach, and its hospitality is greatly appreciated. We are grateful to Mikhail Karpukhin, David Sher and the anonymous referee for helpful remarks. We are also thankful to Dorin Bucur, Fedor Nazarov, Leonid Parnovski, Raphaël Ponge, Alexander Strohmaier and John Toth for useful discussions, as well as to Bartek Siudeja for letting us use his FEniCS eigenvalues computation code.

The research of AG was partially supported by NSERC and FRQNT New Researchers Start-up program. The research of IP was partially supported by NSERC, FRQNT and Canada Research Chairs program.

Richard S. Laugesen and Bartłomiej A. Siudeja

6 Triangles and Other Special Domains

6.1 Introduction

Special domains such as triangles and rectangles can reveal interesting phenomena and suggest conjectures for general domains. For example, the first three Dirichlet eigenvalues of the Laplacian on a triangle of given diameter are all minimal for the equilateral (Theorem 6.16), which suggests an analogous conjecture for general domains with the disk as the minimizing domain (Conjecture 6.61).

Conversely, symmetry problems for N-gons can be more difficult to handle than the corresponding problems for general domains. For example, Pólya and Szegő conjectured that the regular N-gon minimizes the first Dirichlet eigenvalue among all N-gons of given area (Conjecture 6.3), and this problems remains unsolved for $N \geq 5$, ninety years after Faber and Krahn proved that the disk is the minimizer among general domains.

In this chapter, we report on known and conjectured spectral properties of the Laplacian on families of special domains in the plane, including sharp lower bounds and sharp upper bounds, as well as inverse problems, hot spots, and nodal domains. A few three-dimensional results arise too. At the end of the chapter we raise conjectures for general domains, motivated by the results on special domains.

We concentrate on the Dirichlet and Neumann spectra. Much less is known about extremal problems for the Robin and Steklov spectra on special domains. For example, is the second Steklov eigenvalue maximal for the equilateral triangle, among all triangles of given perimeter? If this conjecture were true, it would provide an elegant triangular analogue of Weinstock's result for general domains (Chapter 5), in which the disk is the maximizer.

6.2 Variation, notation, normalization, majorization

We begin with some basic facts and definitions.

Richard S. Laugesen: Department of Mathematics, University of Illinois, Urbana, Illinois 61801, U.S.A., E-mail: Laugesen@illinois.edu
Bartłomiej A. Siudeja: Eugene, Oregon 97403, U.S.A., E-mail: Siudej@gmail.com

Variational principles

The Rayleigh Principle asserts that the first Dirichlet eigenvalue of the Laplacian on a bounded domain Ω is characterized variationally as

$$\lambda_1 = \min\Big\{ \int_\Omega |\nabla u|^2\,dx : u \in H_0^1(\Omega),\ \|u\|_{L^2(\Omega)} = 1 \Big\},$$

where $H_0^1(\Omega)$ denotes the Sobolev space of L^2-functions that have one derivative in L^2 and vanish on the boundary.

The Rayleigh–Poincaré Principle extends the characterization to the *sum* of the eigenvalues:

$$\lambda_1 + \cdots + \lambda_n = \min\Big\{ \int_\Omega |\nabla u_1|^2\,dx + \cdots + \int_\Omega |\nabla u_n|^2\,dx : u_j \in H_0^1(\Omega),$$
$$\langle u_j, u_k\rangle_{L^2(\Omega)} = \delta_{jk} \text{ for } 1 \le j,k \le n \Big\}$$

whenever $n \ge 1$. In other words, one minimizes the sum of Rayleigh quotients over all choices of orthonormal trial functions.

For the Neumann spectrum, corresponding variational characterizations hold using the space H^1 instead of H_0^1.

Notation

Write $j_{n,m}$ for the m-th positive root of the Bessel function J_n. In particular, $j_{0,1} \simeq 2.4048$ and $j_{1,1} \simeq 3.8317$ are the first roots of $J_0(r)$ and $J_1(r)$, respectively.

Given positive numbers $a_1, a_2, \ldots, a_n$, define their

$$\text{arithmetic mean} = M(a_1,\ldots,a_n) = \frac{a_1 + \cdots + a_n}{n},$$
$$\text{geometric mean} = G(a_1,\ldots,a_n) = \sqrt[n]{a_1\cdots a_n},$$
$$\text{harmonic mean} = H(a_1,\ldots,a_n) = 1 \Big/ \Big(\frac{1/a_1 + \cdots + 1/a_n}{n}\Big),$$

so that the arithmetic–geometric–harmonic mean inequality says

$$M \ge G \ge H.$$

Normalizing factors

The most important geometric quantities on triangles and other polygons are:

$$A = \text{area},$$
$$D = \text{diameter},$$
$$L = \text{perimeter} = \text{sum of side lengths},$$
$$S^2 = \text{sum of squares of side lengths}.$$

The diameter of a triangle equals its largest sidelength. The other quantities are related as follows:

Lemma 6.1. *For all triangles,*

$$12\sqrt{3}A \leq L^2 \leq 3S^2.$$

Equality holds (in each inequality) if and only if the triangle is equilateral.

The inequality on the left is the triangular isoperimetric inequality, while the inequality on the right is the arithmetic–geometric mean inequality applied to the sidelengths.

Majorization

The majorization technique of Hardy, Littlewood and Pólya [450, Theorem 108] says that if partial sums of one series are bounded by partial sums of another, then the same holds true for concave images of those series. (Later in this chapter we apply majorization to get from eigenvalue sum inequalities to spectral zeta and heat trace inequalities.) Specifically, majorization says that if

$$a_1 + \cdots + a_n \leq b_1 + \cdots + b_n$$

for all $n \geq 1$, where $\{a_j\}$ and $\{b_j\}$ are increasing sequences of positive numbers, then

$$C(a_1) + \cdots + C(a_n) \leq C(b_1) + \cdots + C(b_n)$$

for all $n \geq 1$, for each concave increasing function $C : \mathbb{R}_+ \to \mathbb{R}$. Special choices of C yield that:

$$\begin{aligned} (a_1^s + \cdots + a_n^s)^{1/s} &\leq (b_1^s + \cdots + b_n^s)^{1/s}, && 0 < s \leq 1, \\ a_1 \cdots a_n &\leq b_1 \cdots b_n, \\ \sum_{j=1}^n a_j^s &\geq \sum_{j=1}^n b_j^s, && s < 0, \\ \sum_{j=1}^n \exp(-ta_j) &\geq \sum_{j=1}^n \exp(-tb_j), && t > 0, \end{aligned}$$

as one sees by taking $C(a) = a^s$ when $0 < s \leq 1$, $C(a) = \log a$ when $s = 0$, $C(a) = -a^s$ when $s < 0$, and $C(a) = -e^{-ta}$ when $t > 0$. The last two expressions can be regarded as partial sums of generalized zeta functions and heat traces respectively.

Even stronger majorization results are available, due to Karamata and Ostrowski [115, §1.31], and these have recently been applied to graph eigenvalues [454]. The monograph of Marshall, Olkin and Arnold [692] is a good reference for the substance, history and applications of majorization.

6.3 Lower bounds by symmetrization

This section presents a variety of symmetrization techniques (Schwarz, Steiner, continuous Steiner, polarization) to prove eigenvalue bounds for special classes of domains, including triangles and quadrilaterals.

Symmetrization methods fix the area of a domain and decrease its perimeter: if $\Omega^\star$ denotes a symmetrized version of a domain Ω, then

$$A(\Omega) = A(\Omega^\star),$$
$$L(\Omega) \geq L(\Omega^\star).$$

Classically, symmetrization arose as a technique to prove that the disk has minimal perimeter among all domains of given area (the isoperimetric theorem).

When symmetrization is applied to the level sets of a function u in order to create a symmetrized function $u^\star$, the Sobolev norm decreases provided the original function is nonnegative and vanishes on the boundary (see Bandle [97], Kawohl [579] or Pólya–Szegő [770]). More precisely, for any nonnegative $u \in H_0^1(\Omega)$ one has

$$\int_\Omega u^2 \, dA = \int_{\Omega^\star} (u^\star)^2 \, dA,$$
$$\int_\Omega |\nabla u|^2 \, dA \geq \int_{\Omega^\star} |\nabla u^\star|^2 \, dA,$$

and hence the Rayleigh quotient decreases under symmetrization. Thus by the Rayleigh principle, the first Dirichlet eigenvalue also decreases under symmetrization.

This powerful tool for proving lower bounds on eigenvalues is restricted to cases where the eigenfunction is nonnegative in the region of interest, and so only the first eigenvalue can be studied by symmetrization, generally speaking.

6.3.1 Dirichlet eigenvalues

The Faber–Krahn inequality says that among all domains with fixed area, the disk has the smallest first Dirichlet eigenvalue (see Chapters 2 and 7). For triangles and quadrilaterals one has the following natural analogue.

Theorem 6.2 (Pólya–Szegő bound on first eigenvalue). *Among triangles, $\lambda_1 A$ is minimal for equilateral triangles. Among quadrilaterals, $\lambda_1 A$ is minimal for squares.*

Before discussing the proof, we mention two unsolved Faber–Krahn type problems for polygons: first for polygons with more sides under Dirichlet boundary conditions, and then for triangles under the Robin boundary condition.

Conjecture 6.3 (Polygonal Faber–Krahn problem, due to Pólya–Szegő [770]). *Among regular polygons with $N \geq 5$ sides, the regular N-gon minimizes $\lambda_1 A$.*

Conjecture 6.4. *The equilateral triangle minimizes the first Robin eigenvalue, among all triangles of given area, when the Robin parameter is a given positive constant. (For the corresponding result on general domains, see Chapter 4.)*

The Faber–Krahn inequality on general domains has many different proofs. The original proof used Schwarz symmetrization, also known as symmetric decreasing rearrangement. For triangles and quadrilaterals as in Theorem 6.2, one needs a different type of symmetrization.

Definition 6.5 (Steiner symmetrization). *Consider a planar domain Ω, and fix a line l in the plane. Given a point $x \in l$, write Ω_x for the intersection of Ω with the line passing through x and perpendicular to l, so that the sets Ω_x foliate Ω into cross-sections perpendicular to l. Let $\Omega_x^\star$ be the segment with the same length measure as Ω_x that is centered at x and perpendicular to l. Define the Steiner symmetrization of Ω to be the set $\Omega^\star = \cup_{x \in l} \Omega_x^\star$.*

Clearly the area of a domain remains unchanged under Steiner symmetrization, while it is well known that the perimeter and the first Dirichlet eigenvalue decrease [579].

Proof of Theorem 6.2 for quadrilaterals: $\lambda_1 A$ is minimal for the square. A triangle that is Steiner symmetrized with respect to any line perpendicular to a side remains a triangle. Hence a quadrilateral symmetrized with respect to any line perpendicular to a diagonal is again a quadrilateral. (The good news stops there though, because symmetrizing a generic pentagon yields a polygon with more than 5 sides, and this difficulty has obstructed every attempt so far to prove Conjecture 6.3 by symmetrization when $N \geq 5$.)

Let Ω be any quadrilateral (not necessarily convex). Let l_1 be a line perpendicular to the diagonal l_2 contained inside Ω. Perform Steiner symmetrization with respect to l_1 to obtain a kite (a convex quadrilateral with a line of symmetry). Now Steiner symmetrize with respect to l_2. The resulting quadrilateral is a rhombus. The final symmetrization uses a line l_3 perpendicular to one of the sides and produces a rectangle, as shown in Figure 6.1. The eigenvalue decreases under each symmetrization.

Now one recalls the formula for the smallest eigenvalue of a rectangle with side lengths a and b (see Section 6.8):

$$\lambda_1(a, b) = \pi^2 \left(\frac{1}{a^2} + \frac{1}{b^2} \right) = \pi^2 \left(\frac{1}{2ab} + \left(\frac{1}{a} - \frac{1}{b} \right)^2 \right).$$

Assuming fixed area (ab = const), we see the square ($a = b$) has the smallest first Dirichlet eigenvalue.

A similar proof works for triangles. One repeatedly symmetrizes with respect to a line perpendicular to the second-longest side of the triangle. This procedure continues indefinitely, but one can show the sequence of triangles converges to an equilateral triangle. Alternatively, one can first show existence of a minimizing triangle (by

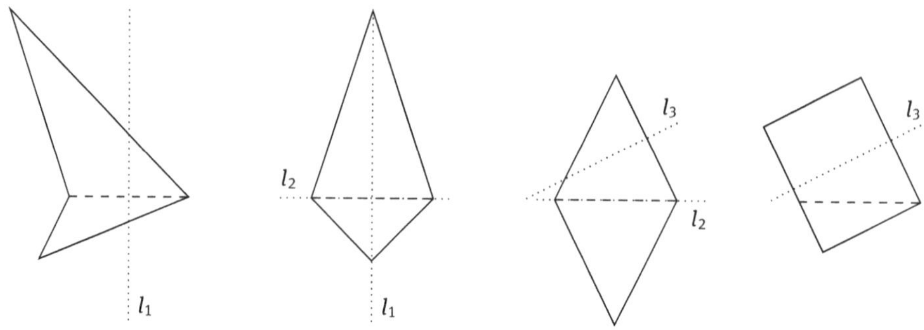

Fig. 6.1. Symmetrization of a quadrilateral into a rectangle.

compactness and domain monotonicity) and then use symmetrization to prove that the minimizer must be equilateral or else the eigenvalue could have been decreased further. □

Steiner symmetrization remains a valuable tool to this day, and has been used recently by Fusco, Maggi and Pratelli [411], Barchiesi, Cagnetti and Fusco [108], and Siudeja [818], among others.

A partial refinement of the lower bound on λ_1 was provided by Hooker and Protter [539].

Theorem 6.6 (Improved inequality for rhombi). *Let $a \geq 1$ be the ratio of the lengths of the diagonals of a rhombus. Then*

$$\lambda_1 A \geq 2\pi^2 \cdot \frac{(1+a)^2}{4a},$$

with equality holding for squares ($a = 1$).

Their proof is not based on symmetrization. Instead, they derive their bound from an explicit perturbation of the quadratic form in the Rayleigh quotient (Picard's method [758]), and find even stronger but more complicated bounds. Later, Freitas and Siudeja noted that if one replaces the final symmetrization in the proof of Theorem 6.2 with the Hooker-Protter inequality, then one gets an improved bound for all quadrilaterals [401, Theorem 1.1].

In contrast with quadrilaterals, no improved inequality is known for triangles. Siudeja [817, Conjecture 1.3] conjectured on the basis of numerical evidence provided by Freitas and Antunes [47] that:

Conjecture 6.7 (Improved bound on first eigenvalue for triangles). *The quantity*

$$\lambda_1 A - \frac{\pi^2}{16} \frac{L^2}{A}$$

is minimal among triangles for the equilateral, with minimum value $7\sqrt{3}\pi^2/12$.

Note that each term in the conjecture is individually minimal for the equilateral triangle. This conjecture would improve for triangles on Makai's inequality [688], which states that for convex domains

$$\lambda_1 A - \frac{\pi^2}{16} \frac{L^2}{A} \geq 0.$$

For triangles one can, however, use symmetrization techniques to provide an intermediate step (Siudeja [818]). For an isosceles triangle, its *aperture* is the angle between the two equal sides.

Theorem 6.8 (Isosceles triangles by symmetrization).

(i) Among all triangles with a given area and smallest angle α, the first eigenvalue λ_1 is minimized by the isosceles triangle with aperture α and the given area.

(ii) For isosceles triangles, $\lambda_1 A$ is decreasing as a function of the aperture $\alpha \in (0, \pi/3)$ and increasing for $\alpha \in (\pi/3, \pi)$.

For more on eigenvalues of isosceles triangles, see Section 6.10.

Proof of Theorem 6.8(ii) for $\alpha \in (0, \pi/3)$. The key tool is "continuous Steiner symmetrization", which is constructed as follows. The shifting of the cross-section Ω_x from its original position to the final segment $\Omega_x^\star$ can be viewed as a time dependent process. Assuming the cross-section Ω_x is connected, we slide it across to $\Omega_x^\star$ at some fixed velocity v_x as time t increases from 0 to 1. This procedure was first described by Blaschke [151, p. 58] and Pólya and Szegő [770, Note B], and later studied by Solynin [824, 827], Brock [194, 195], and Bucur and Henrot [220], and its properties have been summarized by Henrot [505].

The continuous Steiner symmetrization applied to triangles produces an intermediate triangle at each $t \in [0, 1]$. Indeed, each cross-section interval Ω_x is centered at a point on the median of the side to which symmetrization line is perpendicular, and the velocity v_x is assigned so that the points initially on the median remain on a line as t increases.

Let $I(b_1)$ and $I(b_2)$ be isosceles triangles with the same area, and apertures $0 < \alpha_1 < \alpha_2 < \pi/3$. Write $b_1 < b_2$ for the bases of the triangles, where by the base we mean the shortest side. Perform continuous Steiner symmetrization of $I(b_1)$ with respect to a line perpendicular to one of the equal-length sides, as indicated by the blue arrow on the left side of Figure 6.2. The length of the base increases and λ_1 decreases, under this procedure. Either the length of the base reaches b_2 or else the triangle becomes a fully Steiner symmetrized triangle with two sides of length $b_3 < b_2$. In either case, we then perform a standard Steiner symmetrization with respect to the new, elongated base (see the right side of Figure 6.2), yielding either $I(b_2)$ or $I(b_3)$. In the latter case we repeat the procedure and eventually reach $I(b_2)$, as Siudeja showed [818]. That finishes the proof.

□

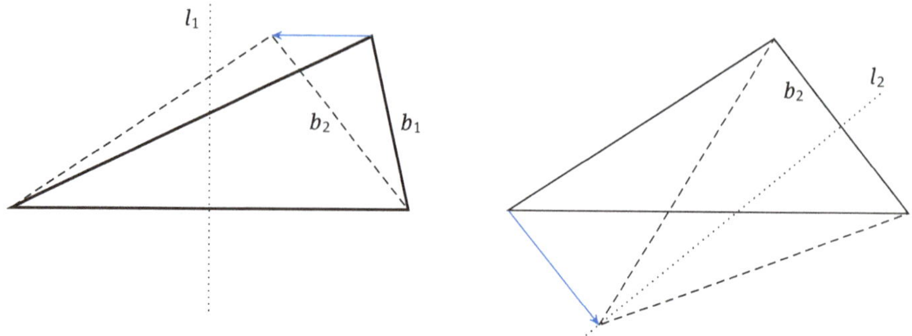

Fig. 6.2. Continuous symmetrization of a triangle into an isosceles triangle with given base length.

Another interesting lower bound for triangles was obtained by Freitas [397], using a version of the Unknown Trial Function Method (see Section 6.4). Let D denote the diameter of the triangle and h its shortest altitude. Then

$$\lambda_1 \geq \pi^2 \left(\frac{4}{D^2} + \frac{1}{h^2} \right). \tag{6.1}$$

The same bound can also be obtained using polarization (another kind of symmetrization described, in the next subsection), as shown by Siudeja [818].

Using the improved bound on λ_1 for quadrilaterals [401, Theorem 1.1], Freitas and Siudeja proved that

$$\lambda_1 \geq \pi^2 \left(\frac{1}{D} + \frac{1}{h} \right)^2. \tag{6.2}$$

Further insight into the lowest eigenvalue of general triangles can be obtained using symmetrization to isosceles triangles as in Theorem 6.8(i), together with the exact eigenvalues for sectors.

Theorem 6.9 (First eigenvalue.). *Let α be the smallest angle in a triangle and let $\gamma = 2\arctan(h/D)$. Then*

$$\lambda_1 \geq \max \left\{ \frac{\alpha j_{\pi/\alpha,1}^2}{hD}, \frac{4(j'_{\pi/\gamma,1})^2}{4h^2 + D^2} \right\},$$

where $j_{\nu,1}$ and $j_{\nu,1}{}'$ denote the first positive roots of the Bessel function J_ν and its derivative, respectively.

The proof of the first inequality involves symmetrization of isosceles triangles into sectors (Siudeja [818]). The second bound, obtained by Freitas [398], follows from an intricate application of a domain monotonicity for mixed Dirichlet-Neumann eigenvalues, due to Harrell [453].

This last theorem together with (6.2) provides the best lower bound for most triangles, except for nearly-equilateral triangles, where the Pólya–Szegő bound in Theorem 6.2 is still the best. Note that Freitas and Siudeja carried out extensive comparisons between the various results [401, Section 2.1]. It is also worth noting that better, much more complicated bounds exist [401, Section 2.2], although even these bounds compare poorly with the Pólya–Szegő inequality for nearly-equilateral triangles. Further bounds for triangles can be deduced from results on wedge-like domains by Hasnaoui and Hermi [466]. Their paper also extensively compares various bounds from the literature in the case of right triangles.

Returning now to the polygonal Faber–Krahn problem in Conjecture 6.3, notice it would imply in particular that $\lambda_1 A$ decreases as one passes from the regular N-gon P_N to the regular $(N+1)$-gon P_{N+1}, since the regular N-gon can itself be regarded as a degenerate $(N+1)$-gon. It seems plausible that:

Conjecture 6.10 (Regular polygons [47]).

$$\frac{\lambda_1(P_{N+1})A(P_{N+1})}{\lambda_1(P_N)A(P_N)} \leq 1$$

for all $N \geq 3$. *Further, the left side is an increasing function of N that converges to* 1 *as* $N \to \infty$.

Remarkably, even this special case of the polygonal Faber–Krahn problem remains unsolved. Some partial results are due to Nitsch [727].

6.3.2 Mixed Dirichlet-Neumann eigenvalues

Even though symmetrization usually requires Dirichlet boundary conditions, it is possible to develop generalizations allowing a certain amount of Neumann boundary. The simplest example is to take a triangle with two Dirichlet sides and one Neumann side, and reflect this triangle along the Neumann side to give a quadrilateral with purely Dirichlet boundary. Due to symmetry of the reflected domain, its ground state must be symmetric, and hence satisfies the Neumann condition on the line of symmetry. Therefore, the lowest Dirichlet eigenvalue of the quadrilateral is the same as the lowest Dirichlet-Dirichlet-Neumann eigenvalue on the original triangle. Now we can apply symmetrization to the quadrilateral to get the next result.

Theorem 6.11 (First eigenvalue, mixed boundary conditions). *Among all triangles of a given area, the first eigenvalue of the mixed problem with two Dirichlet sides and one Neumann side is minimal for the right isosceles triangle having Neumann condition on its longest side.*

One cannot apply the same argument with two Neumann sides, except on right triangles, where four reflections would give a rhombus. In that case again the right isosceles triangle is a minimizer (with Neumann conditions on the short sides).

The preceding theorem concerns optimization across a family of domains with mixed boundary conditions. Another interesting challenge is to optimize the placement of the mixed conditions on a fixed domain. For example, fix a triangle and apply Dirichlet conditions to one, two, or three sides. Denote by λ_1^X the smallest eigenvalue when the Dirichlet condition is imposed on a combination of sides indicated by the symbol X, which is some combination of the letters L, M, S, for the longest, middle and shortest sides. Put Neumann conditions on any side not listed in X. Then Siudeja [820] proved:

Theorem 6.12 (Fixed triangle with Dirichlet conditions on different sides). *On any triangle,*

$$\min\left\{\lambda_1^S, \lambda_1^M, \lambda_1^L\right\} < \lambda_1^{MS} \le \lambda_1^{LS} \le \lambda_1^{LM} < \lambda_1^{LMS},$$

and the inequalities are sharp when the relevant sides have different lengths. Further, for right triangles with smallest angle at least $\pi/6$ we have

$$\lambda_1^S < \lambda_1^M < \mu_2 < \lambda_1^L < \lambda_1^{MS}, \tag{6.3}$$

where μ_2 is the second Neumann eigenvalue of the triangle.

In general, having more Dirichlet boundary should increase the eigenvalue, and so we conjecture that the second part of Theorem 6.12 holds not just for right triangles but for general triangles.

Conjecture 6.13. *For all triangles, $\lambda_1^S < \lambda_1^M < \lambda_1^L < \lambda_1^{MS}$, unless two of the involved sides have equal length.*

The inequality $\lambda_1^{MS} \le \lambda_1^{LS}$ in Theorem 6.12 provides a perfect opportunity to introduce one more type of symmetrization.

Definition 6.14 (Polarization of domains; see Figure 6.3.). *Let Ω be a planar domain. Fix a line l intersecting this domain, and write H_l for one of the halfplanes determined by the line. The polarized domain Ω^P is formed from Ω by reflecting each point in Ω to the H_l-side unless the reflected point already belongs to Ω.*

Polarization clearly preserves area, and can be shown to decrease perimeter, just like other symmetrizations. It is the most fundamental of the symmetrization methods, because all the others can be obtained as limits of suitably chosen sequences of polarizations (see Brock–Solynin [196], Solynin [827] and Burchard–Fortier [226]).

This simple yet powerful transformation was introduced in complex analysis by Wolontis [883] and Ahlfors [14, Lemma 2.2], and later studied in potential theory in

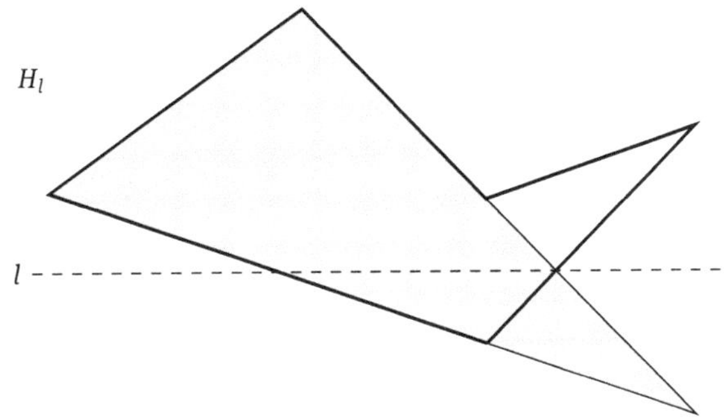

Fig. 6.3. Polarization applied to a triangle.

higher dimensions by Baernstein and Taylor [92]. Then Dubinin [346], and Brock and Solynin [191, 196, 826, 827], developed the theory further. Polarization was used also by Draghici [345] to study heat kernels, Burchard and Hajaiej [227] to study a broad class of functionals, and Siudeja [818, 820] to study the eigenvalues of triangles.

One can polarize functions as well as domains. Given a function φ on Ω, we define the polarized function φ^P as follows. Write Ω_l for the reflection of the domain in the line l, and φ_l for the reflection of the function. On the intersection $\Omega \cap \Omega_l$ we define $\varphi^P = \max(\varphi, \varphi_l)$ for points in H_l and $\varphi^P = \min(\varphi, \varphi_l)$ for points outside H_l. On the rest of Ω^P we define φ^P to equal φ or φ_l, whichever makes sense at the point in question. It is easy to see that this transformation of φ to φ^P preserves continuity, and hence is suitable for creating trial functions.

Proof of Theorem 6.12 for two Dirichlet sides. We will prove the inequality $\lambda_1^{MS} \le \lambda_1^{LS}$. Let φ be the nonnegative eigenfunction for λ_1^{LS} on the triangle at the left of Figure 6.4. Decompose φ into three functions: u (the part above l) and v (the part below l and in $\Omega \cap \Omega_l$) and w (the rest). Recompose the pieces and their reflections u_l and v_l into a trial function for the Rayleigh quotient on the same triangle but with different boundary conditions, related to λ_1^{MS}, as shown at the right of Figure 6.4. The horizontal line l in the figure is the bisector of the angle AOB and we take it as the polarization axis, with H_l lying below the dotted line.

Continuity is preserved at the bisector (the interface of u and v), and the same is true on the interface of $\max(u_l, v)$ and w, since on that interface $u_l = 0 \le v$ and so $\max(u_l, v) = v = w$. Finally, the Dirichlet boundary conditions hold in each part of Figure 6.4 on the solid lines. In fact we get an unnecessary piece of Dirichlet boundary condition on the long side, on the right of Figure 6.4. Hence by the Rayleigh Principle, the first eigenvalue λ_1^{LS} on the left of the figure is greater than the eigenvalue λ_1^{MS} on the right side (noting that the Rayleigh quotients are equal and the function on the right is a trial function but not eigenfunction for λ_1^{MS}). □

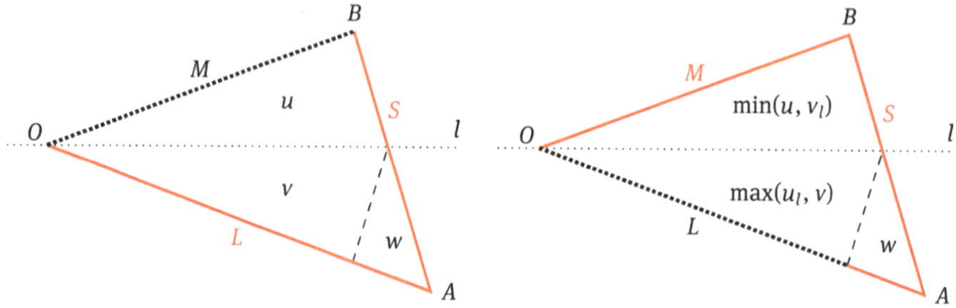

Fig. 6.4. Polarization applied to a triangle with mixed boundary conditions (Dirichlet solid, Neumann dotted).

In the last proof we applied polarization with mixed boundary conditions, and constructed a trial function equal to zero on the appropriate parts of the boundary (where we want to satisfy the Dirichlet boundary condition). In the fully Dirichlet case, sequences of polarizations have been used to prove Theorem 6.8 (see [818]).

One can also compare the first eigenvalue of a triangle having two Neumann sides and one Dirichlet side with that of the corresponding circular sector with Dirichlet condition on the curved boundary by Brandolini–Chiacchio–Trombetti [176, p. 1330].

Theorem 6.15. *Consider a triangle with Neumann boundary conditions on two sides and Dirichlet condition on the third, and denote the angle between the two Neumann sides by α.*

Then the first eigenvalue for this triangle is greater than the first eigenvalue of a sector having aperture α and the same area as the triangle, with Neumann boundary condition on the two sides of the sector and Dirichlet condition on its arc. Furthermore, this sector eigenvalue equals the smallest eigenvalue of the disk having the same radius as the sector.

Proof. When $\alpha = \pi/n$ for a positive integer n, we may reflect the triangle repeatedly to get a polygon with Dirichlet condition on its boundary, then apply the Faber–Krahn theorem. In general, given a function u on the triangle one defines its α-symmetrization in such a way as to show the L^2 norm is preserved and the Dirichlet integral decreases, thus mimicking the proof of the Faber–Krahn theorem. For a full definition of α-symmetrization, see the work of Bandle [97] and Lions–Pacella–Tricarico [674]; in particular, the desired inequality between eigenvalues follows from [674, Proposition 1.2]. For more applications of this technique see Ashbaugh–Chiacchio [75]. □

Given that the first nonconstant Neumann eigenfunction of a rhombus is antisymmetric with respect to the shorter diagonal (Atar–Burdzy [86], Siudeja [820], for some rhombi Bañuelos–Burdzy [104] and Brandolini–Chiacchio–Trombetti [176]), the last theorem also gives a lower bound on the first nonzero Neumann eigenvalue of an arbitrary rhombus.

6.4 Lower bounds by unknown trial functions

Lower bounds on eigenvalues are difficult to establish because eigenvalues are characterized by the *minimization* of a Rayleigh quotient. The Faber–Krahn theorem, that the fundamental tone is minimal for the disk among domains of given area, is generally proved by symmetrization methods (rearrangements). The same is true for the triangular analogue. Symmetrization methods cannot generally be applied for higher eigenvalues, though, because the higher modes change sign.

An alternative method yielding lower bounds for eigenvalues on triangles, which has the advantage of applying to Dirichlet, mixed, and Neumann boundary conditions and to sums of eigenvalues as well as the first eigenvalue, is the *Method of the Unknown Trial Function.* It was developed in a series of papers by Laugesen and Siudeja [633, 635, 642, 820]; the method transplants the (unknown) eigenfunctions of an arbitrary triangle to yield trial functions for the (known) eigenvalues of certain equilateral and right triangles. Transplantation distorts the derivatives in the Rayleigh quotient, but by "interpolating" the various transplantations, one can eliminate the distortions and arrive at sharp lower bounds on the eigenvalues of the arbitrary triangle.

6.4.1 Illustration of the method

As a first example, we prove the inequality $\lambda_1^S < \lambda_1^M$ from (6.3), for right triangles with smallest angle $\alpha \in [\pi/6, \pi/4)$. Equivalently, we will show that μ_2 for an obtuse isosceles triangle with aperture 2β is smaller than μ_A (the smallest Neumann eigenvalue with antisymmetric mode) for an acute isosceles triangle with aperture 2α, where $\pi/6 \leq \alpha = \pi/2 - \beta < \pi/4$. See Figure 6.5. Note that these isosceles triangles, which we denote by $T(2\beta)$ and $T(2\alpha)$ respectively, can be built from two copies of a right triangle with angles $\alpha, \beta, \pi/2$. And $\mu_A = \mu_2$ for the acute triangle $T(2\alpha)$ provided $\alpha \geq \pi/6$, by Theorem 6.40 below.

To obtain an initial upper bound on $\mu_2(T(2\beta))$, we substitute the antisymmetric trial function

$$\varphi = \sin\left(\frac{\pi x}{2\cos\beta}\right)\sin\left(\frac{\pi y}{2\sin\beta}\right)$$

(which is a stretched version of the eigenfunction of the right isosceles triangle with $\alpha = \beta = \pi/4$) into the Rayleigh quotient for $\mu_2(T(2\beta))$, getting that

$$\mu_2(T(2\beta)) \leq \frac{\pi^2 - 8}{4\cos^2\beta} + \frac{\pi^2 + 8}{4\sin^2\beta} < \frac{\pi^2}{\sin^2(2\beta)}. \tag{6.4}$$

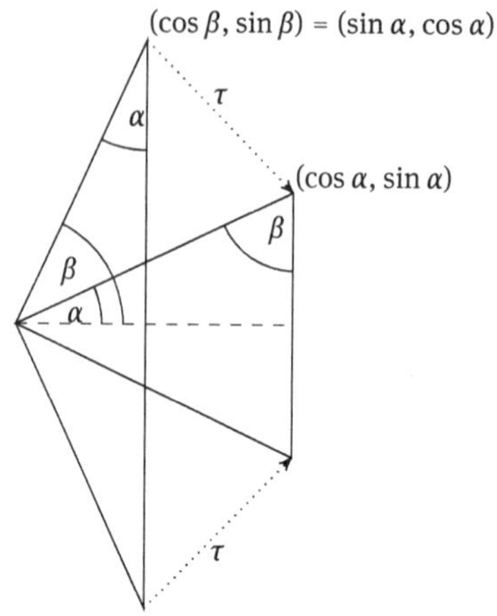

Fig. 6.5. Acute isosceles with aperture 2α and obtuse isosceles with aperture 2β, where $\alpha + \beta = \pi/2$ and $\alpha \le \pi/4 \le \beta$.

Now we are ready to compare $\mu_2\big(T(2\beta)\big)$ and $\mu_A\big(T(2\alpha)\big)$. We start with a direct transplantation, which has little chance of success and yet is enlightening. The linear transformation $\tau(x,y) = (x\cot\alpha, y\tan\alpha)$ transforms $T(2\beta)$ into $T(2\alpha)$. Suppose u is an antisymmetric eigenfunction belonging to $\mu_A\big(T(2\alpha)\big)$. Then $u \circ \tau$ has mean value zero and hence is a valid trial function for $\mu_2\big(T(2\beta)\big)$, and we get

$$\begin{aligned}\mu_2\big(T(2\beta)\big) &\le \frac{\int_{T(2\beta)} |\nabla(u\circ\tau)|^2\,dA}{\int_{T(2\beta)} (u\circ\tau)^2\,dA}\\ &= \frac{\int_{T(2\alpha)} (u_x^2\cot^2\alpha + u_y^2\tan^2\alpha)\,dA}{\int_{T(2\alpha)} u^2\,dA}\\ &\overset{??}{<} \frac{\int_{T(2\alpha)} (u_x^2 + u_y^2)\,dA}{\int_{T(2\alpha)} u^2\,dA} = \mu_A\big(T(2\alpha)\big).\end{aligned}$$

The last inequality is generally false — it holds true if and only if

$$\int_{T(2\alpha)} u_x^2\,dA < \tan^2\alpha \int_{T(2\alpha)} u_y^2\,dA \tag{6.5}$$

(where we use in the derivation that $\sin\alpha < \cos\alpha$). Unfortunately, we do not know anything about u (other than it is antisymmetric), and so we cannot check whether this last inequality holds. In general, it might not.

To obtain additional information, we seek a different domain whose eigenvalue can be estimated using the unknown eigenfunction of the acute isosceles $T(2\alpha)$. This other domain could in principle be a different triangle, or a circular sector, or something else; we will use a one-dimensional interval. Note that u is antisymmetric in the y-variable, and therefore is a valid trial function for the second Neumann eigenvalue on each vertical cross-section of $T(2\alpha)$. That eigenvalue is π^2/ℓ^2 where ℓ denotes the length of the interval, and since the vertical cross-sections of $T(2\alpha)$ have length at most $\ell = 2\sin\alpha$, we compute

$$\begin{aligned}\int_{T(2\alpha)} u_y^2\,dA &= \int_0^{\cos\alpha}\int_{-x\tan\alpha}^{x\tan\alpha} u_y^2\,dydx\\ &\geq \frac{\pi^2}{4\sin^2\alpha}\int_0^{\cos\alpha}\int_{-x\tan\alpha}^{x\tan\alpha} u^2\,dydx = \frac{\pi^2}{4\sin^2\alpha}\int_{T(2\alpha)} u^2\,dA. \qquad (6.6)\end{aligned}$$

Here we have obtained another lower bound on the integral of u_y^2, and our task is now to combine it somehow with the lower bound (6.5). Incidentally, the comparison with an interval was first used by Bañuelos and Burdzy [104, Proposition 2.3].

We are ready for the final computation. If (6.5) holds then we are done. If (6.5) fails then we use it to eliminate the u_x^2 integral from the Rayleigh quotient and then call on (6.6) to eliminate u, as follows:

$$\begin{aligned}\mu_A(T(2\alpha)) = \frac{\int_{T(2\alpha)}(u_x^2+u_y^2)\,dA}{\int_{T(2\alpha)} u^2\,dA} &\geq (\tan^2\alpha+1)\frac{\int_{T(2\alpha)} u_y^2\,dA}{\int_{T(2\alpha)} u^2\,dA}\\ &\geq (\tan^2\alpha+1)\frac{\pi^2}{4\sin^2\alpha} = \frac{\pi^2}{\sin^2(2\beta)} > \mu_2(T(2\beta)),\end{aligned}$$

where the last inequality follows from (6.4).

To summarize: we tried a direct transplantation, and it worked provided that we assumed an uncheckable condition. To handle the remaining cases we assumed the condition to be false, and again proved the required comparison by considering a different transformation. For some of the results in the section below, the method requires more than one other transformation. Two may be needed if a mixed derivative $u_x u_y$ appears in the deformed Rayleigh quotient; note the mixed derivative did not appear in the example above, because the linear transformation was diagonal.

6.4.2 Dirichlet eigenvalues

To obtain lower bounds on triangle eigenvalues we normalize the diameter D, which is simply the largest side length.

Theorem 6.16 (First, second and third eigenvalues). *Among triangular domains, the quantities $\lambda_1 D^2, \lambda_2 D^2, \lambda_3 D^2$ are all minimal for the equilateral triangle.*

The most difficult part of the theorem is the inequality for λ_2. Laugesen and Siudeja [642] prove it by writing $\lambda_2 = (\lambda_1+\lambda_2)-\lambda_1$ and then combining a known upper bound on λ_1 with a new lower bound on $\lambda_1+\lambda_2$ proved by the method of the unknown trial function. The inequality for λ_3 follows immediately from the one for λ_2, since the second and third eigenvalues agree on the equilateral triangle.

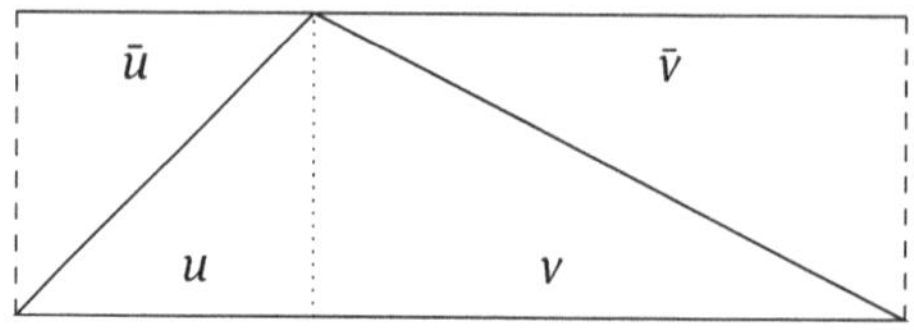

Fig. 6.6. Transplantation from a triangle to a flat cylinder, as used by Freitas.

Conceivably *every* eigenvalue might be minimal for the equilateral triangle.

Conjecture 6.17 (All eigenvalues). *The equilateral triangle minimizes each Dirichlet eigenvalue, among triangles of given diameter. That is, $\lambda_j D^2$ is minimal when the triangle is equilateral, for each $j \geq 1$.*

Note that the analogous conjecture is false for rectangles: the square is not always the minimizer, by the following explicit calculation. The rectangle with sides of length $\cos\phi$ and $\sin\phi$ has diameter $D = 1$, and eigenvalues $\lambda_1 = \pi^2(\sec^2\phi + \csc^2\phi)$ and $\lambda_2 = \pi^2(2^2\sec^2\phi + \csc^2\phi)$ when $0 < \phi \leq \pi/4$; one can show that λ_2 is minimal for some $\phi < \pi/4$, in other words, for some non-square rectangle, and similarly for the sum $\lambda_1 + \lambda_2$. Thus the preceding theorem and conjecture do not extend to rectangles.

To go beyond the third eigenvalue, and hence obtain support for Conjecture 6.17, Laugesen and Siudeja [642] examined sums of eigenvalues, or equivalently, arithmetic means of eigenvalues.

Theorem 6.18 (Sum of n eigenvalues). *Among triangles,*

$$C(\lambda_1 D^2) + \cdots + C(\lambda_n D^2) \tag{6.7}$$

is minimal for the equilateral triangle, for each $n \geq 1$ and each concave increasing function $C : \mathbb{R}_+ \to \mathbb{R}$.

The case $n = 1$ for the fundamental tone follows from the isodiametric inequality (that D^2/A is minimal among triangles for the equilateral) together with Pólya and Szegő's result that $\lambda_1 A$ is minimal for the equilateral triangle [770, Note A]. For $n = 2$ this argument fails, because $(\lambda_1 + \lambda_2)A$ is *not* minimal for the equilateral triangle. Thus the improvement from $n = 1$ to $n \geq 1$ in Theorem 6.18 is possible because we have weakened the geometric normalization from area to diameter.

An earlier manifestation of the the unknown trial function method was discovered by Freitas [397] to prove (6.1). He transplanted the unknown eigenfunction of a triangle into a rectangle with mixed Dirichlet-periodic boundary conditions (a flat cylinder).

The solid lines in Figure 6.6 indicate Dirichlet conditions while the dashed lines indicate periodic conditions on the rectangle. Finally, the dotted line splits the triangle in such a way that $\bar{u}$ and $\bar{v}$ (the rotations of $-u$ and $-v$ around the midpoints of the sloped sides) preserve Dirichlet and periodic boundary conditions on the ends of the rectangle. The resulting trial function is orthogonal to the first eigenvalue of the flat cylinder. Hence Freitas could estimate the second eigenvalue of the flat cylinder from above with the smallest Dirichlet eigenvalue of the triangle.

6.4.3 Neumann eigenvalues

The minimizing domains for the second and third Neumann eigenvalues are quite different from each other, as we shall see in the next two theorems.

Theorem 6.19 (Second eigenvalue, or optimal Poincaré inequality). $\mu_2 D^2$ *is minimal for the degenerate acute isosceles triangle, among all triangles. That is,*

$$\mu_2 D^2 > j_{1,1}^2$$

with this lower bound being attained in the limit for an acute isosceles triangle whose aperture degenerates to 0. *Hence also* $\mu_2 L^2 > 4j_{1,1}^2$.

The first step in Laugesen and Siudeja's proof [635] was to apply domain monotonicity for eigenvalues under one dimensional stretching ([635, Proof of Prop. 8.1] or [726, Lemma 6.6]). Namely:

Lemma 6.20. *Let* Ω *be a Lipschitz domain and write* $\Omega_t = \{(x, ty) : (x, y) \in \Omega\}$ *for* $t > 1$, *so that* Ω_t *is a vertically stretched copy of* Ω. *Then* $\mu_k(\Omega) \geq \mu_k(\Omega_t)$ *for each* $k \geq 1$, *and similarly for the Dirichlet eigenvalues.*

Thus one may stretch a general triangle to an isosceles triangle while reducing the eigenvalues. The second step in the proof is to apply the unknown trial function method to isosceles triangles.

Laugesen *et al.* [633] treated the third eigenvalue by a similar technique.

Theorem 6.21 (Third eigenvalue). $\mu_3 D^2$ *is minimal for the equilateral triangle, among all triangles.*

The fourth eigenvalue is *not* minimal for the equilateral, as seen in the numerical work in Figure 6.17.

Sums of Neumann eigenvalues behave like the third eigenvalue rather than the second, and are known to be minimal for the equilateral [633], as the next theorem explains.

Theorem 6.22 (Sum of n eigenvalues). *Among triangles,*

$$\mu_2 D^2 + \cdots + \mu_n D^2 \tag{6.8}$$

is minimal for the equilateral triangle, for each $n \geq 3$.

The theorem is fully rigorous except when $n = 4, 5, 7, 8, 9$. For those values of n, the proof relies on numerical estimation of the eigenvalues $\mu_2, \ldots, \mu_9$ for one specific isosceles triangle. Note one cannot apply majorization to Theorem 6.22 to get inequalities for $C(\mu_2 D^2) + \cdots + C(\mu_n D^2)$, because (6.8) does not hold for $n = 2$.

Rectangles behave differently, and the square does not always minimize the sum of the first n Neumann eigenvalues under diameter normalization. For example, by plotting the first 12 eigenvalues one finds that the square fails to minimize $(\mu_2 + \cdots + \mu_n)D^2$ when $n = 5, 6, 7, 9, 10, 12$.

6.5 Lower bounds by other methods

6.5.1 Spectral gap for triangles

Antunes and Freitas [40] conjectured that the diameter-normalized spectral gap

$$(\lambda_2 - \lambda_1)D^2$$

is minimal for the equilateral among all triangles. This conjecture was a little surprising, since the spectral gap for general convex domains is minimized by the degenerate rectangle (and not the disk or square), by work of Andrews and Clutterbuck [36].

Lu and Rowlett [680] have given a computer-assisted proof of the gap conjecture for triangles, relying in part on numerical evaluation of the gap functional for a large number of triangles.

Open problem 6.23 (Lower bound on the spectral gap). *Find a purely analytical proof showing that $(\lambda_2 - \lambda_1)D^2$ is minimal for the equilateral triangle, among all triangles.*

6.5.2 High eigenvalues for rectangles

The first Dirichlet eigenvalue λ_1 is minimal for the square among all rectangles of given area, and indeed among all quadrilaterals of given area, by Pólya and Szegö's analogue of the Faber–Krahn result in Theorem 6.2. The square is not the minimizer for the normalized second eigenvalue $\lambda_2 A$, which is minimal instead for the rectangle with side ratio $\sqrt{2} : 1$, by direct computation. (One could hardly expect the minimizer to be square, since among general domains the minimizer is a not a disk but rather a union of two disjoint disks, as discussed in Chapters 2 and 7.)

For high eigenvalues, Antunes and Freitas [43] proved a lovely analogue of the Faber–Krahn result: they showed that the shape of the rectangle minimizing $\lambda_j A$ converges to the square as $j \to \infty$. More precisely, they show that if the rectangle

$(0, a_j) \times (0, b_j)$ is chosen to minimize $\lambda_j A$ among rectangles, then $a_j/b_j \to 1$ as $j \to \infty$. Their tools include a non-sharp two-term Weyl lower bound on λ_j for rectangles, used to prove boundedness of the sidelength ratio, and a sharp two-term Weyl asymptotic formula whose remainder term is controlled uniformly on each family of ellipses having bounded semi-axis ratio. The corresponding result for Neumann eigenvalues is due to van den Berg and Gittins [861]: the shape of the rectangle maximizing $\mu_j A$ will converge to the square as $j \to \infty$.

For general domains, numerical evidence in Chapter 11 suggests that the domain minimizing $\lambda_j A$ might approach a disk as $j \to \infty$. Tools for proving such a conjecture are lacking so far.

6.6 Application to sharp Poincaré inequality and rigorous numerics

Theorem 6.19 provides a sharp lower bound on the second Neumann eigenvalue of an arbitrary triangle in terms of its diameter, which can be interpreted as an optimal Poincaré inequality:

$$\int_T u^2\, dA < \left(\frac{D^2}{j_{1,1}^2}\right) \int_T |\nabla u|^2\, dA$$

for all $u \in H^1(T)$ having mean value zero over the triangle T. This inequality is the best possible bound on the L^2 norm of a mean-zero function on an arbitrary triangle in terms of its Sobolev seminorm. Hence, one obtains the optimal Sobolev embedding inequality for triangular domains.

The optimal Poincaré inequality on triangles is significant for finite element methods, as we now explain. Recent work by Armentano–Duran [52], Liu–Oishi [676] and Carstensen–Gedicke [252] has led to a method of rigorously proving lower bounds for Dirichlet eigenvalues through finite element methods (as opposed to the traditional use of finite elements to obtain numerical approximations). Nigam, Siudeja and Young [726] noticed that the same approach applies to mixed Dirichlet–Neumann eigenvalues, although not to the purely Neumann case.

Suppose Ω is a polygonal domain triangulated into a mesh with longest edge H. Let λ_1^{CR} be the exact eigenvalue of the matrix M resulting from the nonconforming Crouzeix–Raviart approximation scheme [306]. The following result due to Carstensen and Gedicke [252] provides an explicit lower bound for the first Laplace eigenvalue, a bound which converges to the exact value when the mesh parameter H tends to 0.

Theorem 6.24 (Lower bound via numerical approximation). *Let λ_1 be the smallest Dirichlet eigenvalue for the polygonal domain Ω. Then*

$$\lambda_1 \geq \frac{\lambda_1^{CR}}{1 + \kappa^2 H^2 \lambda_1^{CR}},$$

where $\kappa^2 \approx 0.1931$ is a constant obtained from the optimal Poincaré inequality, Theorem 6.19.

The theorem was extended to the mixed Dirichlet–Neumann case by Siudeja, Nigam and Young [726, 815].

Note that λ_1^{CR} is the smallest eigenvalue of a matrix that is usually extremely large, and one cannot hope to evaluate it exactly. Nevertheless, since $f(x) = x/(1 + cx)$ is an increasing function, any lower bound $\lambda_1^{CR} \geq \Lambda$ leads to

$$\lambda_1 \geq \frac{\Lambda}{1 + \kappa^2 H^2 \Lambda} = \Lambda_H.$$

Or in spectral terms, if $M - \Lambda$ is a positive operator (where M is the Crouzeix-Raviart approximation matrix) then $-\Delta - \Lambda_H$ is positive too.

It is rather complicated to find a lower bound for the smallest eigenvalue of a matrix, as is needed for the method above. Many accurate approximation schemes exist for matrix eigenvalues, and yet here one needs a guaranteed lower bound Λ for λ_1^{CR}. The extreme sparsity of the matrices involved allows one to use a mix of Sturm sequences and validated numerical approximations to obtain a lower bound Λ, at least for matrices having few thousand rows and columns. Another way to obtain Λ is to first produce an approximation, then subtract it from the diagonal of the matrix and show that the resulting matrix is positive definite. This last step can be done quickly using sparse LDL' or LU decompositions and interval arithmetic. Both approaches were employed by Nigam, Siudeja and Young [726] to find bounds on mixed eigenvalues of some polygonal domains. Their ultimate goal in that paper was to rigorously study the position of the nodal curve for the second eigenfunction on a right triangle; for more on nodal curves, see Section 6.13.

We should point out that complementary upper bounds are easy to obtain, since the classical conforming finite element approximations provide continuous and piecewise smooth trial functions, which can be substituted into the Rayleigh quotient to yield rigorous upper bounds for the smallest eigenvalue. For example, Bucur and Fragalà [214] followed this approach to get upper bounds on the smallest eigenvalue of a family of octagons. In earlier work, Jakobson *et al.* [556] used both conforming and nonconforming methods to approximate the eigenvalue of a certain domain from below and above, but their approach was not validated, and hence was not rigorous.

6.7 Upper bounds: trial functions

6.7.1 Dirichlet eigenvalues

Upper bounds on Dirichlet eigenvalues will obviously require more than just area normalization, since (for example) λ_1 can be made arbitrarily large for a triangle with area 1 just by taking a sufficiently long, thin triangle. That is why the normalizing factors in this section involve the sidelengths of the triangle: sidelengths penalize long thin domains.

In particular we consider $\lambda_1 A^2/S^2$, which can be written as a product of two scale invariant factors:

$$\lambda_1 A \cdot \frac{A}{S^2}.$$

The first factor, $\lambda_1 A$, can be thought of as the "Faber–Krahn" term (although the Faber–Krahn theorem says this expression is minimal for the disk, not maximal). The second factor, A/S^2, is purely geometric and measures the "deviation from roundness" of the domain. This factor is small when the domain is elongated, and hence it balances the largeness of the first factor on such domains.

Theorem 6.25 (First eigenvalue). *$\lambda_1 A^2/L^2$ and $\lambda_1 A^2/S^2$ are maximal for the equilateral triangle, among all triangles. Further, $\lambda_1 A^2/S^2$ is maximal for rectangles, among all quadrilaterals.*

Among the class of N-gons ($N \geq 3$) having an inscribed circle, $\lambda_1 A^2/L^2$ is maximal for the regular N-gon.

The results on $\lambda_1 A^2/L^2$ are equivalent to results on $\lambda_1 R^2$ where R is the inradius, since $A = \frac{1}{2}LR$ for triangles and other N-gons that have an inscribed circle.

The bound on $\lambda_1 A^2/L^2$ for triangles in Theorem 6.25 (which is due to Solynin [825, 828], see also Siudeja [817]) is stronger than the bound on $\lambda_1 A^2/S^2$, since L/S is maximal for the equilateral by Lemma 6.1. That result on $\lambda_1 A^2/S^2$ for triangles was found by Pólya [772], [769, p. 308,328] and later rediscovered with a different proof by Freitas [397]. The analogous inequality for parallelograms was found by Pólya and later rediscovered for the special case of rhombi by Hooker and Protter [539, §5] and for parallelograms by Hersch [517, formula (5)], again with different proofs, and then extended to all quadrilaterals by Freitas and Siudeja [401]. The final result of the theorem, about N-gons with an inscribed circle, is due to Solynin [825, 828]; note in particular that each triangle ($N = 3$) has an inscribed circle.

Sketch of Solynin's method in Theorem 6.25. The sharp upper bound on $\lambda_1 R^2$ for triangles, in Theorem 6.25, was given a beautiful geometric proof by Solynin [825] using Dubinin's dissymmetrization technique. Let us sketch the main idea. Suppose u is the eigenfunction on the equilateral triangle with inradius 1. This function is symmetric with respect to any isometry of the equilateral triangle. One can cut u into pieces and

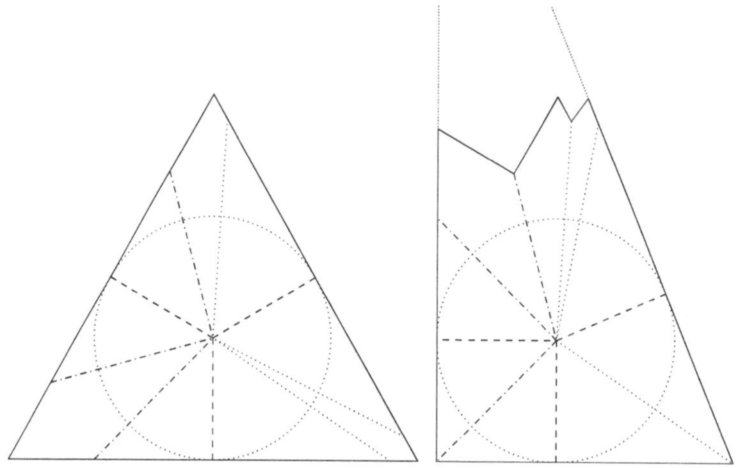

Fig. 6.7. Dissymetrization as applied by Solynin to the equilateral triangle.

glue them together in a different order, provided one makes sure continuity is preserved by ensuring cut lines are mapped to each other under appropriate isometries.

As shown in Figure 6.7, all the pieces from the equilateral triangle are moved inside a new triangle to create a trial function there, with some parts of the new triangle uncovered. Extend the trial function to equal 0 in the uncovered regions, and then substitute the trial function into the Rayleigh Principle to estimate the first eigenvalue on the new triangle. □

One can extend Pólya's result in the previous theorem to eigenvalue sums (or arithmetic means) of arbitrary length, by Laugesen and Siudeja's "Method of Rotations and Tight Frames" [636, Corollary 3.2]:

Theorem 6.26 (Sum of n eigenvalues). *Among triangles,*

$$C(\lambda_1 A^2/S^2) + \cdots + C(\lambda_n A^2/S^2) \tag{6.9}$$

is maximal for the equilateral triangle, while among parallelograms it is maximal for the square, for each $n \geq 1$ and each concave increasing function $C : \mathbb{R}_+ \to \mathbb{R}$.

Further, among all images of the regular N-gon under linear transformations, a functional analogous to (6.9) (except involving moment of inertia instead of the sum of squares of sidelengths) is maximized by the regular N-gon. In the limiting case as $N \to \infty$ one gets maximality of the disk among ellipses. Further, the regular N-gon can be replaced by any domain having N-fold rotational rotational symmetry [636]. Note that tight frames and averaging have also been used in a slightly different context by El Soufi, Harrell, Ilias and Joachim [360].

To give the flavor of all these results, we prove Theorem 6.26 for triangles.

Proof of Theorem 6.26 for triangles. Every triangle can be written (after translation) as the image under a linear transformation T of an equilateral triangle E centered at the origin. The idea of the proof is to construct trial functions on the triangle $T(E)$ by linearly transplanting eigenfunctions of E, and then to average with respect to the rotations of the equilateral triangle.

Let $u_1, u_2, u_3, \dots$ be orthonormal Dirichlet eigenfunctions on E corresponding to the eigenvalues $\lambda_1(E), \lambda_2(E), \lambda_3(E), \dots$. Fix an orthogonal 2×2 matrix U that preserves E. (Later we choose U to represent rotation by a multiple of $2\pi/3$.) Define trial functions

$$v_j = \frac{1}{|\det T|^{1/2}} u_j \circ U \circ T^{-1}$$

on the triangle $T(E)$, noting that $v_j = 0$ on the boundary of the triangle because u_j vanishes on ∂E. Note the functions v_j are orthonormal, since the change of variable formula gives

$$\int_{T(E)} v_j v_k \, dA = \int_E u_j u_k \, dA = \delta_{jk}.$$

Applying the Rayleigh–Poincaré principle (**??**) to the trial functions $v_1, \dots, v_n$, we find

$$\sum_{j=1}^n \lambda_j(T(E)) \le \sum_{j=1}^n \int_{T(E)} |\nabla v_j|^2 \, dA = \sum_{j=1}^n \int_E |(\nabla u_j) U T^{-1}|^2 \, dA, \tag{6.10}$$

where the gradient ∇u_j is regarded as a row vector.

We may choose $U = U_m$, the matrix representing rotation by angle $2\pi m/3$, for $m = 1, 2, 3$. Obviously this matrix fixes the equilateral triangle. By averaging (6.10) over these three rotations we find

$$\begin{aligned}
\sum_{j=1}^n \lambda_j(T(E)) &\le \sum_{j=1}^n \int_E \Big\{ \frac{1}{3} \sum_{m=1}^3 |(\nabla u_j) U_m T^{-1}|^2 \Big\} \, dA \\
&= \sum_{j=1}^n \int_E \Big\{ \frac{1}{2} |\nabla u_j|^2 \|T^{-1}\|_2^2 \Big\} \, dA \qquad \text{by the Tight Frame Lemma 6.27 below} \\
&= \frac{1}{2} \|T^{-1}\|_2^2 \sum_{j=1}^n \lambda_j(E).
\end{aligned}$$

The quantity $\|T^{-1}\|_2^2$ can be substituted in terms of A^2/S^2 by Lemma 6.28 below, and doing so completes the proof of the theorem when the concave function C is the identity function. The general case then follows from majorization as in Section 6.2. □

Tight frames.

A Plancherel formula holds for any orthonormal system of vectors, but also for certain non-orthonormal systems known as *tight frames*. In the preceding proof we needed a special "Mercedes–Benz" tight frame in the plane, which we now develop.

For each nonzero vector $y \in \mathbb{R}^2$, the rotation matrices U_1, U_2, U_3 generate a rotationally symmetric system $\{U_1y, U_2y, U_3y\}$. The tight frame identity says that the inner products against these three vectors recapture the norm of an arbitrary column vector $x \in \mathbb{R}^2$, according to the formula

$$\sum_{m=1}^{3} |x \cdot (U_m y)|^2 = \frac{3}{2}|x|^2|y|^2. \tag{6.11}$$

To prove this identity, one may suppose x and y have length 1 and lie at angles θ and ϕ to the positive horizontal axis, respectively. Then by trigonometric identities and the geometric sum we find

$$\begin{aligned}\sum_{m=1}^{3} |x \cdot (U_m y)|^2 &= \sum_{m=1}^{3} \cos^2(\theta - \phi - 2\pi m/3) \\ &= \frac{1}{2}\sum_{m=1}^{3} \big(1 + \cos 2(\theta - \phi - 2\pi m/3)\big) \\ &= \frac{3}{2} + \frac{1}{2}\operatorname{Re}\left(e^{i2(\theta-\phi)} \sum_{m=1}^{3} (e^{-i4\pi/3})^m\right) = \frac{3}{2}.\end{aligned}$$

Figure 6.8 illustrates the last formula: it shows the projection formula $\sum_{m=1}^{3}(x \cdot U_m y)U_m y = \frac{3}{2}x$ for a typical $x \in \mathbb{R}^2$, where $y = \binom{0}{1}$ is the vertical unit vector and U_m denotes rotation by $2\pi m/3$. Taking the dot product of x with the projection formula yields the Plancherel-type identity in (6.11).

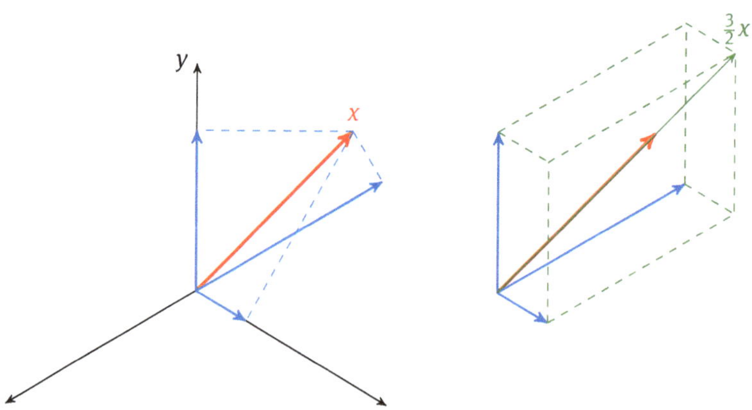

Fig. 6.8. The “Mercedes–Benz” tight frame in the plane, as stated in (6.11).

We need a tight frame identity in which the vector y is generalized to a matrix. Recall that the Hilbert–Schmidt matrix norm of a matrix M is defined by $\|M\|_2 = (\sum_{j,\ell} M_{k\ell}^2)^{1/2}$.

Lemma 6.27 (Tight frame, or Plancherel formula).

$$\frac{1}{3}\sum_{m=1}^{3} |xU_m Y|^2 = \frac{1}{2}|x|^2\, \|Y\|_2^2$$

for all row vectors $x \in \mathbb{R}^2$ and $2 \times \ell$ real matrices Y, where $\ell \geq 1$ is arbitrary.

Proof of Lemma 6.27. Write $y_1, \ldots, y_\ell \in \mathbb{R}^2$ for the column vectors of Y. Then $|xU_m Y|^2 = \sum_{k=1}^{\ell} |xU_m y_k|^2$, and we simply apply (6.11). □

Let us further apply the lemma to express the Hilbert–Schmidt norm of the inverse in terms of A/S.

Lemma 6.28. *For every invertible 2×2 matrix T and equilateral triangle E centered at the origin, one has*

$$\frac{1}{\sqrt{2}}\|T^{-1}\|_2 = \frac{A/S|_E}{A/S|_{T(E)}}.$$

Proof of Lemma 6.28. The explicit formula for the inverse of a 2×2 matrix implies that $\|T^{-1}\|_2 = \|T\|_2/|\det T|$, and of course $|\det T| = A(T(E))/A(E)$. Thus it remains to show that

$$\frac{1}{\sqrt{2}}\|T\|_2 = \frac{S(T(E))}{S(E)}.$$

Write the vertices of E as $(xU_m)^t$ for $m = 1, 2, 3$, where x is some row vector and "t" denotes the transpose. By counting each sidelength twice, we have

$$\begin{aligned} S(T(E))^2 &= \frac{1}{2}\sum_{\ell,m=1}^{3} |T(xU_\ell)^t - T(xU_m)^t|^2 \\ &= 3\sum_{\ell=1}^{3} |xU_\ell T^t|^2 - \sum_{\ell,m=1}^{3} (xU_\ell T^t)\cdot(xU_m T^t) \\ &= \frac{9}{2}|x|^2\, \|T^t\|_2^2 \end{aligned}$$

by Lemma 6.27 and using that $\sum_{m=1}^{3} U_m$ equals the zero matrix. Lastly, T^t and T have the same Hilbert–Schmidt norm, while each side of E has length $\sqrt{3}|x|$ and so $9|x|^2 = S(E)^2$. The lemma now follows. □

More information on tight frames and eigenvalue inequalities can be found in Laugesen's paper [632]. More recently, Siudeja has investigated how higher order tight frame identities can yield spectral inequalities for higher order operators such as the bi-Laplacian [816].

Higher dimensions

In higher dimensions one obtains results analogous to Theorem 6.26 for which the maximizing domains are regular tetrahedra, cubes, and other Platonic solids[637].

Second eigenvalue and spectral gap

Turning now to the spectral gap and second eigenvalue, we state upper sharp bounds by Siudeja [817, 818].

Theorem 6.29 (Second eigenvalue and spectral gap). *$\lambda_2 A^2/L^2$ and $(\lambda_2-\lambda_1)A^2/L^2$ are maximal for the equilateral, among all triangles. Hence $\lambda_2 A^2/S^2$ and $(\lambda_2-\lambda_1)A^2/S^2$ are maximal for the equilateral, among all triangles.*

Siudeja stated his results with an inradius normalization, but for triangles the inradius is proportional to our factor A/L.

The PPW conjecture (proved by Ashbaugh and Benguria) says that the ratio of the first two Dirichlet eigenvalues is maximal for the disk. Siudeja proved an analogous statement for acute triangles [818].

Theorem 6.30 (Ratio of first two eigenvalues). *λ_2/λ_1 is maximal among acute triangles for the equilateral triangle.*

The result should hold for all triangles.

Conjecture 6.31 (Ratio of first two eigenvalues). *λ_2/λ_1 is maximal among all triangles for the equilateral triangle.*

6.7.2 Neumann eigenvalues

We begin with a bound on the second eigenvalue under diameter normalization, due to Cheng [267, Theorem 2.1]. It holds for all domains and achieves equality in the limiting case of a degenerate triangle.

Theorem 6.32 (Second eigenvalue). *$\mu_2 D^2 < 4j_{0,1}^2$ for all convex planar domains. This estimate saturates for degenerate obtuse isosceles triangles (and thin rhombi).*

The idea of the proof is to construct a trial function using two copies of the Bessel function J_0, centered at points distance D apart and rescaled to meet halfway, and with one of the functions multiplied by -1. For the saturation claim, see Proposition 6.42; this limiting case has been known for some time [104, p. 10].

Under area and perimeter normalizations (rather than diameter normalization) we have upper bounds found by Laugesen and Siudeja [634] using a variety of explicit trial functions.

Theorem 6.33 (Means of second and third eigenvalues). *Each of the following scale-invariant functionals is maximized in the class of triangles by the equilateral triangle.*

i: $\mu_2 S^2, \mu_2 L^2, \mu_2 A$
ii: $H(\mu_2, \mu_3)A$
iii: $G(\mu_2, \mu_3)A^{3/2}/S$
iv: $M(\mu_2, \mu_3)A^2/S^2$

The result on $\mu_2 S^2$ implies the results on $\mu_2 L^2$ and $\mu_2 A$, thanks to the inequalities relating S^2, L^2, A in Lemma 6.1. Thus part (i) improves for triangles on Szegö and Weinberger's upper bound saying that $\mu_2 A$ is maximal for the disk among general domains. A further improvement for triangles follows from including an "isoperimetric excess" term [634, Theorem 3.3]. Such improvements are in the spirit of the stability results in Chapter 7.

The second and fourth parts of the theorem are not comparable with each other, since the harmonic mean in part (ii) is smaller than the arithmetic mean in part (iv), while the geometric factor A in part (ii) is "larger" than the geometric functional A^2/S^2 in part (iv) (meaning that S^2/A is minimal for the equilateral). In other words, in those parts of the theorem we trade off the strength of the eigenvalue functional against the strength of the geometric functional.

Part (iii) for the geometric mean follows simply from multiplying parts (ii) and (iv) and taking the square root.

Conjecture 6.34 (Means of second and third eigenvalues). $H(\mu_2, \mu_3)L^2$ *and* $G(\mu_2, \mu_3)A$ *are maximal for the equilateral triangle, among all triangles.*

This conjecture would improve on Theorem 6.33 parts (ii) and (iii), since A/L^2 and A/S^2 are maximal for the equilateral, by Lemma 6.1. A partial result for the geometric mean (also applicable to polygons with arbitrary number of sides) was recently proved by Enache and Philippin [364].

Part (iv) of the theorem extends to arithmetic means of arbitrary length, and so together with majorization we obtain the following result:

Theorem 6.35 (Sum of n eigenvalues). *Among triangles,*

$$C(\mu_2 A^2/S^2) + \cdots + C(\mu_n A^2/S^2)$$

is maximal for the equilateral triangle, while among parallelograms it is maximal for the square, for each $n \geq 2$ and each concave increasing function $C : \mathbb{R}_+ \to \mathbb{R}$.

Indeed, this theorem extends to not just triangles but linear images of regular N-gons [636, Corollary 3.2], in the same manner as for the Dirichlet result Theorem 6.26. Robin eigenvalues can be treated similarly [636, Corollary 3.4], as can eigenvalues of the magnetic Laplacian [638].

For arbitrary polygons, Laugesen (in [84]) raised an analogue of the Szegő–Weinberger result that the ball maximizes $\mu_2 A$ among general domains.

Conjecture 6.36 (Polygonal Szegő–Weinberger problem for the second eigenvalue). *Among N-gons, the regular N-gon maximizes $\mu_2 A$. In particular, the square maximizes $\mu_2 A$ among quadrilaterals.*

The only case proved so far seems to be that of triangles ($N = 3$), in Theorem 6.33, but Antunes and Henrot [45] provided extensive numerical support for the conjecture.

6.8 Rectangles

Rectangles enjoy simple formulas for their eigenvalues and eigenfunctions, and so make the best case study for many general problems. On the other hand, certain problems are difficult even on rectangles; for example, see Section 6.13.

Let a and b denote the lengths of the sides of a rectangle. Separation of variables leads to the eigenvalue formula

$$\mu_{m,n} = \lambda_{m,n} = \pi^2 \left(\frac{m^2}{a^2} + \frac{n^2}{b^2} \right) . \tag{6.12}$$

The only difference between the Neumann and Dirichlet cases are the ranges of the parameters (m, n). For Dirichlet one takes $m, n \geq 1$, while for Neumann $m, n \geq 0$. The eigenfunctions have a common form: $\varphi(\pi m x/a)\varphi(\pi n y/b)$ with $\varphi = \sin$ for Dirichlet and $\varphi = \cos$ for Neumann.

The eigenvalues are simple when the sidelength-squared ratio $(a/b)^2$ is irrational. No other easy examples are known with purely simple eigenvalues, although multiplicities seem unlikely for generic polygons [525–527].

When $(a/b)^2$ is rational, one can have eigenvalues with arbitrarily high multiplicities. It can be difficult to find the exact multiplicity of higher eigenvalues. Indeed, (6.12) shows this problem is equivalent to counting lattice points on an ellipse with semiaxes proportional to a and b.

The simplest case of a multiple eigenvalue on a non-square rectangle occurs when $a/b = \sqrt{8/3}$. In that case the modes $(1, 3)$ and $(2, 1)$ belong to the same eigenvalue, leading to interesting nodal patterns for the eigenfunctions as in Figure 6.9, which otherwise would simply be straight lines (horizontal and/or vertical).

Even more interesting patterns can arise for a square ($a = b = 1$), as illustrated in Figure 6.10. All modes have multiplicity except perhaps those ones with $m = n$, and the

Fig. 6.9. Nodal patterns for the smallest multiple eigenvalue on a rectangle with $a/b = \sqrt{8/3}$.

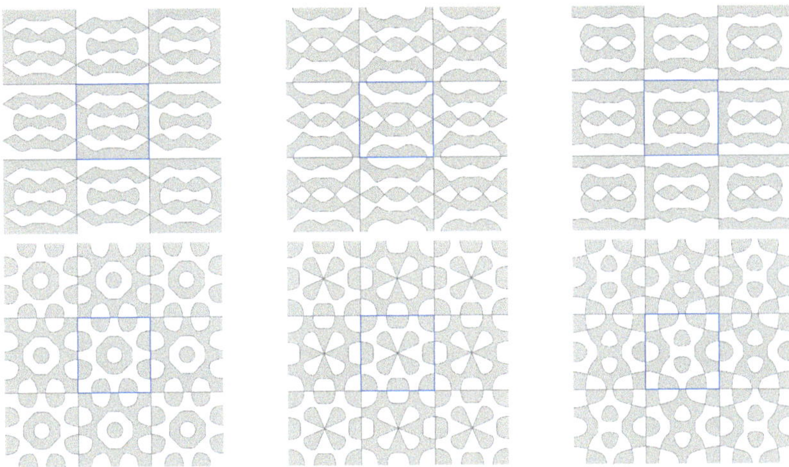

Fig. 6.10. Various nodal patterns belonging to the 31st Dirichlet eigenvalue of a square. This eigenvalue has multiplicity three, with pure modes (1, 7), (7, 1) and (5, 5). Different linear combinations of these modes generate rather diverse nodal domains, including the ones shown here.

multiplicity can arise in interesting ways, such as for $(0, 5)$ and $(3, 4)$ in the Neumann case, or $(1, 7)$ and $(5, 5)$ in the Dirichlet case (which in fact gives triple multiplicity). In Section 6.13 we describe nodal problems that have been solved recently in a few special cases, including squares.

6.9 Equilateral triangles

Explicit formulas for the equilateral triangle are developed in this section, thus allowing the qualitative max/min theorems in this chapter for triangles to be rephrased as computable estimates. For example, the maximality of $\mu_2 A$ for the equilateral triangle (as stated in Theorem 6.33) is equivalent to saying

$$\mu_2 A \le \frac{4\pi^2}{3\sqrt{3}},$$

as one sees by evaluating $\mu_2 A$ for the equilateral using the values stated below.

The modes and frequencies of the equilateral triangle were derived roughly 150 years ago by Lamé. For a modern treatment of the Dirichlet case, see Mathews and Walker's text [695, pp. 237–239], and for a comprehensive treatment under all boundary conditions see McCartin's monograph [701] or his earlier papers upon which the monograph is based.

Consider now the equilateral triangle with vertices at $(0,0)$, $(\sqrt{3}/2, 1/2)$ and $(\sqrt{3}/2, -1/2)$. This triangle has sidelength 1, and

$$A = \sqrt{3}/4, \qquad D = 1, \qquad L = 3, \qquad S^2 = 3.$$

We state below the Dirichlet and Neumann spectra for this triangle.

6.9.1 Dirichlet spectrum of equilateral triangle with unit sidelength

The first three eigenvalues are

$$\lambda_1 = \frac{3 \cdot 16\pi^2}{9}, \qquad \lambda_2 = \lambda_3 = \frac{7 \cdot 16\pi^2}{9}$$

with eigenfunctions

$$u_1(x,y) = \sin\left(\frac{4}{\sqrt{3}}\pi x\right) - 2\cos(2\pi y)\sin\left(\frac{2}{\sqrt{3}}\pi x\right) \tag{6.13}$$

$$u_2(x,y) = \sin\left(\frac{6}{\sqrt{3}}\pi x\right)\cos\left(\frac{2}{3}\pi y\right) - \sin\left(\frac{4}{\sqrt{3}}\pi x\right)\cos\left(\frac{8}{3}\pi y\right) - \sin\left(\frac{2}{\sqrt{3}}\pi x\right)\cos\left(\frac{10}{3}\pi y\right) \tag{6.14}$$

$$u_3(x,y) = \sin\left(\frac{6}{\sqrt{3}}\pi x\right)\sin\left(\frac{2}{3}\pi y\right) - \sin\left(\frac{4}{\sqrt{3}}\pi x\right)\sin\left(\frac{8}{3}\pi y\right) + \sin\left(\frac{2}{\sqrt{3}}\pi x\right)\sin\left(\frac{10}{3}\pi y\right) \tag{6.15}$$

Figure 6.11 displays the nodal patterns of these first three Dirichlet eigenfunctions of the equilateral triangle, extended by odd reflection across the sides of the triangle.

The higher eigenfunctions are more complicated, although the corresponding eigenvalues retain a simple form:

$$\frac{16\pi^2}{9}(j_1^2 + j_1 j_2 + j_2^2), \qquad j_1, j_2 \geq 1. \tag{6.16}$$

Choosing $(j_1, j_2) = (1,1)$ gives the first eigenvalue $48\pi^2/9$ stated above. Note that modes $(j_1, j_2) = (5,6)$ and $(j_1, j_2) = (1,9)$ belong to the same eigenvalue (see Figure 6.11 for some linear combinations).

6.9.2 Neumann spectrum of equilateral triangle with unit sidelength

We have $\mu_1 = 0$, with eigenfunction $u_0 \equiv 1$, and

$$\mu_2 = \mu_3 = \frac{16\pi^2}{9}$$

Fig. 6.11. *Top left:* nodal domain for the first Dirichlet eigenfunction of an equilateral triangle (the central triangle), extended by odd reflection across the boundaries. The shading indicates the sign of the eigenfunction. *Middle left:* nodal pattern of a second Dirichlet eigenfunction (6.14) of the same triangle, extended by odd reflection. *Middle right:* nodal pattern of a third Dirichlet eigenfunction (6.15), extended by odd reflection. The third eigenvalue equals the second, in fact, but the two nodal patterns are different: one eigenfunction is symmetric about the horizontal bisector of the triangle, while the other is antisymmetric. *Top right:* nodal pattern for a generic second eigenfunction obtained from a linear combination of (6.14) and (6.15) . *Bottom left and right:* generic higher eigenfunctions, obtained from linear combinations of modes $(j_1, j_2) = (1, 9)$ and $(5, 6)$ (which have the same eigenvalue, according to (6.16)).

with eigenfunctions

$$u_2(x, y) = 2\cos\left(\frac{2}{\sqrt{3}}\pi x\right)\cos\left(\frac{2}{3}\pi y\right) + \cos\left(\frac{4}{3}\pi y\right) \tag{6.17}$$

$$u_3(x, y) = 2\cos\left(\frac{2}{\sqrt{3}}\pi x\right)\sin\left(\frac{2}{3}\pi y\right) - \sin\left(\frac{4}{3}\pi y\right) \tag{6.18}$$

Fig. 6.12. *Top left:* nodal domain for the (constant) first Neumann eigenfunction of an equilateral triangle, extended by even reflection across the boundaries. *Middle left:* nodal pattern of the second Neumann eigenfunction (6.17) of the same triangle, extended by even reflection. *Middle right:* nodal pattern of the third Neumann eigenfunction (6.18), extended by even reflection. The second and third eigenvalues are equal, but the nodal patterns are different: the second eigenfunction is symmetric about the horizontal bisector of the triangle, while the third eigenfunction is antisymmetric. *Top right:* nodal pattern for a generic second eigenfunction obtained from a linear combination of (6.17) and (6.18) . *Bottom left and right:* generic higher eigenfunctions, obtained from linear combinations of modes $(j_1, j_2) = (0, 7)$ and $(3, 5)$ (which have the same eigenvalue, according to (6.19)).

Clearly u_2 is symmetric about the x-axis, whereas u_3 is antisymmetric about that axis. See Figure 6.12.

Again, the higher eigenfunctions are more complicated, yet the eigenvalues have the same form as in the Dirichlet case:

$$\frac{16\pi^2}{9}(j_1^2 + j_1 j_2 + j_2^2), \qquad j_1, j_2 \geq 0. \tag{6.19}$$

Choosing $(j_1, j_2) = (1, 0)$ gives the first eigenvalue $16\pi^2/9$ stated above. Note that in the Neumann case, modes $(j_1, j_2) = (3, 5)$ and $(j_1, j_2) = (0, 7)$ belong to the same eigenvalue.

6.10 Isosceles triangles

Consider an isosceles triangle $T(\alpha)$ having aperture $\alpha \in (0, \pi)$ and equal sides of length l, oriented so that the vertex sits at the origin and the triangle is symmetric about the positive x-axis. That is,

$$T(\alpha) = \{(x, y) : 0 < x < l\cos(\alpha/2), |y| < x\tan(\alpha/2)\}.$$

Definition 6.37 (Subequilateral and superequilateral). Call the isosceles triangle *subequilateral* it has aperture $\alpha \in (0, \pi/3)$, and *superequilateral* if $\alpha \in (\pi/3, \pi)$. Examples are shown in Figure 6.13.

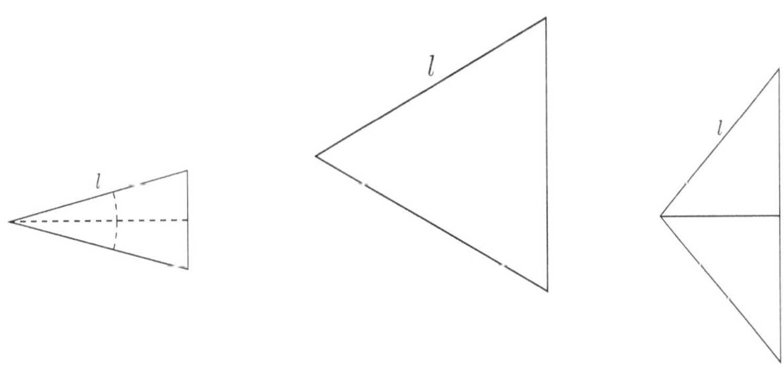

(a) subequilateral triangle **(b)** equilateral triangle **(c)** superequilateral triangle

Fig. 6.13. Plausible nodal curves (solid) for second Dirichlet eigenfunctions of isosceles triangles: see Conjecture 6.39. The solid curves indicate where the eigenfunction vanishes, and the dashing indicates a line of symmetry. The equilateral triangle has a double second eigenvalue, with both patterns present.

6.10.1 Dirichlet eigenvalues

Write $\lambda_1(\alpha)$, $\lambda_A(\alpha)$ and $\lambda_S(\alpha)$ for the smallest eigenvalue of $T(\alpha)$, the smallest eigenvalue with an eigenfunction antisymmetric in the horizontal axis, and the smallest eigenvalue greater than λ_1 with an eigenfunction symmetric in the horizontal axis, respectively. These eigenvalues are plotted numerically in Figure 6.14a, normalized by the square of the sidelength. Figure 6.14c plots them again, this time normalized by the square of the diameter; notice the corners appearing at $\alpha = \pi/3$, due to the diameter switching from l (the length of the two equal sides) to $2l\sin(\alpha/2)$ (the length of the third side) as the aperture passes through $\pi/3$. Figure 6.14b normalizes with perimeter, and Figure 6.14d with area.

The figures are due to Laugesen and Siudeja [642] with eigenvalues computed numerically by the PDE Toolbox in Matlab, using a finite element mesh of about 1 million triangles. To ensure good precision the figures avoid degenerate cases, restricting to apertures in the range $\pi/6 < \alpha < 2\pi/3$.

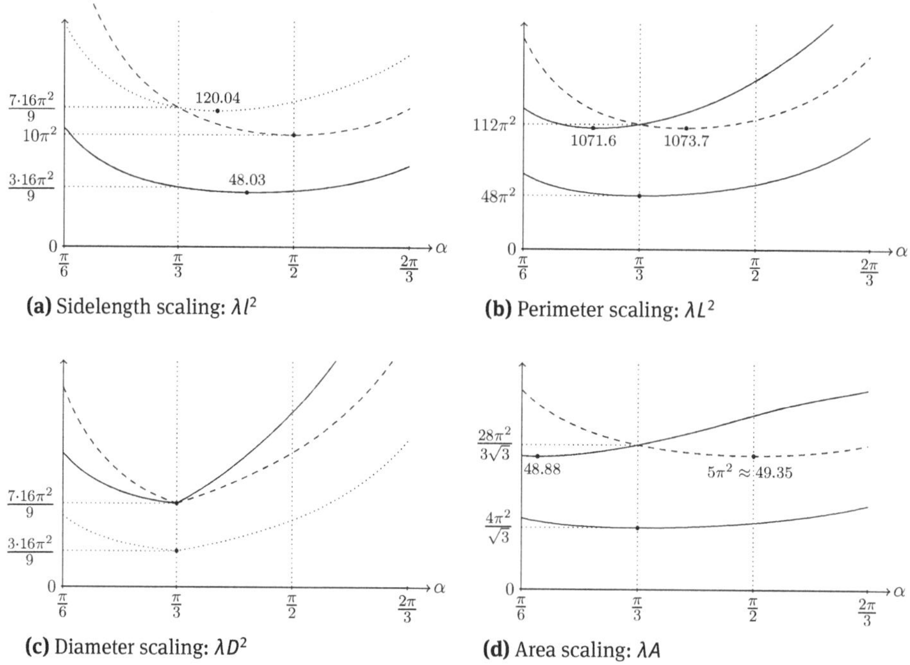

Fig. 6.14. Plots of the Dirichlet fundamental tone λ_1, the smallest antisymmetric tone λ_A (dashed curve), and the smallest symmetric tone λ_S larger than the fundamental tone, for the isosceles triangle $T(\alpha)$ with aperture α. Global minimum points are indicated with dots.

A number of monotonicity relations are suggested by Figure 6.14. Some of them have been proved ([818, Theorem 1.3] and [642, Propositions 7.1 and 7.2]):

Proposition 6.38 (Dirichlet eigenvalue monotonicity for isosceles triangles).

$\lambda_1(\alpha)l^2$ is decreasing for $0 < \alpha \le \pi/3$ and increasing for $\pi/2 \le \alpha < \pi$, and it tends to infinity at the endpoints $\alpha = 0, \pi$.

$\lambda_1(\alpha)D^2, \lambda_1(\alpha)L^2, \lambda_1(\alpha)A$ are decreasing for $0 < \alpha \le \pi/3$ and increasing for $\pi/3 \le \alpha < \pi$.

$\lambda_A(\alpha)l^2$ and $\lambda_A(\alpha)A$ are decreasing for $0 < \alpha \le \pi/2$ and increasing for $\pi/2 \le \alpha < \pi$.

$\lambda_A(\alpha)D^2$ is decreasing for $0 < \alpha \le \pi/3$ and increasing for $\pi/3 \le \alpha < \pi$.

The monotonicity claims on $\lambda_1(\alpha)A$ were explained already in Theorem 6.8 and its proof.

The growth rate to infinity of the first and second eigenvalues as the aperture degenerates to 0 or π has been investigated by Freitas [398]. He obtained asymptotics for degenerating sequences of non-isosceles triangles too.

Certain basic symmetry relations have not yet been proved.

Conjecture 6.39 (Symmetries of second Dirichlet mode [642]).

$\lambda_2(\alpha) = \lambda_S(\alpha)$ when $0 < \alpha < \pi/3$, so that the second Dirichlet mode of a subequilateral triangle is symmetric.

$\lambda_2(\alpha) = \lambda_A(\alpha)$ when $\pi/3 < \alpha < \pi$, so that the second Dirichlet mode of a superequilateral triangle is antisymmetric

The conjecture is plausible because the oscillation of the second Dirichlet mode should take place in the "long" direction of the triangle.

Incidentally, Figures 6.14b and 6.14d show that $\lambda_2 A$ and $\lambda_2 L^2$ are *not* minimal among isosceles triangles at the equilateral triangle. This fact is analogous to the situation for convex domains, where the minimizers of the second eigenvalue are certain "stadium-like" sets rather than disks (Henrot *et al.* [211]).

6.10.2 Neumann eigenvalues

In this section we denote by $\mu_2(\alpha)$, $\mu_A(\alpha)$ and $\mu_S(\alpha)$ the smallest nonzero Neumann eigenvalue, the smallest antisymmetric eigenvalue, and the smallest nonzero symmetric eigenvalue for the isosceles triangle $T(\alpha)$. These eigenvalues are plotted numerically in Figure 6.15a and Figure 6.15b, and the plots suggest the following symmetry properties which have indeed been proved [635].

Theorem 6.40 (Symmetries of second Neumann mode).

$\mu_2(\alpha) = \mu_S(\alpha)$ when $0 < \alpha < \pi/3$, so that the second Neumann mode of a subequilateral triangle is symmetric.

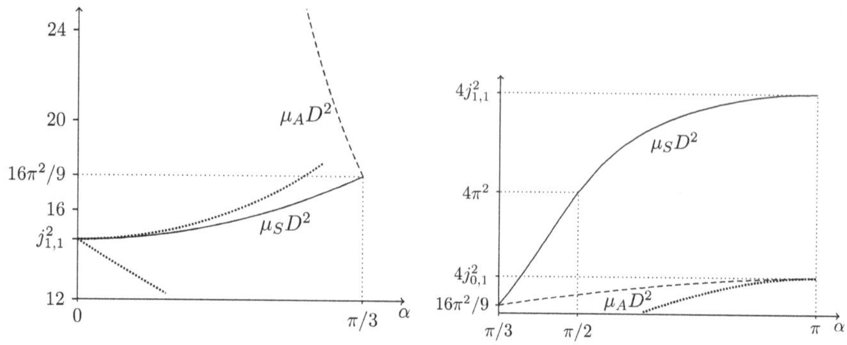

(a) Subequilateral case **(b)** Superequilateral case

Fig. 6.15. Numerical plot of the smallest symmetric Neumann eigenvalue μ_S and smallest antisymmetric Neumann eigenvalue μ_A (dashed curve), for the isosceles triangles $T(\alpha)$ with aperture α. The eigenvalues are normalized by the square of the diameter. The dotted curves show bounds from Proposition 6.42, which converge to the appropriate asymptotic values for the degenerate isosceles triangles.

$\mu_2(\alpha) = \mu_A(\alpha)$ *when* $\pi/3 < \alpha < \pi$, *so that the second Neumann mode of a superequilateral triangle is antisymmetric*

As in the Dirichlet case (Conjecture 6.39), this theorem is plausible because the oscillation of the second Neumann mode should take place in the "long" direction of the triangle.

These symmetry properties of the second Neumann eigenfunction are illustrated in Figure 6.16.

Miyamoto [709, Theorems B and C] refined the previous theorem by proving monotonicity of the second Neumann eigenfunction in the x- and y-directions, as explained in Theorem 6.51 below.

Next we raise some spectral monotonicity questions suggested by Figure 6.15a and Figure 6.15b.

Conjecture 6.41 (Neumann eigenvalue monotonicity for isosceles triangles).
$\mu_S(\alpha)D^2$ *is increasing for* $0 < \alpha < \pi$.
$\mu_A(\alpha)D^2$ *is decreasing for* $0 < \alpha \le \pi/3$ *and increasing for* $\pi/3 \le \alpha < \pi$.

Near the endpoint cases, meaning $\alpha \to 0$ and $\alpha \to \pi$, one has the following reasonably tight bounds [635].

Proposition 6.42. *For subequilateral triangles*

$$\frac{1}{1 + \tan(\alpha/2) + \tan^2(\alpha/2)} < \frac{\mu_2(\alpha)D^2}{j_{1,1}^2} \le \frac{1}{\cos^2(\alpha/2)}, \qquad 0 < \alpha \le \frac{\pi}{3},$$

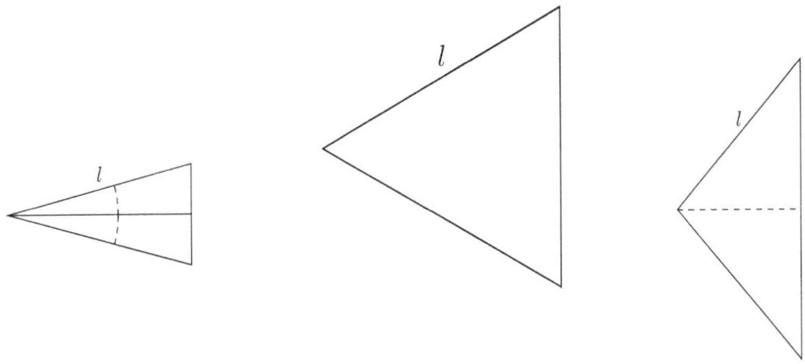

(a) subequilateral triangle **(b)** equilateral triangle **(c)** superequilateral triangle

Fig. 6.16. Nodal curves (dashed) for the second Neumann mode of an isosceles triangle. The mode satisfies a Neumann condition on each solid line, and a Dirichlet condition on each dashed curve. See Theorem 6.40.

with equality holding on both sides in the limit $\alpha \to 0$, that is, for the degenerate subequilateral triangle. For superequilateral triangles

$$\sin^2(\alpha/2) \le \frac{\mu_2(\alpha)D^2}{4j_{0,1}^2} < 1, \qquad \frac{\pi}{3} \le \alpha < \pi,$$

with equality holding on both sides in the limit $\alpha \to \pi$, that is, for the degenerate superequilateral triangle.

Let us close this section by plotting some of the higher Neumann eigenvalues of isosceles triangles, using numerical investigations of Laugesen *et al.* [633] as shown in Figure 6.17.

6.11 Right triangles

Consider the right triangle $R(\gamma)$ having vertices $(0,0)$, $(1,0)$ and $(\cos\gamma, \sin\gamma)$, where $\gamma \in (0, \pi/4]$ is the smallest angle of the triangle.

6.11.1 Dirichlet eigenvalues

We will write $\lambda_1(\gamma)$ for the first Dirichlet eigenvalue of $R(\gamma)$. The corresponding eigenfunction can be reflected oddly across the horizontal axis to form an eigenfunction on the isosceles triangle $T(2\gamma)$. Unfortunately, it is not yet known whether the resulting

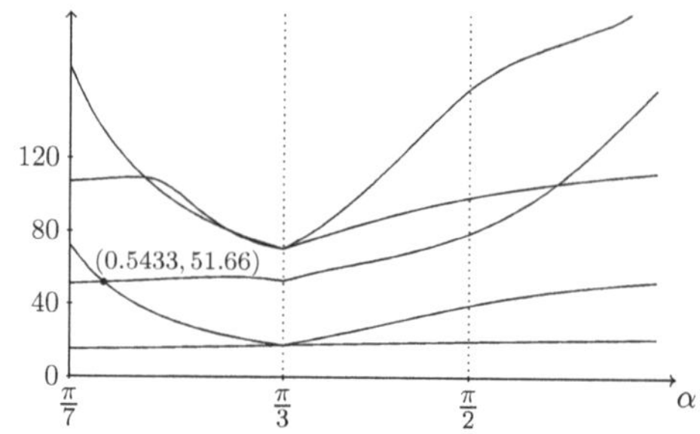

Fig. 6.17. Numerical plot of the first five nonzero Neumann eigenvalues normalized by diameter ($\mu_j D^2$ for $j = 2, \ldots, 6$) for the isosceles triangle $T(\alpha)$, plotted as a function of the aperture angle α. The minimum value of $\mu_4 D^2$ is approximately 51.66, occurring at $\alpha \simeq 0.5433$ (to 4 significant figures). The value at the equilateral triangle ($\alpha = \pi/3$) is larger: $\mu_4 D^2 = 3 \cdot 16\pi^2/9 \simeq 52.64$.

eigenfunction belongs to the second or the third eigenvalue on that isosceles triangle (see Conjecture 6.39). Nonetheless, Siudeja and Laugesen [642, Lemma 7.1] showed by continuous Steiner symmetrization that the eigenvalue of the right triangle behaves monotonically when normalized by area:

Proposition 6.43. *$\lambda_1(\gamma)A$ is decreasing for $\gamma \in (0, \pi/4]$.*

6.11.2 Neumann eigenvalues

The second Neumann eigenvalue of $R(\gamma)$ coincides with the second Neumann eigenvalue of the isosceles triangle $T(2\gamma)$, when $\gamma \in (0, \pi/6)$, in view of the symmetry result in Theorem 6.40. Therefore, any result for subequilateral triangles can be translated into a corresponding bound for narrow right triangles. In particular, Miyamoto's hot spots result (Theorem 6.51) can be translated for right triangles with $\gamma < \pi/6$. The same result, for any right triangle, can be deduced from the last paragraph on page 244 of Atar–Burdzy [86].

Furthermore, Miyamoto [709, Lemma 3.1] showed that a kite built from two copies of a right triangle will have the same second Neumann eigenvalue as the right triangle. Hence bounds on μ_2 for these kites are equivalent to bounds for right triangles. This relation can be extended to more general kites and triangles by work of Siudeja [819, Lemma 1]. Lemma 3 in the same paper gives the following simple bound.

Proposition 6.44 (Second eigenvalue on right triangle). *Let h be the shortest altitude of a right triangle. Then $\mu_2 h^2 \leq \pi^2$, with equality for the right isosceles triangle.*

6.11.3 Mixed Dirichlet–Neumann

We saw in Theorem 6.12 how to compare the mixed eigenvalues of a given right triangle, for many possible combinations of Dirichlet and Neumann conditions on the sides of that triangle. Generally, the longer the Dirichlet part, the higher the eigenvalue. In particular, imposing the Dirichlet condition only on the middle side M gives a smaller eigenvalue than imposing it on the longest side L, and the second Neumann eigenvalue falls between the two: $\lambda_1^M < \mu_2 < \lambda_1^L$.

Nitsch [727] studied monotonicity of λ_1^S assuming fixed length of side M. He found a bound for the derivative of that eigenvalue with respect to γ, and derived some bounds for the eigenvalue and for the Dirichlet eigenvalue on regular polygons.

Finally, any mixed or purely Dirichlet eigenvalue of any right triangle can be bounded with virtually arbitrary precision using the validated numerical approach of Nigam, Siudeja and Young [726], as described in Section 6.6.

6.12 Inverse problem — can one hear the shape of a triangular drum?

The question of whether the eigenvalues determine the domain was raised long ago by Kac [567]. For the special case of triangular domains, Durso [350] found that, indeed, the Dirichlet spectrum $\{\lambda_j\}$ determines the triangle.

Theorem 6.45 (Hearing the shape of a triangle). *If two triangles possess exactly the same Dirichlet spectrum, then the triangles are congruent.*

Durso's proof involves the wave trace. Grieser and Maronna [437] discovered a much simpler approach via the heat trace, and we will sketch their proof. The heat trace for triangular domains has the following asymptotic formula (whose somewhat complicated history is discussed by Mazzeo and Rowlett [698]): as $t \to 0$ one has

$$\sum_{j=1}^{\infty} e^{-\lambda_j t} = \frac{A}{4\pi t} - \frac{L}{4\sqrt{4\pi t}} + \frac{1}{24}\left(-1 + \sum_{k=1}^{3} \frac{\pi}{\theta_k}\right) + o(1) \tag{6.20}$$

where $\theta_1, \theta_2, \theta_3$ are the interior angles of the triangle. The area A and perimeter L can be read off from the expansion (6.20), and one also obtains the reciprocal angle sum $\sum_{k=1}^{3} \theta_k^{-1}$. Of course the angle sum $\sum_{k=1}^{3} \theta_k$ equals π. The cotangent sum $\sum_{k=1}^{3} \cot(\theta_k/2) = \prod_{k=1}^{3} \cot(\theta_k/2)$ is known too, because it equals $L^2/4A$ by a standard formula for triangles. From the last three sums, Grieser and Maronna obtain the angles θ_k by an elementary convexity and monotonicity argument in the parameter space of triangles. The angles and area of the triangle determine it up to congruence, thus completing the proof.

The inverse spectral result (Theorem 6.45) is beautiful, and yet ridiculously overdetermined: it requires infinitely many items of spectral information (the eigenvalues in the heat trace) to determine the three parameters on which the triangle depends. Can one do better?

Conjecture 6.46 (Hearing the shape of a triangle from its first three tones). *If two triangles have the same first three Dirichlet eigenvalues $\lambda_1, \lambda_2, \lambda_3$, then the triangles are congruent.*

Numerical investigations by Antunes and Freitas [41] demonstrate convincingly that the conjecture is true (although interesting it fails if one uses $\lambda_1, \lambda_2, \lambda_4$). The only partial result in this direction is due to Chang and DeTurck [257], saying that finitely many eigenvalues suffice to determine the triangle; unfortunately the required number of eigenvalues is not fixed, but rather depends on the first two eigenvalues. This partial result generalizes to other finite dimensional families of domains.

Incidentally, one can hear the shape of a rectangle from its first two Dirichlet eigenvalues, since $\lambda_1 = \pi^2(l_1^{-2} + l_2^{-2})$ and $\lambda_2 = \pi^2(l_1^{-2} + 4l_2^{-2})$ together determine the side lengths $l_1 \le l_2$.

6.13 Structure of eigenfunctions on special domains

The maxima, minima and nodal curves of eigenfunctions can be extremely hard to study, even for special domains such as rectangles whose eigenvalues and eigenfunctions are explicitly known. In this section we survey recent results on the structure of eigenfunctions on special domains. One of the complications of the analysis is the possibility of multiple eigenvalues, so we begin with observations on the multiplicity of eigenvalues, before turning to maxima and minima (that is, the "hot spots" problem) and nodal curves.

6.13.1 Multiplicity

The first eigenvalue on a connected domain is always simple, and its eigenfunction can be chosen to be nonnegative. The simplicity of the second eigenvalue is not so trivial, and one has a double eigenvalue on highly symmetric domains in dimension 2, or a triple eigenvalue in dimension 3. Such multiplicity occurs, for example, for the equilateral triangle, square, regular polygons, disk, ball, and regular solids in higher dimensions.

One can construct a non-simply connected domain with a second Dirichlet eigenvalue that has multiplicity three [531], but not higher multiplicity [268]. In the other direction, Nadirashvili [716] showed that the second Neumann eigenvalue μ_2 of a simply connected planar domain can have multiplicity at most 2.

For triangles the situation is simpler, according to the following conjecture and theorem.

Conjecture 6.47 (Simplicity of λ_2 on triangles). *The second Dirichlet eigenvalue is simple, on every non-equilateral triangle.*

Theorem 6.48 (Simplicity of μ_2 on triangles). *The second Neumann eigenvalue is simple, on every non-equilateral triangle.*

The theorem is due to Siudeja [819]. The more general Neumann simplicity criterion below, due to Atar and Burdzy [86], covers all obtuse triangles, but not acute triangles.

Theorem 6.49 (Simplicity of μ_2 on 1-Lipschitz domains). *Consider a planar domain contained between the graphs of two Lipschitz functions with constant at most* 1. *The second Neumann eigenvalue on this domain must be simple, unless the domain is a square.*

The 1-Lipschitz criterion includes all rectangles, although in that case it is simpler to use the exact formula for the second eigenvalue. The criterion also covers parallelograms, but does not cover all kites (quadrilaterals symmetric with respect to a diagonal). Indeed, a family of kites exists with double second Neumann eigenvalue – see a dotted line on Figure 3 of Siudeja [819]. Interestingly, this family contains both the equilateral triangle and the square as extreme cases.

6.13.2 Hot spots

The “hot spots” problem due to Rauch [784] claims that:

Conjecture 6.50 (Hot spots on convex domains). *If u is a Neumann eigenfunction with eigenvalue μ_2 on a convex domain, then u attains its maximum value at a point on the boundary.*

The conjecture gains its name from considering the hottest point in a perfectly insulated room: the solution of the heat equation is generically dominated for large times by $c_1 + c_2 e^{-\mu_2 t} u$, and so if Rauch’s conjecture holds true, then for large t the hottest spot should occur close to the boundary.

Notice the conjecture is somewhat ambiguously stated. First, the eigenvalue might not be simple: should the conjecture hold for each eigenfunction with eigenvalue μ_2, or just for one of them? Second, should the boundary maximum point be strict, or could the eigenfunction also attain its maximum at some point inside the domain? These variants have been discussed thoroughly by Bañuelos and Burdzy [104].

It is widely believed that the strongest possible form of the conjecture should hold for every convex domain. The simply connected case might hold true, but the conjecture can fail for a domain with two holes (Burdzy and Werner [230]) or even just one hole (Burdzy [228]).

Few positive results are known. Kawohl [579] treated product type domains, Bañuelos and Burdzy [104] handled obtuse triangles by probabilistic coupling arguments, and Atar and Burdzy [86] tackled a class of Lipschitz domains with constant 1 (including right triangles) by "mirror coupling". Jerison and Nadirashvili [562] proved the conjecture for domains with two lines of symmetry, while Pascu [742] gave a probabilistic proof assuming the eigenfunction is antisymmetric. Without any symmetry assumptions, Miyamoto [708] gave an analytical proof for nearly round domains (diameter-squared to area ratio not much larger than that for a disk).

The conjecture can be proved for isosceles triangles by combining work of Atar and Burdzy [86] and Pascu [742] with eigenfunction (anti-)symmetry results due to Laugesen and Siudeja [635]. A more direct proof for isosceles triangles was given by Miyamoto [709, Theorems B and C]. We state his result in detail below, to show the kind of structural information on the shape of the eigenfunction that seems needed to prove the hot spots conjecture. Write T_+ and T_- for the upper and lower halves, respectively, of the isosceles triangle $T = T(\alpha)$.

Theorem 6.51 (Hot spots on isosceles triangles). *If $0 < \alpha < \pi/3$ (the subequilateral case) then the second Neumann eigenfunction $u(x, y)$ is even with respect to y and can be chosen to satisfy*

$$u_x > 0 \text{ in } T, \text{ with } u_y > 0 \text{ in } T_+ \text{ and } u_y < 0 \text{ in } T_-.$$

If $\pi/3 < \alpha < \pi$ (the superequilateral case) then the second Neumann eigenfunction $u(x, y)$ is odd with respect to y and can be chosen to satisfy

$$u_x > 0 \text{ in } T_+ \text{ and } u_x < 0 \text{ in } T_-, \text{ with } u_y > 0 \text{ in } T.$$

Hence on every subequilateral or superequilateral triangle, the second Neumann eigenfunction achieves its maximum and minimum at vertices of the triangle.

Acute triangles are not covered by any of the results above, and indeed they became the subject of the online collaborative Polymath7 project managed by Terence Tao [848], with the goal of proving:

Conjecture 6.52 (Hot spots on acute triangles). *The second Neumann eigenfunction on an acute triangle attains its maximum at a vertex.*

The conjecture was proved by Siudeja [819] for the class of acute triangles having an angle less than $\pi/6$. The problem remains open for acute triangles all of whose angles lie in the range $(\pi/6, \pi/2)$.

6.13.3 Number of nodal domains

The famous Courant nodal domain theorem states that the j-th Dirichlet or Neumann eigenvalue can have at most j nodal domains. We call the j-th eigenvalue *Courant-sharp* if it possesses an eigenfunction having j nodal domains. Clearly the first two eigenvalues ($j = 1, 2$) are Courant-sharp, because their eigenfunctions have 1 and 2 nodal domains respectively. The existence of other such eigenvalues is not at all obvious.

Further, the Courant upper bound seems rather crude for higher eigenvalues. As an example, the bottom plots on Figure 6.12 show two eigenfunctions for the 36-th eigenvalue of the equilateral triangle, which has multiplicity four. These particular examples have 10 and 13 nodal domains, respectively. Not only are these numbers different, they are much smaller than the Courant upper bound of 36.

Our knowledge of nodal domains advanced significantly in recent years. In particular, Helffer and others have determined all Courant-sharp cases for certain special domains, and we discuss many of these results below. For more on the theory of counting nodal domains, see Chapter 10.

Theorem 6.53 (Courant-sharp nodal domains [131, 132, 498, 501]). *The j-values for which λ_j or μ_j (Dirichlet or Neumann, respectively) possesses an eigenfunction having j nodal domains are:*
- *equilateral triangle (Dirichlet), square (Dirichlet), disk (Dirichlet and Neumann): $j = 1, 2, 4$;*
- *hemi-equilateral triangle, right isosceles triangle, cube (all Dirichlet): $j = 1, 2$;*
- *square (Neumann): $j = 1, 2, 4, 5, 9$.*

See Figure 6.18 for the plots of the nodal domains on the above-mentioned triangles, and Figure 6.19 for the plots of the nodal domains on squares. This theorem leaves the following cases open.

Open problem 6.54. *Find all Courant-sharp Neumann eigenvalues for equilateral, hemi-equilateral, and right isosceles triangles.*

Note that the third eigenvalue is never Courant-sharp on the domains in Theorem 6.53 or the associated figures. In fact, Ashbaugh and Benguria [67] proved that $\mu_2 = \mu_3$ assuming the domain has N-fold symmetry with $N \geq 3$, and so μ_3 cannot be Courant-sharp on such domains. We are not aware of a similar result for the Dirichlet case.

Fig. 6.18. Courant's nodal domain theorem for Dirichlet eigenfunctions of special triangles. The j-th eigenfunction is called *sharp* if it has j nodal domains. Equilateral triangle (first two rows, left to right): $j = 2, 4, 5, 5, 7, 7$; here 2 and 4 are sharp. Right isosceles triangle (third row): $j = 2, 3, 4$; only 2 is sharp. Hemi-equilateral triangle (fourth row): $j = 2, 3, 4$; only 2 is sharp.

In the other direction, Helffer, Hoffmann-Ostenhof and Terracini [477] noticed that rectangles can have arbitrarily many Courant-sharp eigenvalues, as follows.

Fig. 6.19. Courant's nodal domain theorem for eigenfunctions of squares. The j-th eigenfunction is called *sharp* if it has j nodal domains. Neumann (first two rows, left to right): $j = 2, 4, 5, 5, 5, 9$; all are sharp, except that $j = 5$ has two additional non-sharp patterns. Dirichlet (last two rows): $j = 2, 4, 5, 5, 5, 11$; only 2 and 4 are sharp.

Fig. 6.20. Numerical investigation of boundary sign-changing of Neumann eigenfunctions on regular polygons; see results in Section 6.13.4. Equilateral triangle: every mode changes sign; the pictured mode minimizes the number of sign changes among fully symmetric modes. Square: every mode must vanish somewhere on the boundary, yet the pictured modes do not actually change sign. Each regular polygon with five or more sides possesses at least one mode that does not vanish on the boundary. Numerical investigations suggest that exactly two such modes exist on the regular pentagon, and three on the regular hexagon, as illustrated. All plots created by Siudeja [813].

Proposition 6.55 (Courant-sharp rectangles). *The j-th Dirichlet eigenvalue of a rectangle with side ratio $n : 1$ is Courant-sharp if $j = 1, 2, \dots, \lfloor\sqrt{3n^2 + 1}\rfloor$, and similarly for the Neumann eigenvalues if $j = 1, 2, \dots, n + 1$.*

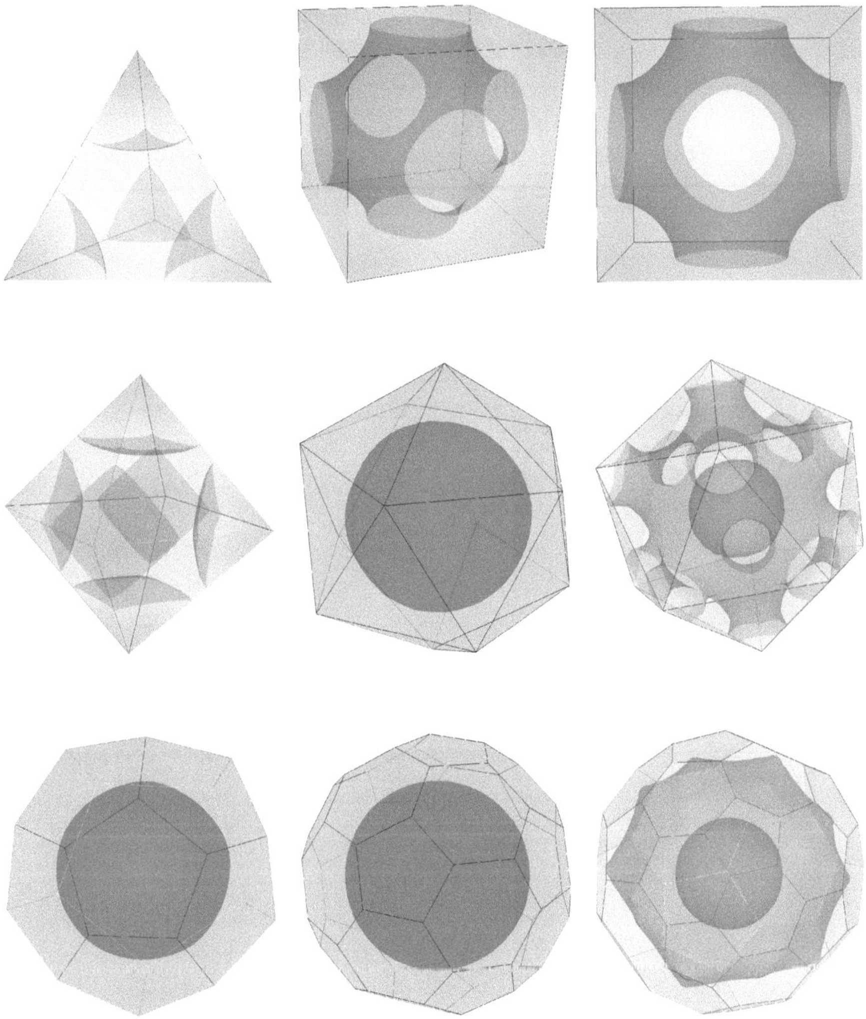

Fig. 6.21. Numerical investigation of boundary sign-changing of Neumann eigenfunctions on some classical solids; see results in Section 6.13.4. Tetrahedron, cube (two views of same mode), octahedron: the pictured modes minimize the number of boundary sign changes among fully symmetric modes. Icosahedron: one mode does not change sign on the boundary; also pictured is the next-highest mode, which does change sign. Dodecahedron: one mode has no boundary sign changes. Truncated icosahedron (football): two modes do not change sign on the boundary. All plots created by Siudeja [813]. Observations are purely numerical except for the cubes, which are treated in Theorem 6.56.

Proof. For the Neumann rectangle, we only need to make sure that the eigenvalue $\mu_{0,k} = \pi^2(0^2/1^2 + k^2/n^2)$ does not exceed the eigenvalue $\mu_{1,0} = \pi^2/1^2$, remembering that $k = 0$ is a valid choice in the Neumann case. This condition holds for $k \leq n$, and ensures that the first $n+1$ eigenfunctions have straight nodal lines that subdivide the long side of the rectangle into smaller rectangles. Then the next eigenvalue will have one subdivision on the short side.

The Dirichlet case is similar, except now $k = 1$ is the smallest valid choice and we need to compare eigenvalues corresponding to $(1, k)$ and $(2, 1)$. □

6.13.4 Boundary sign-changing for eigenfunctions.

Another interesting structural problem involves sign changing on the boundary for Neumann eigenfunctions. The first eigenfunction is constant and so has fixed sign, for every domain. The disk has many radial eigenfunctions, and hence has many eigenfunctions with fixed sign on the boundary. What other domains have eigenfunctions that do not change sign on the boundary?

The following theorem summarizes results due to Hoffmann–Ostenhoff [532] and Siudeja, Nigam and Young [726, 815] for special domains.

Theorem 6.56 (Boundary sign-changing of Neumann eigenfunctions).

Equilateral triangles and cubes: every Neumann eigenfunction that is nonnegative on the boundary must be constant inside the domain.

Squares: every Neumann eigenfunction that is positive on the boundary must be constant inside the domain.

Regular polygons with $N \geq 5$ sides: there exists an eigenfunction that is positive on the boundary.

Comparing the result for squares with that for equilaterals, we note the stronger assumption of positivity on the boundary. This restriction is necessary because on the boundary of the square $[-1, 1]^2$, the nonconstant eigenfunction $u(x, y) = -\cos(\pi x) - \cos(\pi y)$ is positive except for equalling zero when (x, y) equals $(\pm 1, 0)$ or $(0, \pm 1)$. Surprisingly, this positivity restriction is unnecessary on cubes.

Proof sketch for Theorem 6.56. For a regular hexagon we only need look at the middle left image in Figure 6.12 (the symmetric second eigenfunction of the equilateral triangle) and imagine a hexagon centered at the left vertex of the equilateral triangle. Alternatively, we could use the top right image from the same figure. For regular polygons with more than six sides we can use the same construction, except with Theorem 6.40 used to ensure that the symmetric mode has the same general shape as in the equilateral case.

This approach fails for the regular pentagon, since the symmetric mode on the corresponding superequilateral isosceles triangle yields the third eigenvalue rather than the second, and it is difficult to show that a third eigenfunction has just two nodal domains. A more complicated approach was developed to handle this case [726].

Incidentally, to show that non-sign-changing eigenfunctions do not exist on equilateral triangles, squares and cubes, one needs to use exact formulas for the eigenfunctions together with either a combinatorial argument for cubes and equilateral triangles [726] or a number theoretic argument for squares [532]. □

The preceding theorem is connected to an old conjecture.

Conjecture 6.57 (Schiffer's conjecture). *If a Neumann eigenfunction is constant on the boundary, then either the eigenfunction must be constant everywhere or else the domain must be a ball.*

For more information we refer to a recent paper on Schiffer's conjecture [336]. Note that the conjecture is a special case of the Pompeiu problem.

Existence of such a special eigenfunction would imply existence of an eigenfunction having the full symmetry of the underlying domain. One could call these eigenfunctions "nearly radial"; see Figure 6.20 for examples in two dimensions and Figure 6.21 for examples in three dimensions. The first figure shows exactly three such eigenfunctions on the regular hexagon, and we conjecture that no other fully symmetric examples exists on the regular hexagon.

We also raise conjectures for triangles, quadrilaterals and regular solids, as below.

Conjecture 6.58. *(Boundary sign-changing on triangles, quadrilaterals, and regular solids).*

Triangle, non-rectangular quadrilateral, tetrahedron and octahedron: every Neumann eigenfunction that is nonnegative on the boundary must be constant inside the domain.

Dodecahedron and icosahedron: there exists an eigenfunction that is positive on the boundary.

The intuition is that domains "close enough" to the disk or ball will have an eigenfunction that is positive on the boundary.

One can ask a similar question for other boundary conditions too.

Open problem 6.59. *Do there exist eigenfunctions for the Robin boundary condition that are nonnegative/positive on the boundary of highly symmetric domains? Similarly one can examine the sign of the normal derivative of Dirichlet eigenfunctions.*

The difficult pentagonal case in Theorem 6.56 depends on a single case of a right triangle in the following general conjecture.

Conjecture 6.60. *On any triangle, the nodal curve for the second Neumann eigenfunction connects the two longest sides of the triangle, except on superequilateral triangles where it connects the longest side and the opposite vertex.*

So far we know the nodal curve cannot start and end on the same side (Siudeja [819]), but no-one has yet been able to prove which sides it should connect, or to prove that it avoids vertices.

6.14 Conjectures for general domains

How might the results for triangles generalize to arbitrary domains? Let us start by generalizing the Dirichlet results in Section 6.4.

Conjecture 6.61 (All eigenvalues under diameter normalization [642]). *$\lambda_j D^2$ is minimal when the domain is a disk, for each $j \geq 1$.*

For $j = 1$, the conjecture holds by the Faber–Krahn theorem that $\lambda_1 A$ is minimal for the disk together with the isodiametric theorem that D^2/A is minimal for the disk. For $j = 2$, the minimality of $\lambda_2 D^2$ for the disk was conjectured by Bucur, Buttazzo and Henrot [211].

Next, write I for the moment of inertia of the planar domain Ω. That is,

$$I(\Omega) = \int_\Omega |x - \overline{x}|^2 \, dx$$

where the centroid is $\overline{x} = \frac{1}{A} \int_\Omega x \, dx$. We start with a two-sided conjecture on the first Dirichlet eigenvalue, in terms of area and moment of inertia.

Conjecture 6.62 (First eigenvalue [636]). *Suppose Ω is a bounded convex plane domain. Then*

$$\frac{9}{2}\pi^2 < \lambda_1 \frac{A^3}{I} \bigg|_\Omega \leq 12\pi^2$$

with equality on the right for equilateral triangles and all rectangles, and asymptotic equality on the left for degenerate acute isosceles triangles and sectors.

The right side of the conjecture would extend Pólya's bounds on $\lambda_1 A^2/S^2$ in Theorem 6.25 for triangles and parallelograms, since $A/I = (\text{const.})/S^2$ in both those cases. The convexity assumption is necessary on the right side of the conjecture because otherwise one could drive the eigenvalue to infinity without affecting the area or moment of inertia, by removing sets of measure zero (such as curves) from the domain.

Normalizing instead with A^2/L^2, Pólya [767] proved for the first eigenvalue that:

Theorem 6.63 (First eigenvalue). *Among bounded convex planar domains, $\lambda_1 A^2/L^2$ is maximal for the degenerate rectangle.*

This result differs from Theorem 6.25 for triangles with the same normalization and from Conjecture 6.62 for general domains with A^3/I normalization, since in both those cases the equilateral triangle should be a maximizer.

Now we turn to Neumann eigenvalues on general domains. Our result for triangles in Theorem 6.22 suggests that:

Conjecture 6.64 (Sum of n eigenvalues [633]). *Among bounded convex planar domains,*

$$(\mu_2 + \cdots + \mu_n)D^2$$

is minimal when the domain is a disk, for each $n \geq 3$.

The conjecture fails for $n = 2$, because Payne and Weinberger proved $\mu_2 D^2$ is minimal for the degenerate rectangle (and not the disk) among all convex domains [749]. In other words, they proved that the optimal Poincaré inequality for convex domains is saturated by the degenerate rectangle.

The next conjecture would be the analogue of Theorem 6.35, which applied to triangles and parallelograms.

Conjecture 6.65 (Sum of n eigenvalues [636]). *Among bounded convex planar domains,*

$$(\mu_2 + \cdots + \mu_n)\frac{A^3}{I}$$

is maximal for the disk, for each $n \geq 2$.

Open problem 6.66 (Second eigenvalue). *Determine the maximizers for $\mu_2 L^2$, among all bounded convex domains in the plane.*

Notice $\mu_2 L^2$ is *not* maximal for the disk, because the equilateral triangle and the square both have $\mu_2 L^2 = 16\pi^2 \simeq 158$, which exceeds the value $4\pi^2 (j'_{1,1})^2 \simeq 133$ for the disk. The convexity hypothesis in this problem eliminates domains with fractal boundary, for which L is infinite and μ_2 can be positive.

Finally we mention a problem in curved spaces.

Open problem 6.67 (Hyperbolic space and the sphere). *Which spectral results on Euclidean triangles possess analogues for triangles in the hyperbolic disk and in the sphere? For example, among hyperbolic triangles of given area, does the regular triangle minimize the first eigenvalue of the hyperbolic Laplacian?*

The only result that we are aware of is Karp and Peyerimhoff's theorem that among hyperbolic triangles of given perimeter, the regular triangle minimizes the first Dirichlet eigenvalue [572]. Interesting also is Solynin and Zalgaller's investigation of curvilinear polygons in the Euclidean plane [828].

Acknowledgment: The development of this chapter was supported by grants from the Simons Foundation (#204296 to Richard Laugesen) and the Polish National Science Centre (2012/07/B/ST1/03356 to Bartłomiej Siudeja). We also appreciate travel support from the workshop on "Shape Optimization and Spectral Geometry" at the International Centre for Mathematical Sciences, Edinburgh, in June 2015, and from the conference "Fifty Years of Hearing Drums: Spectral Geometry and the Legacy of Mark Kac" at the Pontificia Universidad Católica de Chile, in May 2016.

Figures taken from our earlier papers:
Figure 6.8 from [636], Figure 6.14 from [642], Figure 6.15 from [635], Figure 6.17 from [633].

Lorenzo Brasco and Guido De Philippis

7 Spectral inequalities in quantitative form

7.1 Introduction

7.1.1 The problem

Let $\Omega \subset \mathbb{R}^d$ be an open set, and consider the Laplacian operator $-\Delta$ on Ω under various boundary conditions. When the relevant spectrum happens to be discrete, it is an interesting issue to provide sharp geometric estimates on associated spectral quantities like the *ground state energy* (or first eigenvalue), the *fundamental gap*, or more general functions of the eigenvalues. In this chapter we will consider the following eigenvalue problems for the Laplacian:

Dirichlet conditions

$$\begin{cases} -\Delta u = \lambda\, u, & \text{in } \Omega, \\ u = 0, & \text{on } \partial\Omega, \end{cases}$$

Robin conditions $(\alpha > 0)$

$$\begin{cases} -\Delta u = \lambda\, u, & \text{in } \Omega, \\ \alpha\, u + \dfrac{\partial u}{\partial \nu} = 0, & \text{on } \partial\Omega, \end{cases}$$

Neumann conditions

$$\begin{cases} -\Delta u = \mu\, u, & \text{in } \Omega, \\ u = 0, & \text{on } \partial\Omega, \end{cases}$$

Steklov conditions

$$\begin{cases} -\Delta u = 0, & \text{in } \Omega, \\ \dfrac{\partial u}{\partial \nu} = \sigma\, u, & \text{on } \partial\Omega. \end{cases}$$

We denote by $\lambda_1(\Omega)$, $\lambda_1(\Omega, \alpha)$, $\mu_2(\Omega)$ and $\sigma_2(\Omega)$ the corresponding first (or first non-trivial[7.1]) eigenvalue. We refer to the next sections for the precise definitions of these eiegnvalues and their properties. For these spectral quantities, we have the following well-known sharp inequalities:

Dirichlet case

$$|\Omega|^{2/d}\, \lambda_1(\Omega) \geq |B|^{2/d}\, \lambda_1(B), \qquad (\textit{Faber Krahn inequality}) \tag{7.1}$$

7.1 Observe that in the Neumann and Steklov cases, 0 is always the first eigenvalue, associated to constant eigenfunctions. Thus we use the convention that $\sigma_1(\Omega) = \mu_1(\Omega) = 0$. Also observe that the Robin case can be seen as an interpolation between Neumann (corresponding to $\alpha = 0$) and Dirichlet conditions (when $\alpha = +\infty$), see Chapter 4 for more details.

Lorenzo Brasco: Dipartimento di Matematica e Informatica, Universita degli Studi di Ferrara, Via Machiavelli 35, 44121 Ferrara, Italy
and Institut de Mathématiques de Marseille, Aix-Marseille Université 39, Rue Frédéric Joliot Curie, 13453 Marseille, France, E-mail: lorenzo.brasco@unife.it
Guido De Philippis: SISSA, Via Bonomea 265, 34136 Trieste, Italy, E-mail: guido.dephilippis@sissa.it

Robin case

$$|\Omega|^{2/d}\lambda_1(\Omega,\alpha) \geq |B|^{2/d}\lambda_1(B,\alpha), \qquad \textit{(Bossel-Daners inequality)} \tag{7.2}$$

Neumann case

$$|B|^{2/d}\mu_2(B) \geq |\Omega|^{2/d}\mu_2(\Omega), \qquad \textit{(Szegő-Weinberger inequality)} \tag{7.3}$$

Steklov case

$$|B|^{1/d}\sigma_2(B) \geq |\Omega|^{1/d}\sigma_2(\Omega), \qquad \textit{(Brock-Weinstock inequality)} \tag{7.4}$$

where B denotes an N–dimensional open ball. In all the previous estimates, equality holds only if Ω is a ball.

The fact that balls can be characterized as the only sets for which equality holds in (7.1)-(7.4) naturally leads one to consider the question of the *stability* of these inequalities. More precisely, one would like to improve (7.1), (7.2), (7.3) and (7.4), by adding in the right-hand sides a remainder term that measures the deviation of a set Ω from spherical symmetry.

For example, as for inequality (7.1), a typical quantitative Faber-Krahn inequality would read as follows

$$|\Omega|^{2/d}\lambda_1(\Omega) - |B|^{2/d}\lambda_1(B) \geq g(\delta(\Omega)), \tag{7.5}$$

where:

- $g : [0,+\infty) \to [0,+\infty)$ is some modulus of continuity, i. e. a positive continuous increasing function, vanishing at 0 only;
- $\Omega \mapsto \delta(\Omega)$ is some scaling invariant *asymmetry functional*, i. e. a functional defined over sets such that

$$\delta(t\,\Omega) = \delta(\Omega), \text{ for every } t > 0 \quad \text{and} \quad \delta(\Omega) = 0 \text{ if and only if } \Omega \text{ is a ball.}$$

Moreover, it would desirable to derive quantitative estimates which are "the best possible", in a sense. This means that we not only have (7.5) for every set Ω, but that it is possible to find a sequence of open sets $\{\Omega_n\}_{n\in\mathbb{N}} \subset \mathbb{R}^d$ such that

$$\lim_{n\to\infty}\left(|\Omega_n|^{2/d}\lambda_1(\Omega_n) - |B|^{2/d}\lambda_1(B)\right) = 0,$$

and

$$|\Omega_n|^{2/d}\lambda_1(\Omega_n) - |B|^{2/d}\lambda_1(B) \simeq g(d(\Omega_n)), \qquad \text{as } n \to \infty.$$

In this case, we would say that (7.5) is *sharp*. In other words, the quantitative inequality (7.5) is sharp if it becomes an equality asymptotically, at least for particular shapes having small *deficits*.

The quest for quantitative improvements of spectral inequalities attracted increasing interest in the last years. To the best of our knowledge, this research began with the papers [447] by Hansen and Nadirashvili and [706] by Melas. Both papers concern the Faber-Krahn inequality, which is indeed the most studied case. The reader is invited to consult Section 7.7 for more bibliographical references and comments.

The aim of this chapter is to give a complete picture on recent results about quantitative improvements of sharp inequalities for eigenvalues of the Laplacian. Apart from the inequalities for the first eigenvalues presented above, we will also consider other inequalities involving the second eigenvalue in the Dirichlet case, as well as the *torsional rigidity*. We warn the reader from the very beginning that the presentation will be limited to the Euclidean case. We comment briefly on the case of manifolds in Section 7.7.

7.1.2 Plan of the Chapter

Each section is as self-contained as possible. Where it has not been possible to provide all the details, we have tried to provide precise references.

In Section 7.2 we consider the case of the Faber-Krahn inequality (7.1), while the stability of the Szegő-Weinberger and Brock-Weinstock inequalities is treated in Section 7.4 and 7.5, respectively. For each of these sections, we first present the relevant stability result and then discuss its sharpness.

Section 7.3 is a sort of *divertissement*, which shows some applications of the quantitative Faber-Krahn inequality to estimates for the so called *harmonic radius*. This part of the chapter is essentially new and is placed there because some of the results that are presented will be used in Section 7.4.

Section 7.6 is devoted to presenting the proofs of other spectral inequalities, involving the second Dirichlet eigenvalue λ_2 as well. Namely, we consider the Hong-Krahn-Szego inequality for λ_2 and the Ashbaugh-Benguria inequality for the ratio λ_2/λ_1.

Then in Section 7.7 we present some comments on further bibliographical references, applications and miscellaneous stability results on some particular classes of Riemannian manifolds.

The work is complemented by 4 appendices, that contain technical results which are used throughout the chapter.

7.1.3 An open issue

We conclude the introduction by pointing out that at present no quantitative stability results are available for the case of the Bossel-Daners inequality, see Chapter 4 for more details. We thus start by formulating the following[7.2]

Open problem 7.1. *Prove a quantitative stability estimate of the type (7.5) for the Bossel-Daners inequality for the first eigenvalue of the Robin Laplacian $\lambda_1(\Omega, \alpha)$.*

7.2 Stability for the Faber-Krahn inequality

7.2.1 A quick overview of the Dirichlet spectrum

We briefly recall some useful facts about the Dirichlet spectrum, the reader can refer to Section 2.2 of Chapter 2 for more details. For an open set $\Omega \subset \mathbb{R}^d$, we indicate by $H_0^1(\Omega)$ the closure of $C_0^\infty(\Omega)$ in $H^1(\Omega)$. The first eigenvalue of the Dirichlet Laplacian is defined by

$$\lambda_1(\Omega) := \inf_{u \in H_0^1(\Omega)\setminus\{0\}} \frac{\displaystyle\int_\Omega |\nabla u|^2\,dx}{\displaystyle\int_\Omega |u|^2\,dx}.$$

In other words, this is the sharp constant in the Poincaré inequality

$$c \int_\Omega |u|^2\,dx \le \int_\Omega |\nabla u|^2\,dx, \qquad u \in H_0^1(\Omega).$$

Of course, it may happen that $\lambda_1(\Omega) = 0$ if Ω does not support such an inequality.

The infimum above is attained on $H_0^1(\Omega)$ whenever the embedding $H_0^1(\Omega) \hookrightarrow L^2(\Omega)$ is compact. In this case, the Dirichlet Laplacian has a discrete spectrum $\{\lambda_1(\Omega), \lambda_2(\Omega), \lambda_3(\Omega), \dots\}$ and successive Dirichlet eigenvalues can be defined accordingly. Namely, $\lambda_k(\Omega)$ is obtained by minimizing the Rayleigh quotient above, among functions orthogonal (in the $L^2(\Omega)$ sense) to the first $k-1$ eigenfunctions. Dirichlet eigenvalues have the following scaling property

$$\lambda_k(t\,\Omega) = t^{-2}\,\lambda_k(\Omega), \qquad t > 0.$$

Compactness of the embedding $H_0^1(\Omega) \hookrightarrow L^2(\Omega)$ holds in the case when $\Omega \subset \mathbb{R}^d$ is an open set with finite measure.

7.2 Note added in proof: while completing the editing of this chapter, on ArXiv it has appeared the preprint "The quantitative Faber-Krahn inequality for the Robin Laplacian" (arXiv:1611.06704)" by D. Bucur, V. Ferone, C. Nitsch, and C. Trombetti, where Open Problem 1.1 has been solved.

In this case, it is possible to provide a sharp lower bound on $\lambda_1(\Omega)$ in terms of the measure of the set: this is the celebrated *Faber-Krahn inequality* (7.1) recalled in the Introduction. The usual proof of this inequality relies on *Schwarz symmetrization* (see [505, Chapter 2]). The latter consists in associating to each positive function $u \in H_0^1(\Omega)$ a radially symmetric decreasing function $u^* \in H_0^1(\Omega^*)$, where Ω^* is the ball centered at the origin such that $|\Omega^*| = |\Omega|$. The function u^* is *equimeasurable* with u, that is

$$|\{x \, : \, u(x) > t\}| = |\{x \, : \, u^*(x) > t\}|, \qquad \text{for every } t \geq 0,$$

so that in particular every L^q norm of the function u is preserved. One also has the *Pólya-Szegő principle* (see the Subsection 7.2.3)

$$\int_{\Omega^*} |\nabla u^*|^2 \, dx \leq \int_{\Omega} |\nabla u|^2 \, dx, \tag{7.6}$$

from which the Faber-Krahn inequality easily follows.

For a connected set Ω, the first eigenvalue $\lambda_1(\Omega)$ is simple. In other words, there exists $u_1 \in H_0^1(\Omega) \setminus \{0\}$ such that every solution to

$$-\Delta u = \lambda_1(\Omega)\, u, \quad \text{in } \Omega, \qquad u = 0, \quad \text{on } \partial\Omega,$$

is proportional to u_1. For a ball B_r of radius r, the value $\lambda_1(B_r)$ can be explicitly computed, together with its corresponding eigenfunction. The latter is given by the radial function (see [505])

$$u(x) := |x|^{\frac{2-d}{2}} \, J_{\frac{d-2}{2}} \left(\frac{j_{(d-2)/2,1}}{r} \, |x| \right).$$

Here J_α is a Bessel function of the first kind, solving the ODE

$$g''(t) + \frac{1}{t}\, g'(t) + \left(1 - \frac{\alpha^2}{t^2}\right) g(t) = 0\,,$$

and $j_{\alpha,1}$ denotes the first positive zero of J_α. We have

$$\lambda_1(B_r) = \left(\frac{j_{(d-2)/2,1}}{r} \right)^2. \tag{7.7}$$

7.2.2 Semilinear eigenvalues and torsional rigidity

More generally, for an open set $\Omega \subset \mathbb{R}^d$ with finite measure, we will consider its *first semilinear eigenvalue* of the Dirichlet Laplacian

$$\lambda_1^q(\Omega) = \min_{u \in H_0^1(\Omega) \setminus \{0\}} \frac{\displaystyle\int_\Omega |\nabla u|^2 \, dx}{\left(\displaystyle\int_\Omega |u|^q \, dx \right)^{\frac{2}{q}}} = \min_{u \in H_0^1(\Omega)} \left\{ \int_\Omega |\nabla u|^2 \, dx \, : \, \|u\|_{L^q(\Omega)} = 1 \right\}, \tag{7.8}$$

where the exponent q satisfies

$$1 \le q < 2^* := \begin{cases} \dfrac{2\,d}{d-2}, & \text{if } d \ge 3, \\ +\infty, & \text{if } d = 2. \end{cases} \tag{7.9}$$

For every such an exponent q the embedding $H_0^1(\Omega) \hookrightarrow L^q(\Omega)$ is compact, thus the above minimization problem is well-defined. The shape functional $\Omega \mapsto \lambda_1^q(\Omega)$ satisfies the scaling law

$$\lambda_1^q(t\,\Omega) = t^{d-2-\frac{2}{q}\,d}\,\lambda_1^q(\Omega),$$

with the exponent $d-2-(2\,d)/q$ being negative. By means of Schwarz symmetrization, the following general family of Faber-Krahn inequalities can be derived

$$|\Omega|^{\frac{2}{d}+\frac{2}{q}-1}\,\lambda_1^q(\Omega) \ge |B|^{\frac{2}{d}+\frac{2}{q}-1}\,\lambda_1^q(B), \tag{7.10}$$

where B is any d–dimensional ball. Again, equality in (7.10) is possible if and only if Ω is a ball up to a set of zero capacity. Of course, when $q = 2$ we are back to $\lambda_1(\Omega)$ defined above. We also point out that the quantity

$$T(\Omega) := \frac{1}{\lambda_1^1(\Omega)} = \max_{u\in H_0^1(\Omega)\setminus\{0\}} \frac{\left(\displaystyle\int_\Omega u\,dx\right)^2}{\displaystyle\int_\Omega |\nabla u|^2\,dx},$$

is usually referred to as the *torsional rigidity* of the set Ω. In this case, we can write (7.10) in the form

$$|B|^{-\frac{d+2}{d}}\,T(B) \ge |\Omega|^{-\frac{d+2}{d}}\,T(\Omega). \tag{7.11}$$

This is sometimes called *Saint-Venant inequality*. We recall that for a ball of radius $R > 0$ we have

$$T(B_R) = \frac{1}{\lambda_1^1(B_R)} = \frac{\omega_d}{d\,(d+2)}\,R^{d+2}. \tag{7.12}$$

Remark 7.2 (Torsion function). *By recalling the definition* (2.9) *of torsion energy from Chapter 2*

$$E(\Omega) = \min_{u\in H_0^1(\Omega)} \left\{\frac{1}{2}\int_{\mathbb{R}^d} |\nabla u|^2\,dx - \int_{\mathbb{R}^d} u\,dx\right\},$$

we easily see that the torsional rigidity $T(\Omega)$ *can be equivalently defined through an unconstrained convex problem, i.e.*

$$-\,T(\Omega) = 2\,E(\Omega). \tag{7.13}$$

Indeed, it is sufficient to observe that for every $u \in H_0^1(\Omega)$ and $t > 0$, the function $t\,u$ is still admissible and thus by Young's inequality

$$2\,E(\Omega) = \min_{u\in H_0^1(\Omega)} \left\{ \int_\Omega |\nabla u|^2\,dx - 2\int_\Omega u\,dx \right\}$$

$$= \min_{u\in H_0^1(\Omega)} \min_{t>0} \left\{ t^2 \int_\Omega |\nabla u|^2\,dx - 2\,t\int_\Omega u\,dx \right\} = \min_{u\in H_0^1(\Omega)} -\frac{\left(\int_\Omega u\,dx\right)^2}{\int_\Omega |\nabla u|^2\,dx},$$

which proves (7.13). *The unique solution w_Ω of the problem on the right-hand side in* (7.13) *is called torsion function and it satisfies*

$$-\Delta w_\Omega = 1, \quad \text{in } \Omega, \qquad w_\Omega = 0, \quad \text{on } \partial\Omega.$$

From (7.13) *and the equation satisfied by w_Ω, we thus also get*

$$T(\Omega) = \int_\Omega w_\Omega\,dx.$$

7.2.3 Some pioneering stability results

In this subsection we recall the quantitative estimates for the Faber-Krahn inequality by Hansen & Nadirashvili [447] and Melas [706].

Since the proof of the Faber-Krahn inequality is based on the Pólya-Szegő principle (7.6), it is appropriate to recall how (7.6) can be proved. By following Talenti (see [842, Lemma 1]), the proof combines the *coarea formula*, the *convexity* of the function $t \mapsto t^2$ and the Euclidean *isoperimetric inequality*

$$|\Omega|^{\frac{1-d}{d}}\,P(\Omega) \geq |B|^{\frac{1-d}{d}}\,P(B). \tag{7.14}$$

Here $P(\cdot)$ denotes the perimeter of a set. If $u \in H_0^1(\Omega)$ is a smooth positive function and we set

$$\Omega_t := \{x \in \Omega\,:\,u(x) > t\} \qquad \text{and} \qquad \mu(t) := |\Omega_t|,$$

by using the above mentioned tools, one can infer

$$\begin{aligned}\int_\Omega |\nabla u|^2\,dx &\overset{\text{coarea}}{=} \int_0^\infty \left(\int_{\{u=t\}} |\nabla u|^2\,\frac{d\mathcal{H}^{d-1}}{|\nabla u|} \right) dt \\ &\overset{\text{Jensen}}{\geq} \int_0^\infty P(\Omega_t)^2\,\frac{1}{\int_{\{u=t\}} |\nabla u|^{-1}\,d\mathcal{H}^{d-1}}\,dt = \int_0^\infty \frac{P(\Omega_t)^2}{-\mu'(t)}\,dt \\ &\overset{\text{Isoperimetry}}{\geq} \int_0^\infty \frac{P(\Omega_t^\star)^2}{-\mu'(t)}\,dt = \int_{\Omega^\star} |\nabla u^\star|^2\,dx,\end{aligned} \tag{7.15}$$

where Ω_t^* is the ball centered at the origin such that $|\Omega_t^*| = |\Omega_t|$. For a smooth function, the equality

$$-\mu'(t) = \int_{\{u=t\}} \frac{1}{|\nabla u|}\, d\mathcal{H}^{d-1}, \qquad \text{for a. e. } t > 0,$$

follows from Sard's Theorem, but all the passages in (7.15) can indeed be justified for a genuine H_0^1 function. We refer the reader to [273, Section 2] for more details.

By taking u to be a first eigenfunction of Ω with unit L^2 norm and observing that u^* is admissible for the variational problem defining $\lambda_1(\Omega^*)$, from (7.15) one easily gets the Faber-Krahn inequality

$$\lambda_1(\Omega) \geq \lambda_1(\Omega^*),$$

as desired.

The idea of Hansen & Nadirashvili [447] and Melas [706] was to replace in (7.15) the classical isoperimetric statement (7.14) with an improved quantitative version. At the time of [447] and [706], quantitative versions of the isoperimetric inequality were available only for particular sets, and were known as *Bonnesen inequalities*. These cover simply connected sets in dimension $d = 2$ (see Bonnesen's paper [165], generalized in [409, Theorem 2.2]) and convex sets in every dimension (see [408, Theorem 2.3]).

For this reason, both papers treat simply connected sets in dimension $d = 2$ or convex sets in general dimensions. We now present their results, without entering at all into the details of the proofs. Rather, in the next subsection we will explain the ideas by Hansen and Nadirashvili and use them to prove a more general result (see Theorem 7.11 below).

Theorem 7.3 (Melas). *For every open bounded set $\Omega \subset \mathbb{R}^d$, we define the asymmetry functional*

$$\delta_{\mathcal{M}}(\Omega) := \min \left\{ \max \left\{ \frac{|\Omega \setminus B_1|}{|\Omega|}, \frac{|B_2 \setminus \Omega|}{|B_2|} \right\} : B_1 \subset \Omega \subset B_2 \text{ balls} \right\}. \tag{7.16}$$

Then we have:

- *if $d = 2$, for every Ω open bounded simply connected set, there exists a disc $B_\Omega \subset \Omega$ such that*

$$|\Omega|\, \lambda_1(\Omega) - |B|\, \lambda_1(B) \geq \frac{1}{C} \left(\frac{|\Omega \setminus B_\Omega|}{|\Omega|} \right)^4,$$

for some universal constant $C > 0$;

- *if $d \geq 2$, for every open bounded convex set $\Omega \subset \mathbb{R}^d$ we have*

$$|\Omega|^{2/d}\, \lambda_1(\Omega) - |B|^{2/d}\, \lambda_1(B) \geq \frac{1}{C}\, \delta_{\mathcal{M}}(\Omega)^{2\,d},$$

for some universal constant $C > 0$.

Remark 7.4. *In dimension $d = 2$ Melas proved a slightly more general result. If $\Omega \subset \mathbb{R}^2$ is an open bounded set, not necessarily simply connected, then there exists an open disc B_Ω such that*

$$|\Omega|\,\lambda_1(\Omega) - |B|\,\lambda_1(B) \geq \frac{1}{C}\left(\frac{|\Omega \Delta B_\Omega|}{|\Omega \cup B_\Omega|}\right)^4,$$

for some universal constant $C > 0$.

For every open set $\Omega \subset \mathbb{R}^d$ we denote

$$r_\Omega = \sup\{r > 0 \,:\, \exists x_0 \in \Omega \text{ such that } B_r(x_0) \subset \Omega\} \qquad \text{and} \qquad R_\Omega = \left(\frac{|\Omega|}{\omega_d}\right)^{\frac{1}{d}}.$$

The first quantity is usually called *inradius of Ω*. This is the radius of the largest ball contained in Ω.

Theorem 7.5 (Hansen-Nadirashvili). *For every open set $\Omega \subset \mathbb{R}^d$ with finite measure, we define the asymmetry functional*

$$\delta_{\mathcal{N}}(\Omega) := 1 - \frac{r_\Omega}{R_\Omega}. \tag{7.17}$$

Then we have:
- *if $d = 2$ and Ω is simply connected,*

$$|\Omega|\,\lambda_1(\Omega) - |B|\,\lambda_1(B) \geq \left(\frac{\pi j_{0,1}^2}{250}\right)\delta_{\mathcal{N}}(\Omega)^3;$$

- *if $d \geq 3$, there exist $0 < \varepsilon < 1$ and $C > 0$ such that for every $\Omega \subset \mathbb{R}^d$ open bounded convex set satisfying $\delta_{\mathcal{N}}(\Omega) < \varepsilon$, we have*

$$|\Omega|^{2/d}\,\lambda_1(\Omega) - |B|^{2/d}\,\lambda_1(B) \geq \frac{1}{C}\begin{cases} \dfrac{\delta_{\mathcal{N}}(\Omega)^3}{|\log \delta_{\mathcal{N}}(\Omega)|}, & \text{if } d = 3,\\[2ex] \delta_{\mathcal{N}}(\Omega)^{\frac{d+3}{2}}, & \text{if } d \geq 4.\end{cases}$$

Remark 7.6 (The role of topology). *It is easy to see that the stability estimates of Theorems 7.3 and 7.5 with $\delta_{\mathcal{M}}$ and $\delta_{\mathcal{N}}$ can not hold true without some topological assumptions on the sets. For example, by taking the perforated ball*

$$\Omega_\varepsilon = \{x \in \mathbb{R}^d \,:\, \varepsilon < |x| < 1\}, \qquad 0 < \varepsilon < 1,$$

we have

$$\lim_{\varepsilon \searrow 0}\left(|\Omega_\varepsilon|^{2/d}\,\lambda(\Omega_\varepsilon) - |B|^{2/d}\,\lambda(B)\right) = 0, \tag{7.18}$$

while

$$\lim_{\varepsilon \searrow 0} \delta_{\mathcal{N}}(\Omega_\varepsilon) = \frac{1}{2} \qquad \text{and} \qquad \lim_{\varepsilon \searrow 0} \delta_{\mathcal{M}}(\Omega_\varepsilon) \geq \frac{1}{2}.$$

For the limit (7.18) see for example [387, Theorem 9]. These contradict Theorems 7.3 and 7.5. Observe that for $d = 2$ the set Ω_ε is not simply connected, while for $d \geq 3$ it is. Thus in higher dimensions simple connectedness is still not sufficient to have stability with respect to $\delta_{\mathcal{N}}$ or $\delta_{\mathcal{M}}$.

If we want to obtain a quantitative Faber-Krahn inequality for general open sets in every dimension, a more flexible notion of asymmetry is the so called *Fraenkel asymmetry*, defined by

$$\mathcal{A}(\Omega) = \inf\left\{\frac{|\Omega \Delta B|}{|\Omega|} \;:\; B \text{ ball such that } |B| = |\Omega|\right\}.$$

Observe that for every ball B such that $|B| = |\Omega|$, we have $|\Omega \Delta B| = 2\,|\Omega \setminus B| = 2\,|B \setminus \Omega|$. This simple fact will be used repeatedly.

It is not difficult to see that this is a weaker asymmetry functional than $\delta_{\mathcal{N}}$ and $\delta_{\mathcal{M}}$ above. Indeed, we have the following.

Lemma 7.7 (Comparison between asymmetries). *Let $\Omega \subset \mathbb{R}^d$ be an open bounded set. Then we have*

$$\delta_{\mathcal{N}}(\Omega) \geq \frac{1}{2\,d}\,\mathcal{A}(\Omega), \qquad \delta_{\mathcal{M}}(\Omega) \geq \frac{1}{2}\,\mathcal{A}(\Omega) \qquad \text{and} \qquad \delta_{\mathcal{M}}(\Omega) \geq \frac{d}{2^d}\,\delta_{\mathcal{N}}(\Omega). \tag{7.19}$$

If Ω is convex, we also have

$$\mathcal{A}(\Omega) \geq \frac{1}{d}\left(\frac{\omega_d}{d}\right)^{1/d} \frac{|\Omega|^{1/d}}{\operatorname{diam}(\Omega)}\,\delta_{\mathcal{M}}(\Omega)^d.$$

Proof. By using the elementary inequality

$$a^d - b^d \leq d\,a^{d-1}\,(a-b), \qquad \text{for } 0 \leq b \leq a,$$

and the definitions of r_Ω, R_Ω and $\delta_{\mathcal{N}}(\Omega)$, we have

$$|\Omega| - \omega_d\, r_\Omega^d \leq d\,|\Omega|^{\frac{d-1}{d}}\left(|\Omega|^{\frac{1}{d}} - \omega_d^{\frac{1}{d}}\, r_\Omega\right) = d\,|\Omega|\,\delta_{\mathcal{N}}(\Omega). \tag{7.20}$$

We then consider a ball $B_{r_\Omega}(x_0) \subset \Omega$ and take a concentric ball $\widetilde{B}$ such that $|\widetilde{B}| = |\Omega|$. By definition of Fraenkel asymmetry and estimate (7.20), we obtain

$$\mathcal{A}(\Omega) \leq 2\,\frac{|\Omega \setminus \widetilde{B}|}{|\Omega|} \leq 2\,\frac{|\Omega \setminus B_{r_\Omega}(x_0)|}{|\Omega|} \leq 2\,d\,\delta_{\mathcal{N}}(\Omega),$$

and thus we get the first estimate in (7.19).

For the second one, we take a pair of balls $B_1 \subset \Omega \subset B_2$ and consider the ball $\widetilde{B}_1$ concentric with B_1 and such that $|\Omega| = |\widetilde{B}_1|$. Then we get

$$\mathcal{A}(\Omega) \leq 2\,\frac{|\Omega \setminus \widetilde{B}_1|}{|\Omega|} \leq 2\,\frac{|\Omega \setminus B_1|}{|\Omega|} \leq 2\,\max\left\{\frac{|\Omega \setminus B_1|}{|\Omega|}, \frac{|B_2 \setminus \Omega|}{|B_2|}\right\}.$$

By taking the infimum over the admissible pairs (B_1, B_2) we get the second inequality in (7.19).

Finally, for the third inequality we take again a pair of balls $B_1 \subset \Omega \subset B_2$ and observe that if $r_\Omega \le R_\Omega/2$, then we have

$$\max\left\{\frac{|\Omega \setminus B_1|}{|\Omega|}, \frac{|B_2 \setminus \Omega|}{|B_2|}\right\} \ge \frac{|\Omega \setminus B_1|}{|\Omega|} \ge 1 - \left(\frac{1}{2}\right)^d \ge \left[1 - \left(\frac{1}{2}\right)^d\right] \delta_{\mathcal{N}}(\Omega).$$

In the last inequality we used the fact that $\delta_{\mathcal{N}}(\Omega) < 1$. By taking the infimum over admissible pairs of balls, we obtain the conclusion. If on the contrary Ω is such that $r_\Omega > R_\Omega/2$, then by definition of R_Ω and r_Ω we get

$$\begin{aligned}\delta_{\mathcal{N}}(\Omega) = \frac{R_\Omega - r_\Omega}{R_\Omega} = \frac{|\Omega|^{1/d} - |B_{r_\Omega}|^{1/d}}{|\Omega|^{1/d}} &\le \frac{1}{d}\,\frac{|B_{r_\Omega}|^{1/d}}{|\Omega|^{1/d}}\,\frac{|\Omega| - |B_{r_\Omega}|}{|B_{r_\Omega}|}\\ &\le \frac{1}{d}\,\frac{|\Omega| - |B_1|}{|B_{r_\Omega}|} \le \frac{2^d}{d}\,\frac{|\Omega \setminus B_1|}{|\Omega|}\\ &\le \frac{2^d}{d}\max\left\{\frac{|\Omega \setminus B_1|}{|\Omega|}, \frac{|B_2 \setminus \Omega|}{|B_2|}\right\},\end{aligned}$$

where we used that $|B_1| \le |B_{r_\Omega}| \le |\Omega|$. Thus we get the conclusion in this case as well. Observe that $1 - 2^{-d} \ge d\,2^{-d}$ for $d \ge 2$.

Let us now assume Ω to be convex. We take a ball $|B| = |\Omega|$ such that $|\Omega \setminus B|/|\Omega| = \mathcal{A}(\Omega)/2$. We can assume that $\mathcal{A}(\Omega) < 1/2$, otherwise the estimate is trivial by using the isodiametric inequality

$$|\Omega| \le \omega_d \left(\frac{\operatorname{diam}(\Omega)}{2}\right)^d,$$

and the fact that $\delta_{\mathcal{M}} < 1$. Then from [368, Lemma 4.2] we know that

$$|\Omega \setminus B| \ge \frac{1}{2\,d}\,\frac{|\Omega|}{\operatorname{diam}(\Omega)^d}\,\operatorname{Haus}(\Omega, B)^d.$$

Here $\operatorname{Haus}(E_1, E_2)$ denotes the Hausdorff distance between sets, defined by

$$\operatorname{Haus}(E_1, E_2) = \max\left\{\sup_{x\in E_1}\inf_{y\in E_2}|x-y|,\ \sup_{y\in E_2}\inf_{x\in E_1}|x-y|\right\}. \tag{7.21}$$

We then observe that the ball $B_2 := \gamma B$ contains Ω, provided

$$\gamma = \frac{R_\Omega + \operatorname{Haus}(\Omega, B)}{R_\Omega}.$$

On the other hand, by Lemma 7.74 we have

$$r_\Omega \ge R_\Omega - \operatorname{Haus}(\Omega, B).$$

Let us assume that $\mathrm{Haus}(\Omega, B) < R_\Omega$. From the definition of $\delta_{\mathcal{M}}$, we thus obtain

$$\begin{aligned}\delta_{\mathcal{M}}(\Omega) &\le \max\left\{\frac{\gamma^d - 1}{\gamma^d}, 1 - \frac{r_\Omega^d}{R_\Omega^d}\right\} \le \max\left\{1 - \frac{1}{\gamma^d}, 1 - \left(1 - \frac{\mathrm{Haus}(\Omega, B)}{R_\Omega}\right)^d\right\} \\ &= \max\left\{1 - \left(1 - \frac{\mathrm{Haus}(\Omega, B)}{R_\Omega + \mathrm{Haus}(\Omega, B)}\right)^d, 1 - \left(1 - \frac{\mathrm{Haus}(\Omega, B)}{R_\Omega}\right)^d\right\} \\ &\le d \max\left\{\frac{\mathrm{Haus}(\Omega, B)}{R_\Omega + \mathrm{Haus}(\Omega, B)}, \frac{\mathrm{Haus}(\Omega, B)}{R_\Omega}\right\} \\ &= d\,\frac{\mathrm{Haus}(\Omega, B)}{R_\Omega} \le d\left(\frac{d}{\omega_d}\right)^{\frac{1}{d}} \frac{\mathrm{diam}(\Omega)}{|\Omega|^{1/d}}\,\mathcal{A}(\Omega)^{1/d}.\end{aligned}$$

This concludes the proof for $\mathrm{Haus}(\Omega, B) < R_\Omega$. If on the contrary $\mathrm{Haus}(\Omega, B) \ge R_\Omega$, with similar computations we get

$$\begin{aligned}\delta_{\mathcal{M}}(\Omega) &\le \max\left\{\frac{\gamma^d - 1}{\gamma^d}, 1 - \frac{r_\Omega^d}{R_\Omega^d}\right\} \\ &\le \max\left\{1 - \left(1 - \frac{\mathrm{Haus}(\Omega, B)}{R_\Omega + \mathrm{Haus}(\Omega, B)}\right)^d, 1\right\} \\ &\le \max\left\{d\,\frac{\mathrm{Haus}(\Omega, B)}{R_\Omega + \mathrm{Haus}(\Omega, B)}, 1\right\} \le \frac{d}{2}\,\frac{\mathrm{Haus}(\Omega, B)}{R_\Omega},\end{aligned}$$

and we can conclude as before. □

Remark 7.8. *For general open sets, the asymmetries $\mathcal{A}$, $\delta_{\mathcal{N}}$ and $\delta_{\mathcal{M}}$ are not equivalent. We first observe that if $\Omega_0 = B \setminus \Sigma$, where B is a ball and $\Sigma \subset B$ is a non-empty closed set with $|\Sigma| = 0$, then*

$$\mathcal{A}(\Omega_0) = 0 \qquad \textit{while} \qquad \delta_{\mathcal{N}}(\Omega_0) > 0, \quad \textit{and} \quad \delta_{\mathcal{M}}(\Omega_0) > 0.$$

Moreover, there exists a sequence of open sets $\{\Omega_n\}_{n\in\mathbb{N}} \subset \mathbb{R}^d$ such that

$$\lim_{n\to\infty} \delta_{\mathcal{N}}(\Omega_n) = 0 \qquad \textit{and} \qquad \lim_{n\to\infty} \delta_{\mathcal{M}}(\Omega_n) > 0.$$

Such a sequence $\{\Omega_n\}_{n\in\mathbb{N}}$ can be constructed by attaching a long tiny tentacle to a ball, for example.

7.2.4 A variation on a theme of Hansen and Nadirashvili

We will now show how to adapt the ideas developed by Hansen and Nadirashvili, to get a (non sharp) stability estimate for the general Faber-Krahn inequality (7.10) and for general open sets with finite measure.

First of all, one needs a quantitative improvement of the isoperimetric inequality which is valid for generic sets and dimensions. Such a (sharp) quantitative isoperimetric inequality has been proved by Fusco, Maggi and Pratelli in [411, Theorem 1.1] (see also [275, 381] for different proofs and [412] for an exhaustive review of quantitative forms of the isoperimetric inequality). This reads as follows

$$|\Omega|^{\frac{1-d}{d}}\, P(\Omega) - |B|^{\frac{1-d}{d}}\, P(B) \geq \beta_d\, \mathcal{A}(\Omega)^2. \tag{7.22}$$

An explicit value for the dimensional constant $\beta_d > 0$ can be found in [381, Theorem 1.1]. By inserting this information in the proof (7.15) of Pólya-Szegő inequality, one would get an estimate of the type

$$\int_\Omega |\nabla u|^2\, dx - \int_{\Omega^*} |\nabla u^*|^2\, dx \gtrsim \int_0^\infty \mathcal{A}(\Omega_t)^2\, dt.$$

The difficulty here is to estimate the "propagation of asymmetry" from the whole domain Ω to the superlevel sets Ω_t of the optimal function u. In other words, we would need to know that

$$\mathcal{A}(\Omega) \simeq \mathcal{A}(\Omega_t), \qquad \text{for } t > 0.$$

Unfortunately, in general it is difficult to exclude that

$$\mathcal{A}(\Omega_t) \ll \mathcal{A}(\Omega), \qquad \text{for } t \simeq 0.$$

This means that the graph of u "quickly becomes round" when it detaches from the boundary $\partial\Omega$. This may happen for example if u has a small normal derivative. For these reasons, improving this idea is very delicate, which usually results in a (non sharp) estimate like the ones of Theorems 7.3 and 7.5 and the one of Theorem 7.11 below. We refer to the discussion of Section 7.7.1 for other results of this type, previously obtained by Bhattacharya [146] and Fusco, Maggi and Pratelli [413].

The following expedient result is sometimes useful for stability issues. It states that if the measure of a subset $U \subset \Omega$ differs from that of Ω by an amount comparable to the asymmetry $\mathcal{A}(\Omega)$, then the asymmetry of U can not decrease too much. This is essentially taken from [447, Section 5].

Lemma 7.9 (Propagation of asymmetry). *Let $\Omega \subset \mathbb{R}^d$ be an open set with finite measure. Let $U \subset \Omega$ be such that $|U| > 0$ and*

$$\frac{|\Omega \setminus U|}{|\Omega|} \leq \frac{1}{4}\, \mathcal{A}(\Omega). \tag{7.23}$$

Then there holds

$$\mathcal{A}(U) \geq \frac{1}{2}\, \mathcal{A}(\Omega). \tag{7.24}$$

Proof. Let B be a ball achieving the minimum in the definition of $\mathcal{A}(U)$, by the triangle inequality we get

$$\mathcal{A}(U) = \frac{|U\Delta B|}{|U|} \geq \frac{|\Omega|}{|U|}\left(\frac{|\Omega\Delta B|}{|\Omega|} - \frac{|U\Delta\Omega|}{|\Omega|}\right)$$
$$\geq \frac{|\Omega|}{|U|}\left(\frac{|\Omega\Delta B'|}{|\Omega|} - \frac{|B\Delta B'|}{|\Omega|} - \frac{|U\Delta\Omega|}{|\Omega|}\right),$$

where B' is a ball concentric with B and such that $|B'| = |\Omega|$. By using that

$$|U\Delta\Omega| = |\Omega \setminus U| = |\Omega| - |U| = |B'\Delta B|,$$

and the hypothesis (7.23), we get the conclusion by further noticing that $|\Omega| \geq |U|$. □

By relying on the previous simple result, we can prove a sort of Pólya-Szegő inequality with remainder term. The remainder term depends on the asymmetry of Ω and on the level s of the function, whose corresponding superlevel set $\{x : u(x) > s\}$ has a *measure defect* comparable to the asymmetry $\mathcal{A}(\Omega)$, i.e. it satisfies (7.23).

Lemma 7.10 (Boosted Pólya-Szegő principle). *Let $\Omega \subset \mathbb{R}^d$ be an open set with finite measure, such that $\mathcal{A}(\Omega) > 0$. Let $u \in H_0^1(\Omega)$ be such that $u > 0$ in Ω. For every $t > 0$ we still denote*

$$\Omega_t = \{x \in \Omega : u(x) > t\} \qquad \text{and} \qquad \mu(t) = |\Omega_t|.$$

Let $s > 0$ be the level defined by

$$s = \sup\left\{t : \mu(t) \geq |\Omega|\left(1 - \frac{1}{4}\mathcal{A}(\Omega)\right)\right\}. \tag{7.25}$$

Then we have

$$\int_\Omega |\nabla u|^2\,dx \geq \int_{\Omega^*} |\nabla u^*|^2\,dx + c_d\,\mathcal{A}(\Omega)\,|\Omega|^{1-\frac{2}{d}}\,s^2. \tag{7.26}$$

The dimensional constant $c_d > 0$ is given by

$$c_d = \frac{4^{1/d}\,\beta_d\,d\,\omega_d^{1/d}}{2},$$

where β_d is the same constant appearing in (7.22).

Proof. We first observe that the level s defined by (7.25) is not 0. Indeed, the function $t \mapsto \mu(t)$ is right-continuous, thus we get

$$\lim_{t\to 0^+} \mu(t) = \mu(0) = |\{x \in \Omega : u(x) > 0\}| = |\Omega| > |\Omega|\left(1 - \frac{1}{4}\mathcal{A}(\Omega)\right),$$

where we used the hypothesis on u and the fact that $\mathcal{A}(\Omega) > 0$.

By using the sharp quantitative isoperimetric inequality (7.22), we have

$$P(\Omega_t) \geq P(\Omega_t^*) + \beta_d\,\mu(t)^{\frac{d-1}{d}}\,\mathcal{A}(\Omega_t)^2, \tag{7.27}$$

while by convexity of the map $\tau \mapsto \tau^2$ we get

$$
\begin{aligned}
P(\Omega_t)^2 &\geq P(\Omega_t^\star)^2 + 2\,P(\Omega_t^\star)\left(P(\Omega_t) - P(\Omega_t^\star)\right) \\
&= P(\Omega_t^\star)^2 + 2\left(d\,\omega_d^{1/d}\,\mu(t)^{\frac{d-1}{d}}\right)\left(P(\Omega_t) - P(\Omega_t^\star)\right).
\end{aligned}
$$

By collecting the previous two estimates and reproducing the proof of (7.15), we can now infer

$$
\int_\Omega |\nabla u|^2\,dx \geq \int_{\Omega^*} |\nabla u^\star|^2\,dx + c\int_0^s \mathcal{A}(\Omega_t)^2\,\frac{\left(\mu(t)^{\frac{d-1}{d}}\right)^2}{-\mu'(t)}\,dt, \tag{7.28}
$$

where we set

$$
c = 2\,\beta_d\,d\,\omega_d^{1/d}.
$$

We now observe that μ is a decreasing function, thus we have

$$
\mu(t) > \mu(s) \geq |\Omega|\left(1 - \frac{1}{4}\,\mathcal{A}(\Omega)\right), \qquad 0 < t < s.
$$

This implies that the set Ω_t satisfies the hypothesis of Lemma 7.9 for $0 < t < s$, since

$$
\frac{|\Omega \setminus \Omega_t|}{|\Omega|} = 1 - \frac{\mu(t)}{|\Omega|} \leq 1 - \left(1 - \frac{1}{4}\,\mathcal{A}(\Omega)\right) = \frac{1}{4}\,\mathcal{A}(\Omega), \qquad 0 < t < s.
$$

Thus from (7.24) we get

$$
\mathcal{A}(\Omega_t) \geq \frac{1}{2}\,\mathcal{A}(\Omega), \qquad 0 < t < s.
$$

By inserting the previous information in (7.28) and using that

$$
\mu(t) > \mu(s) \geq |\Omega|\left(1 - \frac{1}{4}\,\mathcal{A}(\Omega)\right) \geq \frac{|\Omega|}{2}, \qquad 0 < t < s,
$$

we get

$$
\int_\Omega |\nabla u|^2\,dx \geq \int_{\Omega^*} |\nabla u^\star|^2\,dx + \frac{c}{4}\,\mathcal{A}(\Omega)^2\left(\frac{|\Omega|^{\frac{d-1}{d}}}{2^{\frac{d-1}{d}}}\right)^2 \int_0^s \frac{1}{-\mu'(t)}\,dt.
$$

We then observe that by convexity of the function $\tau \mapsto \tau^{-1}$, Jensen's inequality gives[7.3]

$$
\int_0^s \frac{1}{-\mu'(t)}\,dt \geq \frac{s^2}{\displaystyle\int_0^s -\mu'(t)\,dt} \geq \frac{s^2}{|\Omega| - \mu(s)} \geq 4\,\frac{s^2}{\mathcal{A}(\Omega)\,|\Omega|},
$$

thanks to the choice (7.25) of s. This concludes the proof. □

7.3 In the second inequality, we used that for a monotone non-decreasing function f

$$
\int_a^b f'(t)\,dt \leq f(b) - f(a), \qquad \text{for } a < b.
$$

We can now prove the following quantitative version of the general Faber-Krahn inequality (7.10). The standard case of the first eigenvalue of the Dirichlet Laplacian corresponds to taking $q = 2$. Though the exponent on the Fraenkel asymmetry is not sharp, the interesting part of the result lies in the computable constant. Moreover, the proof is quite simple and it is based on the ideas by Hansen & Nadirashvili. We also use the *Kohler-Jobin inequality* (see Appendix 7.8.1) to reduce to the case of the torsional rigidity. This reduction trick was first introduced by Brasco, De Philippis and Velichkov in [181].

Theorem 7.11. *There exists an explicit dimensional constant $\tau > 0$ such that for every open $\Omega \subset \mathbb{R}^d$ with finite measure, we have the following* quantitative Saint-Venant inequality

$$|B|^{-\frac{d+2}{d}}\, T(B) - |\Omega|^{-\frac{d+2}{d}}\, T(\Omega) \geq \tau\, \mathcal{A}(\Omega)^3. \tag{7.29}$$

More generally, for $1 \leq q < 2^*$ there exists an explicit constant $\tau_{d,q} > 0$ such that for every open set $\Omega \subset \mathbb{R}^d$ with finite measure, we have

$$|\Omega|^{\frac{2}{d}+\frac{2}{q}-1}\, \lambda_1^q(\Omega) - |B|^{\frac{2}{d}+\frac{2}{q}-1}\, \lambda_1^q(B) \geq \tau_{d,q}\, \mathcal{A}(\Omega)^3. \tag{7.30}$$

Proof. Since inequalities (7.29) and (7.30) are scaling invariant, we can suppose that $|\Omega| = 1$. By Proposition 7.68 with

$$g(t) = t^3 \qquad \text{and} \qquad \delta(\Omega) = \mathcal{A}(\Omega),$$

it is sufficient to prove (7.29). Of course, we can suppose that $\mathcal{A}(\Omega) > 0$, otherwise there is nothing to prove. Let Ω^* be the ball centered at the origin such that $|\Omega^*| = |\Omega| = 1$. Without loss of generality, we can also suppose that

$$T(\Omega) \geq \frac{T(\Omega^*)}{2}. \tag{7.31}$$

Indeed, if the latter is not satisfied, then (7.29) trivially holds with constant $\tau = T(\Omega^*)/16$, thanks to the fact that $\mathcal{A}(\Omega) < 2$.

Let $w_\Omega \in H_0^1(\Omega)$ be the torsion function of Ω (recall Remark 7.2), then we know

$$T(\Omega) = \int_\Omega w_\Omega\, dx = \int_\Omega |\nabla w_\Omega|^2\, dx.$$

Moreover, by standard elliptic regularity we know that $w_\Omega \in C^\infty(\Omega) \cap L^\infty(\Omega)$. By recalling that $|\Omega| = 1$ and (7.31), we get

$$\frac{T(\Omega^*)}{2} \leq T(\Omega) = \int_\Omega w_\Omega\, dx \leq \|w_\Omega\|_{L^\infty(\Omega)}.$$

We now take s as in (7.25), from (7.26) and the definition of torsional rigidity we get

$$T(\Omega) \leq \frac{\left(\displaystyle\int_{\Omega^*} w_\Omega^*\, dx\right)^2}{\displaystyle\int_{\Omega^*} |\nabla w_\Omega^*|^2\, dx + c_d\, \mathcal{A}(\Omega)\, s^2} \leq T(\Omega^*) \left(1 + \frac{c_d\, \mathcal{A}(\Omega)\, s^2}{\displaystyle\int_{\Omega^*} |\nabla w_\Omega^*|^2\, dx}\right)^{-1},$$

that is

$$\frac{T(\Omega^*)}{T(\Omega)} - 1 \geq \frac{c_d\, \mathcal{A}(\Omega)\, s^2}{\displaystyle\int_{\Omega^*} |\nabla w_\Omega^*|^2\, dx}.$$

With simple manipulations, by using $\int_{\Omega^*} |\nabla w_\Omega^*|^2 \leq \int_\Omega |\nabla w_\Omega|^2 = T(\Omega)$, we get

$$T(\Omega^*) - T(\Omega) \geq c_d\, \mathcal{A}(\Omega)\, s^2. \tag{7.32}$$

We then set

$$s_0 = \frac{T(\Omega^*)}{16}\, \frac{d+2}{3\,d+2}\, \mathcal{A}(\Omega), \tag{7.33}$$

and observe that $s_0 < T(\Omega^*)/4$. We have to distinguish two cases.

First case: $s \geq s_0$. This is the easy case, as from (7.32) we directly get (7.29), with constant

$$\tau = \frac{c_d}{256}\, T(\Omega^*)^2 \left(\frac{d+2}{3\,d+2}\right)^2,$$

and we recall that c_d is as in Lemma 7.10.

Second case: $s < s_0$. In this case, by definition (7.25) of s we get

$$\mu(s_0) < \left(1 - \frac{1}{4}\, \mathcal{A}(\Omega)\right). \tag{7.34}$$

We also observe that $\mu(s_0) > 0$, since $s_0 < T(\Omega^*)/4 \leq 1/2\, \|w_\Omega\|_{L^\infty}$ by the discussion above. We now want to work with this level s_0. We have

$$\int_{\Omega_{s_0}} (w_\Omega - s_0)_+\, dx = \int_\Omega (w_\Omega - s_0)_+\, dx \geq \int_\Omega w_\Omega\, dx - s_0 = T(\Omega) - s_0. \tag{7.35}$$

Observe that the right-hand side is strictly positive, since $s_0 < T(\Omega)$ thanks to (7.33) and (7.31). We have $(w_\Omega - s_0)_+ \in H_0^1(\Omega_{s_0})$, thus from the variational characterization of $T(\Omega)$, the Saint-Venant inequality and (7.35)

$$\begin{aligned} T(\Omega) = \frac{\left(\displaystyle\int_\Omega w_\Omega\, dx\right)^2}{\displaystyle\int_\Omega |\nabla w_\Omega|^2\, dx} &\leq \frac{\left(\displaystyle\int_\Omega w_\Omega\, dx\right)^2}{\displaystyle\int_\Omega |\nabla (w_\Omega - s_0)_+|^2\, dx} \leq T(\Omega_{s_0}) \left(\frac{\displaystyle\int_\Omega w_\Omega\, dx}{\displaystyle\int_\Omega (w - s_0)_+\, dx}\right)^2 \\ &\leq T(\Omega^*)\, \mu(s_0)^{\frac{d+2}{d}} \left(1 - \frac{s_0}{T(\Omega)}\right)^{-2}. \end{aligned}$$

Since s_0 satisfies (7.34), from the previous estimate and again (7.31) we can infer

$$\left(1 - \frac{1}{4}\, \mathcal{A}(\Omega)\right)^{-\frac{d+2}{d}} \left(1 - \frac{s_0}{T(\Omega^*)}\right)^2 - 1 \leq \frac{T(\Omega^*)}{T(\Omega)} - 1 \leq \frac{2}{T(\Omega^*)} \left(T(\Omega^*) - T(\Omega)\right). \tag{7.36}$$

We then observe that

$$\left(1-\frac{1}{4}\,\mathcal{A}(\Omega)\right)^{-\frac{d+2}{d}} \ge 1+\frac{d+2}{4\,d}\,\mathcal{A}(\Omega) \qquad \text{and} \qquad \left(1-\frac{s_0}{T(\Omega^*)}\right)^2 \ge 1-\frac{2\,s_0}{T(\Omega^*)}.$$

Thus from (7.36) we get

$$T(\Omega^*)-T(\Omega) \ge \frac{T(\Omega^*)}{2}\left[\left(1+\frac{d+2}{4\,d}\,\mathcal{A}(\Omega)\right)\left(1-\frac{2\,s_0}{T(\Omega^*)}\right)-1\right]. \tag{7.37}$$

We now recall the definition (7.33) of s_0 and finally estimate

$$\left(1+\frac{d+2}{4\,d}\,\mathcal{A}(\Omega)\right)\left(1-\frac{2\,s_0}{T(\Omega^*)}\right)-1 \ge \frac{d+2}{8\,d}\,\mathcal{A}(\Omega).$$

By inserting this in (7.37) and recalling that $\mathcal{A}(\Omega)<2$, we get (7.29) with

$$\tau = T(\Omega^*)\,\frac{d+2}{64\,d}.$$

This concludes the proof of (7.30) and thus of the theorem. □

Remark 7.12 (Value of the constant $\tau_{d,q}$). *In the previous proof Ω^* is a ball with measure 1, then from* (7.12)

$$T(\Omega^*) = \frac{\omega_d^{-2/d}}{d\,(d+2)}.$$

Thus a possible value for the constant τ in the quantitative Saint-Venant inequality (7.29) *is*

$$\tau = \frac{\omega_d^{-2/d}}{16\,d\,(d+2)}\,\min\left\{1,\ \frac{c_d}{16}\,\frac{\omega_d^{-2/d}}{d}\,\frac{d+2}{(3\,d+2)^2},\ \frac{d+2}{4\,d}\right\},$$

with c_d as in Lemma 7.10. Consequently, from Proposition 7.68 we get

$$\tau_{d,q} = (2^{\vartheta}-1)\,|B|^{\frac{2}{d}+\frac{2}{q}-1}\,\lambda_1^q(B)\,\min\left\{\tau\,\frac{|B|^{\frac{d+2}{d}}}{T(B)},\ \frac{1}{8}\right\}, \qquad \vartheta = \frac{2+\dfrac{2}{q}\,d-d}{d+2} < 1.$$

for the constant appearing in (7.30).

Let us make some comments about the dependence of $\tau_{d,q}$ on the parameter q. It is well-known that for $d \ge 3$ we have

$$\lim_{q\nearrow 2^*} \lambda_1^q(\Omega) = \inf_{u\in H_0^1(\Omega)} \left\{\int_{\mathbb{R}^d} |\nabla u|^2\,dx\ :\ \|u\|_{L^{2^*}(\mathbb{R}^d)}=1\right\}.$$

The latter coincides with the best constant in the Sobolev inequality on $\mathbb{R}^d$, a quantity which does not depend on the open set Ω. This implies that the constant $\tau_{d,q}$ must converge to 0 as q goes to 2^*. From the explicit expression above, we have

$$\gamma_{d,q} \simeq (2^{\vartheta}-1) \simeq (2^*-q), \qquad \text{as } q \text{ goes to } 2^*.$$

The conformal case $d = 2$ is a little bit different. In this case we have (see [790, Lemma 2.2])

$$\lim_{q\to+\infty} \lambda_1^q(\Omega) = 0 \qquad \textit{and} \qquad \lim_{q\to+\infty} q\,\lambda_1^q(\Omega) = 8\,\pi\,e,$$

for every open bounded set Ω. By observing that for $d = 2$ we have $\vartheta = 1/q$, the asymptotic behaviour of the constant $\gamma_{2,q}$ is then given by

$$\gamma_{2,q} \simeq \left(2^{\frac{1}{q}} - 1\right) \lambda_1^q(B) \simeq \frac{1}{q^2}, \qquad \textit{as } q \textit{ goes to } +\infty.$$

7.2.5 The Faber-Krahn inequality in sharp quantitative form

As simple and general as it is, the previous result is not sharp. Indeed, Bhattacharya and Weitsman [147, Section 8] and Nadirashvili [719, page 200] independently conjectured the following:

There exists a dimensional constant $c > 0$ such that

$$|\Omega|^{2/d}\,\lambda(\Omega) - |B|^{2/d}\,\lambda(B) \geq c\,\mathcal{A}(\Omega)^2.$$

After some attempts and intermediate results, this has been proved by Brasco, De Philippis and Velichkov in [181]. This follows by choosing $q = 2$ in the statement below, which is again valid in the more general case of the torsional rigidity and the first semilinear eigenvalues.

Theorem 7.13. *There exists a dimensional constant $\gamma_d > 0$ such that for every open $\Omega \subset \mathbb{R}^d$ with finite measure, we have*

$$|B|^{-\frac{d+2}{d}}\,T(B) - |\Omega|^{-\frac{d+2}{d}}\,T(\Omega) \geq \gamma_d\,\mathcal{A}(\Omega)^2. \tag{7.38}$$

More generally, for $1 \leq q < 2^*$ there exists a constant $\gamma_{d,q} > 0$, depending only on the dimension d and q, such that for every open set $\Omega \subset \mathbb{R}^d$ with finite measure we have

$$|\Omega|^{\frac{2}{d}+\frac{2}{q}-1}\,\lambda_1^q(\Omega) - |B|^{\frac{2}{d}+\frac{2}{q}-1}\,\lambda_1^q(B) \geq \gamma_{d,q}\,\mathcal{A}(\Omega)^2. \tag{7.39}$$

The proof of this result is quite long and technical. We will briefly describe the main ideas and steps of the proof, referring the reader to the original paper [181] for all the details. We remark that this time the constant appearing in the estimate *is not explicit*. However, we can trace its dependence on q, which is the same as that of $\gamma_{d,q}$ in Remark 7.12.

Let us stress that in contrast with the previous results, *the proof of Theorem 7.13 does not rely on quantitative versions of the Pólya-Szegő principle*, since this technique seems very hard to implement in sharp form (as explained at the beginning of Subsection 7.2.4). On the contrary, it is based on the *selection principle* introduced by Cicalese and Leonardi in [275] to give a new proof of the aforementioned quantitative isoperimetric inequality (7.27).

Let us now explain the main steps of the proof of Theorem 7.13.

Step 1: reduction to the torsional rigidity

We start by observing that by Proposition 7.68 it is sufficient to prove (7.38). In other words, it is sufficient to prove

$$T(B_1) - T(\Omega) \geq \frac{1}{C}\,\mathcal{A}(\Omega)^2, \qquad \text{for } \Omega \subset \mathbb{R}^d \text{ such that } |\Omega| = |B_1|, \tag{7.40}$$

where C is a dimensional constant and B_1 is the ball of radius 1 centered at the origin.

Step 2: sharp stability for nearly spherical sets

One then observes that if Ω is a sufficiently smooth perturbation of B_1, then (7.40) can be proved by means of a second order expansion argument. More precisely, in this step we consider the following class of sets.

Definition 7.14. *An open bounded set $\Omega \subset \mathbb{R}^d$ is said nearly spherical of class $C^{2,\gamma}$ parametrized by φ, if there exists $\varphi \in C^{2,\gamma}(\partial B_1)$ with $\|\varphi\|_{L^\infty} \leq 1/2$ and such that $\partial\Omega$ is represented by*

$$\partial\Omega = \left\{x \in \mathbb{R}^d \,:\, x = (1+\varphi(y))\,y,\ \textit{for } y \in \partial B_1\right\}.$$

For nearly spherical sets, we then have the following quantitative estimate. The proof relies on a second order Taylor expansion for the torsional rigidity, see [316] and [181, Appendix A].

Proposition 7.15. *Let $0 < \gamma \leq 1$. Then there exists $\delta_1 = \delta_1(d,\gamma) > 0$ such that if Ω is a nearly spherical set of class $C^{2,\gamma}$ parametrized by φ with*

$$\|\varphi\|_{C^{2,\gamma}(\partial B_1)} \leq \delta_1, \qquad |\Omega| = |B_1| \qquad \textit{and} \qquad x_\Omega := \fint_\Omega x\,dx = 0,$$

then

$$T(B_1) - T(\Omega) \geq \frac{1}{32\,d^2}\,\|\varphi\|^2_{L^2(\partial B_1)}. \tag{7.41}$$

Remark 7.16. *It is not difficult to see that (7.41) implies (7.40) for the class of sets under consideration. Indeed, we have*

$$\|\varphi\|^2_{L^2(\partial B_1)} \geq \frac{1}{d\,\omega_d}\left(\int_{\partial B_1} |\varphi|\,d\mathcal{H}^{d-1}\right)^2,$$

and

$$\mathcal{A}(\Omega) \leq \frac{|\Omega \Delta B_1|}{|\Omega|} = \frac{1}{d\,\omega_d}\int_{\partial B_1} |1-(1+\varphi)^d|\,d\mathcal{H}^{d-1} \simeq \frac{1}{\omega_d}\int_{\partial B_1} |\varphi|\,d\mathcal{H}^{d-1}.$$

Let us note for the record that actually inequality (7.41) holds true in a stronger form, where the L^2 norm of φ is replaced by its $H^{1/2}(\partial\Omega)$ norm, see [181, Theorem 3.3].

Step 3: reduction to the small asymmetry regime

This simple step permits us to reduce the task to proving (7.40) for sets that have suitably small Fraenkel asymmetry. We have the following result.

Proposition 7.17. *Let us suppose that there exist $\varepsilon > 0$ and $c > 0$ such that*

$$T(B_1) - T(\Omega) \geq c\,\mathcal{A}(\Omega)^2, \qquad \text{for } \Omega \text{ such that} \quad |\Omega| = |B_1| \text{ and } \mathcal{A}(\Omega) < \varepsilon.$$

Then (7.40) holds true with

$$\frac{1}{C} = \min\{c,\ \varepsilon\,\tau\}$$

where $\tau > 0$ is the dimensional constant appearing in (7.29).

Proof. Once we have Theorem 7.11 at our disposal, the proof is straightforward. Indeed, if $\mathcal{A}(\Omega) \geq \varepsilon$, then by (7.29) we get

$$T(B_1) - T(\Omega) \geq \tau\,\mathcal{A}(\Omega)^3 \geq \varepsilon\,\tau\,\mathcal{A}(\Omega)^2,$$

as desired. However, let us point out that for the proof of this result the power law relation given by (7.29) is not really needed, it would be sufficient to know that $\mathcal{A}(\Omega) \to 0$ as $T(B_1) - T(\Omega) \to 0$. □

Step 4: reduction to bounded sets

We can still make a further reduction, i.e. we can restrict ourselves to proving (7.40) for sets with uniformly bounded diameter. This is a consequence of the following expedient result.

Lemma 7.18. *There exist positive constants $C = C(d)$, $\mathcal{T} = \mathcal{T}(d)$ and $D = D(d)$ such that for every open set $\Omega \subset \mathbb{R}^d$ with*

$$|\Omega| = |B_1| \qquad \text{and} \qquad T(B_1) - T(\Omega) \leq \mathcal{T},$$

we can find another open set $\widetilde{\Omega} \subset \mathbb{R}^d$ with

$$|\widetilde{\Omega}| = |B_1| \qquad \text{and} \qquad \operatorname{diam}(\widetilde{\Omega}) \leq D,$$

such that

$$\mathcal{A}(\Omega) \leq \mathcal{A}(\widetilde{\Omega}) + C\left(T(B_1) - T(\Omega)\right) \quad \text{and} \quad T(B_1) - T(\widetilde{\Omega}) \leq C\left(T(B_1) - T(\Omega)\right). \tag{7.42}$$

The proof of this result is quite tricky and we refer the reader to [181, Lemma 5.3]. It is, however, quite interesting to remark that one of the key ingredients of the proof is the knowledge of some suitable non-sharp quantitative Saint-Venant inequality, where the deficit $T(\Omega) - T(B_1)$ controls a power of the Fraenkel asymmetry. For example, in [181] a prior result by Fusco, Maggi and Pratelli is used, with exponent 4 on the asymmetry (see Section 7.7.1 below for more comments on their result).

By using the previous Lemma, the main achievement of this step is the following result.

Proposition 7.19. *Let D be the same constant as in Lemma 7.18. Let us suppose that there exist $c > 0$ such that*

$$T(B_1) - T(\Omega) \geq c\,\mathcal{A}(\Omega)^2, \qquad \textit{for } \Omega \textit{ such that } \quad |\Omega| = |B_1| \textit{ and } \operatorname{diam}(\Omega) \leq D. \tag{7.43}$$

Then (7.40) *holds true.*

Proof. We suppose that $\operatorname{diam}(\Omega) > D$, otherwise there is nothing to prove. Let $\mathcal{T}$ be as in the statement of Lemma 7.18, we observe that if $T(B_1) - T(\Omega) > \mathcal{T}$, then (7.40) trivially holds true with constant $\mathcal{T}/4$.

We can thus suppose that Ω satisfies the hypotheses of Lemma 7.18 and find a new open set $\widetilde{\Omega}$ for which (7.43) holds true. By using (7.42) and (7.43) we get

$$\begin{aligned}\sqrt{T(B_1) - T(\Omega)} &\geq \sqrt{\frac{1}{C}\left(T(B_1) - T(\widetilde{\Omega})\right)}\\ &\geq \sqrt{\frac{c}{C}}\,\mathcal{A}(\widetilde{\Omega}) \geq \sqrt{\frac{c}{C}}\,\left(\mathcal{A}(\Omega) - C\left(T(B_1) - T(\Omega)\right)\right).\end{aligned}$$

Since we can always suppose that $T(B_1) - T(\Omega) \leq 1$, this shows (7.40) for Ω, as desired. □

Step 5: sharp stability for bounded sets with small asymmetry

This is the core of the proof and the most delicate step. Thanks to **Step 1**, **Step 3** and **Step 4**, in order to prove Theorem 7.13, we have to prove the following.

Theorem 7.20. *For every $R \geq 2$, there exist two constants $\widehat{c} = \widehat{c}(d, R) > 0$ and $\widehat{\varepsilon} = \widehat{\varepsilon}(d, R) > 0$ such that*

$$T(B_1) - T(\Omega) \geq \widehat{c}\,\mathcal{A}(\Omega)^2, \qquad \textit{for } \Omega \subset B_R \textit{ such that } \quad |\Omega| = |B_1| \textit{ and } \mathcal{A}(\Omega) \leq \widehat{\varepsilon}. \tag{7.44}$$

The idea of the proof is to proceed by contradiction. Indeed, let us suppose that (7.44) is false. Thus we may find a sequence of open sets $\{\Omega_j\}_{j\in\mathbb{N}} \subset B_R(0)$ such that

$$|\Omega_j| = |B_1|, \quad \varepsilon_j := \mathcal{A}(\Omega_j) \to 0 \quad \text{and} \quad T(B_1) - T(\Omega_j) \leq c\,\mathcal{A}(\Omega_j)^2, \tag{7.45}$$

with $c > 0$ as small as we wish. The idea is to use a variational procedure to replace the sequence $\{\Omega_j\}_{j\in\mathbb{N}}$ with an “improved” one $\{U_j\}_{j\in\mathbb{N}}$ which still contradicts (7.44) and enjoys some additional smoothness properties.

In the spirit of the celebrated *Ekeland's variational principle*, we want to select such a sequence through some penalized minimization problem. Roughly speaking we look for sets U_j which solve the following

$$\min\left\{T(B_1) - T(\Omega) + \sqrt{\varepsilon_j^2 + \eta\,(\mathcal{A}(\Omega) - \varepsilon_j)^2}\ :\ \Omega \subset B_R,\ |\Omega| = |B_1|\right\}, \tag{7.46}$$

where $\eta > 0$ is a suitably small parameter, which will allow to get the final contradiction.

One can easily show that the sequence U_j still contradicts (7.44) and that $\mathcal{A}(U_j) \to 0$. Relying on the minimality of U_j, one then would like to show that the L^1 convergence to B_1 can be improved to a $C^{2,\gamma}$ convergence. If this is the case, then the stability result for smooth nearly spherical sets Proposition 7.15 applies and shows that (7.45) cannot hold true if c in (7.45) is sufficiently small.

The key point is thus to prove (uniform) regularity estimates for sets that solve (7.46). For this, first one would like to get rid of volume constraints applying some sort of Lagrange multiplier principle to show that U_j solves

$$\min\left\{T(B_1) - T(\Omega) + \sqrt{\varepsilon_j^2 + \eta\,(\mathcal{A}(\Omega) - \varepsilon_j)^2} + \Lambda\,|\Omega|\ :\ \Omega \subset B_R\right\}. \tag{7.47}$$

If we recall the formulation (7.13) for $-T(\Omega)$, then we can take advantage of the fact that we are considering a "min–min" problem. Thus the previous problem is equivalent to require that the torsion function $w_j := w_{U_j}$ of U_j minimizes

$$\int_{\mathbb{R}^d} |\nabla v|^2\,dx - 2\int_{\mathbb{R}^d} v\,dx + \Lambda\,\big|\{v > 0\}\big| + \sqrt{\varepsilon_j^2 + \eta\,(\mathcal{A}(\{v > 0\}) - \varepsilon_j)^2}, \tag{7.48}$$

among all functions with compact support in B_R. Since we are now facing a perturbed free boundary type problem, we aim to apply the techniques of Alt and Caffarelli [25] (see also [187, 189] and Chapter 3) to show the regularity of $\partial U_j = \partial\{u_j > 0\}$ and to obtain the smooth convergence of U_j to B_1.

This is the general strategy, but several non-trivial modifications have to be done to the above sketched proof. A first technical difficulty is that no global Lagrange multiplier principle is available. Indeed, since

$$-T(t\,\Omega) - - -t^{d+2}\,T(\Omega) \qquad \text{and} \qquad |t\,\Omega| = t^d\,|\Omega|, \qquad t > 0,$$

by a simple scaling argument one sees that the infimum of the energy in (7.47) would be identically $-\infty$ in the uncostrained case. This can be fixed by following [13] and replacing the term $\Lambda\,|\Omega|$ with a term of the form $f(|\Omega| - |B_1|)$, for a suitable strictly increasing function f vanishing at 0 only (see Remark 3.26 of Chapter 3).

A more serious obstruction is due to the lack of regularity of the Fraenkel asymmetry. Although solutions to (7.48) enjoy some mild regularity properties, we cannot expect $\partial\{u_j > 0\}$ to be smooth. Indeed, by formally computing the optimality condition[7.4] of (7.48) and assuming that B_1 is the unique optimal ball for the Fraenkel asymmetry of $\{w_j > 0\}$, one gets that w_j should satisfy

$$\left|\frac{\partial w_j}{\partial \nu}\right|^2 = \Lambda + \frac{\eta\,(\mathcal{A}(\{w_j > 0\}) - \varepsilon_j)}{\sqrt{\varepsilon_j^2 + \eta\,(\mathcal{A}(\{w_j > 0\}) - \varepsilon_j)^2}}\left(1_{\mathbb{R}^d\setminus\overline{B}_1} - 1_{B_1}\right) \quad \text{on } \partial\{w_j > 0\},$$

where 1_A denotes the characteristic function of a set A and ν is the outer normal versor. This means that the normal derivative of w_j is discontinuous at points where $U_j = \{w_j > 0\}$ crosses ∂B_1. Since classical elliptic regularity implies that if ∂U_j is $C^{1,\gamma}$ then $u_j \in C^{1,\gamma}(\overline{U_j})$, it is clear that the sets U_j can not be smooth enough. In particular, it seems difficult to obtain the $C^{2,\gamma}$ regularity needed to apply Proposition 7.15.

To overcome this difficulty, we replace the Fraenkel asymmetry with a new asymmetry functional, which behaves like a squared L^2 distance between the boundaries and whose definition is inspired by [24]. For a bounded set $\Omega \subset \mathbb{R}^d$, this is defined by

$$\alpha(\Omega) = \int_{\Omega\Delta B_1(x_\Omega)} \big|1 - |x - x_\Omega|\big|\, dx,$$

where x_Ω is the barycenter of Ω. Notice that $\alpha(\Omega) = 0$ if and only if Ω is a ball of radius 1. This asymmetry is differentiable with respect to the perturbations needed to compute the optimality conditions (this is different from the case of the Fraenkel asymmetry), moreover it enjoys the following crucial properties:

(i) there exists a constant $C_1 = C_1(d) > 0$ such that for every Ω
$$C_1\, \alpha(\Omega) \geq |\Omega\Delta B_1(x_\Omega)|^2; \tag{7.49}$$

(ii) there exist two constants $\delta_2 = \delta_2(d) > 0$ and $C_2 = C_2(d) > 0$ such that for every nearly spherical set Ω parametrized by φ with $\|\varphi\|_{L^\infty} \leq \delta_2$, we have
$$\alpha(\Omega) \leq C_2\, \|\varphi\|^2_{L^2(\partial B_1)}. \tag{7.50}$$

By using the strategy described above and replacing $\mathcal{A}(\Omega)$ with $\alpha(\Omega)$, one can obtain the following.

Proposition 7.21 (Selection Principle). *Let $R \geq 2$ then there exists $\widetilde{\eta} = \widetilde{\eta}(d, R) > 0$ such that if $0 < \eta \leq \widetilde{\eta}(d, R)$ and $\{\Omega_j\}_{j\in\mathbb{N}} \subset \mathbb{R}^d$ verify*

$$|\Omega_j| = |B_1| \quad \textit{and} \quad \varepsilon_j := \alpha(\Omega_j) \to 0, \quad \textit{while} \quad T(B_1) - T(\Omega_j) \leq \eta^4\, \varepsilon_j,$$

then we can find a sequence of smooth open sets $\{U_j\}_{j\in\mathbb{N}} \subset B_R$ satisfying:

7.4 That is, by differentiating the functional along perturbations of the form $v_t = u_j \circ (\mathrm{Id} + tV)$ where V is a smooth vector field.

(i) $|U_j| = |B_1|$;
(ii) $x_{U_j} = 0$;
(iii) ∂U_j *are converging to* ∂B_1 *in* C^k *for every* k;
(iv) *there holds*

$$\limsup_{j\to\infty} \frac{T(B_1) - T(U_j)}{\alpha(U_j)} \le C_3\, \eta,$$

for some constant $C_3 = C_3(d, R) > 0$.

In turn, this permits to prove the following alternative version of Theorem 7.20, by following the contradiction scheme sketched above. Indeed, we can apply Proposition 7.15 to the sets U_j and (7.50) in order to get

$$\frac{1}{32\, d^2} \le \limsup_{j\to\infty} \frac{T(B_1) - T(U_j)}{\|\varphi_j\|_{L^2(\partial B)}} \le C_2 \limsup_{j\to\infty} \frac{T(B_1) - T(U_j)}{\alpha(U_j)} \le C_2\, C_3\, \eta,$$

where φ_j is the parametrization of ∂U_j. By choosing $\eta > 0$ suitably small, we obtain a contradiction and this proves the following result.

Theorem 7.20 bis. *For every* $R \ge 2$, *there exist* $\widetilde{c} = \widetilde{c}(d, R) > 0$ *and* $\widetilde{\varepsilon} = \widetilde{\varepsilon}(d, R) > 0$ *such that*

$$T(B_1) - T(\Omega) \ge \widetilde{c}\, \alpha(\Omega), \qquad \textit{for } \Omega \subset B_R \textit{ such that } |\Omega| = |B_1| \textit{ and } \alpha(\Omega) \le \widetilde{\varepsilon}.$$

Finally, Theorem 7.20 can now be obtained as a consequence of the previous results, by appealing to the properties of $\alpha(\Omega)$. Indeed, by (7.49) we can ensure that $\alpha(\Omega)$ dominates the Fraenkel asymmetry raised to power 2.

Open problem 7.22 (Sharp quantitative Faber-Krahn with explicit constant). *Prove inequality* (7.39) *with a computable constant. Again, it would be sufficient to prove it for the torsional rigidity, still thanks to Proposition 7.68.*

We conclude this subsection by remarking that the Fraenkel asymmetry $\mathcal{A}(\Omega)$ is not affected by removing from Ω a set with positive capacity and zero d–dimensional Lebesgue measure, while this is the case for the Faber-Krahn deficit

$$|\Omega|^{2/d}\, \lambda_1(\Omega) - |B|^{2/d}\, \lambda_1(B).$$

In particular, if $\lambda_1(\Omega) = \lambda_1(B)$ and $|\Omega| = |B|$, from Theorem 7.13 we can only infer that Ω is a ball *up to a set of zero measure*. It could be interesting to have a stronger version of Theorem 7.13, where the Fraenkel asymmetry is replaced by a stronger notion of asymmetry, coinciding on sets which differ by a set with zero capacity. Observe that the two asymmetries $\delta_{\mathcal{M}}$ and $\delta_{\mathcal{N}}$ suffer from the opposite problem, i.e. they are too rigid and affected by removing sets with zero capacity (like points, for example).

Open problem 7.23 (Sharp quantitative Faber-Krahn with capacitary asymmetry). *Prove a quantitative Faber-Krahn inequality with a suitable* capacitary asymmetry *d, i.e. a scaling invariant shape functional $\Omega \mapsto \delta(\Omega)$ vanishing on balls only and such that*

$$\delta(\Omega') = \delta(\Omega) \qquad \textit{if} \quad \mathrm{cap}\,(\Omega\Delta\Omega') = 0.$$

7.2.6 Checking the sharpness

The heuristic idea behind the sharpness of the estimate

$$|\Omega|^{2/d}\,\lambda_1(\Omega) - |B|^{2/d}\,\lambda_1(B) \geq \gamma_{d,2}\,\mathcal{A}(\Omega)^2,$$

is quite easy to understand. It is just the standard fact that a smooth function behaves quadratically near a non degenerate minimum point.

Indeed, λ_1 is twice differentiable in the sense of the *shape derivative* (see [510]). Then any perturbation of the type $\Omega_t := X_t(B)$, where X_t is a measure preserving smooth vector field, should provide a Taylor expansion of the form

$$\lambda_1(\Omega_t) \simeq \lambda_1(B) + O(t^2), \qquad t \ll 1.$$

since the first derivative of λ_1 has to vanish at the "minimum point" B. By observing that the Fraenkel asymmetry satisfies $\mathcal{A}(\Omega_t) \simeq t$, one could prove sharpness of the exponent 2.

Rather than giving the detailed proof of the previous argument, we prefer to give an elementary proof of the sharpness, just based on the variational characterization of λ_1^q and valid for every $1 \leq q < 2^*$. We believe it to be of independent interest.

We still denote by $B_1 \subset \mathbb{R}^d$ the ball with unit radius and centered at the origin. For every $\varepsilon > 0$, consider the $d \times d$ diagonal matrix

$$M_\varepsilon = \mathrm{diag}\Big((1+\varepsilon), (1+\varepsilon)^{-1}, 1, \dots, 1\Big),$$

and take the family of ellipsoids $E_\varepsilon = M_\varepsilon\, B_1$. Observe that by construction we have[7.5]

$$|E_\varepsilon| = |B_1| \qquad \text{and} \qquad \mathcal{A}(E_\varepsilon) = \frac{|E_\varepsilon \Delta B_1|}{|E_\varepsilon|} \simeq \varepsilon. \tag{7.51}$$

Let us fix $q \geq 1$, with a simple change of variables the first semilinear eigenvalue $\lambda_1^q(E_\varepsilon)$ can be written as

$$\lambda_1^q(E_\varepsilon) = \min_{v \in H_0^1(E_\varepsilon)\setminus\{0\}} \frac{\displaystyle\int_{E_\varepsilon} |\nabla v|^2\,dx}{\left(\displaystyle\int_{E_\varepsilon} |v|^q\,dx\right)^{\frac{2}{q}}} = \min_{u \in H_0^1(B_1)\setminus\{0\}} \frac{\displaystyle\int_{B_1} \langle \widetilde{M}_\varepsilon \nabla u, \nabla u\rangle\,dx}{\left(\displaystyle\int_{B_1} |u|^q\,dx\right)^{\frac{2}{q}}}, \tag{7.52}$$

7.5 We recall that for d-dimensional convex sets having d axes of symmetry, the optimal ball for the Fraenkel asymmetry can be centered at the intersection of these axes, see for example [183, Corollary 2 & Remark 6].

where $\widetilde{M}_\varepsilon = M_\varepsilon^{-1} M_\varepsilon^{-1}$. We now observe that

$$\langle \widetilde{M}_\varepsilon \xi, \xi \rangle = \frac{\xi_1^2}{(1+\varepsilon)^2} + (1+\varepsilon)^2 \xi_2^2 + \sum_{i=3}^{d} \xi_i^2, \qquad \xi \in \mathbb{R}^d,$$

and by Taylor's formula

$$\frac{1}{(1+\varepsilon)^2} = 1 - 2\varepsilon + 6 \int_0^\varepsilon \frac{\varepsilon - s}{(1+s)^4}\, ds \le 1 - 2\varepsilon + 3\varepsilon^2.$$

Thus for every $u \in H_0^1(B_1)$ we obtain

$$\int_{B_1} \langle \widetilde{M}_\varepsilon \nabla u, \nabla u \rangle\, dx \le \int_{B_1} |\nabla u|^2\, dx + 2\varepsilon \int_{B_1} \left(|u_{x_2}|^2 - |u_{x_1}|^2 \right) dx$$
$$+ \varepsilon^2 \int_{B_1} \left(3\,|u_{x_1}|^2 + |u_{x_2}|^2 \right) dx. \tag{7.53}$$

We now take $U \in H_0^1(B_1)$ a function which attains the minimum in the definition of $\lambda_1^q(B_1)$, with unit L^q norm. From (7.52) and (7.53) we get

$$\lambda_1^q(E_\varepsilon) \le \lambda_1^q(B_1) + 2\varepsilon \int_{B_1} \left(|U_{x_2}|^2 - |U_{x_1}|^2 \right) dx + \varepsilon^2 \int_{B_1} \left(3\,|U_{x_1}|^2 + |U_{x_2}|^2 \right) dx.$$

By using that U is radially symmetric (this follows again by the Pólya-Szegő principle), it is easy to see that

$$\int_{B_1} \left(|U_{x_2}|^2 - |U_{x_1}|^2 \right) dx = 0,$$

and thus finally

$$\lambda_1^q(E_\varepsilon) \le \lambda_1^q(B_1) + C\,\varepsilon^2.$$

By recalling (7.51), this finally shows sharpness of Theorem 7.13 for every $1 \le q < 2^*$.

7.3 *Intermezzo*: quantitative estimates for the harmonic radius

In this section we present an application of the quantitative Faber-Krahn inequality to estimating the so-called *harmonic radius*. Apart from being interesting in themselves, some of these results will be useful in the next section.

Definition 7.24 (Harmonic radius). *We denote by G_x^Ω the Green function of Ω with singularity at $x \in \Omega$, i.e.*

$$-\Delta G_x^\Omega = \delta_x \quad \text{in } \Omega, \qquad G_x^\Omega = 0 \quad \text{on } \partial\Omega,$$

where δ_x is the Dirac Delta centered at x. We recall that

$$G_x^\Omega(y) = \varsigma_d \left(\Gamma_d(|x-y|) - H_x^\Omega(y) \right),$$

where:

– ς_d *is the following dimensional constant*

$$\varsigma_2 = \frac{1}{2\pi}, \qquad \varsigma_d = \frac{1}{(d-2)\, d\, \omega_d}, \quad \text{for } d \geq 3;$$

– Γ_d *is the function defined on* $(0, +\infty)$

$$\Gamma_2(t) = -\log t, \qquad \Gamma_d(t) = t^{2-d}, \quad \text{for } d \geq 3;$$

– H_x^Ω *is the regular part, which solves*

$$\Delta H_x^\Omega = 0 \quad \text{in } \Omega, \qquad H_x^\Omega = \Gamma_d(|x - \cdot|) \quad \text{on } \partial\Omega.$$

With the notation above, the harmonic radius of Ω *is defined by*

$$\mathfrak{I}_\Omega := \sup_{x \in \Omega} \Gamma_d^{-1}(H_x^\Omega(x)). \tag{7.54}$$

We refer the reader to the survey paper [96] for a comprehensive study of the harmonic radius.

Remark 7.25 (Scaling properties)**.** *It is not difficult to see that* $\mathfrak{I}_\Omega$ *scales like a length. This follows from the fact that for every* $t > 0$

$$G_x^{t\Omega}(y) = t^{2-d}\, G_{x/t}^\Omega\left(\frac{y}{t}\right), \qquad y \neq x \in t\,\Omega. \tag{7.55}$$

Then in dimension $d \geq 3$ *we get*

$$H_x^{t\Omega}(y) = t^{2-d}\, H_{x/t}^\Omega\left(\frac{y}{t}\right)$$

and thus

$$\mathfrak{I}_{t\Omega} := \sup_{x \in t\Omega} \left(t^{2-d}\, H_{x/t}^\Omega\left(\frac{x}{t}\right)\right)^{\frac{1}{2-d}} = t\, \mathfrak{I}_\Omega.$$

In dimension 2 we proceed similarly, by observing that from (7.55)

$$H_x^{t\Omega}(y) = -\log t + H_{x/t}^\Omega\left(\frac{x}{t}\right).$$

For our purposes, it is useful to recall the following spectral inequality.

Theorem 7.26 (Hersch-Pólya-Szegő inequality)**.** *Let* $\Omega \subset \mathbb{R}^d$ *be an open bounded set. Then we have the scaling invariant estimate*

$$\lambda_1(\Omega) \leq \frac{\lambda_1(B_1)}{\mathfrak{I}_\Omega^2}. \tag{7.56}$$

Equality in (7.56) is attained for balls only.

Proof. Under these general assumptions, the result is due to Hersch and is proved by using *harmonic transplantation*, a technique introduced in [518]. The original result by Pólya and Szegő is for $d = 2$ and Ω simply connected, by means of *conformal transplantation*. We present their proof below, by referring to [96, Section 6] for the general case.

Thus, let us take $d = 2$ and Ω simply connected. Without loss of generality, we can assume $|\Omega| = \pi$. For every $x_0 \in \Omega$, we consider the holomorphic isomorphism given by the Riemann Mapping Theorem

$$f_{x_0} : \Omega \to B_1,$$

such that[7.6] $f_{x_0}(x_0) = 0$. Then we have the following equivalent characterization for the harmonic radius

$$\mathfrak{I}_\Omega = \sup_{x_0 \in \Omega} \left|(f_{x_0}^{-1})'(0)\right| = \sup_{x_0 \in \Omega} \frac{1}{|f_{x_0}{}'(x_0)|}. \tag{7.57}$$

Here f' denotes the complex derivative. Indeed, with the notation above the Green function of Ω with singularity at x_0 is given by

$$G^\Omega_{x_0}(y) = -\frac{1}{2\pi} \log |f_{x_0}(y)|, \qquad y \in \Omega \setminus \{x_0\}.$$

We can rewrite it as

$$\begin{aligned} G^\Omega_{x_0}(y) &= -\frac{1}{2\pi} \log |f_{x_0}(y) - f_{x_0}(x_0)| \\ &= -\frac{1}{2\pi} \log |y - x_0| - \frac{1}{2\pi} \log \frac{|f_{x_0}(y) - f_{x_0}(x_0)|}{|y - x_0|}. \end{aligned}$$

By recalling the definition (7.54) of harmonic radius, we get

$$\mathfrak{I}_\Omega = \sup_{x_0 \in \Omega} \left\{ \lim_{y \to x_0} \exp \left(-\log \frac{|f_{x_0}(y) - f_{x_0}(x_0)|}{|y - x_0|} \right) \right\} = \sup_{x_0 \in \Omega} \frac{1}{|f'{}_{x_0}(x_0)|},$$

which proves (7.57).

We now prove (7.56). Let $u \in H^1_0(B_1)$ be the first positive Dirichlet eigenfunction of D, with unit L^2 norm. For $x_0 \subset \Omega$, we consider $f_{x_0} : \Omega \to B_1$ as above, then we set

$$v = u \circ f.$$

By conformality we preserve the Dirichlet integral, i.e.

$$\int_\Omega |\nabla v|^2 \, dx = \int_{B_1} |\nabla u|^2 \, dx = \lambda_1(B_1).$$

7.6 We recall that this is uniquely defined, up to a rotation.

On the other hand, by the change of variable formula we have

$$\int_{\Omega} |v|^2 \, dx = \int_{B_1} |u|^2 \, |(f_{x_0}^{-1})'|^2 \, dx.$$

We now observe that $|(f_{x_0}^{-1})'|^2$ is sub-harmonic, thus the function

$$\Phi(\varrho) = \frac{1}{2\pi\varrho} \int_{\{|x|=\varrho\}} |(f_{x_0}^{-1})'|^2 \, d\mathcal{H}^1, \tag{7.58}$$

is non-decreasing. In particular, we have

$$\Phi(\varrho) \geq \Phi(0) = |(f_{x_0}^{-1})'(0)|^2.$$

Thus we obtain

$$\begin{aligned}\int_{\Omega} |v|^2 \, dx &= \int_{\Omega} |u|^2 \, |(f_{x_0}^{-1})'|^2 \, dx = 2\pi \int_0^1 u^2 \, \varrho \, \Phi(\varrho) \, d\varrho \\ &\geq \left(2\pi \int_0^1 u^2 \, \varrho \, d\varrho\right) |(f_{x_0}^{-1})'(0)|^2 = |(f_{x_0}^{-1})'(0)|^2,\end{aligned}$$

since u has unit L^2 norm. By using the variational characterization of $\lambda_1(\Omega)$, this finally shows

$$|(f_{x_0}^{-1})'(0)|^2 \, \lambda_1(\Omega) \leq \lambda_1(B_1).$$

By taking the supremum over Ω and using (7.57), we get the conclusion. □

Remark 7.27 (Conformal radius). *Historically, the quantity*

$$\max_{x_0 \in \Omega} \left|(f_{x_0}^{-1})'(0)\right| = \max_{x_0 \in \Omega} \frac{1}{|f_{x_0}{}'(x_0)|},$$

was first introduced under the name conformal radius of Ω. The definition of harmonic radius is due to Hersch [518], as we have seen this gives a genuine extension to general sets of the conformal radius.

Among open sets with given measure, the harmonic radius is maximal on balls. By recalling that for a ball the harmonic radius coincides with the radius *tout court*, we thus have the scaling invariant estimate

$$\frac{|\Omega|^{2/d}}{\mathfrak{I}_{\Omega}^2} \geq \omega_d^{2/d}. \tag{7.59}$$

This can be deduced by combining (7.56) and the Faber-Krahn inequality. If we replace the latter by Theorem 7.13, we get a quantitative version of (7.59). This is the content of the next result.

Corollary 7.28 (Stability of the harmonic radius). *Let $\Omega \subset \mathbb{R}^d$ be an open bounded set. Then we have*

$$\frac{|\Omega|^{2/d}}{\mathcal{J}_\Omega^2} - \omega_d^{2/d} \geq c\,\mathcal{A}(\Omega)^2, \tag{7.60}$$

for some constant $c > 0$.

Proof. We multiply both sides of (7.56) by $|\Omega|^{2/d}/\omega_d^{2/d}$, then we get

$$\frac{|\Omega|^{2/d}\,\lambda_1(\Omega)}{\omega_d^{2/d}\,\lambda_1(B_1)} - 1 \leq \frac{1}{\omega_d^{2/d}}\left(\frac{|\Omega|^{2/d}}{\mathcal{J}_\Omega^2} - \omega_d^{2/d}\right).$$

By recalling that $\omega_d^{2/d}\,\lambda_1(B_1)$ is a universal constant and using the sharp quantitative Faber-Krahn inequality of Theorem 7.13, we get the conclusion. □

For simply connected sets in the plane, the previous result has an interesting geometrical consequence, which will be exploited in Section 7.4 in the proof of Theorem 7.31. Indeed, observe that with the notation above we have

$$|\Omega| = \int_{B_1} |(f_{x_0}^{-1})'|^2\,dx \geq \pi\,|(f_{x_0}^{-1})'(0)|^2,$$

where we used again monotonicity of the function (7.58). If we assume for simplicity that $|\Omega| = \pi$, thus we get

$$\frac{1}{|f'_{x_0}(x_0)|} = |(f_{x_0}^{-1})'(0)| \leq 1$$

with equality if Ω is a disc. If Ω is not a disc, then the inequality is strict and we can add a remainder term. In other words, *the local stretching at the origin of the conformal map $f_{x_0}^{-1}$ can tell whether Ω is a disc or not*. This is the content of the next result.

Corollary 7.29. *Let $\Omega \subset \mathbb{R}^2$ be an open bounded simply connected set such that $|\Omega| = \pi$. For every $x_0 \in \Omega$, we consider the holomorphic isomorphism*

$$f_{x_0} : \Omega \to B_1,$$

such that $f_{x_0}(x_0) = 0$. For every $x_0 \in \Omega$ we have

$$\frac{1}{|f'_{x_0}(x_0)|} = |(f_{x_0}^{-1})'(0)| \leq \sqrt{1 - \frac{1}{C}\,\mathcal{A}(\Omega)^2},$$

for some $C > 4$.

Proof. We observe that from (7.60) we get

$$\frac{1}{\mathcal{J}_\Omega^2} \geq 1 + \frac{c}{\pi}\,\mathcal{A}(\Omega)^2,$$

where we used that $|\Omega| = \pi$. From this, with simple manipulations we get

$$\mathcal{J}_\Omega^2 \leq 1 - \frac{1}{C}\,\mathcal{A}(\Omega)^2.$$

It is now sufficient to use the characterization (7.57) to conclude. □

7.4 Stability for the Szegő-Weinberger inequality

7.4.1 A quick overview of the Neumann spectrum

In the case of homogeneous Neumann boundary conditions, the first eigenvalue $\mu_1(\Omega)$ is always 0 and corresponds to constant functions. This reflects the fact that the Poincaré inequality

$$c \int_\Omega |u|^2\,dx \le \int_\Omega |\nabla u|^2\,dx, \qquad u \in H^1(\Omega),$$

can hold only in the trivial case $c = 0$. For an open set $\Omega \subset \mathbb{R}^d$ with finite measure, we define its *first non-trivial Neumann eigenvalue* by

$$\mu_2(\Omega) := \inf_{u \in H^1(\Omega)\setminus\{0\}} \left\{ \frac{\displaystyle\int_\Omega |\nabla u|^2\,dx}{\displaystyle\int_\Omega |u|^2\,dx} \,:\, \int_\Omega u\,dx = 0 \right\}.$$

In other words, this is the sharp constant in the Poincaré-Wirtinger inequality

$$c \int_\Omega \left| u - \fint_\Omega u \right|^2 dx \le \int_\Omega |\nabla u|^2\,dx, \qquad u \in H_0^1(\Omega).$$

When $\Omega \subset \mathbb{R}^d$ has Lipschitz boundary, the embedding $H^1(\Omega) \hookrightarrow L^2(\Omega)$ is compact (see [607, Theorem 5.8.2]) and the infimum above is attained. In this case the Neumann Laplacian has a discrete spectrum $\{\mu_1(\Omega), \mu_2(\Omega), \dots\}$. The successive Neumann eigenvalues can be defined similarly, that is $\mu_k(\Omega)$ is obtained by minimizing the same Rayleigh quotient, among functions orthogonal (in the $L^2(\Omega)$ sense) to the first $k-1$ eigenfunctions.

If Ω has k connected components, we have $\mu_1(\Omega) = \cdots = \mu_k(\Omega) = 0$, with corresponding eigenfunctions given by a constant function on each connected component of Ω. We still have the scaling property

$$\mu_k(t\,\Omega) = t^{-2}\,\mu_k(\Omega), \qquad t > 0,$$

and there holds the Szegő-Weinberger inequality[7.7]

$$|\Omega|^{2/d}\,\mu_2(\Omega) \le |B|^{2/d}\,\mu_2(B), \tag{7.61}$$

with equality if and only if Ω is a ball.

7.7 We point out that Szegő-Weinberger inequality holds for every open set with finite measure, without smoothness assumptions. In other words, the proof does not use neither discreteness of the Neumann spectrum of Ω, nor that the infimum in the definition of $\mu_2(\Omega)$ is attained.

For a ball B_r of radius r, $\mu_2(B_r)$ has multiplicity d, that is $\mu_2(B_r) = \dots = \mu_{d+1}(B_r)$. This value can be explicitly computed, together with its corresponding eigenfunctions. Indeed, these are given by (see [73])

$$\xi_i(x) := |x|^{\frac{2-d}{2}} J_{\frac{d}{2}}\left(\frac{\beta_{d/2,1}|x|}{r}\right)\frac{x_i}{|x|}, \qquad i = 1, \dots, d. \tag{7.62}$$

Here $J_{d/2}$ is still a Bessel function of the first kind, while $\beta_{d/2,1}$ denotes the first positive zero of the derivative of $t \mapsto t^{(2-d)/2} J_{d/2}(t)$, i.e. it satisfies

$$\beta_{d/2,1} J'_{\frac{d}{2}}(\beta_{d/2,1}) + \left(\frac{2-d}{2}\right) J_{\frac{d}{2}}(\beta_{d/2,1}) = 0 .$$

Observe in particular that the radial part of ξ_i

$$\varphi_d(|x|) := |x|^{1-\frac{d}{2}} J_{d/2}\left(\frac{\beta_{d/2,1}|x|}{r}\right), \tag{7.63}$$

satisfies the ODE (of Bessel type)

$$g''(t) + \frac{d-1}{t} g'(t) + \left(\mu_2(B_r) - \frac{d-1}{t^2}\right) g(t) = 0,$$

and one can compute

$$\mu_2(B_r) = \left(\frac{\beta_{d/2,1}}{r}\right)^2 .$$

Finally, we recall that in dimension $d = 2$ inequality (7.61) can be sharpened. Namely, for every $\Omega \subset \mathbb{R}^2$ simply connected open set we have

$$\frac{1}{|\Omega|}\left(\frac{1}{\mu_2(\Omega)} + \frac{1}{\mu_3(\Omega)}\right) \ge \frac{1}{|B|}\left(\frac{1}{\mu_2(B)} + \frac{1}{\mu_3(B)}\right), \tag{7.64}$$

where $B \subset \mathbb{R}^2$ is any open disc. This result has been proved by Szegő in [838] by means of conformal maps, we will recall his proof below. By recalling that for a disc $\mu_2 = \mu_3$, from (7.64) we immediately get (7.61) for simply connected sets in $\mathbb{R}^2$.

Remark 7.30. *The higher dimensional analogue of* (7.64) *would be*

$$\frac{1}{|\Omega|^{2/d}} \sum_{k=2}^{d+1} \frac{1}{\mu_k(\Omega)} \ge \frac{1}{|B|^{2/d}} \sum_{k=2}^{d+1} \frac{1}{\mu_k(B)} .$$

However, the validity of such an inequality is still an open problem, see [505, page 106].

7.4.2 A two-dimensional result by Nadirashvili

One of the first quantitative improvements of the Szegő-Weinberger inequality was due to Nadirashvili, see [719]. Even if his result is limited to simply connected sets in the plane, this is valid for the stronger inequality (7.64). We reproduce the original proof, up to some modifications (see Remark 7.32 below). We will also highlight a quicker strategy suggested to us by Mark S. Ashbaugh (see Remark 7.33 below).

Theorem 7.31 (Nadirashvili). *There exists a constant $C > 0$ such that for every $\Omega \subset \mathbb{R}^2$ smooth simply connected open set we have*

$$\frac{1}{|\Omega|}\left(\frac{1}{\mu_2(\Omega)}+\frac{1}{\mu_3(\Omega)}\right)-\frac{1}{|B|}\left(\frac{1}{\mu_2(B)}+\frac{1}{\mu_3(B)}\right)\geq \frac{1}{C}\,\mathcal{A}(\Omega)^2. \tag{7.65}$$

Here $B \subset \mathbb{R}^2$ is any open disc.

Proof. The proof of (7.65) introduces some quantitative ingredients in the original proof by Szegő. For the reader's convenience, it is useful to begin with the proof of (7.64).

By scale invariance, we can suppose that $|\Omega| = |B| = \pi$ and we may take the disc B to be centered at the origin. From (7.62) above, we know that

$$\xi_1(x) = c\,J_1(\beta_{1,1}\,|x|)\,\frac{x_1}{|x|} \qquad \text{and} \qquad \xi_2(x) = c\,J_1(\beta_{1,1}\,|x|)\,\frac{x_2}{|x|},$$

are two linearly independent Neumann eigenfunctions in B, corresponding to $\mu_2(B) = \mu_3(B)$. The normalization constant c is chosen so to guarantee that ξ_1 and ξ_2 have unit L^2 norm.

Since $\Omega \subset \mathbb{R}^2$ is simply connected, given $x_0 \in \Omega$ by the Riemann Mapping Theorem there exists an analytic isomorphism $f_{x_0} : \Omega \to B$ such that $f_{x_0}(x_0) = 0$. For notational simplicity, we will omit the index x_0 and simply write f. Szegő proved that we can choose $x_0 \in \Omega$ in such a way that if we set $v_i = \xi_i \circ f$ $(i = 1, 2)$ then

$$\int_\Omega v_i\,dx = 0, \qquad i = 1, 2.$$

Then if we set $h = f^{-1}$ we have

$$\int_\Omega |v_i|^2\,dx = \int_B \xi_i^2\,|h'|^2\,dx, \qquad \int_\Omega |\nabla v_i|^2\,dx = \int_B |\nabla \xi_i|^2\,dx, \qquad i = 1, 2, \tag{7.66}$$

where h' denotes the complex derivative. Also observe that by conformality we have

$$\int_\Omega \langle \nabla v_1, \nabla v_2\rangle\,dx = \int_B \langle \nabla \xi_1, \nabla \xi_2\rangle\,dx = 0.$$

By recalling the following variational formulation for the sum of inverses of Neumann eigenvalues (see for example [523, Theorem 1])

$$\frac{1}{\mu_2(\Omega)}+\frac{1}{\mu_3(\Omega)} = \max_{u\in H^1(\Omega)\setminus\{0\}}\left\{\frac{\int_\Omega |u_1|^2\,dx}{\int_\Omega |\nabla u_1|^2\,dx}+\frac{\int_\Omega |u_2|^2\,dx}{\int_\Omega |\nabla u_2|^2\,dx} : \begin{array}{c}\int_\Omega u_1\,dx = \int_\Omega u_2\,dx = 0\\ \int_\Omega \langle \nabla u_1, \nabla u_2\rangle\,dx = 0\end{array}\right\},$$

and using that $\mu_2(B) = \mu_3(B)$, from (7.66) we get

$$\frac{1}{\mu_2(\Omega)} + \frac{1}{\mu_3(\Omega)} \geq \frac{\int_\Omega |v_1|^2\,dx}{\int_\Omega |\nabla v_1|^2\,dx} + \frac{\int_\Omega |v_2|^2\,dx}{\int_\Omega |\nabla v_2|^2\,dx} = \frac{\int_B |\xi_1|^2\,|h'|^2\,dx}{\mu_2(B)} + \frac{\int_B |\xi_2|^2\,|h'|^2\,dx}{\mu_3(B)}$$
$$= \frac{\int_B \left(|\xi_1|^2 + |\xi_2|^2\right)|h'|^2\,dx}{\mu_2(B)}. \tag{7.67}$$

Since h' is holomorphic, the function $|h'|^2$ is subharmonic, thus

$$r \mapsto \Phi(r) := \fint_{\{|x|=r\}} |h'|^2\,d\mathcal{H}^1,$$

is a monotone nondecreasing function. The same is true for the radial function

$$|\xi_1|^2 + |\xi_2|^2 = c^2 J_1(\beta_{1,1}\,|x|)^2,$$

thus by Lemma 7.70 we have

$$\begin{aligned}
\int_B \left(|\xi_1|^2 + |\xi_2|^2\right)|h'|^2\,dx &= 2\pi \int_0^1 \left(|\xi_1|^2 + |\xi_2|^2\right)\Phi(\varrho)\,\varrho\,d\varrho \\
&\geq 2\pi \frac{\int_0^1 \left(|\xi_1|^2 + |\xi_2|^2\right)\varrho\,d\varrho}{\int_0^1 \varrho\,d\varrho} \int_0^1 \Phi(\varrho)\,\varrho\,d\varrho \\
&= 2\int_0^1 \left(|\xi_1|^2 + |\xi_2|^2\right)\varrho\,d\varrho \left(\int_0^1 \int_{\{|x|=\varrho\}} |h'|^2\,d\mathcal{H}^1\,d\varrho\right) \\
&= 2\pi \int_0^1 \left(|\xi_1|^2 + |\xi_2|^2\right)\varrho\,d\varrho = \int_B \left(|\xi_1|^2 + |\xi_2|^2\right)dx = 2,
\end{aligned} \tag{7.68}$$

where we used that

$$\int_0^1 \int_{\{|x|=\varrho\}} |h'|^2\,d\mathcal{H}^1\,d\varrho = \int_B |h'|^2\,dx = \pi.$$

By using the previous estimate in (7.67), we finally get (7.64).

We now come to the proof of (7.65). By using Corollary 7.29 from the previous section, we get

$$|h'(0)|^2 \leq 1 - \frac{1}{C}\,\mathcal{A}(\Omega)^2. \tag{7.69}$$

Since h is analytic, we have

$$h'(z) = \sum_{n=1}^{\infty} n\,a_n\,z^{n-1},$$

and thus

$$\Phi(\varrho) = \fint_{\{|x|=\varrho\}} |h'|^2 \, d\mathcal{H}^1 = \sum_{n=1}^{\infty} n^2 \, a_n^2 \, \varrho^{2\,(n-1)}.$$

The latter can be rewritten as

$$\Phi(\varrho) = \sum_{n=0}^{\infty} \alpha_n \, \varrho^n, \qquad \text{where } \alpha_n = \begin{cases} \left(\frac{n+2}{2}\right)^2 \, a^2_{\frac{n+2}{2}}, & n \text{ even}, \\ 0, & n \text{ odd}, \end{cases}$$

and from (7.69)

$$\alpha_0 = a_1^2 = |h'(0)|^2 \le 1 - \frac{1}{C} \mathcal{A}(\Omega)^2 = 2 \left(1 - \frac{1}{C} \mathcal{A}(\Omega)^2\right) \int_0^1 \Phi(\varrho) \, \varrho \, d\varrho.$$

We can thus apply Lemma 7.71, with the choices

$$\varphi = |\xi_1|^2 + |\xi_2|^2 = c^2 J_1^2, \qquad \psi(\varrho) = \fint_{\{|x|=\varrho\}} |h'|^2 \, d\mathcal{H}^1 \quad \text{and} \quad \gamma = 2 \left(1 - \frac{1}{C} \mathcal{A}(\Omega)^2\right).$$

Thus in place of (7.68) we now obtain

$$\begin{aligned} \int_B \left(|\xi_1|^2 + |\xi_2|^2\right) |h'|^2 \, dx &\ge 2\,\pi \, \frac{\displaystyle\int_0^1 \left(|\xi_1|^2 + |\xi_2|^2\right) \varrho \, d\varrho}{\displaystyle\int_0^1 \varrho \, d\varrho} \int_0^1 \Phi(\varrho) \, \varrho \, d\varrho \\ &\quad + 2\, c' \mathcal{A}(\Omega)^2 \int_0^1 \Phi(\varrho) \, \varrho \, d\varrho = 2 + c' \mathcal{A}(\Omega)^2. \end{aligned}$$

By using this improved estimate in (7.67), we get

$$\frac{1}{\mu_2(\Omega)} + \frac{1}{\mu_3(\Omega)} \ge \frac{2}{\mu_2(B)} + \frac{c'}{\mu_2(B)} \mathcal{A}(\Omega)^2,$$

which concludes the proof. □

Remark 7.32. *The crucial point of the previous proof was to obtain estimate* (7.69) *on* $h'(0) = (f^{-1})'(0)$. *The argument we used to obtain it is slightly different from the original one by Nadirashvili. The latter exploits a stability result of Hansen and Nadirashvili (see* [448, *Corollary 2*]*) for the logarithmic capacity in dimension* $d = 2$, *which assures that*[7.8]

$$\mathrm{Cap}(\Omega) - \mathrm{Cap}(B) \ge c\, \mathcal{A}(\Omega)^2, \qquad \text{if } |\Omega| = |B|. \tag{7.70}$$

Here on the contrary we rely on the stability estimate of Corollary 7.29, which in turn is a consequence of the quantitative Faber-Krahn inequality, as we saw in Section 7.2.

7.8 As explained in the Introduction of [444], for *connected* open sets in $\mathbb{R}^2$ inequality (7.70) follows from an inequality linking capacity and moment of inertia which can be found in the book [770]. This observation is attributed to Keady. In [448] the result is extended to general open sets in $\mathbb{R}^2$.

Remark 7.33 (An overlooked inequality). *Inequality (7.64) in turn can be sharpened. Indeed, in [520] Hersch and Monkewitz have shown that there exists a constant $c > 0$ such that for every $\Omega \subset \mathbb{R}^2$ simply connected open set we have*

$$\frac{1}{|\Omega|}\left(\frac{1}{\mu_2(\Omega)}+\frac{1}{\mu_3(\Omega)}+\frac{c}{\lambda_1(\Omega)}\right) \ge \frac{1}{|B|}\left(\frac{1}{\mu_2(B)}+\frac{1}{\mu_3(B)}+\frac{c}{\lambda_1(B)}\right). \tag{7.71}$$

By using this inequality, we can provide a quicker proof of Theorem 7.31. Indeed, let us suppose for simplicity that $|\Omega| = 1$, from (7.71) we get

$$\left(\frac{1}{\mu_2(\Omega)}+\frac{1}{\mu_3(\Omega)}\right)-\left(\frac{1}{\mu_2(\Omega^*)}+\frac{1}{\mu_3(\Omega^*)}\right) \ge \frac{c}{\lambda_1(\Omega^*)\,\lambda_1(\Omega)}\left(\lambda_1(\Omega)-\lambda_1(\Omega^*)\right),$$

where Ω^* is a disc such that $|\Omega^*| = |\Omega| = 1$. We now observe that if $\lambda_1(\Omega) \ge 2\,\lambda_1(\Omega^*)$, the right-hand side above can be bounded from below as follows

$$\frac{c}{\lambda_1(\Omega^*)\,\lambda_1(\Omega)}\left(\lambda_1(\Omega)-\lambda_1(\Omega^*)\right) \ge \frac{c}{2\,\lambda_1(\Omega^*)} \ge \frac{c}{8\,\lambda_1(\Omega^*)}\,\mathcal{A}(\Omega)^2,$$

where we used that $\mathcal{A}(\Omega) < 2$. If on the contrary $\lambda_1(\Omega) < 2\,\lambda_1(\Omega^*)$, then from the sharp quantitative Faber-Krahn inequality (Theorem 7.13) we get

$$\frac{c}{\lambda_1(\Omega^*)\,\lambda_1(\Omega)}\left(\lambda_1(\Omega)-\lambda_1(\Omega^*)\right) \ge \frac{c'}{\lambda_1(\Omega^*)^2}\,\mathcal{A}(\Omega)^2.$$

In conclusion, we can infer the existence of a constant $c'' > 0$ such that

$$\left(\frac{1}{\mu_2(\Omega)}+\frac{1}{\mu_3(\Omega)}\right)-\left(\frac{1}{\mu_2(\Omega^*)}+\frac{1}{\mu_3(\Omega^*)}\right) \ge c''\mathcal{A}(\Omega)^2,$$

thus proving Theorem 7.31. We thank Mark S. Ashbaugh for kindly pointing out the reference [520].

7.4.3 The Szegő-Weinberger inequality in sharp quantitative form

From Theorem 7.31, one can easily get a quantitative improvement of the Szegő-Weinberger inequality, in the case of simply connected sets in the plane. For general open sets in any dimension, we have the following result proved by Brasco and Pratelli in [178, Theorem 4.1].

Theorem 7.34. *For every $\Omega \subset \mathbb{R}^d$ open set with finite measure, we have*

$$|B|^{2/d}\,\mu_2(B) - |\Omega|^{2/d}\,\mu_2(\Omega) \ge \rho_d\,\mathcal{A}(\Omega)^2, \tag{7.72}$$

where $\rho_d > 0$ is an explicit dimensional constant (see Remark 7.35 below).

Proof. Here as well, we first recall the proof of (7.61). As always, we denote by Ω^* the ball centered at the origin and such that $|\Omega^*| = |\Omega|$. Since (7.72) is scaling invariant, we can suppose that $|\Omega| = \omega_d$, i.e. the radius of Ω^* is 1. Observe that the eigenfunctions ξ_i of Ω^* defined in (7.62) have the following property, which will be crucially exploited:

$$x \mapsto \sum_{i=1}^{d} |\xi_i(x)|^2 \qquad \text{and} \qquad x \mapsto \sum_{i=1}^{d} |\nabla \xi_i(x)|^2 \qquad \text{are monotone radial functions.}$$

Indeed, we have

$$\sum_{i=1}^{d} |\xi_i(x)|^2 = \varphi_d(|x|)^2 \qquad \text{and} \qquad \sum_{i=1}^{d} |\nabla \xi_i(x)|^2 = \varphi'_d(|x|)^2 + (d-1)\,\frac{\varphi_d(|x|)^2}{|x|^2}, \tag{7.73}$$

and the first one is radially increasing, while the second is decreasing. Moreover, since each ξ_i is an eigenfunction of the ball, we have

$$\mu_2(\Omega^*) \int_{\Omega^*} |\xi_i|^2\, dx = \int_{\Omega^*} |\nabla \xi_i|^2\, dx, \qquad i = 1, \dots, d.$$

If we sum the previous identities and use (7.73), we thus end up with

$$\mu_2(\Omega^*) = \frac{\displaystyle\int_{\Omega^*} \left[\varphi'_d(|x|)^2 + (d-1)\,\frac{\varphi_d(|x|)^2}{|x|^2}\right] dx}{\displaystyle\int_{\Omega^*} \varphi_d(|x|)^2\, dx}. \tag{7.74}$$

We then extend φ_d to the whole $[0, +\infty)$ as follows

$$\phi_d(t) = \begin{cases} \varphi_d(t), & 0 \le t \le 1, \\ \varphi_d(1), & t > 1, \end{cases}$$

and consider the new functions defined on $\mathbb{R}^d$

$$\Xi_i(x) = \phi_d(|x|)\,\frac{x_i}{|x|}, \qquad i = 1, \dots, d.$$

Observe that if we define

$$F_d(t) = \int_0^t \phi_d(s)\, ds, \qquad t \ge 0,$$

this is a C^1 convex increasing function, which diverges at infinity. This means that the function

$$x \mapsto \int_\Omega F_d(|x - y|)\, dy,$$

admits a global minimum point $x_0 \in \mathbb{R}^d$ and thus

$$(0, \dots, 0) = \int_\Omega F_d'(|x_0 - y|)\,\frac{x_0 - y}{|x_0 - y|}\, dy = \left(\int_\Omega \Xi_1(x_0 - y)\, dy, \dots, \int_\Omega \Xi_d(x_0 - y)\, dy\right).$$

Thus it is always possible to choose the origin of the coordinate axes in such a way that[7.9]

$$\int_\Omega \Xi_i(x)\,dx = 0,\ i = 1,\dots,d.$$

By making such a choice for the origin, the functions Ξ_i can be used to estimate $\mu_2(\Omega)$ and we can infer

$$\mu_2(\Omega) \le \frac{\displaystyle\int_\Omega |\nabla \Xi_i|^2\,dx}{\displaystyle\int_\Omega \Xi_i^2\,dx}, \qquad i = 1,\dots,d.$$

Again, a summation over $i = 1,\dots,d$ yields

$$\mu_2(\Omega) \le \frac{\displaystyle\sum_{i=1}^d \int_\Omega |\nabla \Xi_i|^2\,dx}{\displaystyle\sum_{i=1}^d \int_\Omega \Xi_i^2\,dx},$$

and the summation trick makes the angular variables disappear and one ends up with

$$\mu_2(\Omega) \le \frac{\displaystyle\int_\Omega \left[\phi'_d(|x|)^2 + (d-1)\,\frac{\phi_d(|x|)^2}{|x|^2}\right] dx}{\displaystyle\int_\Omega \phi_d(|x|)^2\,dx}. \tag{7.75}$$

We set

$$f(t) = \phi'_d(t)^2 + (d-1)\,\frac{\phi_d(t)^2}{t^2} \qquad \text{and} \qquad g(t) = \phi_d(t)^2, \qquad t \ge 0,$$

and recall that f is non-increasing, while g is non-decreasing. Then from (7.74) and (7.75) we get

$$\mu_2(\Omega^*)\int_{\Omega^*} g(|x|)\,dx - \mu_2(\Omega)\int_\Omega g(|x|)\,dx \ge \int_{\Omega^*} f(|x|)\,dx - \int_\Omega f(|x|)\,dx. \tag{7.76}$$

By using the weak Hardy-Littlewood inequality (see Lemma 7.72) and the monotonicity of g, we have

$$\int_\Omega g(|x|)\,dx \ge \int_{\Omega^*} g(|x|)\,dx = \int_{\{|y|\le 1\}} |y|^{2-d} J_{d/2}(\beta_{d/2,1}\,|y|)^2\,dy = \omega_d^{2/d}\,\eta_d,$$

where we used the definition of ϕ_d and that of φ_d, see (7.63). The dimensional constant η_d is defined by

$$\eta_d := d\,\omega_d^{\frac{d-2}{d}} \int_0^1 J_{d/2}(\beta_{d/2,1}\,\varrho)^2\,\varrho\,d\varrho > 0.$$

7.9 We avoid here the original argument based on the Brouwer Fixed Point Theorem.

Thus, by recalling that $|\Omega| = |\Omega^*| = \omega_d$, inequality (7.76) yields

$$|\Omega^*|^{2/d}\,\mu_2(\Omega^*) - |\Omega|^{2/d}\,\mu_2(\Omega) \geq \frac{1}{\eta_d}\left[\int_{\Omega^*} f(|x|)\,dx - \int_{\Omega} f(|x|)\,dx\right]. \tag{7.77}$$

The proof by Weinberger now uses Lemma 7.72 again to ensure that the right-hand side of (7.77) is positive, which leads to (7.61).

If on the contrary we replace Lemma 7.72 by its improved version Lemma 7.73, we can get a quantitative lower bound. Since f is non-increasing, by using (7.149) in (7.77) we get

$$\begin{aligned}|\Omega^*|^{2/d}\,\mu_2(\Omega^*) - |\Omega|^{2/d}\,\mu_2(\Omega) &\geq \frac{d\,\omega_d}{\eta_d}\int_{R_1}^{R_2} |f(\varrho) - f(1)|\,\varrho^{d-1}\,dx \\ &\geq \frac{d\,\omega_d}{\eta_d}\int_{1}^{R_2} [f(1) - f(\varrho)]\,d\varrho.\end{aligned} \tag{7.78}$$

The radii $R_1 < 1 < R_2$ are such that

$$|\Omega^*| - |B_{R_1}| = |\Omega^* \setminus \Omega| \qquad \text{and} \qquad |B_{R_2}| - |\Omega^*| = |\Omega \setminus \Omega^*|.$$

By recalling that $|\Omega| = \omega_d$, they are given by

$$R_1 = \left(\frac{|\Omega \cap \Omega^*|}{\omega_d}\right)^{\frac{1}{d}} \qquad \text{and} \qquad R_2 = \left(\frac{|\Omega \setminus \Omega^*|}{\omega_d} + 1\right)^{\frac{1}{d}}$$

In order to conclude it is now sufficient to observe that

$$f(1) - f(\varrho) \geq (d-1)\,\phi_d(1)^2\left[\frac{\varrho^2 - 1}{\varrho^2}\right] \geq \frac{d-1}{2^{1/d}}\,\phi_d(1)^2\,(\varrho - 1), \qquad \text{for } R_2 \geq \varrho \geq 1,$$

where we also used that $\varrho \leq R_2 \leq 2^{1/d}$. Thus from (7.78) we get

$$|\Omega^*|^{2/d}\,\mu_2(\Omega^*) - |\Omega|^{2/d}\,\mu_2(\Omega) \geq \frac{d\,(d-1)\,\omega_d}{2^{(d+1)/d}\,\eta_d}\,\varphi_d(1)^2\,(R_2 - 1)^2.$$

By using the definition of R_2 we have

$$(R_2 - 1)^2 = \left(\left(1 + \frac{|\Omega \setminus \Omega^*|}{\omega_d}\right)^{\frac{1}{d}} - 1\right)^2 \geq (2^{1/d} - 1)^2\left(\frac{|\Omega \setminus \Omega^*|}{\omega_d}\right)^2, \tag{7.79}$$

thanks to the elementary inequality

$$(1 + t)^{1/d} \geq 1 + (2^{1/d} - 1)\,t, \qquad \text{for every } t \in [0, 1],$$

which follows from concavity. By observing that $|\Omega \Delta \Omega^*| = 2\,|\Omega \setminus \Omega^*|$ and recalling the definition of Fraenkel asymmetry, we get the conclusion. □

Remark 7.35. *An inspection of the proof reveals that a feasible choice for the constant ρ_d appearing in (7.72) is*

$$\rho_d = (d-1)\,\frac{(2^{1/d}-1)^2}{8\cdot 2^{1/d}}\,\frac{\omega_d^{2/d}\,J_{d/2}(\beta_{d/2,1})^2}{\displaystyle\int_0^1 J_{d/2}(\beta_{d/2,1}\,\varrho)^2\,\varrho\,d\varrho}.$$

By observing that $\varrho \mapsto J_{d/2}(\beta_{d/2,1}\,\varrho)^2\,\varrho$ is increasing on $(0,1)$, we can estimate this constant from below by

$$\rho_d \ge (d-1)\,\frac{(2^{1/d}-1)^2}{8\cdot 2^{1/d}}\,\omega_d^{2/d}.$$

7.4.4 Checking the sharpness

The proof of the sharp quantitative Szegő-Weinberger inequality is considerably simpler than that for the Faber-Krahn inequality. But there is a subtlety here: indeed, checking sharpness of Theorem 7.34 is now much more complicated. The argument used for λ_1 can not be applied here: indeed, the shape functional

$$\Omega \mapsto \mu_2(\Omega),$$

is not differentiable at the “maximum point”, i.e. for a ball B. This is due to the fact that $\mu_2(B)$ is a multiple eigenvalue (see [505, Chapter 2]). Thus what now can happen is that $\mu_2(B)-\mu_2(\Omega_n)$ behaves *linearly* along some family $\{\Omega_\varepsilon\}_{\varepsilon>0}$ converging to B, i.e.

$$\mu_2(B)-\mu_2(\Omega_\varepsilon) \simeq \mathcal{A}(\Omega_\varepsilon), \qquad |\Omega_\varepsilon| = |B|.$$

Quite surprisingly, the familiy of ellipsoids $\{E_\varepsilon\}_{\varepsilon>0}$ from the previous section exactly exhibits this behaviour. Indeed, by using the same notation as in Section 7.2.6, we have

$$\mu_2(E_\varepsilon) = \min_{u\in W^{1,2}(B_1)\setminus\{0\}} \left\{ \frac{\displaystyle\int_{B_1}\langle \widetilde{M}_\varepsilon \nabla u, \nabla u\rangle\,dx}{\displaystyle\int_{B_1}|u|^2\,dx} \ :\ \int_{B_1} u\,dx = 0 \right\}.$$

By recalling that

$$\int_{B_1}\langle \widetilde{M}_\varepsilon \nabla u, \nabla u\rangle\,dx = \int_{B_1}\left(\frac{|u_{x_1}|^2}{(1+\varepsilon)^2} + (1+\varepsilon)^2\,|u_{x_2}|^2 + \sum_{i=3}^d |u_{x_i}|^2\right)dx,$$

and

$$\frac{1}{(1+\varepsilon)^2} \le 1 - 2\,\varepsilon + 3\,\varepsilon^2,$$

if we use a L^2–normalized eigenfunction of the ball ξ corresponding to $\mu_2(B_1)$, we obtain

$$\mu_2(E_\varepsilon) \le \mu_2(B_1) + 2\,\varepsilon \int_{B_1}\left(|\xi_{x_2}|^2 - |\xi_{x_1}|^2\right)dx + \varepsilon^2 \int_{B_1}\left(3\,|\xi_{x_1}|^2 + |\xi_{x_2}|^2\right)dx.$$

An important difference with respect to the Dirichlet case now arises. Indeed, ξ is *not* radial and with a suitable choice of ξ we can obtain

$$\int_{B_1} \left(|\xi_{x_2}|^2 - |\xi_{x_1}|^2 \right) \, dx < 0.$$

Thus we finally get for $0 < \varepsilon \ll 1$

$$|B_1|^{2/d} \mu_2(B_1) - |E_\varepsilon|^{2/d} \mu_2(E_\varepsilon) \geq \frac{1}{C}\, \varepsilon \simeq \frac{1}{C}\, \mathcal{A}(E_\varepsilon).$$

This shows that the family of ellipsoids $\{E_\varepsilon\}_{\varepsilon>0}$ has (at most) a linear decay rate and thus it can not be used to show optimality of the estimate (7.72).

The difficulty is to detect families of deformations of a ball such that $\mu_2(B_1) - \mu_2(\Omega_\varepsilon)$ behaves quadratically. In other words, we need to identify directions along which $\Omega \mapsto \mu_2(\Omega)$ is smooth around the maximum point. The next result presents a general way to construct such families. This statement generalizes the one in [178, Section 6] and comes from the analogous discussion for the Steklov case, treated in [180, Section 6].

Theorem 7.36 (Sharpness of the quantitative Szegő-Weinberger inequality). *Let the function $\psi \in C^\infty(\partial B_1)$ satisfy the following assumptions:*
- *for every $a \in \mathbb{R}^d$, there holds*

$$\int_{\partial B_1} \langle a, x\rangle \, \psi \, d\mathcal{H}^{d-1} = 0; \tag{7.80}$$

- *for every $a \in \mathbb{R}^d$, there holds*

$$\int_{\partial B_1} \langle a, x\rangle^2 \, \psi \, d\mathcal{H}^{d-1} = 0. \tag{7.81}$$

Then the corresponding family $\{\Omega_\varepsilon\}_{\varepsilon>0}$ of nearly spherical domains

$$\Omega_\varepsilon = \left\{ x \in \mathbb{R}^d \ : \ x = 0 \ \text{or} \ |x| < 1 + \varepsilon\, \psi\left(\frac{x}{|x|}\right) \right\},$$

is such that

$$\mathcal{A}(\Omega_\varepsilon) \simeq \frac{|\Omega_\varepsilon \Delta B_1|}{|\Omega_\varepsilon|} \simeq \varepsilon \qquad \text{and} \qquad |B_1|^{2/d} \mu_2(B_1) - |\Omega_\varepsilon|^{2/d} \mu_2(\Omega_\varepsilon) \simeq \varepsilon^2,$$

for $\varepsilon \ll 1$.

Remark 7.37. *We may notice that the second condition* (7.81) *implies also*

$$\int_{\partial B_1} \psi \, d\mathcal{H}^{d-1} = 0. \tag{7.82}$$

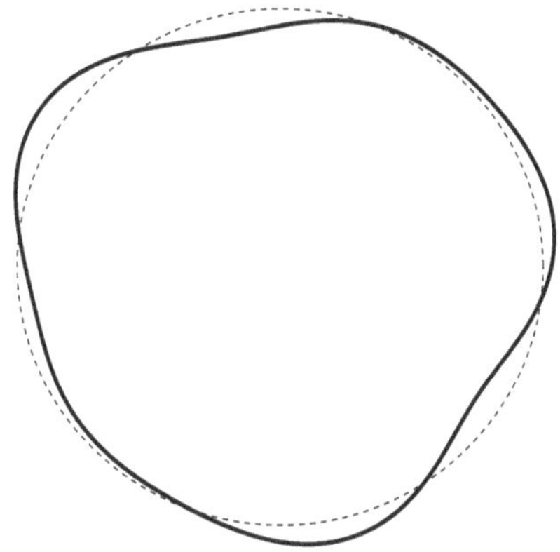

Fig. 7.1. The sets Ω_ε corresponding to the choice $\psi(\vartheta) = 2\sin 3\vartheta + \cos 5\vartheta$. Such a function satisfies (7.80) and (7.81).

Indeed, we have

$$\int_{\partial B_1} \psi \, d\mathcal{H}^{d-1} = \sum_{i=1}^{d} \int_{\partial B_1} \langle x, \mathbf{e}_i\rangle^2 \, \psi \, d\mathcal{H}^{d-1} = 0.$$

Remark 7.38 (Meaning of the assumptions on ψ). *Conditions* (7.80), (7.81) *and* (7.82) *are equivalent to saying that ψ is orthogonal in the $L^2(\partial B_1)$ sense to the first three eigenspaces of the Laplace-Beltrami operator on ∂B_1, i.e. to spherical harmonics of order* 0, 1 *and* 2 *respectively (see [713] for a comprehensive account on spherical harmonics).*

Each of these conditions plays a precise role in the construction: (7.82) *implies that* $|\Omega_\varepsilon| - |B_1| \simeq \varepsilon^2$. *The first condition* (7.80) *implies that Ω_ε has the same barycenter as B_1, still up to an error of order ε^2. Then this order coincides with the magnitude of $\mathcal{A}(\Omega_\varepsilon)^2$.*

In order to understand the second condition (7.81), *one should recall that every Neumann eigenfunction ξ corresponding to $\mu_2(B_1)$ is a linear combination of those defined by* (7.62). *Thus it has the form*

$$\xi(x) = \varphi_d(|x|) \sum_{i=1}^{d} a_i \, x_i = \varphi_d(|x|) \, \langle a, x\rangle, \qquad x \in B_1,$$

where φ_d is the radial function appearing in (7.63) *and $a \in \mathbb{R}^d \setminus \{0\}$. We then obtain for $x \in \partial B_1$*

$$|\xi(x)|^2 = \varphi_d(1)^2 \, \langle a, x\rangle^2,$$

and for the tangential gradient ∇_τ

$$|\nabla_\tau \xi|^2 = |\nabla \xi - \langle \nabla \xi, x\rangle \, x|^2 = |\nabla \xi|^2 - \langle \nabla \xi, x\rangle^2 = -\varphi_d(1)^2 \, \langle a, x\rangle^2 + \varphi_d(1)^2 \, |a|^2.$$

Thus condition (7.81) *implies*

$$\int_{\partial B_1} \psi\,|\xi|^2\,d\mathcal{H}^{d-1} = 0 \qquad \textit{and} \qquad \int_{\partial B_1} \psi\,|\nabla_\tau \xi|^2\,d\mathcal{H}^{d-1} = 0. \tag{7.83}$$

Relations (7.83) *are crucial in order to prove that* $\mu_2(B_1) - \mu_2(\Omega_\varepsilon) \simeq \varepsilon^2$.

We now sketch the proof of Theorem 7.36. In order to compare $\mu_2(\Omega_\varepsilon)$ with $\mu_2(B_1)$, we define an admissible test function in B_1, starting from an eigenfunction u_ε of Ω_ε. First of all, we take any smooth extension of u_ε outside Ω_ε, in order to have it defined in a set containing $\Omega_\varepsilon \cup B_1$. Then we take the test function

$$v_\varepsilon = u_\varepsilon \cdot 1_{B_1} - \delta_\varepsilon, \qquad \text{where } \delta_\varepsilon = \fint_{B_1} u_\varepsilon\,dx = O(\varepsilon).$$

By construction, it is not difficult to see that

$$\left| \int_{B_1} v_\varepsilon^2\,dx - \int_{B_1} u_\varepsilon^2\,dx \right| \le K\,\varepsilon^2, \tag{7.84}$$

By (7.84) and assuming that u_ε has unit L^2 norm on Ω_ε, we can estimate $\mu_2(B_1)$

$$\begin{aligned}
\mu_2(B_1) \le \frac{\displaystyle\int_{B_1} |\nabla v_\varepsilon|^2\,dx}{\displaystyle\int_{B_1} v_\varepsilon^2\,dx} &\le \frac{\displaystyle\int_{B_1\cap\Omega_\varepsilon} |\nabla u_\varepsilon|^2\,dx + \int_{B_1\setminus\Omega_\varepsilon} |\nabla u_\varepsilon|^2\,dx}{\displaystyle\int_{B_1\cap\Omega_\varepsilon} u_\varepsilon^2\,dx + \int_{B_1\setminus\Omega_\varepsilon} u_\varepsilon^2\,dx - K\,\varepsilon^2} \\
&\le \frac{\mu_2(\Omega_\varepsilon) + \mathcal{R}_1(\varepsilon)}{1 + \mathcal{R}_2(\varepsilon) - K\varepsilon^2}.
\end{aligned} \tag{7.85}$$

The two *error terms* $\mathcal{R}_1(\varepsilon)$ and $\mathcal{R}_2(\varepsilon)$ above are given by

$$\mathcal{R}_1(\varepsilon) = \int_{B_1\setminus\Omega_\varepsilon} |\nabla u_\varepsilon|^2\,dx - \int_{\Omega_\varepsilon\setminus B_1} |\nabla u_\varepsilon|^2\,dx,$$

and

$$\mathcal{R}_2(\varepsilon) = \int_{B_1\setminus\Omega_\varepsilon} u_\varepsilon^2\,dx - \int_{\Omega_\varepsilon\setminus B_1} u_\varepsilon^2\,dx.$$

It is not difficult to see that the following rough estimate holds

$$\left|\mathcal{R}_1(\varepsilon)\right| \le K'\,\varepsilon, \qquad \left|\mathcal{R}_2(\varepsilon)\right| \le K'\,\varepsilon, \tag{7.86}$$

for some $K' > 0$. Indeed, as Ω_ε is a small smooth deformation of B_1, u_ε satisfies uniform regularity estimates, thus for example $\|\nabla u_\varepsilon\|_{L^\infty} + \|u_\varepsilon\|_{L^\infty} \le C$ and

$$|\mathcal{R}_1(\varepsilon)| + |\mathcal{R}_2(\varepsilon)| \le 2\,C^2\,|B_1 \Delta \Omega_\varepsilon|,$$

thus giving (7.86). By inserting this in (7.85), one would get

$$\mu_2(B_1) \le \mu_2(\Omega_\varepsilon) + K''\,\varepsilon.$$

This shows that in order to get the correct decay estimate for the deficit, we need to improve (7.86) by replacing ε with ε^2.

We now explain how the assumptions on the function ψ (i.e. on the boundary of $\partial\Omega_\varepsilon$) imply that the rough estimate (7.86) can be enhanced. For ease of readability, we present below the heuristic argument, referring the reader to [178, Section 6] and [180, Section 6] for the rigorous proof. We focus on the term $\mathcal{R}_1(\varepsilon)$, the ideas for $\mathcal{R}_2(\varepsilon)$ being exactly the same. By using polar coordinates

$$\int_{\Omega_\varepsilon\setminus B_1} |\nabla u_\varepsilon|^2\,dx = \int_{\{\psi(y)\geq 0\}} \int_1^{1+\varepsilon\,\psi(y)} \left((\partial_\varrho u_\varepsilon)^2 + \frac{d-1}{\varrho^2}|\nabla_\tau u_\varepsilon|^2\right) \varrho^{d-1}\,d\varrho\,d\mathcal{H}^{d-1},$$

and

$$\int_{B_1\setminus\Omega_\varepsilon} |\nabla u_\varepsilon|^2\,dx = \int_{\{\psi(y)\leq 0\}} \int_{1+\varepsilon\,\psi(y)}^{1} \left((\partial_\varrho u_\varepsilon)^2 + \frac{d-1}{\varrho^2}|\nabla_\tau u_\varepsilon|^2\right) \varrho^{d-1}\,d\varrho\,d\mathcal{H}^{d-1},$$

where we recall that ∇_τ is the tangential gradient and ∂_ϱ is the derivative in the radial direction. The homogeneous Neumann condition of u_ε on $\partial\Omega_\varepsilon$ implies that the gradient ∇u_ε is "almost tangential" in the small sets $B_1\setminus\Omega_\varepsilon$ and $\Omega_\varepsilon\setminus B_1$. In particular

$$\partial_\varrho u_\varepsilon = O(\varepsilon) \qquad \text{for } \varrho = 1 + O(\varepsilon),$$

and

$$|\nabla_\tau u_\varepsilon(\varrho, y)| = |\nabla_\tau u_\varepsilon(1, y)| + O(\varepsilon), \qquad \text{for } \varrho = 1 + O(\varepsilon).$$

By observing that $|B_1\setminus\Omega_\varepsilon| = |\Omega_\varepsilon\setminus B_1| \simeq \varepsilon$, one can compute

$$\begin{aligned}\int_{\Omega_\varepsilon\setminus B_1} |\nabla u_\varepsilon|^2\,dx &= \int_{\{y\in\partial B_1\,:\,\psi(y)\geq 0\}} \int_1^{1+\varepsilon\,\psi(y)} |\nabla_\tau u_\varepsilon|^2\,\varrho^{d-3}\,d\varrho\,d\mathcal{H}^{d-1} + o(\varepsilon^2)\\ &= \varepsilon \int_{\{y\in\partial B_1\,:\,\psi(y)\geq 0\}} \psi\,|\nabla_\tau u_\varepsilon|^2\,d\mathcal{H}^{d-1} + O(\varepsilon^2),\end{aligned}$$

and similarly

$$\begin{aligned}\int_{B_1\setminus\Omega_\varepsilon} |\nabla u_\varepsilon|^2\,dx &= \int_{\{y\in\partial B_1\,:\,\psi(y)\leq 0\}} \int_{1+\varepsilon\psi(y)}^{1} |\nabla_\tau u_\varepsilon|^2\,\varrho^{d-3}\,d\varrho\,d\mathcal{H}^{d-1} + o(\varepsilon^2)\\ &= -\varepsilon \int_{\{y\in\partial B_1\,:\,\psi(y)\leq 0\}} \psi\,|\nabla_\tau u_\varepsilon|^2\,d\mathcal{H}^{d-1} + O(\varepsilon^2).\end{aligned}$$

Hence, recalling the definition of $\mathcal{R}_1(\varepsilon)$, one gets

$$\begin{aligned}\mathcal{R}_1(\varepsilon) &= -\varepsilon \int_{\{y\in\partial B_1\,:\,\psi(y)\leq 0\}} \psi\,|\nabla_\tau u_\varepsilon|^2\,d\mathcal{H}^{d-1}\\ &\quad - \varepsilon \int_{\{y\in\partial B_1\,:\,\psi(y)\geq 0\}} \psi\,|\nabla_\tau u_\varepsilon|^2\,d\mathcal{H}^{d-1} + O(\varepsilon^2)\\ &= -\varepsilon \int_{\partial B_1} \psi\,|\nabla_\tau u_\varepsilon|^2\,d\mathcal{H}^{d-1} + O(\varepsilon^2).\end{aligned} \tag{7.87}$$

It is precisely here that the condition (7.81) on ψ becomes relevant. Indeed, since Ω_ε is smoothly converging to B_1, one can guess that u_ε is sufficiently close to an eigenfunction ξ for $\mu_2(B_1)$. If we assume that we have

$$u_\varepsilon = \xi + O(\varepsilon), \tag{7.88}$$

then substituting u_ε with ξ in (7.87), one would get

$$\mathcal{R}_1(\varepsilon) = O(\varepsilon^2).$$

Indeed, we have seen that (7.81) implies (7.83) and thus

$$\int_{\partial B} \psi\, |\nabla_\tau \xi|^2 \, d\mathcal{H}^{d-1} = 0.$$

This would enhance the rate of convergence to 0 of the term $\mathcal{R}_1(\varepsilon)$ up to an order ε^2. The same arguments can be applied to $\mathcal{R}_2(\varepsilon)$, this time using the first relation in (7.83). By inserting these informations in (7.85), one would finally get

$$\mu_2(B_1) \le \mu_2(\Omega_\varepsilon) + K\,\varepsilon^2,$$

as desired. Of course, the most delicate part of the argument is to prove that the guess (7.88) is correct in a C^1 sense, i.e. that $\|u_\varepsilon - \xi\|_{C^1} = O(\varepsilon)$.

Remark 7.39 (Back to the ellipsoids). *Observe that if on the contrary ψ violates condition (7.81), we can not assure that all the first-order term in the previous estimates cancel out. For example, for the case of the ellipsoids E_ε considered above, their boundaries can be described as follows (let us take $d = 2$ for simplicity)*

$$\partial E_\varepsilon = \left\{ x = \varrho_\varepsilon(y)\, y \, : \, y \in \partial B_1 \text{ and } \varrho_\varepsilon(y) = \left((1+\varepsilon)^2\, y_2^2 + \frac{y_1^2}{(1+\varepsilon)^2} \right)^{-\frac{1}{2}} \right\}.$$

Observe that

$$\varrho_\varepsilon(y) \simeq 1 + \varepsilon\,(y_1^2 - y_2^2), \qquad y \in \partial B_1,$$

and the function $\psi(y) = y_1^2 - y_2^2$ crucially fails to satisfy[7.10] (7.81). *This confirms that*

$$\mu_2(B_1) - \mu_2(E_\varepsilon) \simeq \varepsilon,$$

i.e. ellipsoids do not exhibit the sharp decay rate for the Szegő-Weinberger inequality.

7.10 This function is indeed a spherical harmonic of order 2.

7.5 Stability for the Brock-Weinstock inequality

7.5.1 A quick overview of the Steklov spectrum

In this subsection, we give a brief presentation of the Steklov spectrum, and refer the reader to Section 5.1 of Chapter 5 for more details. Let $\Omega \subset \mathbb{R}^d$ be an open bounded set with Lipschitz boundary. We define its first nontrivial Steklov eigenvalue by

$$\sigma_2(\Omega) := \inf_{u \in H^1(\Omega)\setminus\{0\}} \left\{ \frac{\displaystyle\int_\Omega |\nabla u|^2\, dx}{\displaystyle\int_{\partial\Omega} |u|^2\, d\mathcal{H}^{d-1}} \, : \, \int_{\partial\Omega} u\, d\mathcal{H}^{d-1} = 0 \right\},$$

where the boundary integral in the denominator has to be interpreted in the trace sense. In other words, this is the sharp constant in the Poincaré-Wirtinger trace inequality

$$c \int_{\partial\Omega} \left| u - \fint_{\partial\Omega} u \right|^2 d\mathcal{H}^{d-1} \le \int_\Omega |\nabla u|^2\, dx, \qquad u \in H^1(\Omega).$$

Thanks to the assumptions on Ω, the embedding $H^1(\Omega) \hookrightarrow L^2(\partial\Omega)$ is compact (see [607, Section 6.10.5]) and the infimum above is attained. We have again discreteness of the spectrum of the Steklov Laplacian, that we denote by $\{\sigma_1(\Omega), \sigma_2(\Omega), \dots\}$. The first eigenvalue $\sigma_1(\Omega)$ is 0 and corresponds to constant eigenfunctions. These are the only real numbers σ for which the boundary value problem

$$\begin{cases} -\Delta u = 0, & \text{in } \Omega, \\ \dfrac{\partial u}{\partial \nu} = \sigma u, & \text{on } \partial\Omega, \end{cases}$$

admits nontrivial solutions. Here ν_Ω stands for the exterior normal versor. As always, $\sigma_k(\Omega)$ is obtained by minimizing the same Rayleigh quotient, among functions orthogonal (in the $L^2(\partial\Omega)$ sense, this time) to the first $k-1$ eigenfunctions. The scaling law of Steklov eigenvalues is now

$$\sigma_k(t\,\Omega) = t^{-1}\,\sigma_k(\Omega)$$

and we have the sharp inequality due to Brock

$$|\Omega|^{1/d}\,\sigma_2(\Omega) \le |B|^{1/d}\,\sigma_2(B), \tag{7.89}$$

with equality if and only if Ω is a ball.

As in the case of the Neumann Laplacian, here as well for any ball B_r the first nontrivial eigenvalue has multiplicity d. We have $\sigma_2(B_r) = \cdots = \sigma_{d+1}(B_r)$ and the corresponding eigenfunctions are just the coordinate functions

$$\xi_i(x) = x_i, \qquad i = 1, \dots, d. \tag{7.90}$$

Accordingly, we have

$$\sigma_2(B_r) = r^{-1}.$$

Actually, in dimension $d = 2$ and for simply connected sets, a result stronger than (7.89) holds. Indeed, if we recall the notation $P(\Omega)$ for the perimeter of a set Ω, for every $\Omega \subset \mathbb{R}^2$ open simply connected bounded set with smooth boundary, we know that

$$P(\Omega)\,\sigma_2(\Omega) \le P(B)\,\sigma_2(B), \tag{7.91}$$

where B is any open disc. This is the *Weinstock inequality*, proved in [873] by means of conformal mappings. Observe that by using the planar isoperimetric inequality

$$\frac{P(\Omega)}{\sqrt{|\Omega|}} \ge \frac{P(B)}{\sqrt{|B|}},$$

from (7.91) we get

$$\sqrt{|\Omega|}\,\sigma_2(\Omega) = P(\Omega)\,\sigma_2(\Omega)\,\frac{\sqrt{|\Omega|}}{P(\Omega)} \le P(B)\,\sigma_2(B)\,\frac{\sqrt{|B|}}{P(B)} = \sqrt{|D|}\,\sigma_2(D),$$

thus for simply connected sets in the plane, inequality (7.91) implies (7.89).

Remark 7.40 (The role of topology (again)). *Weinstock inequality is false if we remove the simple connectedness assumption, see Example 4.2.5 of Chapter 5. On the other hand, the quantity*

$$P(\Omega)\,\sigma_2(\Omega),$$

is uniformly bounded from above, but it is still an open problem to compute the sharp bound. We refer to Chapter 5 for more details.

We also recall that it is possible to provide isoperimetric-like estimates for sums of inverses. For example, for simply connected set in the plane Hersch and Payne in [521] showed that (7.91) can be improved as follows

$$\frac{1}{P(\Omega)}\left(\frac{1}{\sigma_2(\Omega)} + \frac{1}{\sigma_3(\Omega)}\right) \ge \frac{1}{P(B)}\left(\frac{1}{\sigma_2(B)} + \frac{1}{\sigma_3(B)}\right).$$

In general dimension and without restrictions on the topology of the sets, in [192] Brock proved that for every $\Omega \subset \mathbb{R}^d$ open bounded set with Lipschitz boundary, we have the stronger inequality

$$\frac{1}{|\Omega|^{1/d}}\sum_{k=2}^{d+1}\frac{1}{\sigma_k(\Omega)} \ge \frac{1}{|B|^{1/d}}\sum_{k=2}^{d+1}\frac{1}{\sigma_k(B)}. \tag{7.92}$$

Equality in (7.92) holds if and only if Ω is a ball. By recalling that for a ball we have $\sigma_2(B) = \cdots = \sigma_{d+1}(B)$, we see that (7.92) implies (7.89).

7.5.2 Weighted perimeters

The proof of (7.89) is similar to Weinberger's proof of (7.61). Namely, one obtains an upper bound on $\sigma_2(\Omega)$ by inserting in the relevant Rayleigh quotient the Steklov eigenfunctions (7.90) of the ball. This would give

$$\sigma_2(\Omega) \le \frac{\displaystyle\int_\Omega |\nabla x_i|^2\,dx}{\displaystyle\int_{\partial\Omega} |x_i|^2\,d\mathcal{H}^{d-1}} = \frac{|\Omega|}{\displaystyle\int_{\partial\Omega} |x_i|^2\,d\mathcal{H}^{d-1}}, \qquad i = 1,\dots,d.$$

Observe that the chosen test functions are admissible, up to translating Ω so that its boundary has its barycenter at the origin. By summing up all the inequalities above, one gets

$$\sigma_2(\Omega) \le \frac{d\,|\Omega|}{\displaystyle\int_{\partial\Omega} |x|^2\,d\mathcal{H}^{d-1}}.$$

Since for a ball we have equality in the previous estimate, in order to finish the proof we need the key ingredient, which is the following *weighted isoperimetric inequality*

$$\int_{\partial\Omega} |x|^2\,d\mathcal{H}^{d-1} \ge \int_{\partial\Omega^*} |x|^2\,d\mathcal{H}^{d-1}, \tag{7.93}$$

where Ω^* is the ball centered at the origin such that $|\Omega^*| = |\Omega|$. Inequality (7.93) has been proved by Betta, Brock, Mercaldo and Posteraro in [145]. If we use the notation

$$P_2(\Omega) = \int_{\partial\Omega} |x|^2\,\mathcal{H}^{d-1},$$

and observe that P_2 scales like a length to the power $d+1$, (7.93) can be rephrased in scaling invariant form as

$$|\Omega|^{-\frac{d+1}{d}}\,P_2(\Omega) \ge |B|^{-\frac{d+1}{d}}\,P_2(B), \tag{7.94}$$

where B is any ball *centered at the origin*. Equality in (7.94) holds if and only if Ω is a ball centered at the origin.

In order to get a quantitative improvement of the Brock-Weinstock inequality, it is sufficient to prove stability of (7.94). This has been done in [180], by means of an alternative proof of (7.94) based on a sort of *calibration technique* (related ideas can be found in the paper [595]).

Theorem 7.41. *For every $\Omega \subset \mathbb{R}^d$ open bounded set with Lipschitz boundary, we have*

$$|\Omega|^{-\frac{d+1}{d}}\,P_2(\Omega) - |\Omega^*|^{-\frac{d+1}{d}}\,P_2(\Omega^*) \ge \mathfrak{c}_d \left(\frac{|\Omega \Delta \Omega^*|}{|\Omega|}\right)^2, \tag{7.95}$$

where $\mathfrak{c}_d > 0$ is an explicit dimensional constant (see Remark 7.42 below).

Proof. As always, by scale invariance we can suppose that $|\Omega| = \omega_d$, so that the radius of the ball Ω^* is 1. We start observing that the vector field $x \mapsto |x|\,x$ is such that

$$\operatorname{div}(|x|\,x) = (d+1)\,|x|, \qquad x \in \mathbb{R}^d.$$

By integrating the previous quantity on Ω, applying the Divergence Theorem and using the Cauchy-Schwarz inequality, we obtain

$$(d+1)\int_\Omega |x|\,dx = \int_{\partial\Omega} |x|^2 \left\langle \frac{x}{|x|}, \nu_\Omega(x) \right\rangle d\mathcal{H}^{d-1} \le P_2(\Omega).$$

On the other hand, by integrating on the ball Ω^* we get

$$(d+1)\int_{\Omega^*} |x|\,dx = \int_{\partial\Omega^*} |x|^2\,d\mathcal{H}^{d-1} = P_2(\Omega^*),$$

since $\nu_{\Omega^*}(x) = x/|x|$. We thus obtain the following lower bound for the isoperimetric deficit

$$P_2(\Omega) - P_2(\Omega^*) \ge (d+1)\left[\int_\Omega |x|\,dx - \int_{\Omega^*} |x|\,dx\right].$$

The proof is now similar to that of the quantitative Szegő-Weinberger inequality. By applying again the quantitative Hardy-Littlewood inequality of Lemma 7.73, we get

$$P_2(\Omega) - P_2(\Omega^*) \ge (d+1)\,d\,\omega_d \int_{R_1}^{R_2} |\varrho - 1|\,\varrho^{d-1}\,d\varrho.$$

The radii $R_1 < 1 < R_2$ are still defined by

$$R_1 = \left(\frac{|\Omega \cap \Omega^*|}{\omega_d}\right)^{\frac{1}{d}} \qquad \text{and} \qquad R_2 = \left(\frac{|\Omega \setminus \Omega^*|}{\omega_d} + 1\right)^{\frac{1}{d}}.$$

With simple manipulations we arrive at

$$P_2(\Omega) - P_2(\Omega^*) \ge (d+1)\,d\,\omega_d \int_1^{R_2} (\varrho - 1)\,d\varrho. \tag{7.96}$$

As in the proof of the quantitative Szegő-Weinberger inequality, we have

$$\int_1^{R_2} (\varrho - 1)\,d\varrho = \frac{(R_2 - 1)^2}{2} \ge \frac{(2^{1/d} - 1)^2}{2}\left(\frac{|\Omega \setminus \Omega^*|}{|\Omega|}\right)^2,$$

where we used again (7.79). By using this in (7.96) and recalling that $|\Omega \setminus \Omega^*| = |\Omega^* \setminus \Omega|$, we get the desired conclusion. □

Remark 7.42. *The previous proof produces the following constant*

$$c_d = \frac{(d+1)\,d}{\omega_d^{1/d}}\,\frac{(2^{1/d} - 1)^2}{8}, \tag{7.97}$$

in inequality (7.95).

Remark 7.43. *The results of [145] and [180] hold for more general weighted perimeters of the form*

$$P_V(\Omega) = \int_{\partial\Omega} V(|x|)\, d\mathcal{H}^{d-1},$$

under suitable assumptions on the weight V. One may also consider anisotropic variants where the Euclidean norm is replaced by a general norm, see [182, Appendix A].

7.5.3 The Brock-Weinstock inequality in sharp quantitative form

By using Theorem 7.41, one can obtain a quantitative improvement of the stronger inequality (7.92) for the sum of inverses. This has been proved by Brasco, De Philippis and Ruffini in [180, Theorem 5.1].

Theorem 7.44. *For every $\Omega \subset \mathbb{R}^d$ open bounded set with Lipschitz boundary, we have*

$$\frac{1}{|\Omega|^{1/d}} \sum_{k=2}^{d+1} \frac{1}{\sigma_k(\Omega)} - \frac{1}{|B|^{1/d}} \sum_{k=2}^{d+1} \frac{1}{\sigma_k(B)} \geq \mathfrak{c}_d \left(\frac{|\Omega \Delta (\Omega^* + x_{\partial\Omega})|}{|\Omega|} \right)^2, \tag{7.98}$$

where Ω^* is the ball centered at the origin such that $|\Omega| = |\Omega^*|$ and the dimensional constant $\mathfrak{c}_d > 0$ is given by (7.97) and $x_{\partial\Omega}$ is the barycenter of the boundary $\partial\Omega$, i.e.

$$x_{\partial\Omega} = \fint_{\partial\Omega} x \, d\mathcal{H}^{d-1}.$$

Proof. We start by reviewing the proof of Brock. The first ingredient is a variational characterization for the sum of inverses of eigenvalues. In the case of Steklov eigenvalues, the following formula holds (see [523, Theorem 1], for example):

$$\sum_{k=2}^{d+1} \frac{1}{\sigma_k(\Omega)} = \max_{(v_2,\dots,v_{d+1}) \in \mathcal{E}} \sum_{k=2}^{d+1} \int_{\partial\Omega} v_k^2 \, d\mathcal{H}^{d-1},$$

where the set of admissible functions is given by

$$\mathcal{E} = \left\{ (v_2, \dots, v_{d+1}) \in (H^1(\Omega))^d \; : \; \begin{array}{c} \int_{\partial\Omega} v_i \, d\mathcal{H}^{d-1} = 0, \\ \int_\Omega \langle \nabla v_i, \nabla v_j \rangle \, dx = \delta_{ij} \end{array} \right\}.$$

The quantities $\sigma_i(\Omega)$ are translation invariant, so without loss of generality we can suppose that the barycenter of $\partial\Omega$ is at the origin, i.e. $x_{\partial\Omega} = 0$. This implies that the eigenfunctions relative to $\sigma_2(\Omega^*) = \dots = \sigma_{d+1}(\Omega^*)$ are admissible in the maximization problem above. More precisely, as admissible trial functions we take

$$v_i(x) = \frac{x_{i-1}}{\sqrt{|\Omega|}}, \qquad i = 2, \dots, d+1.$$

In this way, we obtain

$$\frac{1}{|\Omega|^{1/d}}\sum_{k=2}^{d+1}\frac{1}{\sigma_k(\Omega)} \geq \frac{1}{|\Omega|^{1+1/d}}\int_{\partial\Omega}|x|^2\,d\mathcal{H}^{d-1} = |\Omega|^{-\frac{d+1}{d}}\,P_2(\Omega),$$

which implies

$$\frac{1}{|\Omega|^{1/d}}\sum_{k=2}^{d+1}\frac{1}{\sigma_k(\Omega)} - \frac{1}{|\Omega^*|^{1/d}}\sum_{k=2}^{d+1}\frac{1}{\sigma_k(\Omega^*)} \geq |\Omega|^{-\frac{d+1}{d}}\,P_2(\Omega) - |\Omega^*|^{-\frac{d+1}{d}}\,P_2(\Omega^*). \tag{7.99}$$

In the inequality above we used that (recall that $|\Omega^*|^{1/d}\,\sigma_2(\Omega^*) = \omega_d^{1/d}$)

$$\frac{1}{|\Omega^*|^{1/d}}\sum_{k=2}^{d+1}\frac{1}{\sigma_k(\Omega^*)} = \frac{d}{\omega_d^{1/d}} = |\Omega^*|^{-\frac{d+1}{d}}\,P_2(\Omega^*).$$

It is then sufficient to use the quantitative estimate (7.95) in (7.99) in order to conclude. □

As a corollary, we get the following sharp quantitative version of the Brock-Weinstock inequality.

Theorem 7.45. *For every $\Omega \subset \mathbb{R}^d$ open bounded set with Lipschitz boundary, we have*

$$|B|^{1/d}\,\sigma_2(B) - |\Omega|^{1/d}\,\sigma_2(\Omega) \geq \tilde{c}_d\left(\frac{|\Omega\Delta(\Omega^* + x_{\partial\Omega})|}{|\Omega|}\right)^2, \tag{7.100}$$

where $\tilde{c}_d > 0$ is an explicit constant depending only on d only (see Remark 7.46 below).

Proof. First of all, we can suppose that

$$|\Omega|^{1/d}\,\sigma_2(\Omega) \geq \frac{1}{2}\,|B|^{1/d}\,\sigma_2(B), \tag{7.101}$$

otherwise estimate (7.100) is trivially true with constant $\tilde{c}_d = 1/8\,|B|^{1/d}\,\sigma_2(B)$, just by using the fact that

$$\frac{|\Omega\Delta(\Omega^* + x_{\partial\Omega})|}{|\Omega|} \leq 2.$$

Let us assume (7.101). By recalling that $\sigma_2(\Omega) \leq \sigma_i(\Omega)$ for every $i \geq 3$, from (7.98) we can infer

$$\frac{d}{|\Omega|^{1/d}\,\sigma_2(\Omega)} - \frac{d}{|B|^{1/d}\,\sigma_2(B)} \geq c_d\left(\frac{|\Omega\Delta(\Omega^* + x_{\partial\Omega})|}{|\Omega|}\right)^2.$$

This can be rewritten as

$$\frac{|B|^{1/d}\,\sigma_2(B) - |\Omega|^{1/d}\,\sigma_2(\Omega)}{\left(|\Omega|^{1/d}\,\sigma_2(\Omega)\right)\left(|B|^{1/d}\,\sigma_2(B)\right)} \geq \frac{c_d}{d}\left(\frac{|\Omega\Delta(\Omega^* + x_{\partial\Omega})|}{|\Omega|}\right)^2.$$

By using (7.101), the previous inequality easily implies (7.100). □

Remark 7.46. *By recalling that for every ball* $|B|^{1/d}\,\sigma_2(B)=\omega_d^{1/d}$, *the constant* $\widetilde{\mathfrak{c}}_d$ *above is given by*

$$\widetilde{\mathfrak{c}}_d=\frac{\omega_d^{1/d}}{2}\,\min\left\{\frac{\mathfrak{c}_d}{d}\,\omega_d^{1/d},\frac{1}{4}\right\},$$

and $\mathfrak{c}_d$ *is the same as in* (7.97).

Open problem 7.47 (Stability of the Weinstock inequality). *Prove that for every* $\Omega\subset\mathbb{R}^2$ *simply connected open bounded set with smooth boundary, we have*

$$P(B)\,\sigma_2(B)-P(\Omega)\,\sigma_2(\Omega)\geq c_d\,\mathcal{A}(\Omega)^2,$$

and

$$\frac{1}{P(\Omega)}\left(\frac{1}{\sigma_2(\Omega)}+\frac{1}{\sigma_3(\Omega)}\right)-\frac{1}{P(B)}\left(\frac{1}{\sigma_2(B)}+\frac{1}{\sigma_3(B)}\right)\geq c_d\,\mathcal{A}(\Omega)^2.$$

7.5.4 Checking the sharpness

The discussion here is very similar to that of the quantitative Szegő-Weinberger inequality. Indeed, the ball is the "maximum point" of

$$\Omega\mapsto|\Omega|^{1/d}\,\sigma_2(\Omega),$$

and σ_2 has nontrivial multiplicity for a ball, thus again we do not have differentiability. Then verifying that the exponent 2 on $\mathcal{A}$ is sharp is necessarily involved, exactly like in the Neumann case. In order to check sharpness of (7.34) we can use exactly the same family of domains as in Theorem 7.36. The heuristic ideas are the same as in the Neumann case, we refer the reader to [180, Section 6] for the proof. About the condition (7.81), i.e.

$$\int_{\partial B_1}\langle a,x\rangle^2\,\psi\,d\mathcal{H}^{d-1}=0,\qquad\text{for every } a\in\mathbb{R}^d,$$

we notice that this is still related to the peculiar form of Steklov eigenfunction of a ball. Indeed, from (7.90) we know that each eigenfunction ξ corresponding to $\sigma_2(B)$ has the form

$$\xi=\langle a,x\rangle,\qquad\text{for some } a\in\mathbb{R}^d.$$

Then we get

$$|\xi|^2=\langle a,x\rangle^2\qquad\text{and}\qquad|\nabla_\tau\xi|^2=|a|^2-\langle a,x\rangle^2.$$

Thus condition (7.81) implies again

$$\int_{\partial B_1}\psi\,|\xi|^2\,d\mathcal{H}^{d-1}=0\qquad\text{and}\qquad\int_{\partial B_1}\psi\,|\nabla_\tau\xi|^2\,d\mathcal{H}^{d-1}=0,$$

which are crucial in order to have the sharp decay rate.

Remark 7.48 (Sum of inverses). *Observe that*

$$c_d\,\mathcal{A}(\Omega)^2 \le \frac{1}{|\Omega|^{1/d}} \sum_{k=2}^{d+1} \frac{1}{\sigma_k(\Omega)} - \frac{1}{|B|^{1/d}} \sum_{k=2}^{d+1} \frac{1}{\sigma_k(B)}$$
$$\le \frac{d}{|B|^{1/d}\,\sigma_2(B)} \left(\frac{|B|^{1/d}\,\sigma_2(B)}{|\Omega|^{1/d}\,\sigma_2(\Omega)} - 1 \right).$$

Since the exponent 2 for $\mathcal{A}(\Omega)$ is sharp in the quantitative Brock-Weinstock inequality, this automatically proves the optimality of inequality (7.98) *as well.*

7.6 Some further stability results

7.6.1 The second eigenvalue of the Dirichlet Laplacian

Until now we have only considered isoperimetric-like inequalities for ground state energies of the Laplacian, i.e. for first (or first nontrivial) eigenvalues. In each of the cases previously considered, the optimal set was always a ball. On the contrary, very few facts are known on optimal shapes for successive eigenvalues. In the Dirichlet case, a well-known result states that disjoint pairs of equal balls (uniquely) minimize the second eigenvalue λ_2, among sets with given volume. This is the so-called *Hong-Krahn-Szego inequality*[7.11]. In scaling invariant form this reads

$$|\Omega|^{2/d}\,\lambda_2(\Omega) \ge 2^{2/d}\,|B|^{2/d}\,\lambda_1(B), \tag{7.102}$$

once it is observed that for the disjoint union of two identical balls, the first eigenvalue has multiplicity 2 and coincides with the first eigenvalue of one of the two balls. Equality in (7.102) is attained only for disjoint unions of two identical balls, up to sets of zero capacity.

The proof of (7.102) is quite simple and is based on the following fact.

Lemma 7.49. *Let $\Omega \subset \mathbb{R}^d$ be an open set with finite measure. Then there exist two disjoint open sets Ω_+, $\Omega_- \subset \Omega$ such that*

$$\lambda_2(\Omega) = \max\left\{\lambda_1(\Omega_+),\ \lambda_1(\Omega_-)\right\}. \tag{7.103}$$

7.11 This property of balls was discovered (at least) three times: first by Edgar Krahn ([604]) in the '20s. Then the result has been probably neglected, since in 1955 George Pólya attributes it to Peter Szego (see the final remark of [766]). However, one year before Pólya's paper, there appeared the paper [534] by Imsik Hong, giving once again a proof of this result. We point out that Peter Szego is the son of Gabor Szegő: the difference in the spelling of the surname is due to the fact that Peter passed most of his life in the US and consequently "Americanized" his name. We owe these informations to the kind courtesy of Mark S. Ashbaugh.

For a connected set, the two subsets Ω_+ and Ω_- above are nothing but the *nodal domains* of a second eigenfunction. In this case we have

$$\lambda_1(\Omega_+) = \lambda_2(\Omega) = \lambda_1(\Omega_-).$$

By using property (7.103) and the Faber-Krahn inequality, we get

$$|\Omega|^{2/d}\lambda_2(\Omega) \ge |B|^{2/d}\lambda_1(B)\max\left\{\left(\frac{|\Omega|}{|\Omega_+|}\right)^{\frac{2}{d}}, \left(\frac{|\Omega|}{|\Omega_-|}\right)^{\frac{2}{d}}\right\}. \tag{7.104}$$

By observing that

$$\max\left\{\left(\frac{|\Omega|}{a}\right)^{\frac{2}{d}}, \left(\frac{|\Omega|}{b}\right)^{\frac{2}{d}}\right\} \ge 2^{2/d}, \qquad \text{for every } a, b > 0,\ a + b \le |\Omega|, \tag{7.105}$$

we obtain inequality (7.102). As for equality cases, we observe that if equality holds in (7.102), then we must have equality in (7.104) and (7.105). The first one implies that Ω_+ and Ω_- above must be balls (by using equality cases in the Faber-Krahn inequality). But the lower bound in (7.105) is uniquely attained by the pair $a = b = |\Omega|/2$, thus we finally get that $|\Omega_+| = |\Omega_-| = |\Omega|/2$ and Ω is a disjoint union of two identical balls.

Remark 7.50. *We recall that the Hong-Krahn-Szego inequality is valid in the Robin case, as well (see Theorem 4.36 in Chapter 4).*

As before, we are interested in improving (7.102) by means of a quantitative stability estimate. This has been done in [178, Theorem 3.5]. To present this result, we first need to introduce a suitable variant of the Fraenkel asymmetry. This is the *Fraenkel 2–asymmetry*, which measures the L^1 distance of a set from the collection of disjoint pairs of equal balls. It is given by

$$\mathcal{A}_2(\Omega) := \inf\left\{\frac{|\Omega\Delta(B_+\cup B_-)|}{|\Omega|} \ :\ B_+, B_- \text{ balls s. t. } \begin{array}{c} B_+\cap B_- = \emptyset, \\ |B_+| = |B_-| = |\Omega|/2 \end{array}\right\}.$$

We then have the following quantitative version of the Hong-Krahn-Szego inequality. We point out that the exponent on the Fraenkel 2–asymmetry $\mathcal{A}_2$ in (7.106) is smaller than that in the original statement contained in [178], due to the use of the sharp Faber-Krahn inequality of Theorem 7.13.

Theorem 7.51. *Let $\Omega \subset \mathbb{R}^d$ be an open set with finite measure. Then*

$$|\Omega|^{2/d}\lambda_2(\Omega) - 2^{2/d}|B|^{2/d}\lambda_1(B) \ge \frac{1}{C_d}\mathcal{A}_2(\Omega)^{d+1}, \tag{7.106}$$

for a constant $C_d > 0$ depending on d only.

Proof. Let us set for simplicity

$$\mathcal{K}(\Omega) := |\Omega|^{2/d}\,\lambda_2(\Omega) - 2^{2/d}\,|B|^{2/d}\,\lambda_1(B).$$

The idea of the proof is to insert quantitative elements in (7.104) and (7.105), so as to obtain an estimate of the type

$$\mathcal{K}(\Omega) \geq \frac{1}{C_d}\max\left\{\mathcal{A}(\Omega_+)^2 + \left|\frac{1}{2} - \frac{|\Omega_+|}{|\Omega|}\right|,\ \mathcal{A}(\Omega_-)^2 + \left|\frac{1}{2} - \frac{|\Omega_-|}{|\Omega|}\right|\right\}, \tag{7.107}$$

where Ω_+ and Ω_- are as in Lemma 7.49. Estimate (7.107) would tell that the deficit on the Hong-Krahn-Szego inequality controls how far Ω_+ and Ω_- are from being balls having measure $|\Omega|/2$. Once estimate (7.107) is established, the claimed inequality (7.106) follows from the elementary geometric estimate

$$\mathcal{A}_2(\Omega)^{\frac{d+1}{2}} \leq C_d\left(\mathcal{A}(\Omega_+) + \left|\frac{1}{2} - \frac{|\Omega_+|}{|\Omega|}\right| + \mathcal{A}(\Omega_-) + \left|\frac{1}{2} - \frac{|\Omega_-|}{|\Omega|}\right|\right), \tag{7.108}$$

proved in [178, Lemma 3.3]. Observe that since the quantities appearing in the right-hand side of (7.107) are all bounded by a universal constant, it is not restrictive to prove (7.107) under the further assumption

$$\mathcal{K}(\Omega) \leq 2^{2/d}\,|B|^{2/d}\,\lambda_1(B). \tag{7.109}$$

To obtain (7.107), we need to distinguish two cases.

Case 1. Let us suppose that

$$|\Omega_+| \leq \frac{|\Omega|}{2} \qquad \text{and} \qquad |\Omega_-| \leq \frac{|\Omega|}{2}.$$

In this case, let us apply the quantitative Faber-Krahn inequality of Theorem 7.13 to Ω_+. By recalling (7.103), we find

$$\begin{aligned}
2^{2/d}\,\gamma_{d,2}\,\mathcal{A}(\Omega_+)^2 &\leq (2\,|\Omega_+|)^{2/d}\,\lambda_1(\Omega_+) - 2^{2/d}\,|B|^{2/d}\,\lambda_1(B)\\
&\leq (2\,|\Omega_+|)^{2/d}\,\lambda_2(\Omega) - 2^{2/d}\,|B|^{2/d}\,\lambda_1(B)\\
&= \mathcal{K}(\Omega) + \left((2\,|\Omega_+|)^{2/d} - |\Omega|^{2/d}\right)\lambda_2(\Omega)\\
&= \mathcal{K}(\Omega) + |\Omega|^{2/d}\,\lambda_2(\Omega)\left[\left(\frac{2\,|\Omega_+|}{|\Omega|}\right)^{2/d} - 1\right].
\end{aligned}$$

By concavity of $\tau \mapsto \tau^{2/d}$, we thus get

$$2^{2/d}\,\gamma_{d,2}\,\mathcal{A}(\Omega_+)^2 \leq \mathcal{K}(\Omega) + |\Omega|^{2/d}\,\lambda_2(\Omega)\,\frac{4}{d}\left(\frac{|\Omega_+|}{|\Omega|} - \frac{1}{2}\right). \tag{7.110}$$

By using the hypothesis on $|\Omega_+|$ and the Hong-Krahn-Szego inequality, we thus obtain

$$\mathcal{K}(\Omega) \geq c_d\,\mathcal{A}(\Omega_+)^2 + c_d\left|\frac{1}{2} - \frac{|\Omega_+|}{|\Omega|}\right|. \tag{7.111}$$

Hence, the same computations with Ω_- in place of Ω_+ yield (7.107).

Case 2. Let us suppose that

$$|\Omega_+| > \frac{|\Omega|}{2} \qquad \text{and} \qquad |\Omega_-| \le \frac{|\Omega|}{2}.$$

We still have the estimate (7.110) for both Ω_+ and Ω_-. In particular, for the smaller piece Ω_- we get again (7.111). On the contrary, this time it is no longer true that

$$\left(\frac{1}{2} - \frac{|\Omega_+|}{|\Omega|}\right) = \left|\frac{1}{2} - \frac{|\Omega_+|}{|\Omega|}\right|.$$

Then for Ω_+ the second term in the right-hand side of (7.110) has the wrong sign. The difficulty is that this term could be too big. However, by recalling that $|\Omega_-| + |\Omega_+| \le |\Omega|$ and using (7.111) for $|\Omega_-|$, we have

$$\frac{|\Omega_+|}{|\Omega|} - \frac{1}{2} \le \frac{1}{2} - \frac{|\Omega_-|}{|\Omega|} \le \frac{1}{c_d}\,\mathcal{K}(\Omega). \tag{7.112}$$

Therefore, using this information in (7.110) and recalling (7.109), we immediately get

$$\left(\frac{8}{d\,c_d}\,2^{2/d}\,|B|^{2/d}\,\lambda_1(B) + 1\right)\mathcal{K}(\Omega) \ge 2^{2/d}\,\gamma_{d,2}\,\mathcal{A}(\Omega_+)^2.$$

By combining this and (7.112), we thus obtain estimate (7.111) for Ω_+ as well, possibly with a different constant. Thus we obtain that (7.107) holds in this case as well. □

Remark 7.52. *The following example shows that the dimensional exponent in* (7.108) *is sharp.*

Example 7.53. *Let us fix a small parameter $\varepsilon > 0$ and consider the following set*

$$\Omega^\varepsilon = \{(x_1, x') \in \mathbb{R}^d \,:\, (x_1+1-\varepsilon)^2 + |x'|^2 < 1\} \cup \{(x_1, x') \in \mathbb{R}^d \,:\, (x_1-1+\varepsilon)^2 + |x'|^2 < 1\},$$

which is just the union of two balls of radius 1, with an overlapping region whose volume is of order $\varepsilon^{(d+1)/2}$. We set

$$\Omega_+^\varepsilon = \{(x_1, x') \in \Omega^\varepsilon \,:\, x_1 > 0\} \qquad \textit{and} \qquad \Omega_-^\varepsilon = \{(x_1, x') \in \Omega^\varepsilon \,:\, x_1 < 0\},$$

and by symmetry, we can work only with Ω_+^ε.

Concerning the sharpness of estimate (7.106), some remarks are in order.

Remark 7.54 (Sharpness?). *The proof of* (7.106) *consisted of two steps: the first one is the application of the quantitative Faber-Krahn inequality to the two relevant pieces Ω_+ and Ω_-; the second one is the geometric estimate* (7.108)*, which enables to switch from the error terms of Ω_+ and Ω_- to $\mathcal{A}_2(\Omega)$. Both steps are optimal (for the second one, see*

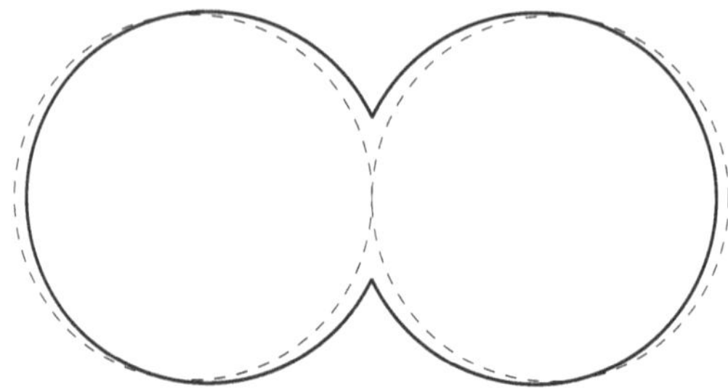

Fig. 7.2. The set Ω_ε of Example 7.55 and the pair of balls achieving $\mathcal{A}_2(\Omega_\varepsilon)$.

[178, Example 3.4]), but unfortunately this is of course not a guarantee of the sharpness of estimate (7.106).

Indeed, we are not able to decide whether the exponent for $\mathcal{A}_2$ in (7.106) *is optimal or not. In any case, we point out that the optimal exponent for the quantitative Hong-Krahn-Szego inequality has to be dimension-dependent. This follows from the next example.*

Example 7.55. *For every $\varepsilon > 0$ sufficiently small, we indicate with B_ε^+ and B_ε^- the open balls of radius 1, centered at $(1-\varepsilon)\,\mathbf{e}_1$ and $(\varepsilon - 1)\,\mathbf{e}_1$ respectively. We also set*

$$\Omega_\varepsilon^+ = B_\varepsilon^+ \cap \{x_1 \geq 0\} \qquad \text{and} \qquad \Omega_\varepsilon^- = B_\varepsilon^- \cap \{x_1 \leq 0\},$$

then we define the set $\Omega_\varepsilon := \Omega_\varepsilon^+ \cup \Omega_\varepsilon^- \subseteq \mathbb{R}^d$, for every $\varepsilon > 0$ sufficiently small. Observe that we have

$$|\Omega_\varepsilon^+| - |B_\varepsilon^+| = O\left(\varepsilon^{\frac{d+1}{2}}\right) \qquad \text{and} \qquad \lambda_2(\Omega_\varepsilon) - \lambda_1(B_\varepsilon^+) \leq O\left(\varepsilon^{\frac{d+1}{2}}\right),$$

for the second estimate see for example [184, Lemma 2.2].

As for asymmetries, it is not difficult to see that

$$\mathcal{A}(\Omega_\varepsilon^+) = \mathcal{A}(\Omega_\varepsilon^-) = O\left(\varepsilon^{\frac{d+1}{2}}\right) \qquad \text{and} \qquad \mathcal{A}_2(\Omega_\varepsilon) = O(\varepsilon).$$

(see Figure 7.2). Then we get

$$\begin{aligned} |\Omega_\varepsilon|^{2/d}\,\lambda_2(\Omega_\varepsilon) - 2^{2/d}\,|B|^{2/d}\,\lambda_1(B) &= 2^{2/d}\left(|\Omega_\varepsilon^+|^{2/d}\,\lambda_2(\Omega_\varepsilon) - |B|^{2/d}\,\lambda_1(B)\right) \\ &= O\left(\varepsilon^{\frac{d+1}{2}}\right) = O\left(\mathcal{A}_2(\Omega_\varepsilon)^{(d+1)/2}\right). \end{aligned}$$

This shows that the sharp exponent in (7.106) *has to depend on the dimension and is in between $(d+1)/2$ and $d+1$.*

Open problem 7.56 (Sharp quantitative Hong-Krahn-Szego inequality). *Prove or disprove that the exponent $d+1$ in* (7.106) *is sharp. If $d+1$ is not sharp, find the optimal exponent.*

7.6.2 The ratio of the first two Dirichlet eigenvalues

Another well-known spectral inequality which involves the second Dirichlet eigenvalue λ_2 is the so-called *Ashbaugh-Benguria inequality*. This asserts that the ratio λ_2/λ_1 is maximal on balls and has been proved in [62, 64]. In other words, for every open set $\Omega \subset \mathbb{R}^d$ with finite measure we have

$$\frac{\lambda_2(\Omega)}{\lambda_1(\Omega)} \le \frac{\lambda_2(B)}{\lambda_1(B)}. \tag{7.113}$$

Remark 7.57 (Equality cases). *Equality cases in* (7.113) *are a bit subtle: indeed, in general it is not true that equality in* (7.113) *is attained for balls only. As a counter-example, it is sufficient to consider any disjoint union of the type*

$$\Omega = B \cup \Omega',$$

with Ω' an open set such that $\lambda_1(\Omega') > \lambda_2(B)$. In general equality in (7.113) *only implies that the connected component of Ω supporting λ_1 and λ_2 is a ball.*

Remark 7.58. *Inequality* (7.113) *is an example of universal inequality. With this name we usually designate spectral inequalities involving eigenvalues only, without any other geometric quantity (see Chapter 8). In particular, inequality* (7.113) *is valid in the larger class of open sets having discrete spectrum, but not necessarily finite measure.*

The first stability result for (7.113) is due to Melas, see [706, Theorem 3.1]. To the best of our knowledge, this is still the best known result on the subject. The original statement was for the asymmetry $\delta_{\mathcal{M}}$ defined in (7.16). Here on the contrary we state the result for the Fraenkel asymmetry.

Theorem 7.59 (Melas). *Let $\Omega \subset \mathbb{R}^d$ be an open bounded convex set. Then we have*

$$\frac{\lambda_2(B)}{\lambda_1(B)} - \frac{\lambda_2(\Omega)}{\lambda_1(\Omega)} \ge \frac{1}{C} \mathcal{A}(\Omega)^m, \tag{7.114}$$

for some $C = C(d) > 0$ and $m = m(d) > 10$ (see Remark (7.64) below) depending on the dimension d only.

We are going to present the core of the proof of Theorem 7.59 below. At first, one needs a handful of technical results.

Lemma 7.60. *Let $\Omega \subset \mathbb{R}^d$ be an open set with finite measure. Let $B \subset \mathbb{R}^d$ be a ball such that*

$$\lambda_1(B) = \lambda_1(\Omega).$$

There exists a constant $C = C(d) > 0$ such that

$$\frac{|\Omega| - |B|}{|\Omega|} \ge \frac{1}{C} \mathcal{A}(\Omega)^2. \tag{7.115}$$

Proof. Observe that thanks to Theorem 7.13, we have

$$\lambda_1(B) = \lambda_1(\Omega) \geq \frac{|B|^{2/d}}{|\Omega|^{2/d}}\,\lambda_1(B) + \frac{\gamma_{d,2}}{|\Omega|^{2/d}}\,\mathcal{A}(\Omega)^2.$$

Thus we get

$$\left(\frac{|\Omega|}{|B|}\right)^{\frac{2}{d}} - 1 \geq \frac{\gamma_{d,2}}{|B|^{2/d}\,\lambda_1(B)}\,\mathcal{A}(\Omega)^2.$$

From the previous inequality, by concavity of the function $\tau \mapsto \tau^{2/d}$ we obtain

$$c'_{d},\mathcal{A}(\Omega)^2 \leq \frac{2}{d}\,\frac{|\Omega| - |B|}{|B|}. \tag{7.116}$$

We now distinguish two possibilities: if $|\Omega| \leq 2\,|B|$, we have

$$\frac{|\Omega| - |B|}{|B|} \leq 2\,\frac{|\Omega| - |B|}{|\Omega|}.$$

By inserting this information in the right-hand side of (7.116), we get (7.115) as desired.

The case $|\Omega| > 2\,|B|$ is even simpler. Indeed, in this case

$$\frac{|\Omega| - |B|}{|\Omega|} > \frac{1}{2} \geq \frac{1}{8}\,\mathcal{A}(\Omega)^2,$$

since the asymmetry of a set does not exceed 2. □

The key ingredient in the proof by Melas is the following result. It asserts that for non degenerating convex sets with given measure, the values of the first Dirichlet eigenfunction control in a quantitative way the measure of the corresponding sublevel sets. Formally:

Lemma 7.61. *Let $\Lambda > 0$, there exist $C = C(d,\Lambda) > 0$ and $\beta = \beta(d,\Lambda) > 1$ such that for every open convex set $\Omega \subset \mathbb{R}^d$ with*

$$|\Omega| = 1 \qquad \text{and} \qquad \lambda_1(\Omega) \leq \Lambda, \tag{7.117}$$

and every $t > 0$ we have

$$\frac{1}{C}\,\Big|\{x \in \Omega\,:\, u_1(x) \leq t\}\Big|^{\beta} \leq t. \tag{7.118}$$

Here u_1 is the first (positive) Dirichlet eigenfunction of Ω with unit L^2 norm.

We omit the proof of the previous result, the interested reader may find it in [706, Lemma 3.5]. We just mention that (7.118) follows by proving the comparison estimate

$$u_1(x) \geq c\,\mathrm{dist}(x,\partial\Omega)^{\beta}, \qquad x \in \Omega. \tag{7.119}$$

Remark 7.62 (The exponent β). *By recalling that, on a convex set, the first eigenfunction u_1 is always globally Lipschitz continuous, we know that the exponent β above can not be smaller than 1. Moreover, it is quite clear that β in (7.119) depends heavily on the regularity of the boundary $\partial\Omega$. To clarify this point, let us stick for simplicity to the case $d = 2$. If $\partial\Omega$ contains a corner at $x_0 \in \partial\Omega$ of opening $\alpha < \pi/2$, classical asymptotic estimates based on comparisons with harmonic homogeneous functions imply that*

$$u_1(x) \simeq \operatorname{dist}(x, \partial\Omega)^{\frac{\pi}{2\alpha}},$$

around the corner x_0. This in particular shows that the smaller the angle α is, the larger the exponent β in (7.119) must be. In particular, without taking any further restriction on the convex sets Ω, it would be impossible to get (7.119).

The hypothesis (7.117) exactly prevents the convex sets considered to become too narrow and permits to have a control like (7.119), with a uniform β.

Finally, one also needs the following interesting result, whose proof can be found in [90, Section 6.1]. This permits to reduce the proof of Theorem 7.59 to the case of convex sets satisfying the hypothesis (7.117) of the previous result. A similar statement was contained in the original paper by Melas (this is essentially [706, Proposition 3.1]), but the proof in [90] is quicker and simpler. We reproduce it here, with some minor modifications.

Lemma 7.63. *Let $\{\Omega_n\}_{n\in\mathbb{N}} \subset \mathbb{R}^d$ be a sequence of open convex sets such that*

$$|\Omega_n| = 1 \qquad \text{and} \qquad \lim_{n\to\infty} \lambda_1(\Omega_n) = +\infty.$$

Then we have

$$\lim_{n\to\infty} \frac{\lambda_2(\Omega_n)}{\lambda_1(\Omega_n)} = 1. \tag{7.120}$$

In particular, for every $\delta > 0$ there exists $\Lambda = \Lambda(\delta) > 0$ such that

$$\sup\left\{\lambda_1(\Omega) \;:\; \Omega \subset \mathbb{R}^d \text{ open convex such that } \frac{\lambda_2(\Omega)}{\lambda_1(\Omega)} \ge (1+\delta)\right\} \le \Lambda. \tag{7.121}$$

Proof. We first observe that (7.121) easily follows from the first part of the statement. Thus we just need to prove (7.120). For every $n \in \mathbb{N}$, we take a pair of points $(a_n, b_n) \in \partial\Omega_n$ such that

$$|a_n - b_n| = \operatorname{diam}(\Omega_n).$$

Up to rigid motions, we can suppose that

$$a_n = (0, \dots, 0, \operatorname{diam}(\Omega_n)) \qquad \text{and} \qquad b_n = (0, \dots, 0).$$

Observe that the hypotheses on the sequence $\{\Omega_n\}_{n\in\mathbb{N}}$ implies that

$$\lim_{n\to+\infty} |a_n - b_n| = \lim_{n\to+\infty} \operatorname{diam}(\Omega_n) = +\infty,$$

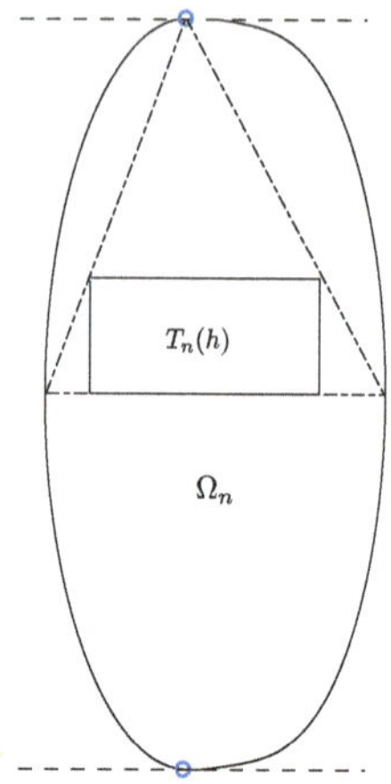

Fig. 7.3. The construction of the cylinder $T_n(h)$ for the proof of Lemma 7.63.

see Remark 7.77. We now need to prove that for every $n \in \mathbb{N}$ there exists $0 < t_n < \operatorname{diam}(\Omega_n)$ such that

$$\lambda_1(\Omega_n) \geq \lambda_1(\Omega_n \cap \{x_d = t_n\}). \tag{7.122}$$

Indeed, let us consider the first (positive) eigenfunction $u_n \in H_0^1(\Omega_n)$ with unit L^2 norm. By Fubini's Theorem we have

$$\begin{aligned}\lambda_1(\Omega_n) = \int_{\Omega_n} |\nabla u_n|^2\, dx &\geq \int_0^{\operatorname{diam}(\Omega_n)} \int_{\Omega_n \cap \{x_d = t\}} |\nabla' u_n|^2\, dx'\, dt \\ &\geq \int_0^{\operatorname{diam}(\Omega_n)} \left(\lambda_1(\Omega_n \cap \{x_d = t\}) \int_{\Omega_n \cap \{x_d = t\}} |u_n|^2\, dx'\right) dt,\end{aligned}$$

where we used the notation $x' = (x_1, \dots, x_{d-1})$ and $\nabla' = (\partial_{x_1}, \dots, \partial_{x_{d-1}})$. Since we assumed

$$\int_0^{\operatorname{diam}(\Omega_n)} \int_{\Omega_n \cap \{x_d = t\}} |u_n|^2\, dx'\, dt = \int_{\Omega_n} |u_n|^2\, dx = 1,$$

from the previous estimate we get (7.122). From the fact that $0 < t_n < |a_n| = \operatorname{diam}(\Omega_n)$, we have

- either
$$|a_n| - t_n = \operatorname{dist}(a_n, \Omega_n \cap \{x_d = t_n\}) \geq \frac{\operatorname{diam}(\Omega_n)}{2};$$
- or
$$t_n = \operatorname{dist}(b_n, \Omega_n \cap \{x_d = t_n\}) \geq \frac{\operatorname{diam}(\Omega_n)}{2}.$$

Without loss of generality we can assume that the first condition is verified, then we consider the cone $\mathfrak{C}_n$ given by the convex hull of $\{a_n\} \cup (\Omega_n \cap \{x_d = t_n\})$. By convexity, we have $\mathcal{C}_n \subset \Omega_n$ and for every $0 < h < |a_n| - t_n$

$$T_n(h) := \frac{|a_n| - t_n - h}{|a_n| - t_n} \left(\Omega_n \cup \{x_d = t_n\}\right) \times (t_n, t_n + h) \subset \mathcal{C}_n \subset \Omega_n.$$

In other words, Ω_n contains a cylinder having height h and with basis a scaled copy of the $(d-1)$–dimensional section $\Omega_n \cap \{x_d = t_n\}$, see Figure 7.3. By monotonicity and scaling properties of Dirichlet eigenvalues and (7.122), we thus obtain[7.12]

$$\begin{aligned}
1 \le \frac{\lambda_2(\Omega_n)}{\lambda_1(\Omega_n)} &\le \frac{\lambda_2(T_n(h))}{\lambda_1(\Omega_n)} \\
&\le \frac{1}{\lambda_1(\Omega_n)} \left[\lambda_1\left(\frac{|a_n| - t_n - h}{|a_n| - t_n} \left(\Omega_n \cup \{x_d = t_n\}\right)\right) + \frac{4\pi^2}{h^2}\right] \\
&\le \left[\left(\frac{|a_n| - t_n}{|a_n| - t_n - h}\right)^2 + \frac{4\pi^2}{h^2} \frac{1}{\lambda_1(\Omega_n)}\right].
\end{aligned}$$

By recalling that $\lambda_1(\Omega_n)$ and $|a_n| - t_n$ are diverging to $+\infty$, we get (7.120) as desired. □

We now come to the proof of the quantitative Ashbaugh-Benguria inequality.

Proof of Theorem 7.59. We first observe that since the functional λ_2/λ_1 is scaling invariant, we can suppose that

$$|\Omega| = 1. \tag{7.123}$$

Moreover, we can always suppose

$$\lambda_2(\Omega) \ge (1 + \delta)\, \lambda_1(\Omega), \tag{7.124}$$

where δ is the dimensional constant

$$\delta := \frac{1}{2} \left(\frac{\lambda_2(B)}{\lambda_1(B)} - 1\right) > 0.$$

Indeed, when (7.124) is not verified, then we have

$$\frac{\lambda_2(B)}{\lambda_1(B)} - \frac{\lambda_2(\Omega)}{\lambda_1(\Omega)} \ge \frac{\lambda_2(B)}{\lambda_1(B)} - (1 + \delta) = \frac{1}{2} \left(\frac{\lambda_2(B)}{\lambda_1(B)} - 1\right) > 0,$$

and the stability estimate is trivially true, with a constant depending on the dimensional constant $\lambda_2(B)/\lambda_1(B)$ only. Finally, thanks to hypothesis (7.124) and Lemma 7.63, we obtain

$$\lambda_1(\Omega) \le \Lambda, \tag{7.125}$$

7.12 We also use the fact that for a cylindric set $\mathcal{O} \times (a, b)$ its Dirichlet eigenfunctions have the form $U(x', x_d) = u(x') \cdot v(x_d)$, with u Dirichlet eigenfunction of $\mathcal{O}$ and v Dirichlet eigenfunction of (a, b). The corresponding eigenvalues take the form

$$\lambda = \lambda' + \frac{n^2 \pi^2}{(b - a)^2},$$

where λ' is an eigenvalue of $\mathcal{O}$ and $n \in \mathbb{N}$.

with Λ depending on δ only and thus only on the dimension d.

We now take the ball B centered at the origin and such that $\lambda_1(\Omega) = \lambda_1(B)$. By (7.7), its radius R is given by

$$R = \frac{j_{d/2-1,1}}{\sqrt{\lambda_1(\Omega)}}.$$

We set u_1 and z for the eigenfunctions corresponding to $\lambda_1(\Omega)$ and $\lambda_1(B)$, normalized by the conditions

$$\int_\Omega u_1^2\, dx = \int_B z^2\, dx = 1.$$

We recall that

$$z(x) = \alpha\, |x|^{\frac{2-d}{2}}\, J_{\frac{d-2}{2}}\left(\frac{j_{(d-2)/2,1}}{R}\,|x|\right),$$

with the normalization constant α given by

$$\begin{aligned}\alpha^2 &= \frac{1}{\displaystyle\int_B |x|^{2-d}\, J_{d/2-1}\left(\frac{j_{d/2-1,1}}{R}|x|\right)^2 dx}\\ &= \frac{1}{\displaystyle R^2 \int_{\{|y|<1\}} |y|^{2-d}\, J_{d/2-1}\left(j_{d/2-1,1}\,|y|\right)^2 dy} =: c_d\, R^{-2}.\end{aligned} \tag{7.126}$$

We then compare $\lambda_2(\Omega)$ and $\lambda_2(B)$: since B and Ω have the same λ_1, we get

$$\lambda_2(B) - \lambda_2(\Omega) = \left(\lambda_2(B) - \lambda_1(B)\right) - \left(\lambda_2(\Omega) - \lambda_1(\Omega)\right). \tag{7.127}$$

We introduce the functions P_i defined as follows

$$P_i(x) = g(|x|)\,\frac{x_i}{|x|}, \qquad i = 1, \dots, d,$$

with g being the ratio of (the radial parts of) the eigenfunctions corresponding to $\lambda_1(B)$ and $\lambda_2(B)$, that is

$$g(t) = \frac{J_{\frac{d}{2}}\left(\frac{j_{d/2,1}}{R}\, t\right)}{J_{\frac{d-2}{2}}\left(\frac{j_{d/2-1,1}}{R}\, t\right)},$$

extended as $g(t) \equiv \lim_{s\to R^-} g(s)$ for $t \geq R$. In this way, the functions P_i are defined over $\mathbb{R}^d$. With a suitable choice of the origin of the axes, one can guarantee (see [505, Lemma 6.2.2])

$$\int_\Omega P_i\, u_1^2\, dx = 0, \qquad i = 1, \dots, d.$$

This implies that the functions $P_i\, u_1$ are L^2–orthogonal to u_1, thus we can use them in the variational problem defining $\lambda_2(\Omega)$. We get

$$\lambda_2(\Omega) \leq \frac{\displaystyle\int_\Omega |\nabla P_i|^2\, u_1^2 dx + \int_\Omega |\nabla u_1|^2\, P_i^2\, dx + 2\int_\Omega \langle \nabla P_i, \nabla u_1\rangle\, P_i\, u_1\, dx}{\displaystyle\int_\Omega P_i^2\, u_1^2\, dx},$$

then we observe that by testing the equation $-\Delta u_1 = \lambda_1(\Omega)\, u_1$ against $P_i^2\, u_1$, we obtain

$$\int_\Omega |\nabla u_1|^2\, P_i^2\, dx + 2\int_\Omega \langle \nabla P_i, \nabla u_1\rangle\, P_i\, u_1\, dx = \lambda_1(\Omega)\int_\Omega P_i^2\, u_1^2\, dx.$$

This permits us to infer

$$\lambda_2(\Omega) - \lambda_1(\Omega) \le \frac{\displaystyle\int_\Omega |\nabla P_i|^2\, u_1^2 dx}{\displaystyle\int_\Omega P_i^2\, u_1^2\, dx}. \tag{7.128}$$

The same computations in the ball B give of course equalities everywhere, since in this case $P_i\, u_1$ would coincide with a second Dirichlet eigenfunction of B. Thus

$$\frac{\displaystyle\int_B |\nabla P_i|^2\, z^2 dx}{\displaystyle\int_B P_i^2\, z^2\, dx} = \lambda_2(B) - \lambda_1(B). \tag{7.129}$$

We then perform the standard trick of adding these (in)equalities for $i = 1, \dots, d$, which let the angular variables disappear, as in the proof of the Szegő-Weinberger inequality. Thus from (7.128) and (7.129), we obtain

$$\left(\int_\Omega g^2\, u_1^2\, dx\right)[\lambda_2(\Omega) - \lambda_1(\Omega)] \le \int_\Omega \left[|g'(|x|)|^2 + (d-1)\left(\frac{g(|x|)}{|x|}\right)^2\right] u_1^2\, dx, \tag{7.130}$$

and

$$\int_B \left[|g'(|x|)|^2 + (d-1)\left(\frac{g(|x|)}{|x|}\right)^2\right] z^2\, dx = \left(\int_B g^2\, z^2\, dx\right)[\lambda_2(B) - \lambda_1(B)]. \tag{7.131}$$

We now use the fact that the function g is monotone non-decreasing on $\mathbb{R}^d$ (see [64, Theorem 3.3]), so that by the Hardy-Littlewood inequality we obtain[7.13]

$$\int_\Omega g^2\, u_1^2\, dx \ge \int_{\Omega^*} g_\star^2\, (u_1^*)^2\, dx \ge \int_{\Omega^*} g^2\, (u_1^*)^2\, dx,$$

where $g_\star$ denotes the spherically *increasing* rearrangement. As always, Ω^* is the ball centered at the origin such that $|\Omega^*| = |\Omega| = 1$. We denote its radius by R_Ω.

7.13 There is a small subtlety here. Indeed, even if g is a radially non-decreasing function, in general we just have

$$g_\star(r) \ge g(r), \qquad \text{for } 0 \le r \le R_\Omega,$$

and inequality could be strict. This is due to the fact that $g_\star$ is constructed by rearranging the super-level sets of g on Ω and not on the whole $\mathbb{R}^d$ (see [64, page 606]).

Another essential ingredient in the proof by Ashbaugh and Benguria is a *comparison result* between z and the spherical rearrangement $u_1^\star$ of u_1. This is due to Chiti, who proved (see [270]) that there exists a radius $r_1 \in (0, R)$ such that

$$\begin{cases} u_1^\star(x) \le z(x), & \text{for } 0 \le |x| \le r_1, \\ u_1^\star(x) \ge z(x), & \text{for } r_1 \le |x| \le R. \end{cases} \tag{7.132}$$

This in turn implies that

$$\int_{\Omega^*} g^2\,(u_1^\star)^2\,dx \ge \int_B g^2\,z^2\,dx.$$

More precisely, by using (7.132), the monotonicity of g and polar coordinates we have

$$\begin{aligned}\int_{\Omega^*} g^2\,[(u_1^\star)^2 - z^2]\,dx &= d\,\omega_d \int_0^{r_1} g^2\,[(u_1^\star)^2 - z^2]\,\varrho^{d-1}\,d\varrho \\ &\quad + d\,\omega_d \int_{r_1}^{R} g^2\,[(u_1^\star)^2 - z^2]\,\varrho^{d-1}\,d\varrho \\ &\quad + d\,\omega_d \int_R^{R_\Omega} g^2\,(u_1^\star)^2\,\varrho^{d-1}\,d\varrho \\ &\ge g(r_1)^2 \int_B [(u_1^\star)^2 - z^2]\,dx + g(R)^2 \int_{\Omega^*\setminus B} (u_1^\star)^2\,dx \\ &= \left(g(R)^2 - g(r_1)^2\right) \int_{\Omega^*\setminus B} u_1^\star(x)^2\,dx.\end{aligned}$$

In the last equality we used that

$$\int_B z^2\,dx = \int_{\Omega^*} (u_1^\star)^2\,dx = 1.$$

We now observe that, using the definition both of z and g, we get

$$\int_B g^2\,z^2\,dx = \alpha^2\,R^2 \int_{\{|y|<1\}} J_{\frac{d}{2}}(j_{d/2,1}\,|y|)^2\,dy := C_d,$$

thanks to (7.126). In this way, we have shown

$$\int_{\Omega^*} g^2\,|u_1^\star|^2\,dx \ge C_d + \left(g(R)^2 - g(r_1)^2\right) \int_{\Omega^*\setminus B} u_1^\star(x)^2\,dx.$$

In the end, by using (7.127), (7.130) and (7.131) we have obtained that

$$\begin{aligned} C_d\left(\lambda_2(B)-\lambda_2(\Omega)\right) &\geq \left(\int_B g^2\, z^2\, dx\right)\left(\lambda_2(B)-\lambda_1(B)\right) \\ &\quad - \left(\int_\Omega g^2\, |u_1^\star|^2\, dx\right)\left(\lambda_2(\Omega)-\lambda_1(\Omega)\right) \\ &\quad + \left(g(R)^2-g(r_1)^2\right)\left(\int_{\Omega^\star\setminus B} u_1^\star(x)^2\, dx\right)\left(\lambda_2(\Omega)-\lambda_1(\Omega)\right) \\ &\geq \int_B \left[|g'|^2+(d-1)\left(\frac{g}{|x|}\right)^2\right] z^2\, dx \\ &\quad - \int_\Omega \left[|g'|^2+(d-1)\left(\frac{g}{|x|}\right)^2\right] u_1^2\, dx \\ &\quad + \delta\left(g(R)^2-g(r_1)^2\right)\left(\int_{\Omega^\star\setminus B} (u_1^\star)^2\, dx\right)\lambda_1(\Omega), \end{aligned}$$

where we also used hypothesis (7.124) for the last term. It is now time to use the monotonicity properties of the function

$$G(t) = g'(t)^2 + (d-1)\left(\frac{g(t)}{t}\right)^2,$$

which is monotone non-increasing (see [64, Corollary 3.4]), so that again by the Hardy-Littlewood inequality

$$\int_\Omega G\, u_1^2\, dx \leq \int_{\Omega^\star} G^\star\,(u_1^\star)^2\, dx \leq \int_{\Omega^\star} G\,(u_1^\star)^2\, dx.$$

and we thus have

$$\begin{aligned} C_d\left(\lambda_2(B)-\lambda_2(\Omega)\right) &\geq \int_B G\, z^2\, dx - \int_{\Omega^\star} G\,(u_1^\star)^2\, dx \\ &\quad + \delta\left(g(R)^2-g(r_1)^2\right)\left(\int_{\Omega^\star\setminus B} (u_1^\star)^2\, dx\right)\lambda_1(\Omega). \end{aligned}$$

Proceeding as before, by using (7.132), the monotonicity of G and indicating as always with R_Ω the radius of $\Omega^\star$, we obtain (we omit the details)

$$\int_B G\, z^2\, dx - \int_{\Omega^\star} G\,(u_1^\star)^2\, dx \geq 0.$$

What we have obtained so far is the following

$$\lambda_2(B)-\lambda_2(\Omega) \geq \frac{\delta}{C_d}\left(g(R)^2-g(r_1)^2\right)\left(\int_{\Omega^\star\setminus B} |u_1^\star|^2\, dx\right)\lambda_1(\Omega).$$

Dividing by $\lambda_1(\Omega)=\lambda_1(B)$, the previous can be rewritten as

$$\frac{\lambda_2(B)}{\lambda_1(B)} - \frac{\lambda_2(\Omega)}{\lambda_1(\Omega)} \geq c\left(g(R)^2-g(r_1)^2\right)\int_{\Omega^\star\setminus B} |u_1^\star|^2\, dx, \tag{7.133}$$

where $c > 0$ depends on d only. We now choose $\widetilde{R} = (R + R_\Omega)/2$, the monotone behaviour of $u_1^\star$ permits us to infer[7.14]

$$\begin{aligned}\int_{\Omega^*\setminus B} |u_1^\star|^2\,dx &\geq \int_{B_{\widetilde{R}}\setminus B} |u_1^\star|^2\,dx \geq |B_{\widetilde{R}} \setminus B|\, u_1^\star(\widetilde{R})^2 \\ &= \frac{\omega_d}{2^d}\left[(R_\Omega + R)^d - (2\,R)^d\right] u_1^\star(\widetilde{R})^2 \\ &\geq \frac{\omega_d}{2^d}\left(R_\Omega^d - R^d\right) u_1^\star(\widetilde{R})^2 = \frac{1}{2^d}\,|\Omega^* \setminus B|\, u_1^\star(\widetilde{R})^2.\end{aligned}$$

In the second inequality above we used the elementary fact

$$(a+h)^d - (b+h)^d \geq a^d - b^d, \qquad a \geq b \geq 0,\ h \geq 0, \tag{7.134}$$

which follows from convexity. Thus from (7.133) we obtain

$$\frac{\lambda_2(B)}{\lambda_1(B)} - \frac{\lambda_2(\Omega)}{\lambda_1(\Omega)} \geq c\left(g(R)^2 - g(r_1)^2\right)\,|\Omega^* \setminus B|\, u_1^\star(\widetilde{R})^2.$$

We now use further properties of g: indeed, in addition to being increasing on $[0, R]$, this is also concave on the same interval, with $g'' < 0$ on $(0, R]$ (see [64, proof of Theorem 3.3]). Thus we get

$$g(R)^2 - g(r_1)^2 \geq g(R)\,(g(R) - g(r_1)) \geq C \int_{r_1}^{R} (-g''(\tau))\,(\tau - r_1)\,d\tau,$$

by Taylor's formula and the fact that $C = g(R) > 0$ is a constant depending on d only. This in particular implies

$$g(R)^2 - g(r_1)^2 \geq c\,(R - r_1)^2,$$

for a possibly different constant $c > 0$. We combine this with the fact that z is a Lipschitz functions. Thus for some $c' > 0$ depending on d only we have

$$(R - r_1)^2 \geq c'\left(z(r_1) - z(R)\right)^2 = c'\,z(r_1)^2 = c'\,u_1^\star(r_1)^2 \geq c'\,u_1^\star(\widetilde{R})^2.$$

By summarizing, we finally obtain

$$\frac{\lambda_2(B)}{\lambda_1(B)} - \frac{\lambda_2(\Omega)}{\lambda_1(\Omega)} \geq c\,|\Omega^* \setminus B|\, u_1^\star(\widetilde{R})^4, \tag{7.135}$$

7.14 The paper [706] contains a misprint in this part of the proof. Indeed, it is claimed that (in our notation)

$$\int_{\Omega^*\setminus B} u_1^\star(x)^2\,dx \geq u_1^\star(R)^2\,|\Omega^* \setminus B|,$$

which is of course not true, since $u_1^\star$ is non-increasing.

for some $c > 0$, still depending on d only. In order to conclude, we now use the key Lemma 7.61. Indeed, by recalling (7.123) and (7.125), we can infer from (7.118)

$$\frac{1}{C}\left|\{x \in \Omega : u_1(x) \le u_1^\star(\widetilde{R})\}\right|^\beta \le u_1^\star(\widetilde{R}).$$

Moreover, by definition of $u_1^\star$ we have

$$\begin{aligned}\left|\{x \in \Omega : u_1(x) \le u_1^\star(\widetilde{R})\}\right| = |\Omega^* \setminus B_{\widetilde{R}}| &= \frac{\omega_d}{2^d}\left((2\,R_\Omega)^d - (R + R_\Omega)^d\right)\\ &\ge \frac{\omega_d}{2^d}\left(R_\Omega^d - R^d\right) = \frac{1}{2^d}\,|\Omega^* \setminus B|,\end{aligned}$$

again thanks to (7.134). By using the last two estimates in (7.135), we get

$$\frac{\lambda_2(B)}{\lambda_1(B)} - \frac{\lambda_2(\Omega)}{\lambda_1(\Omega)} \ge c\,|\Omega^* \setminus B|^{4\,\beta+1},$$

for some constant $c = c(d) > 0$. By recalling that $|\Omega^* \setminus B| = |\Omega| - |B| = 1 - |B|$ and that $\lambda_1(B) = \lambda_1(\Omega)$, we can use Lemma 7.60 and finally get the conclusion by (7.115). □

Remark 7.64. *An inspection of the proof reveals that the exponent m appearing in (7.114) is given by*

$$m = 2\,(4\,\beta + 1) > 10,$$

and β is the exponent coming from Lemma 7.61.

Open problem 7.65 (Sharp quantitative Ashbaugh-Benguria inequality). *Prove that there exists a dimensional constant $c_d > 0$ such that for every open bounded convex set $\Omega \subset \mathbb{R}^d$*

$$\frac{\lambda_2(B)}{\lambda_1(B)} - \frac{\lambda_2(\Omega)}{\lambda_1(\Omega)} \ge c_d\,\mathcal{A}(\Omega)^2.$$

The same family of sets in Theorem 7.36 should give that the exponent 2 is the best possible (recall that λ_2 is multiple for a ball and thus not differentiable).

7.6.3 Neumann vs. Dirichlet

One can immediately see that by combining the Faber-Krahn and the Szegő-Weinberger inequalities, one gets the universal inequality

$$\frac{\mu_2(\Omega)}{\lambda_1(\Omega)} \le \frac{\mu_2(B)}{\lambda_1(B)}. \tag{7.136}$$

Equality in (7.136) holds only for balls. By observing that[7.15]

$$\theta_d := \frac{\mu_2(B)}{\lambda_1(B)} = \left(\frac{\beta_{d/2,1}}{j_{d/2-1,1}}\right)^2 < 1, \tag{7.137}$$

7.15 This can be obtained by direct computation.

one can obtain

$$\mu_2(\Omega) < \lambda_1(\Omega).$$

We have the following quantitative improvement of (7.136).

Corollary 7.66. *For every $\Omega \subset \mathbb{R}^d$ open set with finite measure, we have*

$$\frac{\mu_2(B)}{\lambda_1(B)} - \frac{\mu_2(\Omega)}{\lambda_1(\Omega)} \geq \kappa_d\, \mathcal{A}(\Omega)^2, \tag{7.138}$$

where $\kappa_d > 0$ is an explicit constant depending on d only.

Proof. Let B be a ball such that $|B| = |\Omega|$, then we write

$$\frac{\mu_2(B)}{\lambda_1(B)} - \frac{\mu_2(\Omega)}{\lambda_1(\Omega)} = \left(\frac{\mu_2(B)}{\lambda_1(B)} - \frac{\mu_2(B)}{\lambda_1(\Omega)}\right) + \left(\frac{\mu_2(B)}{\lambda_1(\Omega)} - \frac{\mu_2(\Omega)}{\lambda_1(\Omega)}\right). \tag{7.139}$$

If we suppose that $\lambda_1(\Omega) > 2\,\lambda_1(B)$, by using Szegő-Weinberger for the second term in the right-hand side of (7.139) we get

$$\frac{\mu_2(B)}{\lambda_1(B)} - \frac{\mu_2(\Omega)}{\lambda_1(\Omega)} \geq \mu_2(B)\left(\frac{1}{\lambda_1(B)} - \frac{1}{\lambda_1(\Omega)}\right) \geq \frac{\mu_2(B)}{2\,\lambda_1(B)} \geq \frac{\mu_2(B)}{8\,\lambda_1(B)}\,\mathcal{A}(\Omega)^2.$$

As always, we used that $\mathcal{A}(\Omega) < 2$. Thus the conclusion follows in this case.

If on the contrary $\lambda_1(\Omega) \leq 2\,\lambda_1(B)$, then by using the Faber-Krahn inequality for the first term in the right-hand side of (7.139), we obtain

$$\frac{\mu_2(B)}{\lambda_1(B)} - \frac{\mu_2(\Omega)}{\lambda_1(\Omega)} \geq \left(\frac{\mu_2(B)}{\lambda_1(\Omega)} - \frac{\mu_2(\Omega)}{\lambda_1(\Omega)}\right) \geq \frac{1}{2\,\lambda_1(B)}\left(\mu_2(B) - \mu_2(\Omega)\right).$$

It is now sufficient to use Theorem 7.34 to conclude. □

Remark 7.67. *A feasible value for the constant κ_d appearing in* (7.138) *is*

$$\kappa_d = \frac{1}{2}\min\left\{\frac{\theta_d}{4}, \frac{\rho_d}{|B|^{2/d}\,\lambda_1(B)}\right\}.$$

Here θ_d is defined in (7.137) *and ρ_d is the same constant appearing in* (7.72).

7.7 Notes and comments

7.7.1 Other references

We wish to mention that one of the first paper to introduce quantitative elements in the Pólya-Szegő principle (7.6) was [689] by Makai. This aimed to add some remainder term in order to infer uniqueness of balls as extremals for the Saint-Venant inequality (7.11).

More recently, sophisticated quantitative improvements of the Pólya-Szegő principle were proven in [109] and [273].

The first papers to prove quantitative improvements of the Faber-Krahn inequality for general open sets with respect to the Fraenkel asymmetry have been [839] by Sznitman (for $d = 2$) and [775] by Povel (for $d \geq 3$). It is interesting to notice that both papers prove such estimates for probabilistic purposes. In [839] these estimates are employed to study the asymptotic behaviour of the first eigenvalue of $-\Delta + V_\omega$ of a square $(0, \ell) \times (0, \ell)$ for large ℓ. Here V_ω is a soft repulsive random potential. In [775] a quantitative Faber-Krahn inequality is used to estimate the extinction time of a Brownian motion in presence of (random) absorbing obstacles.

Among the contributions to the subject, it is mandatory to mention the papers [146] by Bhattacharya and [413] by Fusco, Maggi and Pratelli. Both papers consider the more general case of the p–Laplacian operator Δ_p, defined by $\Delta_p u = \operatorname{div}(|\nabla u|^{p-2} \nabla u)$. More precisely, they consider the quantities

$$\min_{u \in W_0^{1,p}(\Omega)} \left\{ \int_\Omega |\nabla u|^p \, dx \; : \; \|u\|_{L^q(\Omega)} = 1 \right\},$$

where

$$1 < q < p^* = \begin{cases} \dfrac{dp}{d-p}, & \text{if } 1 < p < d, \\ +\infty, & \text{if } p \geq d. \end{cases}$$

It is not difficult to generalize the Faber-Krahn inequalities (7.10) to these quantities, again thanks to the Pólya-Szegő principle. Then in [146, 413] some (non sharp) quantitative versions of these Faber-Krahn inequalities are proved, with results similar to Theorem 7.11.

Finally, we wish to cite the recent paper [699] by Mazzoleni and Zucco. There it is shown that quantitative versions of the Faber-Krahn and Hong-Krahn-Szego inequalities can be used to show topological properties of minimizers for a particular spectral optimization problem (see [699, Theorem 1.2]). Namely, the minimization of the convex combination $t\lambda_1 + (1-t)\lambda_2$, with volume constraint.

7.7.2 Nodal domains and Pleijel's Theorem

Let $\Omega \subset \mathbb{R}^2$ be an open bounded set. For every $n \in \mathbb{N}$ let us note by $\mathcal{N}(n)$ the number of nodal domains of the Dirichlet eigenfunction φ_n corresponding to $\lambda_n(\Omega)$. A classical result by Pleijel (see [761] and Theorem 10.21 of Chapter 10) asserts that

$$\limsup_{n \to \infty} \frac{\mathcal{N}(n)}{n} \leq \left(\frac{2}{j_0}\right)^2. \tag{7.140}$$

By observing that $2/j_0 < 1$, this result in particular asymptotically improves the classical *Courant nodal Theorem* (see [505, Theorem 1.3.2]), which asserts that $\mathcal{N}(k) \leq k$.

The proof of (7.140) can be obtained by combining the Faber-Krahn inequality on every nodal domain Ω_i

$$|\Omega|\,\lambda_n(\Omega) = \sum_{i=1}^{\mathcal{N}(n)} |\Omega_i|\,\lambda_1(\Omega_i) \geq \mathcal{N}(n)\,\pi\,j_0^2,$$

and the classical *Weyl's law*, which describes the asymptotic distribution of eigenvalues, i.e.

$$\lim_{t\to\infty} \frac{\#\{\lambda \text{ eigenvalue } : \lambda \leq t\}}{t} = \frac{|\Omega|}{4\pi}.$$

As observed by Bourgain in [174], the estimate (7.140) can be (slightly) improved by using the quantitative Faber-Krahn inequality and the packing density of balls in the Euclidean space. More precisely, the result of [174] gives an explicit improvement in dimension $d = 2$ by appealing to the Hansen-Nadirashvili result of Theorem 7.5, which comes indeed with an explicit constant. On the same problem, we also mention the paper [831] by Steinerberger.

7.7.3 Quantitative estimates in space forms

In this chapter we only discussed the Euclidean case. We briefly mention that some partial results are known for some special classes of manifolds (essentially the so-called *space forms*).

For example, the paper [885] by Xu proves a stability result for the Szegő-Weinberger inequality for smooth (geodesically) convex domains contained in a non-positively curved space form (i.e. the hyperbolic space $\mathbb{H}^d$ or $\mathbb{R}^d$). More precisely, [885, Theorem 4] proves a *pinching result* which shows that if the spectral deficit $\mu_2(B) - \mu_2(\Omega)$ converges to 0, then the Hausdorff distance from the set of geodesic balls with given volume goes to 0.

The paper [91] by Avila proves a quantitative Faber-Krahn inequality for smooth (geodesically) convex domains on the hemisphere $\mathbb{S}^2_+$ ([91, Theorem 0.1]) or on the hyperbolic space $\mathbb{H}^2$ (see [91, Theorem 0.2]). These can be seen as the natural counterparts of Melas' result Theorem 7.3, indeed stability is measured with a suitable variant of his asymmetry $\delta_{\mathcal{M}}$. In [91, Theorem 0.3] the aforementioned Xu's result is extended to a sufficiently narrow polar cap contained in $\mathbb{S}^d_+$.

In [90] Aubry, Bertrand and Colbois prove pinching results for the Faber-Krahn and Ashbaugh-Benguria inequalities for convex sets in $\mathbb{S}^d$, $\mathbb{R}^d$ and $\mathbb{H}^d$. These results show that if the relevant spectral deficit is small, then the set is close to a ball in the Hausdorff metric. As for the hyperbolic space $\mathbb{H}^d$, it should be noted that the inequality

$$\frac{\lambda_2(\Omega)}{\lambda_1(\Omega)} \leq \frac{\lambda_2(B)}{\lambda_1(B)},$$

does not hold true and that the correct replacement of the Ashbaugh-Benguria inequality is

$$\lambda_2(\Omega) \leq \lambda_2(B), \qquad \text{if } \lambda_1(\Omega) = \lambda_1(B), \tag{7.141}$$

where B denotes a geodesic ball (see [121, Theorem 1.1]). Then for $\mathbb{H}^d$ the pinching result of [91, Theorem 1.5] exactly concerns inequality (7.141).

Acknowledgements. *This project was started while L. B. was still at Aix-Marseille Université, during a 6 months CNRS délégation period. He wishes to thank all his former collegues at I2M institution. L. B. has been supported by the Agence Nationale de la Recherche, through the project ANR-12-BS01-0014-01* Geometrya, *G. D. P. is supported by the MIUR SIR-grant* Geometric Variational Problems *(RBSI14RVEZ).*

Both authors wish to warmly thank Mark S. Ashbaugh, Erwann Aubry and Nikolai Nadirashvili for interesting discussions and remarks on the subject, as well as their collaborators Giovanni Franzina, Aldo Pratelli, Berardo Ruffini and Bozhidar Velichkov. Special thanks go to Nicola Fusco, who first introduced L. B. to the realm of stability and quantitative inequalities, during his post-doc position in Naples.

The authors are members of the Gruppo Nazionale per l'Analisi Matematica, la Probabilità e le loro Applicazioni (GNAMPA) of the Istituto Nazionale di Alta Matematica (INdAM).

7.8 Appendix

7.8.1 The Kohler-Jobin inequality and the Faber-Krahn hierarchy

Let $q > 1$ be an exponent satisfying (7.9), for every $\Omega \subset \mathbb{R}^d$ open set with finite measure we still denote by $\lambda_1^q(\Omega)$ its first semilinear Dirichlet eigenvalue, defined in (7.8). The *Kohler-Jobin inequality* states that

$$T(\Omega)^{\vartheta(q,d)}\lambda_1^q(\Omega) \geq T(B)^{\vartheta(q,d)}\,\lambda_1^q(B), \qquad \text{where } \vartheta(q,d) = \frac{2 + \frac{2}{q}d - d}{d+2} < 1. \tag{7.142}$$

The original statement by Marie-Thérèse Kohler-Jobin is for the first eigenvalue of the Laplacian, i.e. for $q = 2$ (see [591, Théorème 1]). This can be equivalently reformulated by saying that balls are the only solutions to the following problem

$$\min\{\lambda_1(\Omega) \,:\, T(\Omega) = c\}.$$

We refer to [593, Theorem 3] and [179, Theorem 1.1] for the general version (7.142) of Kohler-Jobin inequality.

An important consequence of the Kohler-Jobin inequality is that the whole family of Faber-Krahn inequalities (7.10) for the first semilinear eigenvalue λ_1^q can be derived by combining the Saint-Venant inequality (7.11) and (7.142). Indeed, we have

$$\begin{aligned}|\Omega|^{\frac{2}{d}+\frac{2}{q}-1}\lambda_1^q(\Omega) &= \left(|\Omega|^{\frac{2}{d}+\frac{2}{q}-1}T(\Omega)^{-\vartheta(q,d)}\right)\left(T(\Omega)^{\vartheta(q,d)}\lambda_1^q(\Omega)\right)\\ &= \left(|\Omega|^{-\frac{d+2}{d}}T(\Omega)\right)^{-\vartheta(q,d)}\left(T(\Omega)^{\vartheta(q,d)}\lambda_1^q(\Omega)\right)\\ &\geq \left(|B|^{-\frac{d+2}{d}}T(B)\right)^{-\vartheta(q,d)}\left(T(B)^{\vartheta(q,d)}\lambda_1^q(B)\right)\\ &= |B|^{\frac{2}{d}+\frac{2}{q}-1}\lambda_1^q(B).\end{aligned}$$

More interestingly, we can translate every quantitative improvement of the Saint-Venant inequality into a similar statement for λ_1^q. Namely, we have the following expedient result.

Proposition 7.68 (Faber-Krahn hierarchy). *Let $q \geq 1$ be an exponent verifying (7.9). Suppose that there exists $C > 0$ and*

- *$\mathcal{G} : [0,\infty) \to [0,\infty)$ a continuous increasing function vanishing at the origin only,*
- *$\Omega \mapsto \delta(\Omega)$ a scaling invariant shape functional vanishing on balls and bounded by some constant $M > 0$,*

such that for every open set $\Omega \subset \mathbb{R}^d$ with finite measure we have

$$|B|^{-\frac{d+2}{d}}T(B) - |\Omega|^{-\frac{d+2}{d}}T(\Omega) \geq \frac{1}{C}\mathcal{G}(\delta(\Omega)). \tag{7.143}$$

Then we also have

$$|\Omega|^{\frac{2}{d}+\frac{2}{q}-1}\lambda_1^q(\Omega) - |B|^{\frac{2}{d}+\frac{2}{q}-1}\lambda_1^q(B) \geq c\,\mathcal{G}(\delta(\Omega)).$$

The constant $c > 0$ depends on C, d, q and $\mathcal{G}(M)$ only and is given by

$$c = (2^\vartheta - 1)\,|B|^{\frac{2}{d}+\frac{2}{q}-1}\lambda_1^q(B)\min\left\{\frac{1}{C}\frac{|B|^{\frac{d+2}{d}}}{T(B)}, \frac{1}{\mathcal{G}(M)}\right\}.$$

Proof. Without loss of generality, we can suppose that $|\Omega| = 1$ and let B be a ball having unit measure. We also use the shortcut notation $\vartheta = \vartheta(q,d)$. By (7.142) one obtains

$$\frac{\lambda_1^q(\Omega)}{\lambda_1^q(B)} - 1 \geq \left(\frac{T(B)}{T(\Omega)}\right)^\vartheta - 1. \tag{7.144}$$

Since $0 < \vartheta \leq 1$, by concavity we have

$$t^\vartheta - 1 \geq (2^\vartheta - 1)(t-1), \qquad t \in [1,2].$$

Thus from (7.144) and (7.143) we can easily infer that if $T(B) \leq 2\,T(\Omega)$, then

$$\frac{\lambda_1^q(\Omega)}{\lambda_1^q(B)} - 1 \geq (2^\vartheta - 1)\left(\frac{T(B)}{T(\Omega)} - 1\right) \geq \frac{1}{C}\frac{2^\vartheta - 1}{T(B)}\mathcal{G}(\delta(\Omega)).$$

In the last inequality we also used that $T(\Omega) \leq T(B)$ by the Saint-Venant inequality. On the other hand, if $T(B) > 2\,T(\Omega)$, still by (7.144) we get

$$\frac{\lambda_1^q(\Omega)}{\lambda_1^q(B)} - 1 \geq 2^\vartheta - 1 \geq \frac{2^\vartheta - 1}{\mathcal{G}(M)}\,\mathcal{G}(\delta(\Omega)),$$

since $\delta(\Omega) \leq M$ and $\mathcal{G}$ is increasing by hypothesis. □

Remark 7.69. *Examples of shape functionals δ as in the previous statement are: $\delta_{\mathcal{M}}$ defined in (7.16), $\delta_{\mathcal{N}}$ defined in (7.17) and the Fraenkel asymmetry $\mathcal{A}$.*

7.8.2 An elementary inequality for monotone functions

Szegő's proof of the inequality (7.64) is based on the following elementary inequality for monotone functions of one real variable. We give its proof for completeness.

Lemma 7.70 (Chebyshev's sum inequality). *Let $\varphi : [0,1] \to \mathbb{R}_+$ and $\psi : [0,1] \to \mathbb{R}_+$ be two non-decreasing functions, not identically vanishing. Then we have*

$$\int_0^1 \varphi(t)\,\psi(t)\,t\,dt \geq \left(\frac{\displaystyle\int_0^1 \varphi(t)\,t\,dt}{\displaystyle\int_0^1 t\,dt}\right)\int_0^1 \psi(t)\,t\,dt \tag{7.145}$$

Proof. By approximation, we can assume that ψ is C^1. We first observe that if we set

$$\Phi(t) = \int_0^t \varphi(s)\,s\,ds \qquad \text{and} \qquad \overline{\varphi} = \frac{\displaystyle\int_0^1 \varphi(t)\,t\,dt}{\displaystyle\int_0^1 t\,dt},$$

then (7.145) is equivalent to

$$\int_0^1 \left(\Phi'(t) - \overline{\varphi}\,t\right)\psi(t)\,dt \geq 0.$$

If we perform an integration by parts, this in turn is equivalent to

$$\int_0^1 \left(\overline{\varphi}\,\frac{t^2}{2} - \Phi(t)\right)\psi'(t)\,dt \geq 0.$$

Since $\psi' \geq 0$, in order to conclude it is sufficient to prove

$$\Phi(t) \leq \overline{\varphi}\,\frac{t^2}{2}, \qquad \text{i.e.} \quad \frac{2}{t^2}\int_0^t \varphi(s)\,s\,ds \leq 2\int_0^1 \varphi(s)\,s\,ds. \tag{7.146}$$

In turn, in order to show (7.146) it would be enough to prove that the function

$$H(t) = \frac{2}{t^2} \int_0^t \varphi(s)\, s\, ds,$$

is monotone non-decreasing. A direct computation gives

$$H'(t) = -\frac{4}{t^3} \int_0^t \varphi(s)\, s\, ds + \frac{2}{t}\, \varphi(t) = \frac{2}{t} \left[\varphi(t) - \frac{2}{t^2} \int_0^t \varphi(s)\, s\, ds\right] \geq 0,$$

where in the last inequality we used that φ is non-decreasing. This proves (7.146) and thus (7.145). □

The result of Theorem 7.31 is based on the following improvement of the previous inequality.

Lemma 7.71 (Improved Chebyshev's sum inequality). *Let $\varphi : [0, 1] \to \mathbb{R}_+$ be a strictly increasing function. Let $\psi : [0, 1] \to \mathbb{R}_+$ be a non-decreasing function, such that*

$$\psi(t) = \sum_{n=0}^{\infty} \alpha_n\, t^n, \qquad \text{with} \qquad \alpha_n \geq 0 \quad \text{and} \quad \alpha_0 = \psi(0) \leq \gamma \int_0^1 \psi(t)\, t\, dt, \tag{7.147}$$

for some $0 \leq \gamma < 2$. Then we have

$$\int_0^1 \varphi(t)\, \psi(t)\, t\, dt \geq \left(\frac{\displaystyle\int_0^1 \varphi(t)\, t\, dt}{\displaystyle\int_0^1 t\, dt}\right) \int_0^1 \psi(t)\, t\, dt + c\,(2-\gamma) \int_0^1 \psi(t)\, t\, dt, \tag{7.148}$$

for some $c > 0$ depending on φ.

Proof. We use the same notation as in the proof of Lemma 7.70. We then recall that

$$\begin{aligned}
\int_0^1 \varphi(t)\, \psi(t)\, t\, dt - &\left(\frac{\displaystyle\int_0^1 \varphi(t)\, t\, dt}{\displaystyle\int_0^1 t\, dt}\right) \int_0^1 \psi(t)\, t\, dt \\
&= \int_0^1 \left(\overline{\varphi}\, \frac{t^2}{2} - \Phi'(t)\right) \psi'(t)\, dt \\
&= \sum_{n=1}^{\infty} n\, \alpha_n \int_0^1 \left(\overline{\varphi}\, \frac{t^2}{2} - \psi(t)\right) t^{n-1}\, dt \\
&= \sum_{n=1}^{\infty} \frac{n\, \alpha_n}{2} \int_0^1 \Big(H(1) - H(t)\Big)\, t^{n+1}\, dt.
\end{aligned}$$

For the last term, we recall that $H' > 0$ on the interval $[0, 1]$, thus for $n \geq 1$

$$\begin{aligned}
\int_0^1 \Big(H(1) - H(t)\Big)\, t^{n+1}\, dt &\geq \int_0^{1-\frac{1}{2n}} \Big(H(1) - H(t)\Big)\, t^{n+1}\, dt \\
&\geq \left[H(1) - H\left(1 - \frac{1}{2n}\right)\right] \frac{1}{n+2} \left(1 - \frac{1}{2n}\right)^{n+2}.
\end{aligned}$$

Observe that from (7.147) we get

$$\sum_{n=1}^{\infty} \frac{\alpha_n}{n+2} = \int_0^1 \psi(t)\, t\, dt - \alpha_0 \int_0^1 t\, dt \geq \left(1 - \frac{\gamma}{2}\right) \int_0^1 \psi(t)\, t\, dt.$$

Thus in order to conclude the proof of (7.148) we need to prove that for $n \geq 1$

$$n \left[H(1) - H\left(1 - \frac{1}{2n}\right)\right] \left(1 - \frac{1}{2n}\right)^{n+2} \geq c > 0.$$

By observing that

$$\left(1 - \frac{1}{2n}\right)^{n+2} \geq \frac{1}{8}$$

and

$$\lim_{n \to \infty} n \left[H(1) - H\left(1 - \frac{1}{2n}\right)\right] = \frac{H'(1)}{2} = \varphi(1) - 2 \int_0^1 \varphi(s)\, s\, ds > 0,$$

we conclude that this holds true. □

7.8.3 A weak version of the Hardy-Littlewood inequality

The proofs by Weinberger and Brock are based on a test function argument and an isoperimetric-like property of balls with respect to weighted volumes of the type

$$\Omega \mapsto \int_\Omega f(|x|)\, dx,$$

with f a positive monotone function. This is encoded in the following result, which can be seen as a particular case of the Hardy-Littlewood inequality.

Lemma 7.72. *Let $f : \mathbb{R}_+ \to \mathbb{R}_+$ be a non-increasing function and $g : \mathbb{R}_+ \to \mathbb{R}_+$ a non-decreasing function. Let $\Omega \subset \mathbb{R}^d$ be an open set with finite measure, we denote by Ω^* the ball centered at the origin such that $|\Omega| = |\Omega^*|$. Then we have*

$$\int_{\Omega^*} f(|x|)\, dx > \int_\Omega f(|x|)\, dx \qquad \text{and} \qquad \int_{\Omega^*} g(|x|)\, dx \leq \int_\Omega g(|x|)\, dx.$$

The quantitative versions of the Szegő-Weinberger and Brock-Weinstock inequalities are based on the following simple but useful improved version of Lemma 7.72.

Lemma 7.73. *Let $f : \mathbb{R}_+ \to \mathbb{R}_+$ be a nonincreasing function. Let $\Omega \subset \mathbb{R}^d$ be an open set with finite measure, we denote by Ω^* the ball centered at the origin such that $|\Omega| = |\Omega^*|$. Then we have*

$$\int_{\Omega^*} f(|x|)\, dx - \int_\Omega f(|x|)\, dx \geq d\, \omega_d \int_{R_1}^{R_2} |f(\varrho) - f(R_\Omega)|\, \varrho^{d-1}\, d\varrho, \tag{7.149}$$

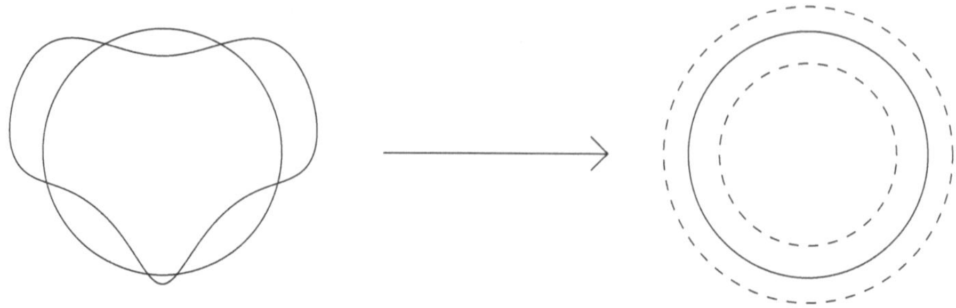

Fig. 7.4. The rearrangement of $\Omega\Delta\Omega^*$ for the proof of (7.149)

where

$$R_\Omega = \left(\frac{|\Omega|}{\omega_d}\right)^{\frac{1}{d}}, \qquad R_1 = \left(\frac{|\Omega \cap \Omega^*|}{\omega_d}\right)^{\frac{1}{d}} \qquad \textit{and} \qquad R_2 = \left(\frac{|\Omega \setminus \Omega^*| + |\Omega|}{\omega_d}\right)^{\frac{1}{d}}. \tag{7.150}$$

Proof. The proof is quite simple, first of all we observe that

$$\int_{\Omega^*} f(|x|)\,dx - \int_{\Omega} f(|x|)\,dx = \int_{\Omega^*\setminus\Omega} f(|x|)\,dx - \int_{\Omega\setminus\Omega^*} f(|x|)\,dx. \tag{7.151}$$

Then the idea is to rearrange the set $\Omega^* \setminus \Omega$ into a spherical shell having radii R_Ω and R_2, and similarly to rearrange the set $\Omega \setminus \Omega^*$ into a spherical shell having radii R_Ω and R_1 (see Figure 7.4). From (7.150), we have that R_1 and R_2 are such that

$$|B_{R_2}| - |\Omega^*| = |\Omega^* \setminus \Omega| \qquad \text{and} \qquad |\Omega^*| - |B_{R_1}| = |\Omega^* \setminus \Omega|,$$

i.e. the two spherical shells mentioned above will preserve the measure. We will prove below that this property and the monotonicity of f entail

$$\int_{\Omega^*\setminus\Omega} f(|x|)\,dx \geq \int_{\Omega^*\setminus B_{R_1}} f(|x|)\,dx \qquad \text{and} \qquad \int_{\Omega\setminus\Omega^*} f(|x|)\,dx \leq \int_{B_{R_2}\setminus\Omega^*} f(|x|)\,dx. \tag{7.152}$$

This means that the worst scenario for the right-hand side of (7.151) is when all the mass is uniformly distributed around $\partial\Omega^*$. Thus from (7.151) and (7.152), we can obtain

$$\int_{\Omega^*} f(|x|)\,dx - \int_{\Omega} f(|x|)\,dx \geq \int_{\Omega^*\setminus B_{R_1}} f(|x|)\,dx - \int_{B_{R_2}\setminus\Omega^*} f(|x|)\,dx.$$

In order to conclude, we observe that since by contruction $|\Omega^* \setminus B_{R_1}| = |B_{R_2} \setminus \Omega^*|$, then we get

$$\int_{\Omega^*\setminus B_{R_1}} f(|x|)\,dx - \int_{B_{R_2}\setminus\Omega^*} f(|x|)\,dx = \int_{\Omega^*\setminus B_{R_1}} [f(|x|) - f(R_\Omega)]\,dx - \int_{B_{R_2}\setminus\Omega^*} [f(|x|) - f(R_\Omega)]\,dx.$$

This finally gives (7.149), by using polar coordinates and since f is nonincreasing.

Let us now prove (7.152). We first observe that we have

$$|(\Omega^* \cap \Omega) \setminus B_{R_1}| = |(\Omega^* \setminus \Omega) \cap B_{R_1}|. \tag{7.153}$$

Indeed, we get

$$\begin{aligned}|(\Omega^* \cap \Omega) \setminus B_{R_1}| + |(\Omega^* \setminus \Omega) \setminus B_{R_1}| &= |\Omega^* \setminus B_{R_1}| = |\Omega^* \setminus \Omega| \\ &= |(\Omega^* \setminus \Omega) \cap B_{R_1}| + |(\Omega^* \setminus \Omega) \setminus B_{R_1}|,\end{aligned}$$

which proves (7.153). By using this and the monotonicity of f, we get

$$\begin{aligned}\int_{\Omega^*\setminus\Omega} f(|x|)\,dx &= \int_{(\Omega^*\setminus\Omega)\cap B_{R_1}} f(|x|)\,dx + \int_{(\Omega^*\setminus\Omega)\setminus B_{R_1}} f(|x|)\,dx \\ &\ge f(R_1)\,|(\Omega^* \setminus \Omega) \cap B_{R_1}| + \int_{(\Omega^*\setminus\Omega)\setminus B_{R_1}} f(|x|)\,dx \\ &= f(R_1)\,|(\Omega^* \cap \Omega) \setminus B_{R_1}| + \int_{(\Omega^*\setminus\Omega)\setminus B_{R_1}} f(|x|)\,dx \\ &\ge \int_{(\Omega^*\cap\Omega)\setminus B_{R_1}} f(|x|)\,dx + \int_{(\Omega^*\setminus\Omega)\setminus B_{R_1}} f(|x|)\,dx \\ &= \int_{\Omega^*\setminus B_{R_1}} f(|x|)\,dx,\end{aligned}$$

which proves the first inequality in (7.152). The second one is proved similarly. □

7.8.4 Some estimates for convex sets

We still denote by r_Ω the inradius of a set and by Haus the Hausdorff distance between sets, defined by (7.21).

Lemma 7.74. *Let $\Omega \subset \mathbb{R}^d$ be an open bounded convex set. For every ball B_R of radius R, we have*

$$\mathrm{Haus}(\Omega, B_R) \ge R - r_\Omega.$$

Proof. We first observe that if $\mathrm{Haus}(\Omega, B) \ge R$ there is nothing to prove. Thus, we set for simplicity $\delta = \mathrm{Haus}(\Omega, B_R)$ and suppose $\delta < R$. By definition of Hausdorff distance, we have that

$$B_R \subset \Omega + \delta\, B_1(0) =: \Omega_\delta, \tag{7.154}$$

where + denotes the Minkowski sum of sets. Let $x \in \Omega$ and $x' \in \partial\Omega$ be such that

$$|x - x'| = \mathrm{dist}(x, \partial\Omega).$$

We also consider the point $x'_\delta = x' + \delta\,(x'-x)/|x-x'| \in \partial\Omega_\delta$, then we obtain for every $x \in \Omega$

$$\operatorname{dist}(x,\partial\Omega) = |x-x'| \geq |x-x'_\delta| - |x'_\delta - x'| \geq \operatorname{dist}(x,\partial\Omega_\delta) - \delta.$$

Since r_Ω coincides with the supremum on Ω of the distance function, this shows

$$r_\Omega + \delta \geq \sup_{x\in\Omega} \operatorname{dist}(x,\partial\Omega_\delta). \tag{7.155}$$

We now want to show that

$$\sup_{x\in\Omega_\delta\setminus\Omega} \operatorname{dist}(x,\partial\Omega_\delta) \leq \delta. \tag{7.156}$$

Let us take $x \in \Omega_\delta \setminus \Omega$, then we know that

$$x = x' + t\,\omega, \qquad \text{for some } x' \in \partial\Omega,\ 0 \leq t < \delta,\ \omega \in \mathbb{S}^{d-1}.$$

The point $x'' = x' + \delta\,\omega$ lies on the boundary of $\partial\Omega_\delta$, thus we get

$$\operatorname{dist}(x,\partial\Omega_\delta) \leq |x-x''| = (\delta - t) < \delta.$$

This shows (7.156). By putting (7.155) and (7.156) together, we thus get

$$r_{\Omega_\delta} = \sup_{x\in\Omega_\delta} \operatorname{dist}(x,\partial\Omega_\delta) = \max\left\{\sup_{\Omega} \operatorname{dist}(x,\partial\Omega_\delta),\ \sup_{\Omega_\delta\setminus\Omega} \operatorname{dist}(x,\partial\Omega_\delta)\right\} \leq r_\Omega + \delta.$$

It is only left to observe that from (7.154), we get $R \leq r_{\Omega_\delta} \leq r_\Omega + \delta$, as desired. □

The following result asserts that for convex sets λ_1 is equivalent to the inradius.

Proposition 7.75. *For every $\Omega \subset \mathbb{R}^d$ open convex set such that $r_\Omega < +\infty$ we have*

$$\frac{1}{4\,r_\Omega^2} \leq \lambda_1(\Omega) \leq \frac{\lambda_1(B_1)}{r_\Omega^2}, \tag{7.157}$$

where B_1 is any d–dimensional ball of radius 1.

Proof. The upper bound easily follows from the monotonicity and scaling properties of λ_1. For the lower bound, we can use the *Hardy inequality* for convex sets

$$\frac{1}{4}\int_\Omega \left|\frac{u}{d_\Omega}\right|^2 dx < \int_\Omega |\nabla u|^2\,dx, \qquad u \in H_0^1(\Omega),$$

where we used the notation $d_\Omega(x) = \operatorname{dist}(x,\partial\Omega)$. By recalling that the inradius r_Ω coincides with the maximum of d_Ω, we get $d_\Omega \leq r_\Omega$ and taking the infimum over $H_0^1(\Omega)$ we get the conclusion. □

Lemma 7.76. *For every $\Omega \subset \mathbb{R}^d$ open bounded convex set, we have*

$$\frac{1}{d\,\omega_d}\,\frac{|\Omega|}{r_\Omega} \leq \operatorname{diam}(\Omega)^{d-1}. \tag{7.158}$$

Proof. By using coarea formula, we obtain

$$|\Omega| = \int_\Omega dx = \int_0^{r_\Omega} P(\{x \in \Omega : d_\Omega(x) = t\})\, dt \le r_\Omega\, P(\Omega),$$

thanks to the convexity of the level sets of the distance function[7.16]. Since Ω is contained in a ball with radius $\mathrm{diam}(\Omega)$, we have

$$P(\Omega) \le d\,\omega_d\, \mathrm{diam}(\Omega)^{d-1}.$$

This concludes the proof. □

Remark 7.77. *By combining* (7.158) *and* (7.157), *we obtain the estimate*

$$\lambda_1(\Omega) \le (d\,\omega_d)^2\, \lambda_1(B_1) \left(\frac{\mathrm{diam}(\Omega)^{d-1}}{|\Omega|}\right)^2.$$

Thus in particular for every sequence of open convex sets $\{\Omega_n\}_{n\in\mathbb{N}} \subset \mathbb{R}^d$ such that

$$|\Omega_n| = 1 \qquad \textit{and} \qquad \lim_{n\to\infty} \lambda_1(\Omega_n) = +\infty,$$

then the diameters diverge to $+\infty$ as well. This fact has been used in the proof of Theorem 7.59.

7.16 We use that on convex sets the perimeter is monotone with respect to inclusion.

Mark S. Ashbaugh

8 Universal Inequalities for the Eigenvalues of the Dirichlet Laplacian

8.1 Introduction

The subject of universal inequalities for the eigenvalues of differential operators was initiated by Payne, Pólya, and Weinberger in 1955 (see [746, 751]). Their work began by focusing on the eigenvalues of the Dirichlet Laplacian, so we begin our discussion there. Payne, Pólya, and Weinberger (PPW) confined their attention to two dimensions, but since their approach works equally well in d dimensions with essentially no extra work, we will consistently address the problem in that generality below.

If Ω is a bounded domain in $\mathbb{R}^d$ it is well-known that the Laplacian $-\Delta$ on Ω with Dirichlet boundary conditions is positive and self-adjoint with spectrum consisting of an infinite sequence of eigenvalues,

$$(0 <)\lambda_1 < \lambda_2 \le \lambda_3 \le \lambda_4 \le \cdots \to \infty, \tag{8.1}$$

where the indexing here takes account of multiplicities, i.e., in this list each eigenvalue is repeated according to its multiplicity. A corresponding orthonormal basis of real eigenfunctions will be denoted

$$u_1, u_2, u_3, \ldots . \tag{8.2}$$

That is, we assume known that, for each $i = 1, 2, \ldots$, u_i is real-valued, $-\Delta u_i = \lambda_i u_i$ throughout Ω, $u_i = 0$ on $\partial\Omega$, and that $\int_\Omega u_i u_j = \delta_{ij}$, where δ_{ij} denotes the Kronecker delta.

In their initial announcement [751] in 1955, Payne, Pólya, and Weinberger presented the *ratio bound*

$$\lambda_2/\lambda_1 \le 3 \tag{8.3}$$

for domains in $\mathbb{R}^2$. Exactly the same proof, promoted to dimension d, proves the general inequality

$$\lambda_2/\lambda_1 \le 1 + \frac{4}{d} \tag{8.4}$$

for domains $\Omega \subset \mathbb{R}^d$.

In their full paper [746] in 1956, they gave the general inequality, again only for the 2-dimensional case,

$$\lambda_{m+1} - \lambda_m \le \frac{2}{m} \sum_{i=1}^{m} \lambda_i \text{ for } m = 1, 2, \ldots \tag{8.5}$$

Mark S. Ashbaugh: Department of Mathematics, University of Missouri, Columbia, MO 65211, USA, E-mail: ashbaughm@missouri.edu (with $m = 1$ giving (8.4)). (Note that the $m = 1$ case of this is simply (8.3).) As in the case of the inequality $\lambda_2/\lambda_1 \leq 3$, this inequality is easily promoted to d dimensions [851], where it becomes

$$\lambda_{m+1} - \lambda_m \leq \frac{4}{dm} \sum_{i=1}^{m} \lambda_i \text{ for } m = 1, 2, \ldots . \tag{8.6}$$

These bounds tell us that the mth spectral gap is controlled by the average of the first m eigenvalues.

These inequalities deserve the name *universal* because they hold in complete generality, requiring no hypotheses on the domain Ω (other than that it is of dimension d). Protter [776, 778] seems to have been the first person to use the term *universal inequalities* in print. Among others who have used it are [66, 67], [463], [55, 57, 58], and [83, 511]. Universal eigenvalue inequalities in the style of PPW are also rather easy to prove and require very few tools. In fact, beyond the basics of the Rayleigh-Ritz inequality and its use in spectral theory, one only needs integration by parts and the Cauchy-Schwarz inequality, two of the most basic tools of analysis.

To give some flavor of how these proofs go, let us prove the inequality $\lambda_2/\lambda_1 \leq 1 + 4/d$ as a warm-up exercise. By the Rayleigh-Ritz inequality for the second eigenvalue we have

$$\lambda_2 \leq \frac{\int_\Omega |\nabla \varphi|^2}{\int_\Omega \varphi^2} \tag{8.7}$$

where φ, called a *trial function* for λ_2 in this context, is any nontrivial function in the Sobolev space $H_0^1(\Omega)$ that is L^2-orthogonal to u_1, the first eigenfunction of $-\Delta$ on Ω (with Dirichlet boundary conditions). That is, we require that $\varphi \not\equiv 0$ in $L^2(\Omega)$, $\varphi = 0$ on $\partial\Omega$, and that $\int_\Omega \varphi u_1 = 0$. In fact, in our work here we will be using trial functions that are C^1 (or more) on Ω and continuous up to the boundary (so in $C(\overline{\Omega})$). Thus we shall not need to worry about technicalities involving Sobolev spaces.

Let us consider Ω as a domain in $\mathbb{R}^d$ with its center of mass (with respect to the mass density u_1^2) located at the origin (this can always be accomplished via a translation of the coordinates, as necessary). That is, with x_ℓ for $\ell = 1, 2, \ldots, d$ denoting Cartesian coordinates for $\mathbb{R}^d$,

$$\int_\Omega x_\ell u_1^2 = 0 \text{ for } \ell = 1, 2, \ldots, d. \tag{8.8}$$

This makes the functions

$$\varphi^{(\ell)} \equiv x_\ell u_1 \text{ for } \ell = 1, 2, \ldots, d \tag{8.9}$$

legitimate trial functions for λ_2 in (8.7). To proceed, we note two facts. The first is that the numerator of the Rayleigh quotient (defined as the right-hand side of (8.7)) may be rewritten as $\int_\Omega \varphi(-\Delta\varphi)$ via an integration by parts (using the fact that φ vanishes on $\partial\Omega$). The other is that

$$-\Delta\varphi^{(\ell)} = -\Delta(x_\ell u_1) = x_\ell(-\Delta u_1) - 2u_{1,x_\ell}$$

$$\begin{aligned} &= x_\ell \lambda_1 u_1 - 2u_{1,x_\ell} \\ &= \lambda_1 \varphi^{(\ell)} - 2u_{1,x_\ell}, \end{aligned} \tag{8.10}$$

where u_{1,x_ℓ} denotes the partial derivative of u_1 with respect to x_ℓ. From these two facts it follows that the numerator in the Rayleigh quotient for λ_2 using $\varphi^{(\ell)}$ is

$$\int_\Omega \varphi^{(\ell)}(\lambda_1 \varphi^{(\ell)} - 2u_{1,x_\ell}) = \lambda_1 \int_\Omega (\varphi^{(\ell)})^2 - 2 \int_\Omega \varphi^{(\ell)} u_{1,x_\ell}. \tag{8.11}$$

Inserting this into the Rayleigh-Ritz bound for λ_2, clearing the denominator, and rearranging terms, we have

$$(\lambda_2 - \lambda_1) \int_\Omega (\varphi^{(\ell)})^2 \le -2 \int_\Omega \varphi^{(\ell)} u_{1,x_\ell}, \tag{8.12}$$

and it only remains to work the right-hand side of this expression into a more tractable form. In fact, there are two useful things we can do with the right-hand side. One is that we can rewrite it in terms of u_1 alone and use integration by parts to evaluate it:

$$\begin{aligned} -2 \int_\Omega \varphi^{(\ell)} u_{1,x_\ell} &= -2 \int_\Omega x_\ell u_1 u_{1,x_\ell} \\ &= - \int_\Omega x_\ell \, (u_1^2)_{x_\ell} = \int_\Omega u_1^2 = 1. \end{aligned} \tag{8.13}$$

The other is that we can use the Cauchy-Schwarz inequality to estimate its square in terms of the L^2 norms of φ_ℓ and u_{1,x_ℓ}. Thus we have

$$\left(-2 \int_\Omega \varphi^{(\ell)} u_{1,x_\ell} \right)^2 \le 4 \left(\int_\Omega (\varphi^{(\ell)})^2 \right) \left(\int_\Omega u_{1,x_\ell}^2 \right), \tag{8.14}$$

or

$$\frac{-2 \int_\Omega \varphi^{(\ell)} u_{1,x_\ell}}{\int_\Omega (\varphi^{(\ell)})^2} \le 4 \frac{\int_\Omega u_{1,x_\ell}^2}{-2 \int_\Omega \varphi^{(\ell)} u_{1,x_\ell}} \tag{8.15}$$

(note that the denominator on the right-hand side here is positive (in fact $= 1$) by (8.13)), and this allows us to rewrite (8.12) as

$$\begin{aligned} \lambda_2 - \lambda_1 &\le \frac{-2 \int_\Omega \varphi^{(\ell)} u_{1,x_\ell}}{\int_\Omega (\varphi^{(\ell)})^2} \\ &\le 4 \frac{\int_\Omega u_{1,x_\ell}^2}{-2 \int_\Omega \varphi^{(\ell)} u_{1,x_\ell}} \\ &= 4 \int_\Omega u_{1,x_\ell}^2. \end{aligned} \tag{8.16}$$

At this point we have an inequality without any explicit φ's, and we are almost done. We only have to note that by averaging these inequalities over ℓ the integrals $\int_\Omega u_{1,x_\ell}^2$ on the right-hand side sum together to give $\int_\Omega |\nabla u_1|^2$, which, by an integration by parts and use of the eigenvalue equation, is just λ_1. Thus we have reached

$$\lambda_2 - \lambda_1 \le \frac{4}{d} \lambda_1, \tag{8.17}$$

which is the inequality we sought,

$$\frac{\lambda_2}{\lambda_1} \le 1 + \frac{4}{d}. \tag{8.18}$$

In obtaining the more general inequalities of Payne-Pólya-Weinberger and others, we shall use similar techniques, as will be seen below. In particular, to prove inequalities such as (8.6) we shall use trial functions for λ_{m+1} that we build from products $x_\ell u_i$ for $i = 1, 2, \ldots, m$. The main difference will be that where we were able to use a center of mass argument above to construct our trial functions appropriately (and thus had some geometric intuition into what we were doing), in our later work we forgo this and simply subtract counter-terms in the eigenfunctions $\{u_j\}_{j=1}^m$ to enforce the necessary orthogonalities. Thus, for φ built from $x_\ell u_i$ $(1 \le i \le m)$ to be L^2-orthogonal to $\{u_k\}_{k=1}^m$ and hence be a legitimate trial function for λ_{m+1} we will take

$$\varphi = x_\ell u_i - \sum_{j=1}^{m} a_{ij} u_j, \tag{8.19}$$

where the a_{ij}'s are chosen to make φ L^2-orthogonal to $u_1, u_2, \ldots, u_m$. We will do all this in detail below. In this further work it will transpire that the symmetry or anti-symmetry of various terms that arise will be important, which is a facet of the work that did not show up above.

We end this introduction by mentioning the other "main-line" universal inequalities for the eigenvalues of the Dirichlet Laplacian on a bounded domain Ω in Euclidean space $\mathbb{R}^d$. In historical order (and also in order of strength, as we shall see later) these are

$$\frac{d}{4} \le \frac{1}{m} \sum_{i=1}^{m} \frac{\lambda_i}{\lambda_{m+1} - \lambda_i}. \tag{8.20}$$

which was published by Hile and Protter in 1980. Note that in the event that $\lambda_{m+1} - \lambda_m$ vanishes, the right-hand side here should be viewed as taking the value ∞ with the inequality holding trivially. The next result is the Yang inequality (Yang1)

$$\sum_{i=1}^{m} (\lambda_{m+1} - \lambda_i)^2 \le \frac{4}{d} \sum_{i=1}^{m} (\lambda_{m+1} - \lambda_i)\lambda_i. \tag{8.21}$$

There is also an important simpler inequality that can be derived from Yang1, which is

$$\lambda_{m+1} \le \left(1 + \frac{4}{d}\right) \frac{1}{m} \sum_{j=1}^{m} \lambda_j, \tag{8.22}$$

an inequality which will be called the Yang2 inequality. Note that the Yang2 inequality controls the $(m+1)$th eigenvalue in terms of the average of the first m eigenvalues.

As will be seen later, for each $m = 1, 2, \ldots,$

$$Yang1 \implies Yang2 \implies HP \implies PPW. \tag{8.23}$$

Also, for each m each of the Yang2, HP, and PPW inequalities can be written as a strict inequality, and, when $m = 1$, each of them and Yang1 as well reduce to $\lambda_2/\lambda_1 \le 1+4/d$, which is known to hold as a strict inequality. In fact, the sharp upper bound for λ_2/λ_1, which was conjectured by PPW to saturate at the disk (and at the ball in d dimensions, as conjectured by Thompson [851]), is known due to work of Ashbaugh and Benguria [62–64] in the early 1990's. The result is indeed as PPW/Thompson conjectured, that the saturating case is the disk/ball in d-dimensional Euclidean space.

It is clear that HP $\Longrightarrow$ PPW, since we can replace the difference $\lambda_{m+1} - \lambda_i$ in the denominator on the right by $\lambda_{m+1} - \lambda_m$ and simplify algebraically to obtain the PPW inequality. Similarly, that Yang1 $\Longrightarrow$ Yang2 follows in an elementary way once one writes the Yang1 inequality explicitly (see (8.31) and following).

From the fact that the Yang1 inequality (8.21) may be written as

$$\sum_{i=1}^{m} (\lambda_{m+1} - \lambda_i)^2 - \frac{4}{d} \sum_{i=1}^{m} (\lambda_{m+1} - \lambda_i)\lambda_i \le 0, \tag{8.24}$$

or

$$\sum_{i=1}^{m} (\lambda_{m+1} - \lambda_i)\left(\lambda_{m+1} - \left(1 + \frac{4}{d}\right)\lambda_i\right) \le 0, \tag{8.25}$$

and that the left-hand member of this inequality is quadratic in λ_{m+1}, it is easy to see how to write an explicit inequality for λ_{m+1} which is essentially equivalent to (8.21) (in fact, at times as we proceed we will use the term "the Yang1 inequality" interchangeably for (8.21) or (8.31) below). Now (8.25) may be written more explicitly (as a quadratic in λ_{m+1}) as

$$m\lambda_{m+1}^2 - 2\left(1 + \frac{2}{d}\right)\left(\sum_{i=1}^{m} \lambda_i\right)\lambda_{m+1} + \left(1 + \frac{4}{d}\right)\left(\sum_{i=1}^{m} \lambda_i^2\right) \le 0. \tag{8.26}$$

Before proceeding further we adopt some useful notation, which will also be used later in the chapter. We define Λ_m to be the average of the first m λ_i's, i.e.,

$$\Lambda_m = \frac{1}{m} \sum_{i=1}^{m} \lambda_i, \tag{8.27}$$

and T_m to be the average of the first m λ_i^2's, i.e.,

$$T_m = \frac{1}{m} \sum_{i=1}^{m} \lambda_i^2. \tag{8.28}$$

In terms of these quantities, (8.26) (divided by m) may be conveniently written simply as

$$\lambda_{m+1}^2 - 2\left(1 + \frac{2}{d}\right)\Lambda_m \lambda_{m+1} + \left(1 + \frac{4}{d}\right) T_m \le 0. \tag{8.29}$$

This inequality implies immediately that λ_{m+1} lies between the two roots of the quadratic on the left-hand side (that these roots are real is a simple consequence of the fact that the sequences of real numbers $\{\lambda_i\}_{i=1}^{\infty}$ are known to exist for any given domain Ω, so that each of these inequalities must have real roots; this implies, in particular, that the discriminants of these quadratics must be nonnegative). We therefore learn that

$$\lambda_{m+1} \leq [\text{ larger root }] = \left(1+\frac{2}{d}\right)\Lambda_m + \sqrt{\left(1+\frac{2}{d}\right)^2 \Lambda_m^2 - \left(1+\frac{4}{d}\right)T_m}. \tag{8.30}$$

This is our second form of the Yang1 inequality. That the lower bound that could be obtained in this way from $\lambda_{m+1} \geq [\text{ smaller root }]$ is of less interest follows from the fact that in fact $[\text{ smaller root }] \leq \lambda_m \leq \lambda_{m+1}$ by virtue of the fact that replacing λ_{m+1} by λ_m in (8.25) simply gives back the same inequality with $m+1$ replaced by m (since the mth term of the sum vanishes). Another way to say this is that the interval $[\lambda_m, \lambda_{m+1}]$ lies within the interval on which the quadratic in (8.25) is nonpositive. Thus the lower bound is weaker than simply $\lambda_{m+1} \geq \lambda_m$, which we've known from the beginning. (But one should know that on occasion useful conclusions follow from making use of the (implicit) lower bound $\lambda_m \geq [\text{ smaller root }]$; in particular, if one is focusing on the difference $\lambda_{m+1} - \lambda_m$. For example, both [886] (1995 version) and [463] make use of the lower bound in this way.)

Finally, for the record, we write out the explicit form of the Yang1 inequality in all details:

$$\lambda_{m+1} \leq \left(1+\frac{2}{d}\right)\left(\frac{1}{m}\sum_{i=1}^{m}\lambda_i\right) + \sqrt{\left(1+\frac{2}{d}\right)^2\left(\frac{1}{m}\sum_{i=1}^{m}\lambda_i\right)^2 - \left(1+\frac{4}{d}\right)\left(\frac{1}{m}\sum_{i=1}^{m}\lambda_i^2\right)}. \tag{8.31}$$

That (8.22) follows from this can be seen by replacing the term $-\frac{1}{m}\sum_{i=1}^{m}\lambda_i^2$ under the radical here by $-(\frac{1}{m}\sum_{i=1}^{m}\lambda_i)^2$, since by the Cauchy-Schwarz inequality $(\sum_{i=1}^{m}\lambda_i)^2 \leq m\sum_{i=1}^{m}\lambda_i^2$. In fact this inequality is strict for all $m \geq 2$, since $\lambda_i > \lambda_1$ for $i > 1$.

8.2 Proof of the Main Inequality: Yang1

In this section we prove the main universal inequality of Yang (the "Yang1 inequality"), but in a somewhat more general context. It will turn out that to give the proof in this more general context requires almost no additional work. Our goal here will be to prove the inequality in the context of a Schrödinger operator with both a scalar and a vector potential. Thus the operator that we consider is

$$H = (\vec{p} - \vec{A}(x))^2 + V(x) = -(\nabla - i\vec{A}(x))^2 + V(x), \tag{8.32}$$

on a bounded domain $\Omega \subset \mathbb{R}^d$, where Dirichlet boundary conditions are imposed on $\partial\Omega$, which generalizes the Dirichlet Laplacian (where $\vec{A} = 0$, $V = 0$) and the "standard Schrödinger operator" (where $\vec{A} = 0$). This could model the quantum mechanics of a particle acted upon by both electric and magnetic forces, where V would give the electric potential and $\vec{A}$ would be the magnetic vector potential (for more on such operators, see [308, 485]; or see the recent volume based on a conference on magnetic systems [190], which contains a number of interesting articles). Our use of the differential operator $\vec{p} = \frac{1}{i}\nabla$ here also comes from quantum mechanics, $\vec{p}$ denoting the (linear) momentum operator in $\mathbb{R}^d$. But we can equally well take the form of our operator just to represent a rather general second order differential operator. Throughout this section we shall typically denote the potentials by simply V and $\vec{A}$, suppressing any explicit mention of their x-dependence. Both are taken to be real-valued on Ω throughout.

Even though our potentials are real-valued, the explicit appearance of $i = \sqrt{-1}$ in H means that here we must be prepared to deal with eigenfunctions which are in general complex-valued, and thus we will need to work explicitly in the (complex) Hilbert space $L^2(\Omega;\mathbb{C})$ and with subspaces of this space, in particular the form domain of H, $H_0^1(\Omega)$. For this purpose we adopt the notation

$$\langle u, v\rangle \equiv \int_\Omega \overline{u}\, v \tag{8.33}$$

as our *inner product* (inner product on $L^2(\Omega;\mathbb{C})$). Note that it is linear in its **second** argument, conjugate linear in its **first** argument (following the usual physicists' convention). In line with this, in our Rayleigh quotients throughout this section we will have numerators given by $\langle\varphi, H\varphi\rangle = \int_\Omega \overline{\varphi} H\varphi$, and denominators $\langle\varphi,\varphi\rangle = \int_\Omega \overline{\varphi}\varphi = \int_\Omega |\varphi|^2$.

To lay the groundwork for the proof, we note that in this context, for reasonable potentials V and $\vec{A}$ (e.g., in $C^\infty(\Omega) \cap C(\overline{\Omega})$), H is a self-adjoint operator which is bounded below and has only discrete spectrum consisting of an infinite sequence of eigenvalues

$$\lambda_1 < \lambda_2 \le \lambda_3 \le \lambda_4 \le \cdots \to \infty, \tag{8.34}$$

where the indexing here takes account of multiplicities, i.e., in this list each eigenvalue is repeated according to its multiplicity. A corresponding L^2-orthonormal basis of eigenfunctions will be denoted

$$u_1, u_2, u_3, \ldots. \tag{8.35}$$

In general these will be complex-valued. Thus, we assume known that, for each $i = 1, 2, \ldots$, $Hu_i = \lambda_i u_i$ throughout Ω, $u_i = 0$ on $\partial\Omega$, and that $\langle u_i, u_j\rangle = \int_\Omega \overline{u_i} u_j = \delta_{ij}$, where δ_{ij} denotes the Kronecker delta.

In our use of the Rayleigh-Ritz inequality (and variants, such as the quadratic form version) in this chapter, we refer the reader to [505], [97], [789], and/or [663].

In the organization of our proof we shall find it useful to employ the notion of the *commutator* of two operators. Given two operators B and C one defines their commutator via $[B, C] = BC - CB$. The commutator bracket $[\cdot,\cdot]$ is then linear in each of its entries, and, moreover, it enjoys the property that $[BC, D] = B[C, D] + [B, D]C$, where D is some third operator. These are in some sense formal properties of the commutator insofar as they apply to the differential operators that we are considering, despite the domain questions that are left to be considered. However, in our context there will not be problems, since our trial functions will always be built out of eigenfunctions and the Cartesian coordinate functions, and hence no delicate issues arise regarding domains.

We are now ready to proceed with the proof. To lend structure to the proof, we begin with a series of lemmas. Throughout our work here we will be considering trial functions (for λ_{m+1})

$$\varphi_i^{(\ell)} \equiv x_\ell u_i - \sum_{j=1}^{m} a_{ij}^{(\ell)} u_j. \tag{8.36}$$

By rights we should hang another index, m, on each of $\varphi_i^{(\ell)}$ and $a_{ij}^{(\ell)}$, but we forgo that here in the interest of economy of notation. Throughout this section the index m will be considered to be fixed, and will denote the index up to which we will combine u_j's to construct our trial functions $\varphi_i^{(\ell)}$ (for λ_{m+1}).

First we make some observations about $\varphi_i^{(\ell)}$. Since $\varphi_i^{(\ell)}$ is to be orthogonal to $\{u_j\}_{j=1}^m$ (i.e., $\varphi_i^{(\ell)} \perp u_j$ for $1 \le j \le m$), we must have

$$a_{ij}^{(\ell)} = \langle u_j, x_\ell u_i \rangle = \int_\Omega x_\ell u_i \overline{u_j}. \tag{8.37}$$

It is easily seen that, while $a_{ij}^{(\ell)}$ is in general complex,

$$a_{ji}^{(\ell)} = \overline{a_{ij}^{(\ell)}}. \tag{8.38}$$

We collect these, and a few more facts concerning $\varphi_i^{(\ell)}$ into a lemma.

Lemma 8.1. *For $\varphi_i^{(\ell)} \equiv x_\ell u_i - \sum_{j=1}^m a_{ij}^{(\ell)} u_j$, we have*
(i) $a_{ij}^{(\ell)} = \langle u_j, x_\ell u_i \rangle = \int_\Omega x_\ell u_i \overline{u_j}$;
(ii) $a_{ji}^{(\ell)} = \overline{a_{ij}^{(\ell)}}$;
(iii) $\varphi_i^{(\ell)} \perp u_j$ for all $1 \le i, j \le m$;
(iv) $\|\varphi_i^{(\ell)}\|^2 = \langle \varphi_i^{(\ell)}, x_\ell u_i \rangle = \|x_\ell u_i\|^2 - \sum_{j=1}^m |a_{ij}^{(\ell)}|^2$.

Proof. For (i) we have

$$\begin{aligned} 0 = \langle u_k, \varphi_i^{(\ell)} \rangle &= \left\langle u_k, x_\ell u_i - \sum_{j=1}^m a_{ij}^{(\ell)} u_j \right\rangle \\ &= \langle u_k, x_\ell u_i \rangle - \sum_{j=1}^m a_{ij}^{(\ell)} \langle u_k, u_j \rangle \end{aligned}$$

$$
\begin{aligned}
&= \langle u_k, x_\ell u_i \rangle - \sum_{j=1}^{m} a_{ij}^{(\ell)} \delta_{jk} \\
&= \langle u_k, x_\ell u_i \rangle - a_{ik}^{(\ell)} \\
&= \int_\Omega x_\ell u_i \overline{u_k} - a_{ik}^{(\ell)},
\end{aligned} \tag{8.39}
$$

and thus $a_{ij}^{(\ell)} = \langle u_j, x_\ell u_i \rangle = \int_\Omega x_\ell u_i \overline{u_j}$, as stated. For (ii), we simply observe that $a_{ji}^{(\ell)} = \int_\Omega x_\ell u_j \overline{u_i} = \overline{\int_\Omega x_\ell u_i \overline{u_j}} = \overline{a_{ij}^{(\ell)}}$. Condition (iii) is simply the condition that we imposed to make $\varphi_i^{(\ell)}$ a suitable trial function for λ_{m+1}, and by which we determined the $a_{ij}^{(\ell)}$'s. Finally, condition (iv) follows easily from the previous results, since, by the orthogonalities in (iii),

$$
\begin{aligned}
\|\varphi_i^{(\ell)}\|^2 = \langle \varphi_i^{(\ell)}, \varphi_i^{(\ell)} \rangle &= \langle \varphi_i^{(\ell)}, x_\ell u_i - \sum_{j=1}^{m} a_{ij}^{(\ell)} u_j \rangle \\
&= \langle \varphi_i^{(\ell)}, x_\ell u_i \rangle \\
&= \langle x_\ell u_i - \sum_{j=1}^{m} a_{ij}^{(\ell)} u_j, x_\ell u_i \rangle \\
&= \langle x_\ell u_i, x_\ell u_i \rangle - \sum_{j=1}^{m} \overline{a_{ij}^{(\ell)}} \langle u_j, x_\ell u_i \rangle \\
&= \langle x_\ell u_i, x_\ell u_i \rangle - \sum_{j=1}^{m} \overline{a_{ij}^{(\ell)}} a_{ij}^{(\ell)} \\
&= \|x_\ell u_i\|^2 - \sum_{j=1}^{m} |a_{ij}^{(\ell)}|^2.
\end{aligned} \tag{8.40}
$$

Lemma 8.2. *For the operator H given by* (8.32), *we have* $[H, x_\ell] = -2(\partial_\ell - iA_\ell)$.

Note. We use the notation ∂_ℓ for $\frac{\partial}{\partial x_\ell}$ here and in what follows.

Proof. We compute

$$
\begin{aligned}
[H, x_\ell] &= \left[-\sum_{k=1}^{d} (\partial_k - iA_k)^2 + V, x_\ell \right] \\
&= -\sum_{k=1}^{d} \left\{ (\partial_k - iA_k)[\partial_k - iA_k, x_\ell] + [\partial_k - iA_k, x_\ell](\partial_k - iA_k) \right\} \\
&= -2(\partial_\ell - iA_\ell),
\end{aligned} \tag{8.41}
$$

since

$$
[\partial_k - iA_k, x_\ell] = \delta_{k\ell}. \tag{8.42}
$$

Here as our first step we used the fact that $[V, x_\ell] = 0$ and the commutator identity $[BC, D] = B[C, D] + [B, D]C$.

Lemma 8.2 leads immediately to the following two results:

Lemma 8.3. *For*

$$\varphi_i^{(\ell)} \equiv x_\ell u_i - \sum_{j=1}^{m} a_{ij}^{(\ell)} u_j, \tag{8.43}$$

$$H\varphi_i^{(\ell)} = \lambda_i x_\ell u_i - 2(\partial_\ell - iA_\ell)u_i - \sum_{j=1}^{m} a_{ij}^{(\ell)} \lambda_j u_j. \tag{8.44}$$

Moreover,

$$\langle \varphi_i^{(\ell)}, H\varphi_i^{(\ell)} \rangle = \lambda_i \langle \varphi_i^{(\ell)}, \varphi_i^{(\ell)} \rangle - 2\langle \varphi_i^{(\ell)}, (\partial_\ell - iA_\ell)u_i \rangle. \tag{8.45}$$

Proof. We have

$$\begin{aligned} H\varphi_i^{(\ell)} &= H\Big(x_\ell u_i - \sum_{j=1}^{m} a_{ij}^{(\ell)} u_j\Big) \\ &= H(x_\ell u_i) - \sum_{j=1}^{m} a_{ij}^{(\ell)} H u_j \\ &= (x_\ell H + [H, x_\ell])u_i - \sum_{j=1}^{m} a_{ij}^{(\ell)} \lambda_j u_j \\ &= x_\ell H u_i - 2(\partial_\ell - iA_\ell)u_i - \sum_{j=1}^{m} a_{ij}^{(\ell)} \lambda_j u_j \\ &= \lambda_i x_\ell u_i - 2(\partial_\ell - iA_\ell)u_i - \sum_{j=1}^{m} a_{ij}^{(\ell)} \lambda_j u_j. \end{aligned} \tag{8.46}$$

For the second part, we apply $\langle \varphi_i^{(\ell)}, \cdot \rangle$ to the equation we just derived, obtaining

$$\begin{aligned} \langle \varphi_i^{(\ell)}, H\varphi_i^{(\ell)} \rangle &= \Big\langle \varphi_i^{(\ell)}, \lambda_i x_\ell u_i - 2(\partial_\ell - iA_\ell)u_i - \sum_{j=1}^{m} a_{ij}^{(\ell)} \lambda_j u_j \Big\rangle \\ &= \lambda_i \langle \varphi_i^{(\ell)}, x_\ell u_i \rangle - 2\langle \varphi_i^{(\ell)}, (\partial_\ell - iA_\ell)u_i \rangle - \sum_{j=1}^{m} a_{ij}^{(\ell)} \lambda_j \langle \varphi_i^{(\ell)}, u_j \rangle \\ &= \lambda_i \langle \varphi_i^{(\ell)}, \varphi_i^{(\ell)} \rangle - 2\langle \varphi_i^{(\ell)}, (\partial_\ell - iA_\ell)u_i \rangle \end{aligned} \tag{8.47}$$

by (iii) and (iv) of Lemma 8.1.

Lemma 8.4. *With notation as above, $-2\langle \varphi_i^{(\ell)}, (\partial_\ell - iA_\ell)u_i \rangle$ is real and given by*

$$-2\langle \varphi_i^{(\ell)}, (\partial_\ell - iA_\ell)u_i \rangle = 1 + \sum_{j=1}^{m} \left\{ \overline{a_{ij}^{(\ell)}} b_{ij}^{(\ell)} + a_{ij}^{(\ell)} \overline{b_{ij}^{(\ell)}} \right\} \tag{8.48}$$

where the coefficients $b_{ij}^{(\ell)}$ are defined by

$$b_{ij}^{(\ell)} \equiv \int_\Omega ((\partial_\ell - iA_\ell)u_i)\,\overline{u_j}. \tag{8.49}$$

Proof. That $-2\langle \varphi_i^{(\ell)}, (\partial_\ell - iA_\ell)u_i\rangle$ is real-valued follows from the identity

$$-2\langle \varphi_i^{(\ell)}, (\partial_\ell - iA_\ell)u_i\rangle = \langle \varphi_i^{(\ell)}, H\varphi_i^{(\ell)}\rangle - \lambda_i\langle \varphi_i^{(\ell)}, \varphi_i^{(\ell)}\rangle \tag{8.50}$$

(the second part of Lemma 8.3) and the self-adjointness of H (from which it also follows that λ_i is real, as already noted). (Indeed, by the Rayleigh-Ritz inequality for λ_{m+1} we have

$$\begin{aligned}(\lambda_{m+1} - \lambda_i)\int_\Omega |\varphi_i^{(\ell)}|^2 &\le -2\langle \varphi_i^{(\ell)}, (\partial_\ell - iA_\ell)u_i\rangle \\ &= -2\int_\Omega \overline{\varphi_i^{(\ell)}}(\partial_\ell - iA_\ell)u_i,\end{aligned} \tag{8.51}$$

so that not only is $-2\langle \varphi_i^{(\ell)}, (\partial_\ell - iA_\ell)u_i\rangle$ real-valued, it is nonnegative, a fact which we shall use later.) Because of this we can write

$$\begin{aligned}-2\langle \varphi_i^{(\ell)}, (\partial_\ell - iA_\ell)u_i\rangle &= -2\int_\Omega \overline{\varphi_i^{(\ell)}}(\partial_\ell - iA_\ell)u_i \\
&= -\int_\Omega \overline{\varphi_i^{(\ell)}}(\partial_\ell - iA_\ell)u_i - \int_\Omega \varphi_i^{(\ell)}\overline{(\partial_\ell - iA_\ell)u_i} \\
&= -\int_\Omega ((\partial_\ell - iA_\ell)u_i)\overline{\Big(x_\ell u_i - \sum_{j=1}^m a_{ij}^{(\ell)}u_j\Big)} \\
&\quad - \int_\Omega \overline{((\partial_\ell - iA_\ell)u_i)}\Big(x_\ell u_i - \sum_{j=1}^m a_{ij}^{(\ell)}u_j\Big) \\
&= -\int_\Omega \{((\partial_\ell - iA_\ell)u_i)\,\overline{(x_\ell u_i)} + \overline{((\partial_\ell - iA_\ell)u_i)}\,(x_\ell u_i)\} \\
&\quad + \sum_{j=1}^m \int_\Omega \{((\partial_\ell - iA_\ell)u_i)\overline{a_{ij}^{(\ell)}u_j} + \overline{((\partial_\ell - iA_\ell)u_i)}a_{ij}^{(\ell)}u_j\} \\
&= -\int_\Omega x_\ell\partial_\ell(u_i\overline{u_i}) \\
&\quad + \sum_{j=1}^m \Big\{\overline{a_{ij}^{(\ell)}}\int_\Omega ((\partial_\ell - iA_\ell)u_i)\overline{u_j} + a_{ij}^{(\ell)}\int_\Omega \overline{(\partial_\ell - iA_\ell)u_i}u_j\Big\} \\
&= \int_\Omega |u_i|^2 \\
&\quad + \sum_{j=1}^m \Big\{\overline{a_{ij}^{(\ell)}}\int_\Omega ((\partial_\ell - iA_\ell)u_i)\overline{u_j} + a_{ij}^{(\ell)}\overline{\int_\Omega ((\partial_\ell - iA_\ell)u_i)\overline{u_j}}\Big\}\end{aligned}$$

$$= 1 + \sum_{j=1}^{m} \{\overline{a_{ij}^{(\ell)}} b_{ij}^{(\ell)} + a_{ij}^{(\ell)} \overline{b_{ij}^{(\ell)}}\} \tag{8.52}$$

where we have defined

$$b_{ij}^{(\ell)} \equiv \int_{\Omega} ((\partial_\ell - iA_\ell)u_i)\overline{u_j}. \tag{8.53}$$

This covers all the preliminary material, except that we can also relate the coefficients $b_{ij}^{(\ell)}$ back to the $a_{ij}^{(\ell)}$'s.

Lemma 8.5. *The coefficients* $b_{ij}^{(\ell)} \equiv \int_\Omega ((\partial_\ell - iA_\ell)u_i)\overline{u_j}$ *satisfy* $b_{ji}^{(\ell)} = -\overline{b_{ij}^{(\ell)}}$ *and are given by*

$$2b_{ij}^{(\ell)} = (\lambda_i - \lambda_j)a_{ij}^{(\ell)}. \tag{8.54}$$

Proof. That $b_{ji}^{(\ell)} = -\overline{b_{ij}^{(\ell)}}$ is easily seen by integrating by parts on the right-hand side of $b_{ij}^{(\ell)} \equiv \int_\Omega ((\partial_\ell - iA_\ell)u_i)\overline{u_j}$, but it also follows from the explicit evaluation, which we now carry out. We have

$$\begin{aligned} -2b_{ij}^{(\ell)} \equiv -2\int_{\Omega} ((\partial_\ell - iA_\ell)u_i)\overline{u_j} &= \int_{\Omega} \overline{u_j}\{-2(\partial_\ell - iA_\ell)\}u_i \\ &= \int_{\Omega} \overline{u_j}[H, x_\ell]u_i \\ &= \langle u_j, [H, x_\ell]u_i\rangle \end{aligned} \tag{8.55}$$

by Lemma 8.2. Expanding out the commutator (using the definition) and by the self-adjointness of H (and the definition of $a_{ij}^{(\ell)}$) we have

$$-2b_{ij}^{(\ell)} = (\lambda_j - \lambda_i)a_{ij}^{(\ell)}, \tag{8.56}$$

which is the result we wished to show.

Using our explicit evaluation of $b_{ij}^{(\ell)}$ we can now compute from Lemma 8.4 that

$$\begin{aligned} -2\langle\varphi_i^{(\ell)}, (\partial_\ell - iA_\ell)u_i\rangle &= -2\int_{\Omega} \overline{\varphi_i^{(\ell)}}(\partial_\ell - iA_\ell)u_i \\ &= 1 + \sum_{j=1}^{m} \left\{\overline{a_{ij}^{(\ell)}} b_{ij}^{(\ell)} + a_{ij}^{(\ell)} \overline{b_{ij}^{(\ell)}}\right\} \\ &= 1 + \frac{1}{2}\sum_{j=1}^{m} (\lambda_i - \lambda_j)\left(\overline{a_{ij}^{(\ell)}} a_{ij}^{(\ell)} + a_{ij}^{(\ell)} \overline{a_{ij}^{(\ell)}}\right) \\ &= 1 + \sum_{j=1}^{m} (\lambda_i - \lambda_j)|a_{ij}^{(\ell)}|^2. \end{aligned} \tag{8.57}$$

We will need this explicit evaluation momentarily.

By the Rayleigh-Ritz inequality we now have

$$(\lambda_{m+1}-\lambda_i)\int_\Omega |\varphi_{ij}^{(\ell)}|^2 \le -2\langle \varphi_i^{(\ell)}, (\partial_\ell - iA_\ell)u_i\rangle$$
$$= -2\int_\Omega \overline{\varphi_i^{(\ell)}}(\partial_\ell - iA_\ell)u_i, \tag{8.58}$$

or

$$\lambda_{m+1}-\lambda_i \le \frac{-2\int_\Omega \overline{\varphi_i^{(\ell)}}(\partial_\ell - iA_\ell)u_i}{\int_\Omega |\varphi_{ij}^{(\ell)}|^2}. \tag{8.59}$$

Next we estimate the numerator on the right-hand side here using the Cauchy-Schwarz inequality. By the CS inequality we have

$$\begin{aligned}\Big(-2\int_\Omega \overline{\varphi_i^{(\ell)}}(\partial_\ell - iA_\ell)u_i\Big)^2 &= (-2\langle \varphi_i^{(\ell)}, (\partial_\ell - iA_\ell)u_i\rangle)^2\\ &= \Big(-2\Big\langle \varphi_i^{(\ell)}, \Big\{(\partial_\ell - iA_\ell)u_i - \sum_{j=1}^m b_{ij}^{(\ell)}u_j\Big\}\Big\rangle\Big)^2\\ &\le 4\Big(\int_\Omega |\varphi_i^{(\ell)}|^2\Big)\Big(\int_\Omega \Big|(\partial_\ell - iA_\ell)u_i - \sum_{j=1}^m b_{ij}^{(\ell)}u_j\Big|^2\Big).\end{aligned} \tag{8.60}$$

Note that here we have introduced counter-terms with the factor $(\partial_\ell - iA_\ell)u_i$ before employing the Cauchy-Schwarz inequality, as this has the potential of getting us a significantly better estimate. Since $\varphi_i^{(\ell)} \perp u_j$ for all $1 \le i, j \le m$ (part (iii) of Lemma 8.1), this costs us nothing. The choice of the coefficients $b_{ij}^{(\ell)}$ here is optimal, since these are precisely the components of $(\partial_\ell - iA_\ell)u_i$ along u_j. Something equivalent to this improvement upon a more straightforward use of the Cauchy-Schwarz inequality is what allowed H. C. Yang [886] to obtain his improved inequality (Yang1), rather than a weaker inequality. We shall go into more details of this in the next section.

For $-2\int_\Omega \overline{\varphi_i^{(\ell)}}(\partial_\ell - iA_\ell)u_i > 0$ (by the Rayleigh-Ritz inequality it is always ≥ 0, as seen above) we can proceed as follows:

$$\begin{aligned}\lambda_{m+1}-\lambda_i &\le \frac{-2\int_\Omega \overline{\varphi_i^{(\ell)}}(\partial_\ell - iA_\ell)u_i}{\int_\Omega |\varphi_{ij}^{(\ell)}|^2}\\ &= \frac{-2\langle \varphi_i^{(\ell)}, \{(\partial_\ell - iA_\ell)u_i - \sum_{j=1}^m b_{ij}^{(\ell)}u_j\}\rangle}{\int_\Omega |\varphi_{ij}^{(\ell)}|^2}\\ &\le 4\,\frac{\int_\Omega |(\partial_\ell - iA_\ell)u_i - \sum_{j=1}^m b_{ij}^{(\ell)}u_j|^2}{-2\int_\Omega \overline{\varphi_i^{(\ell)}}(\partial_\ell - iA_\ell)u_i},\end{aligned} \tag{8.61}$$

which implies that

$$(\lambda_{m+1}-\lambda_i)\Big(-2\int_\Omega \overline{\varphi_i^{(\ell)}}(\partial_\ell - iA_\ell)u_i\Big) \le 4\int_\Omega \Big|(\partial_\ell - iA_\ell)u_i - \sum_{j=1}^m b_{ij}^{(\ell)}u_j\Big|^2, \tag{8.62}$$

or

$$
\begin{aligned}
(\lambda_{m+1}-\lambda_i)\Big[1+\sum_{j=1}^{m}(\lambda_i-\lambda_j)|a_{ij}^{(\ell)}|^2\Big] &\le 4\Big[\int_\Omega |(\partial_\ell - iA_\ell)u_i|^2 \\
&\quad - \sum_{j=1}^{m} b_{ij}^{(\ell)}\int_\Omega \overline{(\partial_\ell - iA_\ell)u_i u_j} \\
&\quad - \sum_{j=1}^{m} \overline{b_{ij}^{(\ell)}}\int_\Omega \overline{u_j}(\partial_\ell - iA_\ell)u_i \\
&\quad + \sum_{j=1}^{m}\sum_{k=1}^{m} \overline{b_{ij}^{(\ell)}}b_{ik}^{(\ell)}\int_\Omega \overline{u_j}u_k\Big] \\
&= 4\Big[\int_\Omega |(\partial_\ell - iA_\ell)u_i|^2 - \sum_{j=1}^{m} b_{ij}^{(\ell)}\overline{b_{ij}^{(\ell)}} \\
&\quad - \sum_{j=1}^{m} \overline{b_{ij}^{(\ell)}}b_{ij}^{(\ell)} + \sum_{j=1}^{m}\sum_{k=1}^{m} \overline{b_{ij}^{(\ell)}}b_{ik}^{(\ell)}\delta_{jk}\Big] \\
&= 4\Big[\int_\Omega |(\partial_\ell - iA_\ell)u_i|^2 - \sum_{j=1}^{m} |b_{ij}^{(\ell)}|^2 \\
&\quad - \sum_{j=1}^{m} |b_{ij}^{(\ell)}|^2 + \sum_{j=1}^{m} |b_{ij}^{(\ell)}|^2\Big] \\
&= 4\int_\Omega |(\partial_\ell - iA_\ell)u_i|^2 - 4\sum_{j=1}^{m} |b_{ij}^{(\ell)}|^2 \\
&= 4\int_\Omega |(\partial_\ell - iA_\ell)u_i|^2 - \sum_{j=1}^{m}(\lambda_i-\lambda_j)^2\,|a_{ij}^{(\ell)}|^2,
\end{aligned}
\tag{8.63}
$$

where we have simplified the expressions appearing on either side of (8.63) using the definition of the $b_{ij}^{(\ell)}$'s and Lemmas 8.4 and 8.5. To clear up a loose end, we note that this final inequality continues to hold even in the event that $-2\int_\Omega \overline{\varphi_i^{(\ell)}}(\partial_\ell - iA_\ell)u_i = 0$ (recall that we began this paragraph by taking this quantity to be positive). In that case $-2\int_\Omega \overline{\varphi_i^{(\ell)}}(\partial_\ell - iA_\ell)u_i = 1+\sum_{j=1}^{m}(\lambda_i-\lambda_j)|a_{ij}^{(\ell)}|^2 = 0$ and the left-hand side of (8.63) is 0, while its right-hand side, $4\int_\Omega |(\partial_\ell - iA_\ell)u_i|^2 - \sum_{j=1}^{m}(\lambda_i-\lambda_j)^2\,|a_{ij}^{(\ell)}|^2 = 4\,\|(\partial_\ell - iA_\ell)u_i - \sum_{j=1}^{m} b_{ij}^{(\ell)}u_j\|^2$, is clearly ≥ 0, confirming our claim. Thus even in the unlikely event that $-2\int_\Omega \overline{\varphi_i^{(\ell)}}(\partial_\ell - iA_\ell)u_i = 0$, i.e., that $\varphi_i^{(\ell)} \perp (\partial_\ell - iA_\ell)u_i$ (which would hold, in particular, if $\varphi_i^{(\ell)}$ vanished identically on Ω), the inequality (8.63) holds, which is necessary to the continuation of our arguments.

At this stage we are almost done. Moving all the terms involving the $a_{ij}^{(\ell)}$'s to the left and simplifying, we have

$$(\lambda_{m+1} - \lambda_i) + \sum_{j=1}^{m} (\lambda_{m+1} - \lambda_j)(\lambda_i - \lambda_j)|a_{ij}^{(\ell)}|^2 \le 4 \int_\Omega |(\partial_\ell - iA_\ell)u_i|^2. \tag{8.64}$$

We note that the integral on the right-hand side here has one of the gradient-squared terms that appear in our operator H (in a quadratic form sense), so we can make it into the quantity $\langle u_i, (H - V)u_i\rangle$ if we simply sum on ℓ from 1 to d. We therefore arrive at

$$d(\lambda_{m+1} - \lambda_i) + \sum_{j=1}^{m} (\lambda_{m+1} - \lambda_j)(\lambda_i - \lambda_j)\left\{\sum_{\ell=1}^{d} |a_{ij}^{(\ell)}|^2\right\} \le 4\langle u_i, (H - V)u_i\rangle. \tag{8.65}$$

Setting $\Theta_{ij} \equiv \sum_{\ell=1}^{d} |a_{ij}^{(\ell)}|^2$, which is easily seen to have the property $\Theta_{ji} = \Theta_{ij} \ge 0$, we have

$$\begin{aligned} d(\lambda_{m+1} - \lambda_i) + \sum_{j=1}^{m} (\lambda_{m+1} - \lambda_j)(\lambda_i - \lambda_j)\Theta_{ij} &\le 4\langle u_i, (H - V)u_i\rangle \\ &= 4\lambda_i - 4\int_\Omega V|u_i|^2. \end{aligned} \tag{8.66}$$

To finish obtaining the Yang1 inequality, we have only to multiply through by $(\lambda_{m+1} - \lambda_i)$ and sum on i from 1 to m. We obtain (after dividing by d)

$$\sum_{i=1}^{m} (\lambda_{m+1} - \lambda_i)^2 \le \frac{4}{d} \sum_{i=1}^{m} (\lambda_{m+1} - \lambda_i)\left[\lambda_i - \int_\Omega V|u_i|^2\right], \tag{8.67}$$

because all the terms in the $a_{ij}^{(\ell)}$'s disappear, due to anti-symmetry (this is why we introduced the factor $(\lambda_{m+1} - \lambda_i)$).

If we introduce the notation $V_i \equiv \int_\Omega V|u_i|^2$ for the matrix element of V in the state u_i, we have simply

$$\sum_{i=1}^{m} (\lambda_{m+1} - \lambda_i)^2 \le \frac{4}{d} \sum_{i=1}^{m} (\lambda_{m+1} - \lambda_i)[\lambda_i - V_i]. \tag{8.68}$$

This is the general inequality of Yang generalized to the case of a Schrödinger operator with a vector potential. In the case of just the Dirichlet Laplacian (Yang's case) the inequality becomes the classical Yang1 inequality.

$$\sum_{i=1}^{m} (\lambda_{m+1} - \lambda_i)^2 \le \frac{4}{d} \sum_{i=1}^{m} (\lambda_{m+1} - \lambda_i)\lambda_i. \tag{8.69}$$

We note that by our work above this inequality also holds for the magnetic Schrödinger operator H as defined by (8.32) whenever the scalar potential V is nonnegative, and for essentially arbitrary magnetic vector potential $\vec{A}$. Thus, the range of validity of the bound (8.21) (which is the same as (8.69)) is extended considerably.

8.3 The Other Main Inequalities and their Proofs: PPW, HP, and Yang2

In this section we quickly run through the proofs of the other main universal inequalities for the eigenvalues of the Dirichlet Laplacian on a bounded domain in Euclidean space $\mathbb{R}^d$. Although our methods (the Rayleigh-Ritz inequality with trial functions based on the lower eigenfunctions, integration by parts, the Cauchy-Schwarz inequality) all come from the original papers of Payne, Pólya, and Weinberger [746, 751], we shall not follow the historical proofs, but rather give the most direct proofs we can using our proof of Yang's first inequality in the last section. We should bear in mind here that we work in the context of the general operator H (see (8.32)), so that the inequalities we give are more general in that respect. We give the inequalities in both the form that contains a term based on the potential V, and one that does not. It may be worthwhile to observe that the results that hold with V included all can be reduced to the V-less results whenever $V \geq 0$ on Ω. Moreover, if it is known that V is not a.e. 0 there, all of the latter inequalities can be written as *strict* inequalities.

At this point, the easiest of the other inequalities to obtain is that of Hile and Protter. To derive it, we observe that if we go back into the proof of Yang1 in the last section and use the Cauchy-Schwarz inequality without subtracting any of the counter-terms based on the $b_{ij}^{(\ell)}$'s we arrive at (8.63), but without the subtracted terms on the right-hand side. Thus we have

$$(\lambda_{m+1} - \lambda_i)\left[1 + \sum_{j=1}^{m} (\lambda_i - \lambda_j)|a_{ij}^{(\ell)}|^2\right] \leq 4\int_\Omega |(\partial_\ell - iA_\ell)u_i|^2. \tag{8.70}$$

If we carry through with our other manipulations up to just before we summed on i to obtain a double sum that vanished by anti-symmetry we have the inequality

$$d(\lambda_{m+1} - \lambda_i) + \sum_{j=1}^{m} (\lambda_{m+1} - \lambda_i)(\lambda_i - \lambda_j)\Theta_{ij} \leq 4\langle u_i, (H - V)u_i\rangle = 4(\lambda_i - V_i), \tag{8.71}$$

the only difference from what we had at (8.66) above being that here there is a factor of $(\lambda_{m+1} - \lambda_i)$ with the sum on the left, in place of a $(\lambda_{m+1} - \lambda_j)$.

We can continue with the same strategy we used before, only now in order to free ourselves of the terms in the Θ_{ij}'s we have to divide through by $\lambda_{m+1} - \lambda_i$ and sum on i from 1 to m, rather than multiply by this factor and sum. We thus arrive at

$$\frac{d}{4} \leq \frac{1}{m}\sum_{i=1}^{m} \frac{\lambda_i - V_i}{\lambda_{m+1} - \lambda_i} \text{ for } m = 1, 2, \ldots, \tag{8.72}$$

which holds for the eigenvalues of the operator H of (8.32), and, in its V-less form,

$$\frac{d}{4} \leq \frac{1}{m}\sum_{i=1}^{m} \frac{\lambda_i}{\lambda_{m+1} - \lambda_i} \text{ for } m = 1, 2, \ldots, \tag{8.73}$$

which holds for H with $V \geq 0$, or for just the Laplacian $-\Delta$. This last form is the classical HP inequality. We note that this inequality should be understood to hold trivially (as $d/4 \leq \infty$) if λ_{m+1} shares its value with any lower eigenvalue, i.e., if it shares its value with any λ_i, $1 \leq i \leq m$.

As already mentioned in our introduction, the PPW inequality follows easily from the HP inequality by replacing the difference $\lambda_{m+1} - \lambda_i$ by $\lambda_{m+1} - \lambda_m$, clearing fractions, and simplifying algebraically. One could equally well go back to our general proof and at one point or another replace the difference $\lambda_{m+1} - \lambda_i$ which occurs on the left-hand side of our inequalities via the Rayleigh-Ritz inequality by $\lambda_{m+1} - \lambda_m$ and continue on with what results, using the same ideas (but with everything being simplified).

One can also reach the Yang2 inequality by simple modifications of what we did in the previous section. One has only to average the inequalities (8.66) and (8.71) to obtain

$$\begin{aligned} d(\lambda_{m+1} - \lambda_i) + \sum_{j=1}^{m} (\lambda_{m+1} - (\lambda_i + \lambda_j)/2)(\lambda_i - \lambda_j)\Theta_{ij} &\leq 4\langle u_i, (H - V)u_i\rangle \\ &= 4(\lambda_i - V_i). \end{aligned} \tag{8.74}$$

Here we have the same inequality as appears in (8.66) and (8.71) except that where (8.66) had the factor $(\lambda_{m+1} - \lambda_j)$ and (8.71) had the factor $(\lambda_{m+1} - \lambda_i)$, here we have the factor $(\lambda_{m+1} - (\lambda_i + \lambda_j)/2)$. The effect of this, following our usual strategy of eliminating the terms in Θ_{ij} using anti-symmetry, is that we can sum this inequality directly on i from 1 to m to obtain the Yang2 inequality:

$$\begin{aligned} \lambda_{m+1} &\leq \frac{1}{m}\sum_{i=1}^{m} \lambda_i + \frac{4}{d}\frac{1}{m}\sum_{i=1}^{m}(\lambda_i - V_i) \\ &= \left(1 + \frac{4}{d}\right)\frac{1}{m}\sum_{i=1}^{m} \lambda_i - \frac{4}{d}\frac{1}{m}\sum_{i=1}^{m} V_i \text{ for } m = 1, 2, \dots. \end{aligned} \tag{8.75}$$

We note that the method of proof used here suggests that (8.22) is in some sense midway between the HP and Yang1 inequalities, a point of view that will also be supported by our work in the next section. Again, the classical version of the Yang1 inequality is obtained by dropping the terms in V:

$$\lambda_{m+1} \leq \left(1 + \frac{4}{d}\right)\frac{1}{m}\sum_{i=1}^{m} \lambda_i \text{ for } m = 1, 2, \dots. \tag{8.76}$$

This holds for arbitrary H with $V \geq 0$ on Ω, or in general for the Dirichlet Laplacian on a bounded domain Ω in $\mathbb{R}^d$.

Furthermore, we have the PPW variants of these,

$$\lambda_{m+1} - \lambda_m \leq \frac{4}{d}\frac{1}{m}\sum_{i=1}^{m}(\lambda_i - V_i), \tag{8.77}$$

and the V-less form (good for $V \geq 0$),

$$\lambda_{m+1} - \lambda_m \leq \frac{4}{d}\frac{1}{m}\sum_{i=1}^{m} \lambda_i. \tag{8.78}$$

That these hold follows immediately from (8.75) and (8.76) because $\frac{1}{m}\sum_{i=1}^{m} \lambda_i \leq \lambda_m$ for all $m \geq 1$ (with strict inequality for $m \geq 2$ since $\lambda_i > \lambda_1$ for $i > 1$). They also follow from the HP inequalities, (8.72) and (8.73).

8.4 The Hierarchy of Inequalities: PPW, HP, Yang

A nice way to understand the classical inequalities of Payne, Pólya, and Weinberger, Hile and Protter, Yang, and more from a unified point of view is provided by the Chebyshev inequality (see, for example, Hardy, Littlewood, and Pólya [449]). Let $\{a_i\}_{i=1}^m$, $\{b_i\}_{i=1}^m$ be two finite sequences of real numbers. These sequences are said to be *similarly ordered* if $(a_i - a_j)(b_i - b_j) \geq 0$ for all $1 \leq i, j \leq m$, and *oppositely ordered* if $(a_i - a_j)(b_i - b_j) \leq 0$ for all $1 \leq i, j \leq m$. We also introduce a system of weights, $\{w_i\}_{i=1}^m$ defined such that each w_i is real and nonnegative, and $\sum_1^m w_i > 0$ (note that this allows some, but not all, of the weights to be 0).

Theorem 8.6 (Chebyshev's inequality). *If the sequences $\{a_i\}_{i=1}^m$ and $\{b_i\}_{i=1}^m$ are similarly ordered, then*

$$\Big(\sum_{i=1}^{m} w_i a_i\Big)\Big(\sum_{j=1}^{m} w_j b_j\Big) \leq \Big(\sum_{i=1}^{m} w_i\Big)\Big(\sum_{j=1}^{m} w_j a_j b_j\Big), \tag{8.79}$$

and if they are oppositely ordered, then

$$\Big(\sum_{i=1}^{m} w_i a_i\Big)\Big(\sum_{j=1}^{m} w_j b_j\Big) \geq \Big(\sum_{i=1}^{m} w_i\Big)\Big(\sum_{j=1}^{m} w_j a_j b_j\Big). \tag{8.80}$$

Remarks. (1) When there is no room for confusion we shall sometimes write these inequalities simply as $(\sum wa)(\sum wb) \leq (\sum w)(\sum wab)$, etc.
(2) Clearly sequences that are both increasing (more precisely, nondecreasing) or both decreasing are similarly ordered, and sequences one of which is decreasing and the other increasing are oppositely ordered. Here all our sequences will be either increasing or decreasing so this remark will suffice for us to determine when any two of them are similarly or oppositely ordered. The underlying concept is not tied to the notions of increasing or decreasing; these just provide us with a convenient standard form in which to view them. One could also get into rearranging sequences, and then the increasing (resp., decreasing) rearrangement of a sequence would be important. But we will not do that here.

(3) In fact, all of the sequences that we shall ever apply the Chebyshev inequality to below will consist of nonnegative elements, with all the sums that appear being positive (or possibly infinity; more on that later). This will allow us to move the sums around in our inequalities essentially at will, dividing or multiplying by them as we see fit. In the process of doing this we shall only comment if there is a possibility that one of these is 0 (or possibly ∞). In such cases we will either exclude the case in question, or argue that the inequality still holds in an extended sense. In our early work here all our sequences will be based on eigenvalues, or eigenvalue differences, of the Dirichlet Laplacian, or of H with potential $V \geq 0$, and all terms will be nonnegative. However, at the very end of this section we will touch just slightly on what can be said for the eigenvalues of the general operator H (so with V and $\vec{A}$, and without the restriction $V \geq 0$).
(4) As will be clear from the proof, when all the weights are positive Chebyshev's inequality is strict unless one (or both) of the two sequences is constant (i.e., for $\{a_i\}$, for example, unless $a_i = a_j$ for all $1 \leq i, j \leq m$). Or see [449]. For us, however, it is not so important to characterize the case of equality, and it is useful to allow some of our weights to be 0.
(5) If the weights are all equal (the case of constant weights), then their value does not matter so long as it is greater than 0. We refer to this as the "unweighted case". Convenient choices are either to take all $w_i = 1$, or all $w_i = 1/m$. Both choices have their uses. In the first case $\sum w = m$, while in the latter $\sum w = 1$.

Proof. One only has to consider the sum $\sum_i \sum_j w_i w_j (a_i - a_j)(b_i - b_j)$ and follow one's nose. We consider only the case of similarly ordered sequences here. The other case just reverses the inequality. Because each term of the sum is nonnegative, one has

$$
\begin{aligned}
0 &\leq \sum_{i=1}^{m} \sum_{j=1}^{m} w_i w_j (a_i - a_j)(b_i - b_j) \\
&= \sum_{i=1}^{m} \sum_{j=1}^{m} w_i w_j (a_i b_i - a_i b_j - a_j b_i + a_j b_j) \\
&= \sum_{i=1}^{m} \sum_{j=1}^{m} (w_i a_i b_i w_j - w_i a_i w_j b_j - w_j a_j w_i b_i + w_i w_j a_j b_j) \\
&= \left(\sum_{i=1}^{m} w_i a_i b_i\right)\left(\sum_{j=1}^{m} w_j\right) - \left(\sum_{i=1}^{m} w_i a_i\right)\left(\sum_{j=1}^{m} w_j b_j\right) \\
&\quad - \left(\sum_{i=1}^{m} w_i b_i\right)\left(\sum_{j=1}^{m} w_j a_j\right) + \left(\sum_{i=1}^{m} w_i\right)\left(\sum_{j=1}^{m} w_j a_j b_j\right) \\
&= 2\left[\left(\sum_{i=1}^{m} w_i a_i b_i\right)\left(\sum_{j=1}^{m} w_j\right) - \left(\sum_{i=1}^{m} w_i a_i\right)\left(\sum_{j=1}^{m} w_j b_j\right)\right]
\end{aligned}
$$

$$= 2\Big[\big(\sum w\big)\big(\sum wab\big) - \big(\sum wa\big)\big(\sum wb\big)\Big], \tag{8.81}$$

which gives us the inequality we want.

For our purpose here, that of understanding the various universal eigenvalue inequalities in a unified way, and, in particular, how they compare to and imply one another, we have only to make the following choices:

$$w_i = \lambda_i(\lambda_{m+1} - \lambda_i), \tag{8.82}$$

$$a_i = \frac{\lambda_{m+1} - \lambda_i}{\lambda_i} \tag{8.83}$$

(note that with the λ_i's the Dirichlet eigenvalues of the Laplacian, each a_i is positive and $\{a_i\}$ is decreasing in i), and

$$b_i = (\lambda_{m+1} - \lambda_i)^{-p} \text{ for } p \in [0, 2] \tag{8.84}$$

(here, with $p \geq 0$, the sequence of b_i's is increasing in i).

Putting these choices into Chebyshev's inequality, we have the following theorem (since our sequences $\{a\}$ and $\{b\}$ are oppositely ordered, we use the reversed form of the inequality).

Theorem 8.7. *For the eigenvalues of the Dirichlet Laplacian on a bounded domain in $\mathbb{R}^d$ or for the operator H of (8.32) with a potential $V \geq 0$, we have, for $p \in [0, 2]$,*

$$\begin{aligned}\big(\sum (\lambda_{m+1} - \lambda_i)^2\big)\big(\sum \lambda_j(\lambda_{m+1} - \lambda_j)^{1-p}\big)& \\ \geq \big(\sum \lambda_i(\lambda_{m+1} - \lambda_i)\big)\big(\sum (\lambda_{m+1} - \lambda_j)^{2-p}\big).&\end{aligned} \tag{8.85}$$

From this follows, due to Yang's first inequality (Yang1),

$$\frac{4}{d} \geq \frac{\sum (\lambda_{m+1} - \lambda_j)^{2-p}}{\sum \lambda_j(\lambda_{m+1} - \lambda_j)^{1-p}} \tag{8.86}$$

for $p \in [0, 2]$, or, equivalently,

$$\frac{d}{4} \leq \frac{\sum \lambda_j(\lambda_{m+1} - \lambda_j)^{1-p}}{\sum (\lambda_{m+1} - \lambda_j)^{2-p}} \tag{8.87}$$

for the same range of p's. The cases $p = 0$, $p = 1$, and $p = 2$ give, respectively, the Yang1 inequality, the Yang2 inequality, and the Hile-Protter inequality.

Remarks. (1) We confine p to the interval $[0, 2]$ even though it might appear that our argument is good for all $p \geq 0$ because of the possibility that some of the differences $\lambda_{m+1} - \lambda_i$ (and, in particular, $\lambda_{m+1} - \lambda_m$) might vanish. If we restrict p not to go above 2 then at least one side of inequality (8.85) will still be meaningful and finite, whereas the other side will be infinity, and in such cases the inequality still holds (as can be

verified directly, if nothing else). For p beyond 2 this will no longer be the case.
(2) The argument here might go more smoothly if we replaced λ_{m+1} by σ, a *strict* upper bound for λ_{m+1}. Then all our weights would be positive, and our a_i's and b_i's would be positive and finite. The use of the Chebyshev inequality would be straightforward, and we could recover all our inequalities for λ_{m+1} by sending σ to λ_{m+1} from above. At that point our interpretations and understandings from Remark (1) would come into play again. A point of view close to this is used in [82, 83, 511]. See also [57].

Indeed, the point of view taken in [82, 83, 511] is to definite quantities σ_{Y1}, σ_{Y2}, σ_{HP}, σ_{PPW} as those values of $\sigma > \lambda_m$ where the respective inequalities (with σ replacing λ_{m+1}) are equalities (this defines these quantities uniquely), and then to show, using Chebyshev, that $\lambda_{m+1} \le \sigma_{Y1} < \sigma_{Y2} < \sigma_{HP} < \sigma_{PPW}$ for all $m \ge 2$ (at $m = 1$ one has $\lambda_2 < (1+\frac{4}{d})\lambda_1 = \sigma_{Y1} = \sigma_{Y2} = \sigma_{HP} = \sigma_{PPW}$). Although not done in [82, 83, 511], the strict inequalities here for $m \ge 2$ follow by using the weaker σ (= larger σ) in Chebyshev's inequality for the individual pairwise comparisons and using the characterization of equality in Chebyshev's inequality. One can even do this in the context of a parameter p, $p \in [0, 2]$, where one shows strict monotonicity of the bound $\sigma(p)$ (cf. our comments at (8.96) below; here $\sigma(p)$ is defined such that $\sigma(0) = \sigma_{Y1}$, $\sigma(1) = \sigma_{Y2}$, $\sigma(2) = \sigma_{HP}$, and all the bounds already listed for these quantities are included).

These inequalities (the strict ones, as listed above) were all originally proved in [57]. There the proofs did not use the Chebyshev inequality, but rather relied on convexity arguments. The notation was $\sigma_{Y1} = G_m^{(Yang1)}(\lambda_1, \dots, \lambda_m)$, $\sigma_{Y2} = G_m^{(Yang2)}(\lambda_1, \dots, \lambda_m)$, etc., and it was proved that, for $m \ge 2$,

$$\lambda_{m+1} \le G_m^{(Yang1)}(\lambda_1, \dots, \lambda_m) < G_m^{(Yang2)}(\lambda_1, \dots, \lambda_m) \tag{8.88}$$

$$< G_m^{(HP)}(\lambda_1, \dots, \lambda_m) < G_m^{(PPW)}(\lambda_1, \dots, \lambda_m) \tag{8.89}$$

for all $(\lambda_1, \dots, \lambda_m)$ arising as the first m eigenvalues, with multiplicities, of the Dirichlet Laplacian $(-\Delta^{(Dir)})$ on a bounded domain $\Omega \in \mathbb{R}^d$. In the context of this chapter, all of these conclusions continue to hold for the eigenvalues of the operator H (8.32) with $V \ge 0$, with strict inequality from the beginning if V is not a.e. 0.
(3) That Yang2 is given by taking $p = 1$, while Yang1 and the Hile-Protter inequality are the cases $p = 0$ and 2, respectively, shows again that Yang2 is in a sense midway between the other two inequalities. We saw this also in the way we proved Yang2 near the end of the last section.

Proof. We substitute our given choices of w_i, a_i, and b_i into the inequality

$$\Big(\sum_i wa\Big)\Big(\sum_j wb\Big) \ge \Big(\sum_i w\Big)\Big(\sum_j wab\Big), \tag{8.90}$$

obtaining

$$\begin{aligned} 0 \le {} & \Big(\sum \lambda_i(\lambda_{m+1} - \lambda_i)\frac{\lambda_{m+1} - \lambda_i}{\lambda_i}\Big)\Big(\sum \lambda_j(\lambda_{m+1} - \lambda_j)(\lambda_{m+1} - \lambda_j)^{-p}\Big) \\ & - \Big(\sum \lambda_i(\lambda_{m+1} - \lambda_i)\Big)\Big(\sum \lambda_j(\lambda_{m+1} - \lambda_j)\frac{\lambda_{m+1} - \lambda_j}{\lambda_j}(\lambda_{m+1} - \lambda_j)^{-p}\Big) \end{aligned}$$

$$= \left(\sum (\lambda_{m+1} - \lambda_i)^2\right)\left(\sum \lambda_j(\lambda_{m+1} - \lambda_j)^{1-p}\right)$$
$$- \left(\sum \lambda_i(\lambda_{m+1} - \lambda_i)\right)\left(\sum (\lambda_{m+1} - \lambda_j)^{2-p}\right). \tag{8.91}$$

Hence,

$$\frac{\sum (\lambda_{m+1} - \lambda_i)^2}{\sum \lambda_i(\lambda_{m+1} - \lambda_i)} \geq \frac{\sum (\lambda_{m+1} - \lambda_j)^{2-p}}{\sum \lambda_j(\lambda_{m+1} - \lambda_j)^{1-p}}, \tag{8.92}$$

or

$$\frac{\sum \lambda_i(\lambda_{m+1} - \lambda_i)}{\sum (\lambda_{m+1} - \lambda_i)^2} \leq \frac{\sum \lambda_j(\lambda_{m+1} - \lambda_j)^{1-p}}{\sum (\lambda_{m+1} - \lambda_j)^{2-p}}. \tag{8.93}$$

Since Yang1 can be written as the inequality

$$\frac{d}{4} \leq \frac{\sum \lambda_i(\lambda_{m+1} - \lambda_i)}{\sum (\lambda_{m+1} - \lambda_i)^2}, \tag{8.94}$$

the statement in the theorem follows. The identifications of the cases $p = 0, 1, 2$ now follow directly and explicitly.

By further exploitation of Chebyshev's inequality one can in fact obtain the further refinements

$$\frac{d}{4} \leq \frac{\sum \lambda_i(\lambda_{m+1} - \lambda_i)}{\sum (\lambda_{m+1} - \lambda_i)^2} \leq \frac{\sum \lambda_j}{\sum (\lambda_{m+1} - \lambda_j)}$$
$$\leq \frac{1}{m}\sum \lambda_j(\lambda_{m+1} - \lambda_j)^{-1}. \tag{8.95}$$

Indeed, one can show that the quantity

$$\frac{\sum \lambda_j(\lambda_{m+1} - \lambda_j)^{1-p}}{\sum (\lambda_{m+1} - \lambda_j)^{2-p}} \tag{8.96}$$

is increasing in p for $p \in [0, 2]$. This can be done by using the Chebyshev inequality to compare this expression with the same expression with p replaced by $q \in [0, 2]$. Then for $q > p$ one can use Chebyshev to compare the two expressions, obtaining the desired result. We forgo giving the details here. Similar conclusions to those we drew above, but in a more abstract setting, and based on the exploitation of Chebyshev's inequality, can be found in Hermi's dissertation [511], and in the papers [82, 83] of Ashbaugh and Hermi. In particular, the monotonicity of the expression in (8.96) is mentioned in [83], with details in [82].

Finally, we conclude this section with a quick summary of what can be done in the general case of the operator H (see (8.32)), where the eigenvalues are unrestricted. There is certainly also the ordering question to address, whether the necessary sequences are similarly or oppositely ordered as they were before. The key to some of what we do here is (8.68), in which V_i is included (and similar observations on the inequalities of PPW, HP, and Yang2 in which the term in V_i is included). One has

$$\sum_{i=1}^{m} (\lambda_{m+1} - \lambda_i)^2 \leq \frac{4}{d}\sum_{i=1}^{m} (\lambda_{m+1} - \lambda_i)[\lambda_i - V_i]. \tag{8.97}$$

We set

$$\tau_i \equiv \lambda_i - V_i = \sum_{\ell=1}^{d} \int_\Omega \left|\left(\partial_\ell - iA_\ell\right)u_i\right|^2 > 0. \tag{8.98}$$

Thus we now can write our inequality as

$$\sum_{i=1}^{m} (\lambda_{m+1} - \lambda_i)^2 \leq \frac{4}{d} \sum_{i=1}^{m} (\lambda_{m+1} - \lambda_i)\tau_i. \tag{8.99}$$

Remarks. (1) The use of the letter τ here is not by accident, since it is supposed to suggest a kinetic energy, which is what the term $\lambda_i - V_i$ represents in physical terms. This has already been employed by Harrell and Stubbe in their first paper [463], as well as in [652] (see also the comments in [57]). In those papers the letter T (often with one or more indices attached to it) was typically associated with kinetic energy terms. Also, Ashbaugh and Hermi [83] use Λ_i for Harrell and Stubbe's $4T_i$.
(2) If one finds that the eigenvalues start off negative because the potential has a deep well (or wells) one can simply introduce an additive constant M into both the λ_i's and the V_i's (so this should adjust all λ's and all V's) in such a way that the eigenvalues all become positive. In particular, this can be done if the potential is bounded below by $-M$. If one does this and renames the eigenvalues and the potential, nothing will have changed (the inequality will continue to hold), but the eigenvalues will be known to be positive. And if the new potential is nonnegative, it can be thrown away as before to reach the classical version of the inequality.
(3) This phenomenon, that a small shift in notation or point of view can bring us back to the classical universal inequalities even when studying a somewhat different or more general problem, occurs several times in the study of universal eigenvalue inequalities, as, for example, when studying the eigenvalues of the Laplacian (Laplace-Beltrami operator) on Riemannian manifolds or those of related more general operators in such more extended settings. For example, much the same thing happens in the case of domains in $\mathbb{S}^d$ and $\mathbb{H}^d$. Interested readers are referred to the final section of [57], or the work in [265] and elsewhere. In $\mathbb{S}^d$ the appropriate shift is by $d^2/4$ (see [55, 57] and other references).

The particular cases of our universal inequalities that include V_i's (or with τ_i for $\lambda_i - V_i$) with $p = 0$, 1, and 2 are all fine, since these results all hold because our earlier proofs establish them. In fact, there are even "p-versions", $p \in [0, 2]$, of these results as in (8.86) or (8.87) where the "free λ_j" is replaced by τ_j. See [463], Theorem 5, p. 1801; the proof does not use the Chebyshev inequality, but rather employs a multiplicative function introduced into certain eigenvalue inequalities. What we cannot necessarily do here is to say that one or another of these inequalities is stronger than another, i.e., the sorts of things we were doing in this section based on the Chebyshev inequality. If one tries to do such things, one finds one needs to check for similar or opposite ordering (as the case may be), and that, while the necessary relations seem like they might have a chance, they do not hold up in general.

Therefore, while the classical universal inequalities of Yang (Yang1 and Yang2), HP, and PPW with V_i's (or τ_i's) included hold up for the general operator H even without any restriction on V, we cannot be sure of the hierarchy (Yang1 $\Longrightarrow$ Yang2 $\Longrightarrow$ HP $\Longrightarrow$ PPW) in general. But for the Dirichlet Laplacian, or for H with $V \geq 0$, one can get essentially everything one could wish for (having dropped the V's, or replaced τ's by λ's, since $\tau_i \leq \lambda_i$ holds then).

In fact, one can sometimes do more than just use $\tau_i \leq \lambda_i$, whether or not $V \geq 0$, to free oneself of the τ_i's (equivalently, the V_i's). Under the subject of "virial theory" physicists have studied the relation between τ_i and λ_i and, under specific conditions on the potentials, it can follow that $\tau_i \leq f\lambda_i$, where f is some proper fraction (i.e., $f \in (0, 1)$). When this applies, we can recover the classical inequalities in an improved form, for now $4/d$ can be replaced by the smaller constant $4f/d$. This applies, for example, to certain power-law potentials as discussed by Harrell and Stubbe in [463]. In particular, they show that by this means one can get sharp inequalities in the case of the isotropic quantum harmonic oscillator in d dimensions. For general background on virial theory in the quantum-mechanical setting see the book by Eastham and Kalf [351]. In this context (no τ_i's or V_i's, $4f/d$ replacing $4/d$ with $f \in (0, 1]$) we can do everything we did above with $4f/d$ replacing $4/d$, even including p-versions, $p \in [0, 2]$, as in Theorem 8.7. See [82, 83, 463, 511] for more in this direction. Also, the material in Section 8.5 extends to this case.

On a final note, in the case of the Dirichlet Laplacian or H with $V \geq 0$, because of the implications that exist there is no real reason to continue to consider the inequalities of PPW or HP, except out of historical interest. One has the main Yang1 inequality, which is better than any of the other inequalities, and is relatively simple and can be written explicitly. One should probably keep the Yang2 inequality in mind because it is even simpler, and is still stronger than the PPW and HP inequalities. Beyond that, the HP inequality is the most complicated, since to use it to get an explicit bound on λ_{m+1} one would have to solve explicitly for the roots of a polynomial of degree m, which is not generally possible in terms of familiar functions.

8.5 Asymptotics and Explicit Inequalities

It is an interesting and technically demanding problem to turn a given set of universal inequalities, in particular, the Yang1 inequality (8.30) or (8.31) for $m = 1, 2, 3, \ldots$, into a set of more explicit bounds on λ_{m+1}. To get a sense of what is wanted here, note that the PPW bound (8.6) can be dropped back to saying simply that $\lambda_{m+1} \leq (1 + \frac{4}{d})\lambda_m$ for $m \geq 1$, and that by simply concatenating these inequalities from m back to 1 one obtains the explicit inequality $\lambda_{m+1} \leq (1 + \frac{4}{d})^m \lambda_1$. This is the sort of inequality that we have in mind when we speak of deriving explicit inequalities from a set of universal inequalities, a bound on λ_{m+1} (or possibly λ_m) in terms of an expression involving

m and a factor of λ_1. That is, we want a bound that is explicitly computable based on m and λ_1. Of course, we would like a bound that is much better than the one we just derived, and one can certainly hope to do that by basing the bound on one of the stronger sets of universal inequalities, such as Yang1 (8.21), although extracting such explicit information from an inequality like (8.31) presents a formidable challenge.

Before going on with this story, we lay the groundwork for it by presenting some background information.

According to Weyl, for $\Omega \subset \mathbb{R}^d$ the *dominant factor* in how λ_m grows as $m \to \infty$ is $m^{2/d}$ (see, e.g., [48, 94, 277, 610, 777, 789]). That is, $\lambda_m/m^{2/d}$ goes to a constant in the limit as $m \to \infty$.

The dominant factor comes from the leading behavior, or leading term, in the asymptotic expansion of the logarithm of λ_m, and as such says nothing about whatever constant C might appear in a formula such as $\lambda_m = Cm^{2/d}[1 + o(1)]$ as $m \to \infty$. In fact, according to Weyl,

$$C = C_d(\Omega) = (2\pi)^2(v_d|\Omega|)^{-2/d} = \frac{(2\pi)^2}{(v_d|\Omega|)^{2/d}} = 4\pi\left(\frac{\Gamma(d/2+1)}{|\Omega|}\right)^{2/d}, \tag{8.100}$$

where $|\Omega|$ is the (Lebesgue) measure of Ω and v_d denotes the volume of the ball of unit radius in $\mathbb{R}^d$, given explicitly by $v_d = \pi^{d/2}/\Gamma(d/2+1)$ (with Γ denoting the usual factorial, or Γ, function).

For the various universal inequalities that have been developed, it is interesting to see how close they come to capturing the dominant factor in the growth of λ_m as $m \to \infty$ (that the constant C cannot be captured is clear from the fact that each of our universal eigenvalue inequalities is *homogeneous* in the λ_i's). If one substitutes the behavior $\lambda_m \sim cm^\alpha$ as $m \to \infty$ into (8.31), one will obtain constraints on the size of α, both from the fact that the discriminant must remain nonnegative and from the overall inequality, and something similar will occur if one substitutes this behavior into the other classical universal inequalities (by which is meant the PPW, HP, and Yang2 inequalities). Leaving aside the Hile-Protter inequality (which is somewhat intractable from this point of view), what one finds here is that the PPW inequality (8.6) does not limit α at all, that Yang2 (8.22) limits it by $\alpha \le 4/d$, and that Yang1 limits it by $\alpha \le 2/d$. Thus, as a bound the Yang1 inequality limits the growth of the λ_m as much as it possibly can, that is, it correctly captures the actual (Weyl) growth of the eigenvalues λ_m. This observation was first made by Yang [886] already in his 1991 Trieste preprint (see also [57] for further discussion).

This means that Yang's first inequality comes close to giving the Weyl asymptotics of the eigenvalues of operators such as the Dirichlet Laplacian. In fact, it comes as close as it can, in a sense, since as an upper bound, homogeneous in all the λ_i's, we cannot hope to establish the constant factor C from these bounds alone, nor can we expect to get more than upper bounds.

The fact that the Yang1 inequality captures the correct Weyl power-law growth of the eigenvalues led to much additional interest in this inequality, from researchers substantially beyond those already working in the field of universal eigenvalue inequalities. Before, since the growth constraints given by universal inequalities were not particularly tight at large indices, there had not been much interest in how these inequalities interacted with eigenvalue asymptotics. It had been understood that universal eigenvalue inequalities were perhaps best viewed as useful constraints on how the various low eigenvalues could be related to each other (i.e., they gave constraints on various combinations of low eigenvalues, and it seemed that these restrictions were the most interesting restrictions coming from the subject of universal inequalities). But now the situation had changed materially, attracting substantial additional interest.

Nevertheless, it was Hongcang Yang, together with Qing-Ming Cheng, who made the greatest strides in the field. In what amounted to a *tour de force*, in 2007 they proved

Theorem 8.8 (Cheng-Yang [266], 2007). *Let $\{\lambda_i\}_{i=1}^{\infty}$ be any sequence of positive real numbers satisfying*

$$\sum_{i=1}^{m} (\lambda_{m+1} - \lambda_i)^2 \le \frac{4}{d} \sum_{i=1}^{m} (\lambda_{m+1} - \lambda_i)\lambda_i. \tag{8.101}$$

Then for all $m \ge 1$ the λ_m's obey

$$\lambda_{m+1} < c\lambda_1 m^{2/d}, \tag{8.102}$$

where the constant c can be taken as $c = 1 + \frac{4}{d}$.

Remarks. (1) Apparently Cheng and Yang had established this theorem already by 2005, as the result is mentioned in [265]. See the remark at the bottom of p. 448 (which concerns a somewhat different problem, but, as far as establishing an explicit asymptotic result from the Yang1 inequality for the problem, the transition from one problem to the other is direct and straightforward; see, for example, our comments about shifting by M toward the end of Section 8.4).
(2) With a bit more work than we do here the constant c can be improved in various ways. In particular, Cheng and Yang [266] give the constant as $c = C_0(d, m) \le K_d \equiv j_{d/2,1}^2/j_{d/2-1,1}^2$, where $j_{\nu,k}$ denotes the kth positive zero of the Bessel function J_ν (see [4]). (It is known that $K_d < 1 + \frac{4}{d}$, see, for example, [62, 64]; in fact, K_d is the best upper bound for λ_2/λ_1 for $\Omega \subset \mathbb{R}^d$, with equality at the d-dimensional ball.) They also give further refinements in the later parts of their paper. Moreover, Chen and Zheng [262] have been able to push these ideas farther, obtaining some further refinements of the basic growth estimate.
(3) The main results of this section extend easily to cases where $4/d$ can be replaced by $4f/d$ (or any other positive constant, for that matter) as covered in Section 8.4 in our discussion of virial theory (or otherwise). One would just have to trace the constant through the computations. Moreover, the main results of this section also apply to cases where all eigenvalues have been shifted by the same amount; the constant giv-

ing the shift just has to be taken account of in the final results. And, of course, these two modifications can be combined to produce even more general results.

The proof of this theorem relies essentially on the following lemma.

Lemma 8.9 (Cheng-Yang, 2007). *Let* $\{\lambda_i\}_{i=1}^{\infty}$ *be any sequence of positive real numbers satisfying*

$$\sum_{i=1}^{m}(\lambda_{m+1}-\lambda_i)^2 \le \frac{4}{d}\sum_{i=1}^{m}(\lambda_{m+1}-\lambda_i)\lambda_i, \tag{8.103}$$

let

$$\Lambda_m \equiv \frac{1}{m}\sum_{i=1}^{m}\lambda_i, \tag{8.104}$$

$$T_m \equiv \frac{1}{m}\sum_{i=1}^{m}\lambda_i^2, \tag{8.105}$$

and define F_m *by*

$$F_m \equiv \left(1+\frac{2}{d}\right)\Lambda_m^2 - T_m. \tag{8.106}$$

It follows that

$$F_{m+1} < \left(\frac{m+1}{m}\right)^{4/d} F_m \qquad \text{for all} \qquad m \ge 1, \tag{8.107}$$

and thus that $F_m/m^{4/d}$ *is strictly decreasing in m for all* $m \ge 1$.

Remark. The definition $F_m \equiv (1+\frac{2}{d})\Lambda_m^2 - T_m$ represents an *inspired* choice, as it lends itself to working out the explicit asymptotic growth of λ_m in an efficient and relatively straightforward way. Nevertheless, the calculations involved are somewhat daunting.

Proof of Lemma 8.9. To get started, we first observe that all our F_m's are positive by (8.115) (below). From (8.115), we have

$$\left[\lambda_{m+1} - \left(1+\frac{2}{d}\right)\Lambda_m\right]^2 + \frac{2}{d}\left(1+\frac{2}{d}\right)\Lambda_m^2 \le \left(1+\frac{4}{d}\right)F_m, \tag{8.108}$$

and since the terms on the left are nonnegative, with $\frac{2}{d}(1+\frac{2}{d})\Lambda_m^2$ surely positive, the conclusion follows. We remark that this inequality is equivalent to Yang's first inequality, and can be used now, even though we will not explicitly derive it until later. See (8.115). In fact, we could get by here using only the fact that the discriminant of the quadratic in the Yang1 inequality (8.29) is nonnegative, since the quadratic has real roots. But we learn a bit more by completing the square and keeping the squared term, as in (8.108).

Note. Another relation from which the same conclusion follows is (8.131). There both terms on the left are nonnegative with the second clearly positive, so that F_m must be positive for all m. Again this is a reformulation of the Yang1 inequality, and may be regarded as known already at this stage in our proof.

That $F_m > 0$ for all $m \geq 1$ will be important for us, because later in our upper bound for F_{m+1} we will be replacing the coefficient of F_m by a larger quantity, and for that it would not do to have $F_m \leq 0$. As we proceed, we shall use this fact without comment.

Following Cheng and Yang, we set

$$p_{m+1} \equiv \Lambda_{m+1} - \left(1 + \frac{2}{d}\frac{1}{m+1}\right)\Lambda_m \tag{8.109}$$

and observe that this gives

$$\lambda_{m+1} = (m+1)\Lambda_{m+1} - m\Lambda_m = (m+1)\left[p_{m+1} + \left(1 + \frac{2}{d}\right)\frac{1}{m+1}\Lambda_m\right] \tag{8.110}$$

We may then compute (to understand our starting equation here, replace the F_m that ends the first line by its definition, $F_m = (1 + \frac{2}{d})\Lambda_m^2 - T_m$, and observe that what remains reduces to the definition of F_{m+1})

$$\begin{aligned} F_{m+1} &= \left(1 + \frac{2}{d}\right)\Lambda_{m+1}^2 - \frac{m}{m+1}\left(1 + \frac{2}{d}\right)\Lambda_m^2 - \frac{1}{m+1}\lambda_{m+1}^2 + \frac{m}{m+1}F_m \\ &= \left(1 + \frac{2}{d}\right)\left[p_{m+1} + \left(1 + \frac{2}{d}\frac{1}{m+1}\right)\Lambda_m\right]^2 - \frac{m}{m+1}\left(1 + \frac{2}{d}\right)\Lambda_m^2 \\ &\quad - (m+1)\left[p_{m+1} + \left(1 + \frac{2}{d}\right)\frac{1}{m+1}\Lambda_m\right]^2 + \frac{m}{m+1}F_m \\ &= \frac{m}{m+1}F_m - \left(m - \frac{2}{d}\right)p_{m+1}^2 + \frac{4}{d}\frac{1 + \frac{2}{d}}{m+1}p_{m+1}\Lambda_m \\ &\quad + \frac{2}{d}\frac{1 + \frac{2}{d}}{m+1}\left(1 + \frac{2}{d}\frac{1}{m+1}\right)\Lambda_m^2. \end{aligned} \tag{8.111}$$

At this point we would like to work (8.111) into a form where, if it were an equality, it would be a linear first order homogeneous recursion relation for the F_m's. In that form it would be easy to "solve" the recursion relation as an explicit inequality for F_m (or F_{m+1}) in terms of F_1 via iteration, and that is what we aim to do here. As we proceed we shall use the language of linear difference equations even though what we deal with will be a linear difference inequality. Because of the close connection between the two, the extended use of this terminology will lend unity to the discussion and should assist with understanding.

On the right-hand side of (8.111) there are terms in F_m (which we want to keep), p_{m+1}^2, $p_{m+1}\Lambda_m$, and Λ_m^2, and we would like to get free of the last three of these using inequalities (replacing them, ultimately, with terms in F_m). We would be prepared to just drop terms that are negative, but we would rather not do this for nothing, but instead would like to use such terms to help cancel positive terms. And, ultimately, if we hope to achieve a good bound, we should try to keep the bound on F_{m+1} (i.e., the right-hand side of (8.111) as we transform it) as small as possible. We shall see that we will almost be able to do this in an optimal way as well. To transform the right-hand side without costing us too much, the first idea we use is to package the term in $p_{m+1}\Lambda_m$ as the middle term in a binomial-squared, with the binomial built as a difference of a term in p_{m+1} and Λ_m and including a free parameter. Because $p_{m+1}\Lambda_m$ comes with a positive coefficient, this will allow us to subtract the binomial-squared (and later drop it, at not too much cost), at the expense of having to add terms in p_{m+1}^2 and Λ_m^2, which we will need to control in some way. But at least we will have fewer different terms to control. Also, since the term in the original inequality in p_{m+1}^2 comes with a negative coefficient, that will help us as we proceed. On the other hand, since the term in Λ_m^2 comes with a positive coefficient, we need to work to control it in some way in terms of F_m. In fact, as we shall see, we will exploit an interplay between such terms that helps us to handle them together.

We now get down to business with the calculations. First, using the binomial-squared idea, we substitute away the $2p_{m+1}\Lambda_m$ as $-(\alpha p_{m+1} - \frac{1}{\alpha}\Lambda_m)^2 + \alpha^2 p_{m+1}^2 + \frac{1}{\alpha^2}\Lambda_m^2$, where α is any positive number. We have, for $\alpha > 0$,

$$\begin{aligned} F_{m+1} &= \frac{m}{m+1}F_m + \left(\frac{2}{d}\frac{1+\frac{2}{d}}{m+1}\alpha^2 - m + \frac{2}{d}\right)p_{m+1}^2 - \frac{2}{d}\frac{1+\frac{2}{d}}{m+1}\left(\alpha p_{m+1} - \frac{1}{\alpha}\Lambda_m\right)^2 \\ &+ \frac{2}{d}\frac{1+\frac{2}{d}}{m+1}\left(\frac{1}{\alpha^2} + 1 + \frac{2}{d}\frac{1}{m+1}\right)\Lambda_m^2 \\ &\le \frac{m}{m+1}F_m + \left(\frac{2}{d}\frac{1+\frac{2}{d}}{m+1}\alpha^2 - m + \frac{2}{d}\right)p_{m+1}^2 \\ &+ \frac{2}{d}\frac{1+\frac{2}{d}}{m+1}\left(\frac{1}{\alpha^2} + 1 + \frac{2}{d}\frac{1}{m+1}\right)\Lambda_m^2. \end{aligned} \tag{8.112}$$

We have one other useful inequality at our disposal, and it is crucial. This is the inequality

$$0 \le -(m+1)^2 p_{m+1}^2 - \frac{2}{d}\left(1 + \frac{2}{d}\right)\Lambda_m^2 + \left(1 + \frac{4}{d}\right)F_m, \tag{8.113}$$

which is essentially just the Yang1 inequality. To see this, we recall that Yang1 reads

$$\lambda_{m+1}^2 - 2\left(1 + \frac{2}{d}\right)\Lambda_m\lambda_{m+1} + \left(1 + \frac{4}{d}\right)T_m \le 0, \tag{8.114}$$

and, if we complete the square in the first two terms and rearrange,

$$\begin{aligned} \left[\lambda_{m+1} - \left(1 + \frac{2}{d}\right)\Lambda_m\right]^2 &\le \left(1 + \frac{2}{d}\right)^2\Lambda_m^2 - \left(1 + \frac{4}{d}\right)T_m \\ &= \left(1 + \frac{4}{d}\right)F_m - \frac{2}{d}\left(1 + \frac{2}{d}\right)\Lambda_m^2 \end{aligned} \tag{8.115}$$

since $F_m \equiv (1+\frac{2}{d})\Lambda_m^2 - T_m$. The last two terms here are the last two terms of (8.113), and thus all that remains is to observe that $[\lambda_{m+1} - (1+\frac{2}{d})\Lambda_m]^2$ is the same as $(m+1)^2 p_{m+1}^2$, which follows from $(m+1)p_{m+1} = \lambda_{m+1} - (1+\frac{2}{d})\Lambda_m$. To see this, we simply observe that the last expression in (8.110) tells us that

$$\lambda_{m+1} = (m+1)\left[p_{m+1} + \left(1 + \frac{2}{d}\right)\frac{1}{m+1}\Lambda_m\right] = (m+1)p_{m+1} + \left(1 + \frac{2}{d}\right)\Lambda_m, \tag{8.116}$$

which is the identity we want.

We will now use (8.113) by taking an appropriate (positive) multiple of it and adding that to our main inequality for F_{m+1}. In order to eliminate the terms in Λ_m^2 we choose the factor to be

$$\frac{1}{m+1}\left(1+\frac{2}{d}\frac{1}{m+1}+\frac{1}{\alpha^2}\right). \tag{8.117}$$

This yields

$$\begin{aligned} F_{m+1} &\leq \left(1+\frac{4}{d}\frac{1}{m+1}+\frac{2}{d}\frac{1+\frac{4}{d}}{(m+1)^2}+\frac{1+\frac{4}{d}}{m+1}\frac{1}{\alpha^2}\right)F_m \\ &\quad+\left(\frac{2}{d}\frac{1+\frac{2}{d}}{m+1}\alpha^2+\frac{2}{d}-m-\left((m+1)+\frac{2}{d}+\frac{m+1}{\alpha^2}\right)\right)p_{m+1}^2 \\ &=\left(1+\frac{4}{d}\frac{1}{m+1}+\frac{2}{d}\frac{1+\frac{4}{d}}{(m+1)^2}+\frac{1+\frac{4}{d}}{m+1}\frac{1}{\alpha^2}\right)F_m \\ &\quad+\left(\frac{2}{d}\frac{1+\frac{2}{d}}{m+1}\alpha^2-(2m+1)-\frac{m+1}{\alpha^2}\right)p_{m+1}^2 \end{aligned} \tag{8.118}$$

Finally, we would like to eliminate the terms in p_{m+1}^2, and we can do that by an appropriate choice of the parameter α ($\alpha > 0$).

Now the coefficient of p_{m+1}^2 increases with increasing α for $\alpha > 0$, ranging from $-\infty$ to $+\infty$ as α ranges from 0 to ∞. Also, the coefficient of F_m decreases with increasing α, so that the optimal choice of α at this point would be to take it as the value where the coefficient of p_{m+1}^2 is 0. We could do that, but it would complicate our further calculations, so we make a choice that's almost as good, and causes no significant loss asymptotically (at large m). The optimal α^2 would be

$$\alpha^2=\frac{(2m+1)+\sqrt{(2m+1)^2+\frac{8}{d}(1+\frac{2}{d})}}{\frac{4}{d}\frac{1+\frac{2}{d}}{m+1}}, \tag{8.119}$$

and obviously if we ignore the final $\frac{8}{d}(1+\frac{2}{d})$ under the radical we have a slightly smaller value of α^2, which also serves our purposes, giving

$$\alpha^2=\frac{2m+1}{\frac{2}{d}\frac{1+\frac{2}{d}}{m+1}}, \tag{8.120}$$

or, equivalently,

$$\alpha^{-2}=\frac{\frac{2}{d}(1+\frac{2}{d})}{(m+1)(2m+1)}. \tag{8.121}$$

Using this choice of α, we have

$$\begin{aligned} F_{m+1} &\leq \left(1+\frac{4}{d}\frac{1}{m+1}+\frac{2}{d}\frac{1+\frac{4}{d}}{(m+1)^2}+\frac{1+\frac{4}{d}}{m+1}\frac{\frac{2}{d}(1+\frac{2}{d})}{(m+1)(2m+1)}\right)F_m \\ &\quad-\left((m+1)\frac{\frac{2}{d}(1+\frac{2}{d})}{(m+1)(2m+1)}\right)p_{m+1}^2 \end{aligned}$$

$$= \left(1 + \frac{4}{d}\frac{1}{m+1} + \frac{2}{d}\frac{1+\frac{4}{d}}{(m+1)^2} + \frac{\frac{2}{d}(1+\frac{2}{d})(1+\frac{4}{d})}{(m+1)^2(2m+1)}\right)F_m - \frac{\frac{2}{d}(1+\frac{2}{d})}{2m+1}p_{m+1}^2$$
$$\leq \left(1 + \frac{4}{d}\frac{1}{m+1} + \frac{2}{d}\frac{1+\frac{4}{d}}{(m+1)^2} + \frac{\frac{2}{d}(1+\frac{2}{d})(1+\frac{4}{d})}{(m+1)^2(2m+1)}\right)F_m. \quad (8.122)$$

This, finally, is the homogeneous difference inequality for F_m that we sought. That this is enough to prove our lemma follows from the following computation (first done by Cheng and Yang [266]; see p. 165):

$$\left(\frac{m+1}{m}\right)^{4/d} = \left(1 - \frac{1}{m+1}\right)^{-4/d}$$
$$= 1 + \frac{4}{d}\frac{1}{m+1} + \frac{1}{2}\frac{4}{d}\frac{1+\frac{4}{d}}{(m+1)^2} + \frac{1}{6}\frac{4}{d}\frac{(1+\frac{4}{d})(2+\frac{4}{d})}{(m+1)^3}$$
$$+ \frac{1}{24}\frac{4}{d}\frac{(1+\frac{4}{d})(2+\frac{4}{d})(3+\frac{4}{d})}{(m+1)^4} + \cdots$$
$$\geq 1 + \frac{4}{d}\frac{1}{m+1} + \frac{1}{2}\frac{4}{d}\frac{1+\frac{4}{d}}{(m+1)^2} + \frac{1}{3}\frac{4}{d}\frac{(1+\frac{4}{d})(1+\frac{2}{d})}{(m+1)^3}$$
$$+ \frac{1}{4}\frac{4}{d}\frac{(1+\frac{4}{d})(1+\frac{2}{d})}{(m+1)^4}, \quad (8.123)$$

where in the last step we have made an obvious simplification in the last displayed term.

Since the first three terms on the right here are the first three terms of the coefficient of F_m in (8.122) we can use the inequality above to write

$$F_{m+1} \leq \left(\left(\frac{m+1}{m}\right)^{4/d} - \frac{1}{3}\frac{4}{d}\frac{(1+\frac{4}{d})(1+\frac{2}{d})}{(m+1)^3} - \frac{1}{d}\frac{(1+\frac{4}{d})(1+\frac{2}{d})}{(m+1)^4}\right.$$
$$\left. + \frac{\frac{2}{d}(1+\frac{2}{d})(1+\frac{4}{d})}{(m+1)^2(2m+1)}\right)F_m$$
$$= \left(\left(\frac{m+1}{m}\right)^{4/d} - \frac{2}{3n}\frac{(1+\frac{2}{d})(1+\frac{4}{d})(m-1)}{(2m+1)(m+1)^3} - \frac{1}{d}\frac{(1+\frac{4}{d})(1+\frac{2}{d})}{(m+1)^4}\right)F_m$$
$$< \left(\frac{m+1}{m}\right)^{4/d} F_m \text{ for all } m \geq 1. \quad (8.124)$$

This is a nice simple linear difference inequality for the F_m's, which we can now solve.

Note that Cheng and Yang do a little more here, estimating the final two terms in the second to last expression in (8.124) and thus obtaining the last expression in (8.124) in the form

$$F_{m+1} \leq C(d,m)\left(\frac{m+1}{m}\right)^{4/d} F_m, \quad (8.125)$$

where

$$C(d,m) = \left[1 - \frac{1}{3n}\left(\frac{m}{m+1}\right)^{4/d}\frac{(1+\frac{2}{d})(1+\frac{4}{d})}{(m+1)^3}\right] < 1 \quad (8.126)$$

This expression for $C(d,m)$ is obtained by combining the final two terms in the second to last expression in (8.124) and using the estimate $\frac{2(m-1)}{2m+1} + \frac{3}{m+1} = \frac{2m^2+6m+1}{(m+1)(2m+1)} =$

$\frac{2m^2+6m+1}{2m^2+3m+1} > 1$ for all $m \geq 1$. We forgo any more detailed analysis here in the interest of finishing our calculation in as simple a way as possible. Further refinements along the lines of what Cheng and Yang do are certainly possible, but here we choose simply not to delve into these matters.

By iteration of the inequality one easily sees that

$$F_{m+1} < F_1(m+1)^{4/d}, \tag{8.127}$$

or

$$F_m < F_1 m^{4/d}, \tag{8.128}$$

and this is the explicit bound on F_m for all $m \geq 1$ that has been our objective throughout.

We also observe from (8.124) that

$$F_{m+1} < \left(\frac{m+1}{m}\right)^{4/d} F_m \qquad \text{for all} \qquad m \geq 1, \tag{8.129}$$

which shows immediately that $F_m/m^{4/d}$ is decreasing in m for $m \geq 1$, and concluding the proof of Lemma 8.9.

Proof of Theorem 8.8. To finish the argument for Theorem 8.8, one needs to manipulate the Yang1 inequality once more, so as to see how to bound λ_{m+1} in terms of F_m. Recalling (8.115), we have

$$\left[\lambda_{m+1} - \left(1+\frac{2}{d}\right)\Lambda_m\right]^2 \leq \left(1+\frac{4}{d}\right)F_m - \frac{2}{d}\left(1+\frac{2}{d}\right)\Lambda_m^2. \tag{8.130}$$

To proceed, we multiply out the left-hand side, transpose the Λ_m^2 term from the right, and complete the square in the terms in Λ_m^2 and $\Lambda_m \lambda_{m+1}$. This leaves us with two positive terms on the left, bounded by $(1+\frac{4}{d})F_m$. In detail, we have

$$\left(1+\frac{2}{d}\right)\left[\left(1+\frac{4}{d}\right)\Lambda_m - \lambda_{m+1}\right]^2 + \frac{2}{d}\lambda_{m+1}^2 \leq \left(1+\frac{4}{d}\right)^2 F_m, \tag{8.131}$$

having moved a factor of $(1+\frac{4}{d})$ to the right.

Since both terms on the left are nonnegative (if the λ_i's form a sequence of Dirichlet eigenvalues, then the first term is positive as well, by virtue of the Yang2 inequality) each is bounded by the right member of the inequality, and it follows that

$$\lambda_{m+1}^2 < \frac{d}{2}\left(1+\frac{4}{d}\right)^2 F_m, \tag{8.132}$$

or, in terms of λ_{m+1},

$$\lambda_{m+1} < \sqrt{\frac{d}{2}}\left(1+\frac{4}{d}\right)\sqrt{F_m}. \tag{8.133}$$

Now by Lemma 8.9 and observing that $F_1 = \frac{2}{d}\lambda_1^2$ from (8.106), we have

$$\lambda_{m+1} < \sqrt{\frac{d}{2}}\left(1+\frac{4}{d}\right)\sqrt{F_1}\,m^{2/d} = \left(1+\frac{4}{d}\right)\lambda_1\, m^{2/d}, \tag{8.134}$$

which is the conclusion of the theorem of Cheng and Yang [266].

Improvements to Cheng and Yang's theorem are possible, since, as already remarked, there are various places where one could hope to improve the estimates that go into their proof. Suffice it to say here that the most leeway for making improvements occurs for low values of m. There it is well-known that the various inequalities we have used (i.e., the Yang1 bound at small values of m) can be improved (see our notes in Section 8.6 for more in this direction). Thus, what one needs to do is to use the general iteration scheme given above, obtaining an upper bound for F_m (for $m \geq k$) in terms of some F_k for a relatively small k, and then do the rest "by hand", using the more specialized (and sharpened) inequalities that are available for low values of k. One obtains an inequality

$$F_m < \left[\frac{F_k}{k^{4/d}}\right] m^{4/d} \qquad \text{for all} \qquad m > k \tag{8.135}$$

(with equality at $m = k$), and then by using the arguments above to get back λ_{m+1} and the better inequalities for F_k one can get the desired improvements. For example, to finish from $k = 2$ one would use the fact that

$$\begin{aligned} F_2 &\leq \left[\frac{d}{4}\left(1 + \frac{2}{d}\right)K_d - \frac{d}{8}\left(1 - \frac{2}{d}\right)(1 + K_d^2)\right] F_1, \\ &= \left[\frac{1}{2}\left(1 + \frac{2}{d}\right)K_d - \frac{1}{4}\left(1 - \frac{2}{d}\right)(1 + K_d^2)\right] \lambda_1^2, \end{aligned} \tag{8.136}$$

since $F_1 = \frac{2}{d}\lambda_1^2$, where K_d is a ratio of squares of zeros of Bessel functions, as given following Theorem 8.8. In particular, for $d = 2$ this simplifies considerably and leads to

$$\begin{aligned} \lambda_{m+1}^2 &< 9\,F_m \\ &\leq 9\,F_2\left(\frac{m}{2}\right)^2 \\ &\leq 9\,K_2\lambda_1^2\left(\frac{m}{2}\right)^2 \qquad \text{for all} \qquad m \geq 2. \end{aligned} \tag{8.137}$$

Since this gives

$$\lambda_{m+1} < \frac{3}{2}\sqrt{K_2}\,\lambda_1\,m \qquad \text{for all} \qquad m \geq 2, \tag{8.138}$$

and $3/2 < 1.593 \approx \sqrt{K_2}$ while $\lambda_2/\lambda_1 \leq K_2$ (with equality at the disk), it follows that

$$\lambda_{m+1} \leq K_2\,\lambda_1\,m \qquad \text{for all} \qquad m \geq 1, \tag{8.139}$$

which is the $d = 2$ case of the Cheng-Yang explicit growth estimate with constant $c = K_2 \approx 2.5387$. Of course, not all of this is yet known in detail, and there could be more improvements as the story develops.

Before closing this section we discuss one further implication of what we have done above. As observed in Lemma 8.9 (or see (8.135)), $F_m/m^{4/d}$ is strictly decreasing in m for all $m \geq 1$. We can make use of this result in two ways: (1) To bound F_m for all sufficiently large m in terms of lower-indexed F_m's, as we have done above. In particular, one might typically obtain bounds going back to F_1, F_2, or some other F_k for a relatively low value of k, and ultimately to λ_1^2 (the last parts accomplished by specialized and strengthened low-order inequalities, if we stop short of F_1 in using the general result). This leads to explicit *upper bounds* for λ_m/λ_1 with sharp order of growth when carried out in detail.

(2) To get explicit *semi-classical lower bounds* for quantities such as F_k, Λ_k, T_k, and even λ_k. Since $F_m/m^{4/d}$ is strictly decreasing, $F_k/k^{4/d} > F_m/m^{4/d}$ for $k < m$, and, because the power of m in the denominator on the right is asymptotically correct (in the sense of Weyl growth), we can take the limit as $m \to \infty$ and obtain a positive constant on the right, with the constant determined appropriately in terms of the Weyl asymptotics. This constant is called a *semi-classical constant* and comes directly from the Weyl growth law (only modified somewhat from the more usual because we start with F_m and not λ_m or Λ_m). If carried out in detail, one arrives at lower bounds for quantities such as Λ_k in terms of the semi-classical constant, with dimensional factors, which have the form of the Li-Yau-Berezin inequality for Λ_k. These have the correct order of growth in k, but probably the multiplicative factor is not sharp (whereas it is in the Li-Yau-Berezin lower bound for Λ_k). Because of that, we do not explore the further consequences and ramifications of (8.135) here. In fact, these are probably best explored by exploiting the Yang1 inequality (8.21) directly using Legendre transforms and Riesz means, as employed by Hermi [512] and Harrell and Hermi [457, 458]. See also our comments near the end of Section 8.7.

8.6 Further Work

There is a large body of further work concerned with the low eigenvalues of the Dirichlet Laplacian and related operators. Here we concentrate only on those results that concern the Dirichlet Laplacian on a bounded domain in Euclidean space. First, beginning with the work of Payne, Pólya, and Weinberger [746, 751] many people have worked to find upper bounds on the ratio λ_2/λ_1, especially for the 2-dimensional case. PPW had obtained the bound 3, and they conjectured that the optimal upper bound was the value of λ_2/λ_1 taken on the disk, approximately 2.5387 (this value is given exactly by the ratio of squares of Bessel function zeros, $j_{1,1}^2/j_{0,1}^2$, where $j_{\nu,k}$ denotes the kth positive zero of the Bessel function J_ν; we follow Abramowitz and Stegun [4] in our notation for Bessel functions and their zeros). Improvements were subsequently made by Brands [177] who obtained 2.686, de Vries [333] who obtained 2.658, and Chiti [272] who obtained 2.586. Also, Thompson, in [851], gave the explicit statement of some of the d-dimensional results that follow straightforwardly from the methods and work of Payne, Pólya, and Weinberger, and he also gave the d-dimensional generalization of their conjecture concerning λ_2/λ_1. This line of development culminated in the work of Ashbaugh and Benguria, who proved that the sharp bound for λ_2/λ_1 was indeed its value for the disk/ball in Euclidean space of any dimension [62–64]. Thus, for $\Omega \subset \mathbb{R}^d$, one has $\lambda_2/\lambda_1 \le j_{d/2,1}^2/j_{d/2-1,1}^2$. The proof was achieved via rearrangement results, building in part on the earlier work of Chiti (see [272] and his foregoing papers) and Talenti [843, 846].

Ashbaugh and Benguria were then able to show that (again in general dimension d) $\lambda_4/\lambda_2 < j_{d/2,1}^2/j_{d/2-1,1}^2$, which is sharp in the limiting case of two equal disks/balls (among other possibilities). This established the next two cases of the conjecture of PPW [746] that λ_{m+1}/λ_m is always bounded above by the value of λ_2/λ_1 at the disk/ball.

Other eigenvalue ratios, or combinations of eigenvalue ratios, have also been considered, especially in dimension 2 (though certainly such questions could be considered in any dimension, and would hold appreciable interest). For example, a number of authors have considered (in 2 dimensions) λ_3/λ_1, λ_4/λ_1, and $(\lambda_2+\lambda_3)/\lambda_1$. Consideration of λ_3/λ_1 was made by PPW [746], and later by Brands [177], Hile-Protter [522], Marcellini [691], and perhaps others. Later considerations are due to Yang [886], Ashbaugh and Benguria [66, 69, 71], with the most recent improvements being in [262]. For a discussion of what's known for λ_3/λ_1 one can consult Section 13 of [73]. We have the impression that improvements will still be made on a regular basis, as the techniques used thus far are not yet exhausted. Indeed, there may be more improvements in the near future by some of the authors listed above.

Another recent paper worthy of mention is [263], by Chen, Zheng, and Yang, where attention is focused on obtaining good upper bounds for the gap $\lambda_{m+1} - \lambda_m$.

One of the interesting developments in this low eigenvalue work has been the study of the *range of values* of λ_2/λ_1 and λ_3/λ_1 for planar domains. Here one attempts to understand the region in the $(\lambda_2/\lambda_1, \lambda_3/\lambda_1)$-plane that could be inhabited by the eigenvalue ratios of some domain Ω. This work was initiated in [69] and continued in [71], with valuable inputs from Yang [886]. The full picture of the range of values was looked at from a variety of perspectives by Levitin and Yagudin [655], who did extensive numerical studies to try to determine the actual range of values (most other workers in the field had worked to put bounds on the possibilities for how large the range of values could be, but had not looked so hard (or had fewer results) for the range that could be realized). Obviously this work has much to say about the bounds we can get on λ_3/λ_1, $(\lambda_2+\lambda_3)/\lambda_1$, $(\lambda_3-\lambda_2)/\lambda_1$, and the like.

The initial results of PPW on $(\lambda_2+\lambda_3)/\lambda_1$ (in 2 dimensions) have been generalized and extended in various ways, including to higher dimensions. In fact, it is straightforward to see that the bound generalizes to $(\lambda_2+\lambda_3+\cdots\lambda_{d+1})/\lambda_1 \le d+4$ for $\Omega \subset \mathbb{R}^d$. Brands obtained an improvement of this (with right-hand side $5+\lambda_1/\lambda_2$ in two dimensions), and this generalizes nicely to d dimensions, yielding the bound

$(\lambda_2+\lambda_3+\cdots\lambda_{d+1})/\lambda_1 \le d+3+\lambda_1/\lambda_2$. This generalization is proved in [66], but certainly Hile and Protter [522] were aware of it earlier.

We note that the constant in $(\lambda_2+\lambda_3+\cdots\lambda_{d+1})/\lambda_1 \le d+4$ is $d+4 = d(1+\frac{4}{d})$, so that the quantity $(1+\frac{4}{d})$ is still with us even here. Thus, this inequality is, in a sense, closely related to the general inequality of PPW and its successors, which characteristically contain the combination $1 + \frac{4}{d}$, or at least $\frac{4}{d}$.

There is also an extension (or improvement) in a different direction, by Levitin and Parnovski [652], to $(\lambda_{\ell+1}+\cdots+\lambda_{\ell+d})/\lambda_\ell \le d+4$ for ℓ arbitrary, $\ell = 1, 2, 3, \ldots$. We do not know how widely this inequality has become known, but thus far, to our knowledge, it has not figured in any other considerations of universal eigenvalue inequalities, even those involving finding explicit upper bounds on λ_m (or λ_{m+1}) in terms of λ_1 and other more accessible factors. Note that from this inequality we can obtain $d\,\lambda_{\ell+1}/\lambda_\ell \le d + 4$, i.e., the inequality $\lambda_{\ell+1}/\lambda_\ell \le 1 + 4/d$ for all ℓ, which also follows from the PPW inequality (8.6).

As realized by Levitin and Parnovski [652], there are connections between the standard universal eigenvalue inequalities of PPW, HP, and Yang and the *sum rules* of quantum mechanics, which were established in the early days by some of the pioneers in the field. The two best-known sum rules are those of Thomas-Reiche-Kuhn and of Bethe. References for these, and a generalization, can be found in a paper of S. Wang [867] which appeared in Phys. Rev. A and can also be found on the internet. Other references for sum rules include the book, **Intermediate Quantum Mechanics** by Bethe and Jackiw [144], and J. J. Sakurai's **Modern Quantum Mechanics** [797]. Some of what is known on sum rules has fed back into developments on universal eigenvalue inequalities; we cite here some of the more recent work of Harrell and Hermi [458] and of Harrell and Stubbe [465].

There has been considerable interest in extending the technology that gives universal eigenvalue inequalities for the Dirichlet Laplacian to operators with other boundary conditions, e.g., the Laplacian with Neumann boundary conditions. However, efforts in that direction have so far proved disappointing, mainly due to the considerable extra problems caused by boundary conditions other than Dirichlet. In the abstract setting these problems concern the domains of the various operators that one considers. In particular, nothing approaching the sort of capture of the leading behavior of the Weyl asymptotics (i.e., the *dominant factor*) as is done by Yang's inequality has been achieved for other boundary conditions. Indeed, one would be quite happy to have a general inequality of the nature of the PPW or HP inequalities for such problems. Work on such problems occurs in the articles of Harrell and Michel [461], and of Levitin and Parnovski [652], who built on the work of Harrell and Michel, and others. A recent preprint of Funano [410] also treats ratios of eigenvalues of the Neumann Laplacian.

There are other types of "universal eigenvalue inequalities" that we have not even touched upon in this chapter. The largest oversight might be that we have not said anything regarding inequalities between the Dirichlet and Neumann eigenvalues of the

Laplacian. These inequalities have a storied history of their own, with the key players being Payne, Levine and Weinberger, Aviles, L. Friedlander, R. Mazzeo, and N. Filonov. A review of the main developments in this area occurs as Section 10 in [73]. A MathSciNet search on "Payne's inequality" or "Payne's conjecture" (or on any of these authors) should turn up the relevant references, and probably some more recent ones as well (see, for example, a relatively recent paper by F. Gesztesy and M. Mitrea; that paper will have all, or almost all, of the relevant references that predate it).

Other kinds of universal eigenvalue inequalities, such as those involving the inradius (defined as the radius of the largest inscribed disk/ball), are discussed in Section 12 of [73]. In particular, this takes you into interesting territory involving the bass note of a drum.

There are also bounds for eigenvalues, and combinations of eigenvalues, involving the *inner conformal radius* of the domain. This work is associated with the names of Pólya, Szegő, Hersch, and others. See, for example, [770]. Some of the later references will be found in [66], Bandle's book [97], or in the paper of Laugesen and Morpurgo [641].

In addition to the books by Henrot [505] and Bandle [97], other general references on the subject of universal and isoperimetric inequalities for eigenvalues are [588, 744, 753, 770]. A few other classics in the field, that we would be remiss if we did not mention, include Weyl's inspiring paper "Ramifications, old and new, of the eigenvalue problem" [877], Kac's seminal paper [567] on "Can one hear the shape of a drum?", the monograph of Pólya and Szegő [770] **Isoperimetric Inequalities of Mathematical Physics**, and Lieb's "The number of bound states of one-body Schrödinger operators and the Weyl problem" [662]. There are also a number of very inspiring writings on the subject by Pólya, many of which can be found in his collected works; see, in particular, **George Pólya: Collected Papers, Vol. III: Analysis** [771]. And finally, Chavel's book [261], **Eigenvalues in Riemannian Geometry**, gives an excellent treatment of eigenvalue problems in the context of Riemannian geometry (and thus for the Laplace-Beltrami operator, which generalizes the Laplacian in that setting).

There is by now an extensive literature of universal eigenvalue inequalities that have been developed for domains on manifolds, or for compact manifolds. Such work began with work of S. Y. Cheng, M. Maeda, P. C. Yang and S.-T. Yau, P. Li, P.-F. Leung, and others, and continued with works of Harrell [456], Ashbaugh and Benguria [68], Harrell and Michel [460], Ashbaugh [55, 57], Hermi [511], and others. In more recent times there have been many works by H. C. Yang and Q.-M. Cheng, Q. Wang and C. Xia, and of their students and collaborators. For some of the more recent literature in this vein, one can search MathSciNet for those names, or check the bibliography of Xia's book [884] (which can be found online).

Many people have considered the generalization of the standard universal eigenvalue inequalities to the case of uniformly elliptic second order partial differential operators. Special cases appear in the geometric settings mentioned above, but Allegretto [23] was most likely the first to realize that these inequalities had something to say about general second order elliptic operators. Also, Ashbaugh and Benguria [64] (see also [68]) developed inequalities that hold for uniformly elliptic second order operators in the context of their work on λ_2/λ_1.

Regarding higher order differential operators, already in 1955-56 Payne, Pólya, and Weinberger considered two higher order problems, the eigenvalues of the vibrating clamped plate, and the eigenvalues for the buckling of a clamped plate. Each of these problems is 4th order, involving at highest order, the square of the Laplacian, called the *biharmonic operator*. In their 1956 paper they obtained results for the vibrating clamped plate that are roughly analogous to what they had done for the eigenvalues of the vibrating membrane (with fixed edges; this corresponds to the Dirichlet Laplacian). However, they were less successful with the buckling problem, and only presented an upper bound for the ratio of the first two eigenvalues. Also, they only presented their results for dimension 2, but essentially everything they did can be generalized to higher (Euclidean) dimensions in a straightforward way (except, perhaps, for their work on the buckling problem; cf. [58]).

In the 1980's there was work by Hile and Yeh [524] and several papers by Zu Chi Chen, sometimes with Chun Lin Qian or others (with their work continuing into the 1990's). Hile and Yeh considered the vibrating clamped plate and buckling problems especially, while Chen and collaborators looked at powers of the Laplacian, polynomials in the Laplacian, and other such higher order problems. At around the same time Hook [535–537] was also looking at certain higher order cases, including powers of and polynomials in, the Laplacian. Most of the literature up to the early 1990's is cited in [64, 68]. On the "classical side", work on the higher order problems on which PPW initiated the direct study, beyond Hile and Yeh one finds [55, 58, 702, 703, 856] working mainly on low eigenvalues (as did Hile and Yeh), and Cheng and Yang working to establish general Yang-type inequalities for the vibrating clamped plate and the buckling of a clamped plate. Hile and Yeh had also done work on improved general inequalities. For discussion of the known inequalities that preceded Cheng and Yang see [55]. There, for example, improvements on what PPW and Hile-Yeh did for the problem of the vibrations of a clamped plate are noted, in particular, a result found independently by Hook, and by Chen and Qian, in 1990.

In the 2000's the literature on the subject has proliferated to the point where it would be prohibitive to give a complete listing here; suffice it to say that one should search especially on such names as Q.-M. Cheng, H. Yang, Q. Wang, C. Xia, and their collaborators (in addition to many of the names mentioned elsewhere in this chapter). The book [884] of Changyu Xia contains many of the relevant references.

We also mention that in recent years interesting connections between the buckling problem and the Krein (or Krein-von Neumann) Laplacian have emerged [77–79]. The Krein Laplacian is the minimal nonnegative operator associated with the formal Laplace operator, $-\Delta$. It has, aside from 0, a discrete spectrum, but its eigenfunctions obey nonlocal boundary conditions. In fact, on the orthogonal complement of its kernel (which is infinite dimensional unless $d = 1$), it is unitarily equivalent to the clamped buckling problem in a sense made precise by [77]. Discussion of the eigenvalue asymptotics for the nonzero eigenvalues of the Krein Laplacian (i.e., the buckling eigenvalues of the corresponding clamped plate) can be found in [76, 418].

In recent years there has been interest in proving *stability results* for various of the known isoperimetric inequalities for eigenvalues. These are certainly interesting and technically quite challenging. We will not enter into a discussion of them here, since they are discussed in Chapter 7 of this book.

There are a variety of one-dimensional results focused on eigenvalues or, especially, eigenvalue ratios. These are associated with the names of M. G. Krein, Mahar and Willner, J. Keller, Gentry and Banks, D. O. Banks, E. R. Barnes, and D. C. Barnes. Most of the works of these authors could be traced through the bibliography in [64], or, better, [505]. Later contributors include Ashbaugh and Benguria, Y. L. Huang and C. K. Law, M. J. Huang, M. Horvath, M. Horvath and M. Kiss, and others.

8.7 History

The original proof of the Hile-Protter inequality was somewhat complicated, since it involved the introduction of certain parameters and optimization with respect to them (or, at least, one of them, after simplifying assumptions had been made). The proof was given in a simpler form in [68]. Later this simplified form was incorporated into a proof of the Yang inequality (much as we did in this chapter; see below for more on this). Also, proofs have been given in the context of the "algebraized" approach to universal eigenvalue inequalities discussed below.

Hongcang Yang originally presented his inequality (Yang1) in a Trieste preprint [886], which circulated in that form for some years. An updated version appeared in 1995, but again circulated only as a preprint. The original proof by Yang was finally published as a part of [266] (see the Appendix, pp. 172–174). A more direct version of this proof had been given in the meantime by Ashbaugh in [55, 57] (here we have followed essentially the same development of the proofs as occur in these two papers). In that proof one sees directly how the use of the counter-terms involving the $b_{ij}^{(\ell)}$'s come into play allowing the "optimal" use of the Cauchy-Schwarz inequality. A foreshadowing of the optimal use of the Cauchy-Schwarz inequality, but only in the context of the first three eigenvalues, occurs already in [71]. Additionally, proofs of the Yang inequality have been given in the context of the algebraic approach to such inequalities, as discussed below.

To our knowledge, the Chebyshev inequality was first used in the context of universal eigenvalue inequalities by Hermi in his Ph.D. dissertation (see [511], and the later paper [83]). Hermi's usage was similar to our treatment in Section 8.4, though he began by using the unweighted Chebyshev inequality to compare the main inequalities, and only later dealt with the cases where the parameter p is introduced in the exponent.

It was realized fairly early on, by E. B. Davies in discussions with Evans Harrell, and then by others, that much of the PPW argument (and subsequent arguments) could be framed in an algebraic setting. These ideas first appeared in print in [452], and only for the PPW inequality (8.6) (see also [456]). Later Hook [535] and Harrell and Michel [460, 461] were able to extend this approach so as to obtain the Hile-Protter inequality. In light of Yang's contribution in 1991, the algebraized approach to universal eigenvalue inequalities was then continued by several authors, including Harrell and Stubbe [463], Hermi [511] (see also [83]), Levitin and Parnovski [652], and others. Harrell and Stubbe introduced a non-variational approach to proving Yang's inequalities, which has led to various further developments. Levitin and Parnovski [652] also followed this approach to obtaining Yang's inequality, while making several useful observations and extensions. With Levitin and Parnovski the connection of the universal eigenvalue inequalities of PPW, HP, and Yang to the "sum rules" explored in the early days of quantum mechanics was made explicit (the connection was made based on a comment from Barry Simon). Thus the algebraization of the treatment of universal eigenvalue inequalities relates back in some sense to the algebraic approach to quantum mechanics introduced by Dirac in the 1920's and subsequently developed by him and others.

The other main approach to the Yang inequality, using Rayleigh-Ritz with the trial functions that Payne, Pólya, and Weinberger had used, Cauchy-Schwarz, and reductions based on (anti-)symmetry, was followed by most others. See [886], [55, 57, 83, 266, 391, 511], with most of these following the simpler approach introduced in [55, 57]. Our approach here, and that in [391], follows that proof. Note that both in [391] and here the method works smoothly even for magnetic Schrödinger operators.

The reader will note that in some circumstances we have used an algebraic approach to the proofs in this chapter. In particular, our use of commutator identities can be mentioned. Of course, one can go much farther in this direction, removing all integrals in favor of (abstract) inner products, and working with general operators (on which we might put further conditions, such as that they be self-adjoint and bounded below, etc.); in fact, the reader might find it productive to study our proofs with this in mind.

Moreover, it will be observed that much of what is done in the abstract versions can be seen in the proofs given above. One could write everything in terms of operators and inner products, and avoid all the integrals altogether. For example, see [83] or the other references listed above. One might worry that the integral way of writing things gives extra scope for dealing with expressions such as $-2\langle x_\ell u_1, u_{1,x_\ell}\rangle$. Recall, for example, that we handled $-2\int_\Omega u_1 x_\ell u_{1,x_\ell}$ by rewriting it as $-\int_\Omega x_\ell (u_1^2)_{x_\ell}$ and integrating by parts to get $\int_\Omega u_1^2 = 1$ (here we have taken u_1 to be real-valued to simplify our computation; the analogous calculation for the general complex-valued case would be somewhat more complicated, but could be done in much the same way). But the operator and inner product point of view can still handle such calculations. The way out, if one wants to give some structure to the calculation, is to use commutators (see, for example, [83], or the earlier papers in this vein). In the example given above, keeping to inner products, we could write

$$\begin{aligned} -2\langle x_\ell u_1, u_{1,x_\ell}\rangle &= 2\langle (x_\ell u_1)_{x_\ell}, u_1\rangle \\ &= 2\langle (u_1 + x_\ell u_{1,x_\ell}), u_1\rangle \\ &= 2\langle u_1, u_1\rangle + 2\langle x_\ell u_{1,x_\ell}, u_1\rangle \\ &= 2 + 2\langle x_\ell u_1, u_{1,x_\ell}\rangle, \end{aligned} \tag{8.140}$$

and thus $-4\langle x_\ell u_1, u_{1,x_\ell}\rangle = 2$, or $-2\langle x_\ell u_1, u_{1,x_\ell}\rangle = 1$. The calculation may not be as efficient when done in this way, but it gets the job done, and offers the possibility of generalization. And if one puts these calculations in terms of commutators and works out the consequences in general, one can maintain efficiency in the applications (one gets the messy calculations out of the way once and for all at the beginning). In those terms, the result here can be seen to rely on the commutator identity, $[\partial_{x_\ell}, x_\ell] = 1$, which is the basic commutator identity of quantum mechanics (usually expressed as $[p, x] = -i\hbar$, where x is a Cartesian position variable, and p is its conjugate momentum, often identified with $\frac{\hbar}{i}\frac{\partial}{\partial x}$ (where $\hbar$ is, of course, Planck's constant)).

The extension of the PPW, HP, and Yang inequalities to Schrödinger operators with nonnegative potentials (but without the magnetic vector potential, $\vec{A}$) occurred to many as the field progressed, so that once a given inequality was established for the eigenvalues of the Laplacian it was soon extended to the more general Schrödinger operator case. See [57, 68, 83, 452, 456, 460, 461, 463–465, 511, 535, 652] and probably several others. In particular, the HP inequality (with V_i's, i.e., (8.68)) for $H = -\Delta + V(\vec{x})$ first appeared in [68, 463]. Recent times have seen a proliferation of results in this area, so here we mention only some further names: El Soufi, Ilias, Cheng and Yang (together and with others), Wang and Xia (see also the book by Xia [884]).

Finally, we mention some of the references that added magnetic (vector) potentials to the picture. For the most part this happened around the time of Yang's Trieste preprint in 1991 or after, but it took a bit of time for the methods to include the Yang inequality within their scope. In 1997 we find Harrell and Stubbe [463] including magnetic potentials in their abstract (algebraized) discussion of universal inequalities, and we find Hermi [511] doing the same in 1999. Hook [535] also treated a special case in 1990, and Harrell and Michel dealt with the HP-extension to magnetic Schrödinger operators in 1995 [461]. Such results are also included, to an extent, in the abstract frameworks found in [83, 535]. The later paper by Frank, Laptev, and Molchanov [391] contains a very complete treatment of magnetic Schrödinger operators in this vein, and much else besides. Further developments can be found in [465]. The works that include Yang-type inequalities for Schrödinger operators with magnetic potentials are [391, 463, 465, 511].

Our proof of the main explicit upper bound in Section 8.5 follows essentially all of the details of the proof given in [266], with perhaps a small improvement or simplification in one or two places. As noted earlier, the result is as mentioned in [265], but without proof.

If one wants to improve the constant appearing in the explicit upper bound one can work harder at low indices, making use of various other inequalities that are known there (such as the best bound for λ_2/λ_1). Then one also starts with a higher-indexed λ, λ_k, say, and works to bound λ_m for $m \geq k$ n terms of λ_k. Work of this type already occurs in Cheng and Yang's first paper on the subject [266], and further work along these lines was done by Chen and Zheng [262].

In our discussion in Section 8.5 we have chosen not to go into the details of these refinements, and have instead tried to emphasize the main ideas of the development. Thus, we have even reverted to simpler versions of the main inequalities of Cheng and Yang when that would strip away some of the complications from their development.

In their work on magnetic Schrödinger operators, Frank, Laptev, and Molchanov [391] make use of the explicit upper bound of Cheng and Yang but do not give an independent proof.

A rather interesting chapter in the subject of eigenvalues of the Laplacian is the story of Riesz means and their connection to various asymptotic estimates, including Weyl asymptotics. But that, together with its connections to sharp asymptotics, Lieb-Thirring inequalities, and the Pólya conjecture for λ_m, could easily be a chapter on its own. Suffice it to say that there are very nice connections between various of the classical inequalities for the eigenvalues of the Dirichlet Laplacian (such as those for the trace of the heat kernel and the spectral zeta function) which result by applying such operations as the Legendre transform, the Laplace transform, and the Mellin transform to the appropriate inequality. Typically the Legendre transform takes you between inequalities for Riesz means and inequalities for eigenvalues or sums of eigenvalues and reverses the sense of the inequality (and applying the Legendre transform twice in succession takes you back where you started), while the Laplace and Mellin transforms both respect the sense of the inequality. The Laplace transform can be used to convert Riesz means into traces of the associated heat kernels, while the Mellin transform is applied to traces of heat kernels to produce spectral zeta functions. Riesz means also afford an efficient and relatively effortless way of obtaining explicit inequalities of the type we rather laboriously found in Section 8.5, following [266]. See [512] and especially [391, 457, 458].

For a start on this subject, one can consult [626], [512], [457], [391], [458], and many other papers. In particular, the subject of Riesz means is developed in the 1952 book of Chandrasekharan and Minakshisundaram [256], and a number of people used them subsequently in studies of eigenvalue asymptotics. Among these are [478, 479, 551, 552, 627–630, 795, 796].

The Berezin and Li-Yau inequalities figure in this story as well, as they turn out to be the same inequality a Legendre transform apart. Thus, many people now refer to those inequalities collectively as the Berezin-Li-Yau inequality. References for the original papers on these inequalities are [135, 136, 660].

Finally, for a discussion of the Pólya conjectures for both the Dirichlet and Neumann eigenvalues of the Laplacian, see the Open Problems section (Section 6) of [68] or Section 11 of [73]. Note, though, that in both those papers the Neumann eigenvalues are indexed from 0, whereas in this book they are indexed from 1. This introduces a shift (of one unit) into the conjectured inequalities for the Neumann eigenvalues.

Acknowledgements. I thank Richard Laugesen and Giuseppe Buttazzo for their conscientious proofreading of this chapter and for their suggestions toward improving its exposition. I am, of course, responsible for any errors and obscurities that remain.

Giuseppe Buttazzo and Bozhidar Velichkov

9 Spectral optimization problems for Schrödinger operators

In this chapter we consider Schrödinger operators of the form $-\Delta + V(x)$ on the Sobolev space $H_0^1(D)$, where D is an open subset of $\mathbb{R}^d$. We are interested in finding optimal potentials for some suitable criteria; the optimization problems we deal with are then written as

$$\min\{F(V) \ : \ V \in \mathcal{V}\}$$

where F is a suitable cost functional and $\mathcal{V}$ is a suitable class of admissible potentials. For simplicity, we consider the case when D is bounded and $V \geq 0$; under these conditions the resolvent operator of $-\Delta + V(x)$ is compact and the spectrum $\lambda(V)$ of the Schrödinger operator is discrete and consists of an increasing sequence of positive eigenvalues

$$\lambda(V) = \big(\lambda_1(V), \lambda_2(V), \dots\big).$$

This allows us to consider as cost functions the so-called *spectral functionals*, of the form

$$F(V) = \Phi\big(\lambda(V)\big),$$

where Φ is a given function. The cases when D is unbounded or V takes on negative values may provide in general a continuous spectrum and are more delicate to treat; some examples in this framework are considered in [171] and in the references therein.

The largest framework in which Schrödinger operators can be considered is the one where the potentials are *capacitary measures*; these ones are nonnegative Borel measures on D, possibly taking on the value $+\infty$ and vanishing on all sets of capacity zero (we refer to Section 2.2 for the definition of capacity). This framework will be considered in Section 9.1 together with the related optimization problems. We want to stress here that the class of capacitary measures μ is very large and contains both the case of standard potentials $V(x)$, in which $\mu = V\,dx$, as well as the case of classical domains Ω, in which $\mu = +\infty_{D\setminus\Omega}$. By this notation, we intend to reference the measure defined in (9.3).

Optimization problems for domains, usually called *shape optimization problems*, are often considered in the literature; the other chapters in the present volume deal with this kind of problem and in particular with *spectral optimization problems*, in

Giuseppe Buttazzo: Dipartimento di Matematica, Università di Pisa, Largo B. Pontecorvo 5, 56127 Pisa - Italy, E-mail: buttazzo@dm.unipi.it
Bozhidar Velichkov: Laboratoire Jean Kuntzmann, Université de Grenoble et CNRS 38041 Grenoble cedex 09 - France, E-mail: bozhidar.velichkov@imag.fr

which the cost functional depends on the spectrum of the Laplace operator $-\Delta$ on $H^1_0(\Omega)$:

$$F(\Omega) = \Phi\big(\lambda(\Omega)\big)$$

being Ω a domain which varies in the admissible class. For further details on shape optimization problems we refer the reader to the other chapters of this book and to [207], [505], [510]; here we simply recall some key facts. The existence of optimal domains for a problem of the form

$$\min\big\{F(\Omega)\ :\ \Omega \subset D,\ |\Omega| \le m\big\} \tag{9.1}$$

has been obtained under some additional assumptions, that we resume below.

- On the admissible domains Ω, some additional geometrical constraints are imposed, including convexity, uniform Lipschitz condition, uniform exterior cone properties, capacitary conditions, Wiener properties, ...; a detailed analysis of these conditions can be found in the book [207].
- No geometrical conditions are required on the admissible domains Ω but the functional F is assumed to satisfy some monotonicity conditions; in particular it is supposed to be decreasing with respect to set inclusion. The first result in this direction has been obtained in [238] and several generalizations, mainly to the cases where the set D is not bounded, have been made in [206] and in [700].

Without the extra assumptions above, the existence of an optimal shape may fail, in general, as several counterexamples show (see for instance [207]); in these cases the minimizing sequences (Ω_n) for the problem (9.1) converge in the γ-convergence sense (see Definition 9.1) to capacitary measures μ. In Section 9.1 we will see that many problems admit a capacitary measure as an optimal solution; this class is very large and only mild assumptions on the cost functional are required to provide the existence of a solution. In Section 9.2 we restrict our attention to the subclass of *Schrödinger potentials* $V(x)$ that belong to some space $L^p(D)$; we call them *integrable potentials* and we will see that suitable assumptions on the cost functional still imply the existence of an optimal potential. Finally, in Section 9.3 we consider the case of *confining potentials* $V(x)$ that are very large out of a bounded set, or more generally fulfill some integral inequalities of the form $\int_D \psi\big(V(x)\big)\,dx \le 1$ for some suitable integrand ψ.

9.1 Existence results for capacitary measures

In this section we consider a bounded open subset D of $\mathbb{R}^d$ and the class $\mathcal{M}_{cap}(D)$ of all capacitary measures on D, that is the Borel nonnegative measures on D, possibly $+\infty$ valued, that vanish on all sets of capacity zero. The analysis of capacitary measures and of their variational properties was made in [314]; the related optimization problems have been first considered in [237].

The key ingredient we need is the notion of γ-convergence. For a given measure $\mu \in \mathcal{M}_{cap}(D)$ we consider the Schrödinger-like operator $-\Delta + \mu$ defined on $H_0^1(D)$ and its resolvent operator R_μ which associates to every $f \in L^2(D)$ the unique solution $u = R_\mu(f)$ of the PDE

$$-\Delta u + \mu u = f, \qquad u \in H_0^1(D) \cap L_\mu^2(D).$$

The PDE above has to be defined in the weak sense

$$\begin{cases} u \in H_0^1(D) \cap L_\mu^2(D) \\ \int_D \nabla u \nabla \phi \, dx + \int_D u\phi \, d\mu = \int_D f\phi \, dx \qquad \forall \phi \in H_0^1(D) \cap L_\mu^2(D). \end{cases} \tag{9.2}$$

Definition 9.1. *We say that a sequence (μ_n) of capacitary measures γ-converges to a capacitary measure μ if and only if*

$$R_{\mu_n}(f) \to R_\mu(f) \text{ weakly in } H_0^1(D) \qquad \forall f \in L^2(D).$$

In the definition above one can equivalently require that the resolvent operators R_{μ_n} converge to the resolvent operator R_μ in the norm of the space of operators $\mathcal{L}\big(L^2(D); L^2(D)\big)$.

We summarize here below the main properties of the class $\mathcal{M}_{cap}(D)$; we refer for the details to [207].

- Every domain Ω can be seen as a capacitary measure, by taking $\mu = \infty_{D\setminus\Omega}$, or more precisely

$$\mu(E) = \begin{cases} 0 & \text{if } \operatorname{cap}(\Omega \setminus E) = 0 \\ +\infty & \text{if } \operatorname{cap}(\Omega \setminus E) > 0. \end{cases} \tag{9.3}$$

- Every capacitary measure is the γ-limit of a suitable sequence (Ω_n) of (smooth) domains; in other words, the class $\mathcal{M}_{cap}(D)$ is the closure with respect to the γ-convergence, of the class of (smooth) domains D.
- For every sequence (μ_n) of capacitary measures there exists a subsequence (μ_{n_k}) which γ-converges to a capacitary measure μ; in other words the class $\mathcal{M}_{cap}(D)$ is compact with respect to the γ-convergence.
- If μ is a capacitary measure, we may consider the PDE formally written as

$$-\Delta u + \mu u = f, \qquad u \in H_0^1(D). \tag{9.4}$$

The meaning of the equation above, as specified in (9.2), is in a weak sense, by considering the Hilbert space $H_\mu^1(D) = H_0^1(D) \cap L_\mu^2(D)$ with the norm

$$\|u\|_{H_\mu^1(D)} = \|u\|_{H_0^1(D)} + \|u\|_{L_\mu^2(D)}$$

and defining the solution in the weak sense (9.2). By Lax-Milgram theory, for every $\mu \in \mathcal{M}_{cap}(D)$ and $f \in L^2(D)$ (actually it would be enough to have f in the dual space of $H^1_\mu(D)$) there exists a unique solution $u_{\mu,f}$ of the PDE above. Moreover, if $\mu_n \to \mu$ in the γ-convergence, we have $u_{\mu_n,f} \to u_{\mu,f}$ weakly in in $H^1_0(D)$, hence strongly in $L^2(D)$.

- In order to have the γ-convergence of μ_n to μ it is enough to have the weak convergence in $H^1_0(D)$ of $R_{\mu_n}(1)$ to $R_\mu(1)$; in other words, we need to test the convergence of solutions of the PDEs related to the operators $-\Delta + \mu_n$ only with $f = 1$.
- The space $\mathcal{M}_{cap}(D)$, endowed with the γ-convergence, is metrizable; more precisely, the γ-convergence on $\mathcal{M}_{cap}(D)$ is equivalent to the distance

$$d_\gamma(\mu, \nu) = \|w_\mu - w_\nu\|_{L^2(D)}$$

where w_μ and w_ν are the solutions of the problems

$$-\Delta w_\mu + \mu w_\mu = 1 \text{ on } H^1_\mu(D)\,, \qquad -\Delta w_\nu + \nu w_\nu = 1 \text{ on } H^1_\nu(D)\,.$$

Remark 9.2. *We notice that the definition of γ-convergence of a sequence of capacitary measures μ_n to μ can be equivalently expressed in terms of the Γ-convergence in $L^2(D)$ of the corresponding energy functionals*

$$J_n(u) = \int_D |\nabla u|^2\,dx + \int_D u^2\,d\mu_n$$

to the limit energy

$$J(u) = \int_D |\nabla u|^2\,dx + \int_D u^2\,d\mu.$$

For all details about Γ-convergence theory we refer to [313].

The γ-convergence is very strong, and so many functionals are γ-lower semicontinuous, or even continuous (see below some important examples). The classes of functionals we are interested in are the following.

Integral functionals. Given a function $f \in L^2(D)$, for every $\mu \in \mathcal{M}_{cap}(D)$ we consider the solution $u_{\mu,f} = R_\mu(f)$ to the elliptic PDE (9.4). The integral cost functionals we consider are of the form

$$F(\mu) = \int_D j(x, u_{\mu,f}, \nabla u_{\mu,f})\,dx, \tag{9.5}$$

where $j(x, s, z)$ is a suitable integrand that we assume measurable in the x variable, lower semicontinuous in the s, z variables, and convex in the z variable. Moreover, the function j is assumed to fulfill bounds from below of the form

$$j(x, s, z) \geq -a(x) - c|s|^2,$$

with $a \in L^1(D)$ and c smaller than the first Dirichlet eigenvalue of the Laplace operator $-\Delta$ in D. In particular, the energy $\mathcal{E}_f(\mu)$ defined by

$$\mathcal{E}_f(\mu) = \inf\left\{\frac{1}{2}\int_D |\nabla u|^2\,dx + \frac{1}{2}\int_D u^2\,d\mu - \int_D fu\,dx \;:\; u \in H_0^1(D)\right\}, \tag{9.6}$$

belongs to this class since, integrating its Euler-Lagrange equation by parts, we have

$$\mathcal{E}_f(\mu) = -\frac{1}{2}\int_D f(x)u_{\mu,f}\,dx,$$

which corresponds to the integral functional above with

$$j(x,s,z) = -\frac{1}{2}f(x)s.$$

Thanks to the assumptions above and to the strong-weak lower semicontinuity theorem for integral functionals (see for instance [235]) all functionals of the form (9.5) are γ-lower semicontinuous on $\mathcal{M}_{cap}(D)$.

Spectral functionals. For every capacitary measure $\mu \in \mathcal{M}_{cap}(D)$ we consider the spectrum $\lambda(\mu)$ of the Schrödinger operator $-\Delta+\mu$ on $H_0^1(D)\cap L_\mu^2(D)$. Since D is bounded (it is enough to consider D to be of finite measure), then the operator $-\Delta + \mu$ has a compact resolvent and so its spectrum $\lambda(\mu)$ is discrete:

$$\lambda(\mu) = \big(\lambda_1(\mu), \lambda_2(\mu), \dots\big),$$

where $\lambda_k(\mu)$ are the eigenvalues of $-\Delta + \mu$, counted with their multiplicity. The same occurs if D is unbounded, and the measure μ satisfies some suitable *confinement* integrability properties (see for instance [208]). The spectral cost functionals we may allow are of the form

$$F(\mu) = \Phi\big(\lambda(\mu)\big),$$

for suitable functions $\Phi : \mathbb{R}^{\mathbb{N}} \to (-\infty, +\infty]$. For instance, taking $\Phi(\lambda) = \lambda_k$ we obtain

$$F(\mu) = \lambda_k(\mu).$$

Since a sequence (μ_n) γ-converges to μ if and only if the sequence of resolvent operators (R_{μ_n}) converges in the operator norm convergence of linear operators on $L^2(D)$ to the resolvent operator R_μ, the spectrum $\lambda(\mu)$ is continuous with respect to the γ-convergence, that is

$$\mu_n \to_\gamma \mu \;\Rightarrow\; \lambda_k(\mu_n) \to \lambda_k(\mu) \quad \forall k \in \mathbb{N}.$$

Therefore, the spectral functionals above are γ-lower semicontinuous, provided that the function Φ is lower semicontinuous, in the sense that

$$\lambda_n \to \lambda \text{ in } \mathbb{R}^{\mathbb{N}} \;\Rightarrow\; \Phi(\lambda) \le \liminf_n \Phi(\lambda_n),$$

where $\lambda_n \to \lambda$ in $\mathbb{R}^{\mathbb{N}}$ is intended in the componentwise convergence.

The relation between γ-convergence and weak*-convergence of measures is given in the proposition below.

Proposition 9.3. *Let $\mu_n \in \mathcal{M}_{cap}(D)$ be capacitary and Radon measures weakly* converging to the measure ν and γ-converging to the capacitary measure $\mu \in \mathcal{M}_{cap}(D)$. Then $\mu \le \nu$ in D.*

Proof. It is enough to show that $\mu(K) \le \nu(K)$ whenever K is a compact subset of D. Let u be a nonnegative smooth function with compact support in D such that $u \le 1$ in D and $u = 1$ on K; we have

$$\mu(K) \le \int_D u^2\,d\mu \le \liminf_{n\to\infty} \int_D u^2\,d\mu_n = \int_D u^2\,d\nu \le \nu(\{u > 0\}).$$

Since u is arbitrary, the conclusion follows from the definition of Borel regularity of the measure ν. □

Remark 9.4. *When $d = 1$, as a consequence of the compact embedding of $H_0^1(D)$ into the space of continuous functions on D, we obtain that any sequence (μ_n) weakly* converging to μ is also γ-converging to μ.*

In several shape optimization problems the class of admissible domains Ω is slightly larger than the class of open sets.

Definition 9.5. *We say that a set $\Omega \subset \mathbb{R}^d$ is quasi-open if for every $\varepsilon > 0$ there exists an open subset $\Omega_\varepsilon \subset \mathbb{R}^d$ such that $\operatorname{cap}(\Omega_\varepsilon \triangle \Omega) < \varepsilon$, where $\triangle$ denotes the symmetric difference of sets.*

Remark 9.6. *It is possible to prove (see for instance [207]) that a set $\Omega \subset D$ is quasi-open if and only if it can be written as*

$$\Omega = \{x \in D \,:\, u(x) > 0\}$$

for a suitable function $u \in H_0^1(D)$. Since Sobolev functions are defined only up to sets of capacity zero, a quasi-open set is defined up to capacity zero sets too.

In many problems the admissible domains Ω are constrained to verify a measure constraint of the form $|\Omega| \le m$; in order to *relax* this constraint to capacitary measures we have to introduce, for every $\mu \in \mathcal{M}_{cap}(D)$, the *set of finiteness* Ω_μ. A precise definition would require the notion of fine topology and finely open sets (see for instance [207]); however, a simpler equivalent definition can be given in terms of the solution $w_\mu = R_\mu(1)$ of the elliptic PDE

$$-\Delta u + \mu u = 1, \qquad u \in H_\mu^1(D).$$

Definition 9.7. *For every $\mu \in \mathcal{M}_{cap}(D)$ we denote by Ω_μ the set of finiteness of μ, defined by*

$$\Omega_\mu = \{w_\mu > 0\}.$$

By definition, the set Ω_μ is quasi-open, being the set where a Sobolev function is positive. Of course, since the function w_μ is defined only up to sets of capacity zero, the set Ω_μ is defined up to sets of capacity zero too.

Proposition 9.8. *The Lebesgue measure $|\Omega_\mu|$ is γ-lower semicontinuous.*

Proof. This follows from the definition of Ω_μ and from the fact that the γ-convergence $\mu_n \to_\gamma \mu$ is equivalent to the convergence of the solutions $w_{\mu_n} = R_{\mu_n}(1)$ to $w_\mu = R_\mu(1)$ in $L^2(D)$. The conclusion then follows by the Fatou's lemma. □

In summary, thanks to the γ-compactness of the class $\mathcal{M}_{cap}(D)$, the following general existence result holds.

Theorem 9.9. *Let $F : \mathcal{M}_{cap}(D) \to \overline{\mathbb{R}}$ be a γ-lower semicontinuous functional (for instance one of the classes above); then the minimization problem*

$$\min\left\{F(\mu)\ :\ \mu \in \mathcal{M}_{cap}(D),\ |\Omega_\mu| \le m\right\}$$

admits a solution $\mu_{opt} \in \mathcal{M}_{cap}(D)$.

In general, the optimal measure μ_{opt} is not unique; however, in the situation described below, the uniqueness occurs. Consider the optimization problem for the integral functional

$$F(\mu) = \int_D j\big(x, u_{\mu,f}, \nabla u_{\mu,f}\big)\,dx$$

where $f \ge 0$ is a given function in $L^2(D)$. We can write the problem as a double minimization, in μ and in u:

$$\min\left\{\int_D j(x, u, \nabla u)\,dx\ :\ \mu \in \mathcal{M}_{cap}(D),\ u \in H_0^1(D),\ -\Delta u + \mu u = f\right\}.$$

Since $f \ge 0$, by the maximum principle we know that $u \ge 0$ and, at least formally (the rigorous justification can be found in [269]),

$$\mu = \frac{f + \Delta u}{u},$$

so that we can eliminate the variable μ from the minimization and the optimization problem can be reformulated in terms of the function u only, as

$$\min\left\{\int_D j(x, u, \nabla u)\,dx\ :\ u \in \mathcal{K}\right\},$$

where $\mathcal{K}$ is the subset of $H_0^1(D)$ given by

$$\mathcal{K} = \left\{u \in H_0^1(D)\ :\ f + \Delta u \ge 0\right\}.$$

The inequality $f + \Delta u \geq 0$ has to be formulated in a weak sense, as

$$\int_D f\phi\, dx - \int_D \nabla u \nabla \phi\, dx \geq 0 \qquad \forall \phi \in H_0^1(D),\ \phi \geq 0. \tag{9.7}$$

The set $\mathcal{K}$ is clearly convex and it is easy to see that it is also closed. Hence, as a consequence, if the function $j(x, s, z)$ is *strictly convex* with respect to the pair (s, z), the solution of (9.7) is unique. Thus the solution μ_{opt}, that exists thanks to Theorem 9.9 is also unique. Note that in this case, no measure constraint of the form $|\Omega_\mu| \leq m$ is imposed.

In several situations the optimal measure μ_{opt} given by Theorem 9.9 has more regularity or summability properties than a general element of $\mathcal{M}_{cap}(D)$.This happens in the cases below:

- If the functional F is monotonically increasing with respect to the usual order of measures, and a constraint $|\Omega_\mu| \leq m$ is added, then an optimal measure μ_{opt} that is actually a domain exists, that is $\mu_{opt} = \infty_{D\setminus\Omega}$ for some quasi-open subset Ω of D. This fact should be rigorously justified (see [238]), but the argument consists in the fact that the measure $\infty_{D\setminus\Omega}$ is smaller than μ and has the same set of finiteness; then it provides an optimum for the minimization problem due to the monotonicity of F and to the constraint on the measure of the set of finiteness.
- In [241] the optimization of the *elastic compliance* for a membrane is considered, with the additional constraint that the measure μ has a prescribed total mass. In this case it is shown that μ_{opt} is actually an $L^1(D)$ function, that is no singular parts with respect to the Lebesgue measure occur.

In general, we should not expect that μ_{opt} is a domain or a function with any summability; the following example shows that even in simple and natural problems this does not occur.

Example 9.10. *Let D be a ball of radius R and let $f = 1$; consider the optimization problem for the integral functional*

$$F(\mu) = \int_D |u_{\mu,1} - c|^2\, dx, \tag{9.8}$$

where c is a given constant and $u_{\mu,1}$ denotes as before the solution of the PDE

$$-\Delta u + \mu u = 1, \qquad u \in H_\mu^1(D).$$

By the argument described above the problem can be reformulated in terms of the function u only, as

$$\min\left\{\int_D |u - c|^2\, dx\ :\ u \in \mathcal{K}\right\}$$

where $\mathcal{K}$ is the convex closed subset of $H_0^1(D)$ given by

$$\mathcal{K} = \left\{u \in H_0^1(D)\ :\ \Delta u + 1 \geq 0\right\}.$$

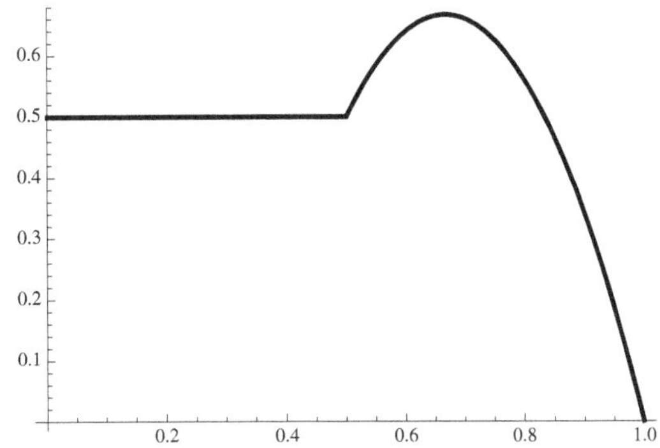

Fig. 9.1. The behavior of an optimal state function $u(r)$.

As we have seen, this auxiliary problem has a unique solution which is radially symmetric. Thus we can write the problem in polar coordinates as

$$\min\left\{\int_0^R |u-c|^2 r^{d-1}\,dx \;:\; u'' + \frac{d-1}{r}u' + 1 \geq 0,\; u(R) = 0\right\}.$$

The minimum problem above can be fully analyzed and its solution is characterized as follows (see [207] for the details).

- *If c is large enough, above a certain threshold $\bar{c}$ that can be computed explicitly, we have for the optimal solution (u, μ)*

$$u(r) = \frac{R^2 - r^2}{2d}, \qquad \textit{hence} \qquad \mu \equiv 0.$$

- *Below the threshold $\bar{c}$ the optimal measure μ is given by*

$$\mu = \frac{1}{c}\mathcal{L}^d \lfloor B_{R_c} + \alpha_c \mathcal{H}^{d-1} \lfloor \partial B_{R_c},$$

  where $\mathcal{L}^d$ denotes the Lebesgue measure in $\mathbb{R}^d$, $\alpha_c > 0$ is a suitable constant, and $R_c < R$ is a suitable radius. The solution u is computed correspondingly, through the equation

$$u'' + \frac{d-1}{r}u' + \mu u = 1.$$

A plot of the behavior of an optimal state function u is given in Figure 9.1.

Note that the functional in (9.8) is not monotonically increasing with respect to μ.

9.2 Existence results for integrable potentials

In this section we consider optimization problems of the form

$$\min\left\{F(V) \;:\; V : D \to [0, +\infty],\; \int_D V^p\,dx \leq 1\right\}, \tag{9.9}$$

where $p > 0$ and $F(V)$ is a cost functional acting on Schrödinger potentials, or more generally on capacitary measures. We assume that F is γ-lower semicontinuous, an assumption that, as we have seen in the previous section, is very mild and verified for most of the functionals of integral or spectral type.

When $p > 1$ a general existence result follows from the following proposition, where we show that the weak $L^1(D)$ convergence (that is the one having $L^\infty(D)$ as the space of test functions) of potentials implies the γ-convergence.

Proposition 9.11. *Let $V_n \in L^1(D)$ converge weakly in $L^1(D)$ to a function V. Then the capacitary measures $V_n\,dx$ γ-converge to $V\,dx$.*

Proof. We have to prove that the solutions $u_n = R_{V_n}(1)$ of the PDE

$$\begin{cases} -\Delta u_n + V_n(x)u_n = 1 \\ u \in H_0^1(D) \end{cases}$$

weakly converge in $H_0^1(D)$ to the solution $u = R_V(1)$ of

$$\begin{cases} -\Delta u + V(x)u = 1 \\ u \in H_0^1(D). \end{cases}$$

Equivalently, as noticed in Remark 9.2, we may prove that the functionals

$$J_n(u) = \int_D |\nabla u|^2\,dx + \int_D V_n(x)u^2\,dx$$

Γ-converge in $L^2(D)$ to the functional

$$J(u) = \int_D |\nabla u|^2\,dx + \int_D V(x)u^2\,dx.$$

Let us prove the Γ-liminf inequality:

$$J(u) \le \liminf_{n\to\infty} J(u_n) \qquad \forall u_n \to u \text{ in } L^2(D).$$

Indeed, if $u_n \to u$ in $L^2(D)$, we have

$$\int_D |\nabla u|^2\,dx \le \liminf_{n\to\infty} \int_D |\nabla u_n|^2\,dx$$

by the lower semicontinuity of the $H^1(D)$ norm with respect to the $L^2(D)$-convergence, and

$$\int_D V(x)u^2\,dx \le \liminf_{n\to\infty} \int_D V_n(x)u_n^2\,dx$$

by the strong-weak lower semicontinuity theorem for integral functionals (see for instance [235]).

Let us now prove the Γ-limsup inequality: there exists $u_n \to u$ in $L^2(D)$ such that

$$J(u) \geq \limsup_{n\to\infty} J(u_n). \tag{9.10}$$

For every $t > 0$ we set

$$u^t = (u \wedge t) \vee (-t);$$

then, by the weak $L^1(D)$ convergence of V_n to V, for every t fixed we have

$$\lim_{n\to\infty} \int_D V_n(x)|u^t|^2\,dx = \int_D V(x)|u^t|^2\,dx.$$

Moreover, letting $t \to \infty$ we have by the monotone convergence theorem

$$\lim_{t\to+\infty} \int_D V(x)|u^t|^2\,dx = \int_D V(x)|u|^2\,dx.$$

Then, by a diagonal argument, we can find a sequence $t_n \to +\infty$ such that

$$\lim_{n\to\infty} \int_D V_n(x)|u^{t_n}|^2\,dx = \int_D V(x)|u|^2\,dx.$$

Taking now $u_n = u^{t_n}$, and noticing that for every $t > 0$

$$\int_D |\nabla u^t|^2\,dx \leq \int_D |\nabla u|^2\,dx,$$

we obtain (9.10) and so the proof is complete. □

The existence of an optimal potential for problems of the form (9.9) is now straightforward.

Theorem 9.12. *Let $F(V)$ be a functional defined for $V \in L^1_+(D)$ (the set of nonnegative functions in $L^1(D)$), lower semicontinuous with respect to the γ-convergence, and let $\mathcal{V}$ be a subset of $L^1_+(D)$, compact for the weak L^1-convergence. Then the problem*

$$\min\left\{F(V)\ :\ V \in \mathcal{V}\right\},$$

admits a solution.

Proof. Let (V_n) be a minimizing sequence in $\mathcal{V}$. By the compactness assumption on $\mathcal{V}$, we may assume that V_n tends to some $V \in \mathcal{V}$ weakly in $L^1(D)$. By Proposition 9.11, we have that V_n γ-converges to V and so, by the semicontinuity of F,

$$F(V) \leq \liminf_{n\to\infty} F(V_n),$$

which gives the conclusion. □

In some cases the optimal potential can be explicitly determined through the solution of a partial differential equation, as for instance in the examples below.

Example 9.13. *Take $F = -\mathcal{E}_f$, where $\mathcal{E}_f$ is the energy functional defined in* (9.6), *with f a fixed function in $L^2(D)$, and*

$$\mathcal{V} = \left\{ V \geq 0,\ \int_D V^p \, dx \leq 1 \right\} \qquad \text{with } p > 1. \tag{9.11}$$

Then, the problem we are dealing with is

$$\max_{V\in\mathcal{V}} \mathcal{E}_f(V) = \max_{V\in\mathcal{V}} \min_{u\in H_0^1(D)} \left\{ \frac{1}{2}\int_D |\nabla u|^2 \, dx + \frac{1}{2}\int_D Vu^2 \, dx - \int_D fu \, dx \right\}. \tag{9.12}$$

As we have already seen above, the energy functional can be written, by an integration by parts, as

$$\mathcal{E}_f(V) = -\frac{1}{2}\int_D f(x) R_V(f) \, dx$$

where R_V is the resolvent operator of $-\Delta + V(x)$. Therefore, the functional F is γ-continuous and the existence Theorem 9.12 applies. In order to compute the optimal potential, interchanging the min and the max in (9.12) *we obtain the inequality*

$$\begin{aligned} &\max_{V\in\mathcal{V}} \min_{u\in H_0^1(D)} \left\{ \frac{1}{2}\int_D |\nabla u|^2 \, dx + \frac{1}{2}\int_D Vu^2 \, dx - \int_D fu \, dx \right\} \\ &\quad \leq \min_{u\in H_0^1(D)} \max_{V\in\mathcal{V}} \left\{ \frac{1}{2}\int_D |\nabla u|^2 \, dx + \frac{1}{2}\int_D Vu^2 \, dx - \int_D fu \, dx \right\}. \end{aligned}$$

The maximization with respect to V is very easy to compute; in fact, for a fixed u, the maximal value is reached at

$$V = \left(\int_D |u|^{2p/(p-1)} \, dx \right)^{-1/p} |u|^{2/(p-1)}, \tag{9.13}$$

so that

$$\begin{aligned} &\min_{u\in H_0^1(D)} \max_{V\in\mathcal{V}} \left\{ \frac{1}{2}\int_D |\nabla u|^2 \, dx + \frac{1}{2}\int_D Vu^2 \, dx - \int_D fu \, dx \right\} \\ &\quad = \min_{u\in H_0^1(D)} \left\{ \frac{1}{2}\int_D |\nabla u|^2 \, dx + \frac{1}{2}\left(\int_D |u|^{2p/(p-1)} \, dx \right)^{(p-1)/p} - \int_D fu \, dx \right\}. \end{aligned}$$

In order to find the optimal potential V_{opt} we have then to solve the auxiliary variational problem

$$\min_{u\in H_0^1(D)} \left\{ \frac{1}{2}\int_D |\nabla u|^2 \, dx + \frac{1}{2}\left(\int_D |u|^{2p/(p-1)} \, dx \right)^{(p-1)/p} - \int_D fu \, dx \right\},$$

and then, by means of its solution $\bar{u}$, recovering V_{opt} from (9.13). *The auxiliary variational problem above can be written, via its Euler-Lagrange equation, as the nonlinear PDE*

$$-\Delta\bar{u} + C(p, \bar{u})|\bar{u}|^{2/(p-1)}\bar{u} = f, \qquad \bar{u} \in H_0^1(D),$$

with the constant $C(p,\bar u)$ given by

$$C(p,\bar u)=\left(\int_D |\bar u|^{2p/(p-1)}\,dx\right)^{-1/p}.$$

The fact that V_{opt} actually solves our optimization problem (9.12) follows from the fact that $\bar u = R_{V_{opt}}(f)$, hence we have

$$\begin{aligned}
\mathcal{E}_f(V_{opt}) &= \min_{u\in H_0^1(D)}\left\{\frac12\int_D |\nabla u|^2\,dx+\frac12\int_D V_{opt}u^2\,dx-\int_D fu\,dx\right\}\\
&=\frac12\int_D |\nabla \bar u|^2\,dx+\frac12\int_D V_{opt}\bar u^2\,dx-\int_D f\bar u\,dx\\
&=\min_{u\in H_0^1(D)}\left\{\frac12\int_D |\nabla u|^2\,dx+\frac12\left(\int_D |u|^{2p/(p-1)}\,dx\right)^{(p-1)/p}-\int_D fu\,dx\right\}\\
&\ge \max_{V\in\mathcal{V}}\mathcal{E}_f(V).
\end{aligned}$$

We notice that, replacing $-\mathcal{E}_f$ by $\mathcal{E}_f$ transforms the maximization problem in (9.12) into the minimization of $\mathcal{E}_f$ on $\mathcal{V}$, which has the only trivial solution $V\equiv 0$.

Example 9.14. *More generally, we may consider the optimization problem*

$$\min_{V\in\mathcal{V}}\ \min_{u\in H_0^1(D)}\left\{\int_D j(x,u)\,dx\ :\ -\Delta u+Vu=f\right\}$$

where the constraint $\mathcal{V}$ is given by (9.11). If $f\ge 0$ and $j(x,\cdot)$ is decreasing, by the maximum principle the best choice for the potential V is on the boundary of the admissible set $\mathcal{V}$ and we may consider the Lagrangian functional

$$\int_D \big(j(x,u)+\nabla u\nabla\phi+Vu\phi-f\phi\big)\,dx\,,$$

where ϕ is the adjoint state function. Optimizing with respect to V provides the optimal potential

$$V=|u\phi|^{1/(p-1)}\left(\int_D |u\phi|^{p/(p-1)}\,dx\right)^{-1/p}$$

which, combined with the Lagrangian functional above, reduces the problem to the minimization of the functional

$$\int_D \big(j(x,u)+\nabla u\nabla\phi-f\phi\big)\,dx+\left(\int_D |u\phi|^{p/(p-1)}\,dx\right)^{(p-1)/p}.$$

Differentiating with respect to ϕ gives the PDE (to let ϕ go up) for the state function u:

$$-\Delta u-f+C(p,u,\phi)|u\phi|^{1/(p-1)}u=0\,,$$

where

$$C(p,u,\phi) = \left(\int_D |u\phi|^{p/(p-1)}\,dx\right),$$

while, differentiating with respect to u gives the equation for the adjoint state function ϕ:

$$-\Delta\phi + j'(x,u) + C(p,u,\phi)|u\phi|^{1/(p-1)}\phi = 0\,.$$

Example 9.15. *Similarly to what done in Example 9.13 we may consider F as the functional $-\lambda_1(V)$, where $\lambda_1(V)$ is the first eigenvalue of the Schrödinger operator $-\Delta + V(x)$, given by the minimization*

$$\lambda_1(V) = \min_{u\in H_0^1(D)} \left\{\int_D \left(|\nabla u|^2 + Vu^2\right)dx \;:\; \int_D u^2\,dx = 1\right\}. \tag{9.14}$$

We are then dealing with the optimization problem

$$\max_{V\in\mathcal{V}} \lambda_1(V) = \max_{V\in\mathcal{V}} \min_{u\in H_0^1(D)} \left\{\int_D \left(|\nabla u|^2 + Vu^2\right)dx \;:\; \int_D u^2\,dx = 1\right\}, \tag{9.15}$$

where the constraint $\mathcal{V}$ is as in (9.11). Arguing as before, we interchange the max and the min above and we end up with the auxiliary problem

$$\min_{u\in H_0^1(D)} \left\{\int_D |\nabla u|^2\,dx + \left(\int_D |u|^{2p/(p-1)}\,dx\right)^{(p-1)/p} \;:\; \int_D u^2\,dx = 1\right\}.$$

In the same way as before, the optimal potential V_{opt} can be recovered through the solution $\bar{u}$ of the auxiliary problem above, by taking

$$V_{opt} = \left(\int_D |\bar{u}|^{2p/(p-1)}\,dx\right)^{-1/p} |\bar{u}|^{2/(p-1)}.$$

Remark 9.16. *In the case $p < 1$ problem (9.12) with the admissible class (9.11) does not admit any solution. Indeed, for a fixed real number $\alpha > 0$, take $V_n(x) = n\chi_{\Omega_n}(x)$, where χ_E denotes the characteristic function of the set E (with value 1 on E and 0 outside E) and $\Omega_n \subset D$ are such that the sequence (V_n) converges weakly in $L^1(D)$ to the constant function α. In particular, we have $n|\Omega_n| \to \alpha$ as $n \to \infty$ and so, since $p < 1$, we have*

$$\int_D V_n^p\,dx = n^p|\Omega_n| \to 0 \qquad \text{as } n \to \infty.$$

Therefore, for n large enough, the potentials V_n belong to the admissible class $\mathcal{V}$. By Proposition 9.11 we have $\mathcal{E}_f(V_n) \to \mathcal{E}_f(\alpha)$ and, since α was arbitrary, we obtain

$$\sup_{V\in\mathcal{V}} \mathcal{E}_f(V) \ge \sup_{\alpha\in\mathbb{R}} \mathcal{E}_f(\alpha) = \lim_{\alpha\to+\infty} \mathcal{E}_f(\alpha).$$

The limit on the right-hand side above is zero; on the other hand we have $\mathcal{E}_f(V) \le 0$ for any V. Thus, if a maximal potential V_{opt} exists, it should verify $\mathcal{E}_f(V_{opt}) = 0$ which is impossible.

It remains to consider the maximization problem (9.12) when $p = 1$. In this case the result of Proposition 9.11 cannot be applied because the unit ball of $L^1(D)$ is not weakly compact. However, the existence of an optimal potential still holds, as we show below. It is convenient to introduce the functionals

$$J_p(u) := \frac{1}{2}\int_D |\nabla u|^2\,dx + \frac{1}{2}\left(\int_D |u|^{2p/(p-1)}\,dx\right)^{(p-1)/p} - \int_D fu\,dx \qquad \text{if } p > 1$$

$$J_1(u) := \frac{1}{2}\int_D |\nabla u|^2\,dx + \frac{1}{2}\|u\|_\infty^2 - \int_D fu\,dx \qquad \text{if } p = 1.$$

Proposition 9.17. *The functionals J_p Γ-converge in $L^2(D)$ to J_1 as $p \to 1$.*

Proof. Let $v_n \in L^2(D)$ be a sequence of positive functions converging in $L^2(D)$ to $v \in L^2(D)$ and let $\alpha_n \to +\infty$. Then we have

$$\|v\|_{L^\infty(D)} \le \liminf_{n\to\infty} \|v_n\|_{L^{\alpha_n}(D)}. \tag{9.16}$$

In fact, suppose first that $\|v\|_{L^\infty} = M < +\infty$ and let $\omega_\varepsilon = \{v > M - \varepsilon\}$, for some $\varepsilon > 0$. Then, we have

$$\liminf_{n\to\infty} \|v_n\|_{L^{\alpha_n}(D)} \ge \lim_{n\to\infty} |\omega_\varepsilon|^{(1-\alpha_n)/\alpha_n} \int_{\omega_\varepsilon} v_n\,dx = |\omega_\varepsilon|^{-1} \int_{\omega_\varepsilon} v\,dx \ge M - \varepsilon,$$

and so, letting $\varepsilon \to 0$, we have

$$\liminf_{n\to\infty} \|v_n\|_{L^{\alpha_n}(\Omega)} \ge M.$$

If $\|v\|_{L^\infty} = +\infty$, then setting $\omega_k = \{v > k\}$, for any $k \ge 1$, and arguing as above, we obtain (9.16).

Now, let $u_n \to u$ in $L^2(D)$. Then, by the semicontinuity of the L^2 norm of the gradient, by (9.16), and by the continuity of the term $\int_D uf\,dx$, we have

$$J_1(u) \le \liminf_{n\to\infty} J_{p_n}(u_n),$$

for any decreasing sequence $p_n \to 1$. On the other hand, for any $u \in L^2(D)$, we have $J_{p_n}(u) \to J_1(u)$ as $n \to \infty$ and so, we have the conclusion. □

Lemma 9.18. *Let $D \subset \mathbb{R}^d$ be a bounded open set. Then for every $p \ge 1$ there is a unique minimizer u_p of the functional $J_p : H_0^1(D) \to \mathbb{R}$. Moreover, the following facts hold.*

(a) *There is a constant $C > 0$ such that for every $p > 1$ we have*

$$\|\nabla u_p\|_{L^2(D)} + \|u_p\|_{L^{2p/(p-1)}(D)} \le C\|f\|_{L^2(D)}.$$

(b) *For every open set $\Omega \subset\subset D$, there is a constant C_Ω such that*

$$\|u_p\|_{H^2(\Omega)} \le C_\Omega \|f\|_{L^2(D)}, \quad \text{for every} \quad p > 1. \tag{9.17}$$

Proof. The existence of a minimizer follows by the direct method in the calculus of variations, while the uniqueness is a consequence of the strict convexity of the functional. Moreover, for every $p > 1$, the minimizer u_p satisfies the Euler-Lagrange PDE

$$-\Delta u_p + c|u_p|^\alpha u_p = f, \qquad u_p \in H_0^1(D) \tag{9.18}$$

where

$$\alpha = \frac{2}{p-1} \qquad \text{and} \qquad c = \left(\int_D |u_p|^{\frac{2p}{p-1}}\right)^{-1/p}.$$

Now (a) follows by multiplying equation (9.18) by u_p and integrating on D. In fact one may simply take the constant C to be the first Dirichlet eigenvalue $\lambda_1(D)$.

In order to prove (b) we use an argument similar to that of the classical elliptic regularity theorem. For $h \in \mathbb{R}$ and $k = 1, \dots, d$, we use the notation

$$\partial_k^h u = \frac{u(x + he_k) - u(x)}{h},$$

and we consider a function $\phi \in C_c^\infty(D)$ such that $\phi \equiv 1$ on Ω. Then we have that for h small enough $\partial_k^h u$ satisfies the following equation on the support of ϕ :

$$-\Delta \partial_k^h u + \frac{c}{h}\left(u(x + he_k)|u(x + he_k)|^\alpha - u(x)|u(x)|^\alpha\right) = \partial_k^h f. \tag{9.19}$$

Multiplying (9.19) by $\phi^2 \partial_k^h u$ and taking into account the inequality

$$(X|X|^\alpha - Y|Y|^\alpha)(X - Y) \geq 0, \qquad \text{for all} \quad X, Y \in \mathbb{R},$$

we obtain

$$\int_D \nabla(\partial_k^h u) \cdot \nabla(\phi^2 \partial_k^h u)\, dx \leq \int_D (\partial_k^h f)\phi^2 \partial_k^h u\, dx.$$

By a change of variables, the Cauchy-Schwartz and the Poincaré inequalities we get that

$$\begin{aligned}\int_D (\partial_k^h f)\phi^2 \partial_k^h u\, dx &= -\int_D f \partial_k^{-h}(\phi^2 \partial_k^h u)\, dx \leq \|f\|_{L^2} \|\partial_k^{-h}(\phi^2 \partial_k^h u)\|_{L^2} \\ &\leq \|f\|_{L^2} \|\nabla(\phi^2 \partial_k^h u)\|_{L^2} \leq C_\phi \|f\|_{L^2} \|\phi \nabla(\partial_k^h u)\|_{L^2},\end{aligned} \tag{9.20}$$

where C_ϕ is a constant depending on ϕ. On the other hand we have

$$\begin{aligned}\int_D \phi^2 |\nabla(\partial_k^h u)|^2\, dx &= \int_D \nabla(\partial_k^h u) \cdot \nabla(\phi^2 \partial_k^h u)\, dx - 2\int_D \phi(\partial_k^h u)\nabla\phi \cdot \nabla(\partial_k^h u)\, dx \\ &\leq C_\phi \|f\|_{L^2} \|\phi \nabla(\partial_k^h u)\|_{L^2} + \frac{1}{2}\int_D \phi^2 |\nabla(\partial_k^h u)|^2\, dx + 2\|\nabla\phi\|_{L^\infty}^2 \|\nabla u\|_{L^2}^2.\end{aligned} \tag{9.21}$$

Thus, there is a constant $C_{D,\phi}$ depending on D and ϕ such that

$$\|\phi \nabla(\partial_k^h u)\|_{L^2}^2 \leq C_{D,\phi} \|f\|_{L^2} \|\phi \nabla(\partial_k^h u)\|_{L^2} + C_{D,\phi}^2 \|f\|_{L^2}^2,$$

which finally gives that

$$\|\partial_k^h(\nabla u)\|_{L^2(\Omega)} \leq \|\phi\nabla(\partial_k^h u)\|_{L^2(D)} \leq 2C_{D,\phi}\|f\|_{L^2(D)},$$

and since this last ineaqulity is true for every $k = 1, \dots, d$ and every h small enough we get that $u \in H^2(\Omega)$ and

$$\|\nabla^2 u\|_{L^2(\Omega)} \leq C_\Omega \|f\|_{L^2(D)},$$

for an appropriate constant C_Ω depending on the function ϕ associated to Ω. □

Proposition 9.19. *Let $D \subset \mathbb{R}^d$ be a bounded open set and $f \in L^2(D)$ a given function. Then there is a unique minimizer $u_1 \in H_0^1(D)$ of the functional $J_1 : H_0^1(D) \to \mathbb{R}$,*

$$J_1(u) = \frac{1}{2}\int_D |\nabla u|^2\,dx + \frac{1}{2}\|u\|_{L^\infty(D)}^2 - \int_D uf\,dx.$$

Setting $M = \|u_1\|_{L^\infty}$, $\omega_+ = \{u_1 = M\}$ and $\omega_- = \{u_1 = -M\}$, we have that
(i) $u_1 \in H_{loc}^2(D)$;
(ii) u_1 is the solution of the equation

$$-\Delta u + \frac{1}{M}(\chi_{\omega_+} f - \chi_{\omega_-} f)\,u = f, \qquad u \in H_0^1(D)\,; \tag{9.22}$$

(iii) $\displaystyle\frac{1}{M}\int_{\omega_+} f\,dx - \frac{1}{M}\int_{\omega_-} f\,dx = 1\,;$
(iv) $f \geq 0$ on ω_+ and $f \leq 0$ on ω_-.

Proof. (i) Let u_p be the minimizer of J_p. By Proposition 9.18 we have that the family $\{u_p\}_{p>1}$ is bounded in $H_0^1(D)$. From the estimate (9.17) we have that for every sequence $p_n \to 1$ the solutions u_{p_n} admit a subsequence converging weakly in $H_0^1(D)$ to some $u \in H_{loc}^2(D) \cap H_0^1(D)$. Since by Proposition 9.17 the functionals J_p Γ-converge in $L^2(D)$ to the functional J_1 as $p \to 1$, and since the minimizer of the functional J_1 is unique, we have $u = u_1$ and so, $u_1 \in H_{loc}^2(D) \cap H_0^1(D)$. Moreover, since this happens for every sequence $p_n \to 1$ we have $u_p \to u_1$ in $L^2(D)$ as $p \to 1$.

(ii) Let us define $\omega = \omega_+ \cup \omega_-$. We claim that u_1 satisfies, on D the PDE

$$-\Delta u_1 + \chi_\omega f = f, \qquad u_1 \in H_0^1(D). \tag{9.23}$$

Since $u_1 \in H_{loc}^2(D)$ (9.23) is equivalent to

$$\int_D (-\Delta u_1)\varphi\,dx = \int_D \chi_{D\setminus\omega} f\varphi\,dx, \quad \text{for every} \quad \varphi \in C_c^\infty(D). \tag{9.24}$$

Let $v_{\varepsilon,\delta} = 1 \wedge \left(\frac{1}{\delta}u\big((M - \varepsilon - u) \vee 0\big)\big((u + M - \varepsilon) \vee 0\big)\right)$. Since $u \in H_0^1(D) \cap L^\infty(D)$, we have that $v_{\varepsilon,\delta} \in H_0^1(D)$. By choosing ε and δ appropriately we may construct a sequence $v_n \in H_0^1(D)$ such that

- $v_n \uparrow \chi_{D\setminus\omega}$ as $n \to \infty$;
- $v_n = 0$ a.e. on the (Lebesgue measurable) set $\left\{-M + \frac{1}{n} \le u_1 \le M - \frac{1}{n}\right\}$.

Now since for $t \in \mathbb{R}$ such that $|t| \le \frac{1}{n\|\varphi\|_{L^\infty}}$ we have that $\|u_1 + t\varphi v_n\| \le M$, the minimality of u_1 gives that

$$\frac{1}{2}\int_D |\nabla u_1|^2\,dx - \int_D u_1 f\,dx \le \frac{1}{2}\int_D |\nabla(u_1 + t\varphi v_n)|^2\,dx - \int_D (u_1 + t\varphi v_n)f\,dx,$$

and taking the derivative with respect to t at $t = 0$, we get that

$$\int_D (-\Delta u_1)\varphi v_n\,dx = \int_D \nabla u_1 \cdot \nabla(\varphi v_n)\,dx = \int_D f\varphi v_n,\,dx,$$

and passing to the limit as $n \to \infty$ we obtain (9.24). We can now obtain (9.22) by (9.23) and the fact that $u_1 = M$ on ω_+ and $u = -M$ on ω_-.

(iii) Since u_1 is the minimizer of J_1, we have

$$J_1\big((1+\varepsilon)u_1\big) - J_1(u_1) \ge 0 \qquad \forall \varepsilon \in \mathbb{R}.$$

Taking the derivative of this difference at $\varepsilon = 0$, we obtain

$$\int_D |\nabla u_1|^2\,dx + M^2 = \int_D f u_1\,dx.$$

On the other hand, by (9.23), we have

$$\int_D |\nabla u_1|^2\,dx + \int_\omega f u_1\,dx = \int_D f u_1\,dx.$$

Now since $u_1 = M$ on ω_+ and $u_1 = -M$ on ω_-, we obtain

$$M^2 = \int_\omega f u_1\,dx = M\int_{\omega_+} f\,dx - M\int_{\omega_-} f\,dx.$$

(iv) Consider the function $w_\varepsilon = \big(M - \varepsilon - u\big) \vee 0$ which vanishes on the set $\{u \ge M - \varepsilon\}$ and is strictly positive on the set $\{u < M - \varepsilon\}$. For any non-negative $\varphi \in C_c^\infty(D)$ we have that for t small enough $\|u_1 + t\varphi\|_{L^\infty} < M$. Therefore, the optimality of u_1 gives

$$\begin{aligned}0 &\le \lim_{t\to 0^+} \frac{J_1(u_1 + t\varphi w_\varepsilon) - J_1(u_1)}{t}\\ &= \int_D \nabla u_1 \cdot \nabla(\varphi w_\varepsilon)\,dx - \int_D f\varphi w_\varepsilon\,dx\\ &= \int_D (-\Delta u_1 - f)\varphi w_\varepsilon\,dx.\end{aligned}$$

Since the last inequality holds for any $\varphi \ge 0$ and any $\varepsilon > 0$ we get that

$$-\Delta u_1 - f \ge 0 \quad \text{almost everywhere on} \quad \{u_1 < M\}.$$

On the other hand, $\Delta u_1 = 0$ almost everywhere on $\omega_- = \{u = -M\}$, and so we obtain that $f \le 0$ on ω_-. Arguing in the same way, and considering test functions supported on $\{u_1 \ge -M + \varepsilon\}$, we can prove that $f \ge 0$ on ω_+. □

Theorem 9.20. *Let $D \subset \mathbb{R}^d$ be a bounded open set, let $p = 1$, and let $f \in L^2(D)$. Then there is a unique solution to problem*

$$\max\left\{\mathcal{E}_f(V)\ :\ V \ge 0,\ \int_D V\,dx \le 1\right\}, \tag{9.25}$$

given by

$$V_1 = \frac{1}{M}\,(\chi_{\omega_+}f - \chi_{\omega_-}f),$$

where $M = \|u_1\|_{L^\infty(D)}$, $\omega_+ = \{u_1 = M\}$, $\omega_- = \{u_1 = -M\}$, being $u_1 \in H_0^1(D) \cap L^\infty(D)$ the unique minimizer of the functional J_1.

Proof. For any $u \in H_0^1(D)$ and any $V \ge 0$ with $\int_D V\,dx \le 1$ we have

$$\int_D u^2 V\,dx \le \|u\|_{L^\infty}^2 \int_D V\,dx \le \|u\|_{L^\infty}^2.$$

Thus we obtain the inequality

$$\frac{1}{2}\int_\Omega |\nabla u|^2\,dx + \frac{1}{2}\int_\Omega u^2 V\,dx - \int_\Omega uf\,dx \le J_1(u), \quad \text{for every} \quad u \in H_0^1(D),$$

and taking the minimum with respect to u we get

$$\mathcal{E}_f(V) \le \min_{u \in H_0^1(D)} J_1(u),$$

which finally gives

$$\max\left\{\mathcal{E}_f(V)\ :\ V \ge 0,\ \int_D V\,dx \le 1\right\} \le \min_{u \in H_0^1(D)} J_1(u) = J_1(u_1),$$

where u_1 is the minimizer of J_1. By Proposition 9.19 we have that u_1 satisfies the equation

$$-\Delta u_1 + V_1 u_1 = f, \qquad u_1 \in H_0^1(D), \tag{9.26}$$

where we set

$$V_1 := \frac{1}{M}\,(\chi_{\omega_+}f - \chi_{\omega_-}f).$$

By Proposition 9.19 (iii) we have that $\int_D V_1 u_1^2\,dx = M^2$ and so

$$J_1(u_1) = \mathcal{E}_f(V_1).$$

Moreover, again by (iii) and (iv) we obtain that $V_1 \ge 0$ and $\int_D V_1\,dx = 1$, which concludes the proof. □

Example 9.21. *When f is a general $H^{-1}(D)$ function, the result of Theorem 9.20 may fail in the sense that problem (9.25) may admit an optimal solution which is merely a capacitary measure. Take for instance $f = \mathcal{H}^{d-1}\lfloor S$ where $S \subset D$ is a regular $(d-1)$-dimensional surface. In this case the energy $\mathcal{E}_f(V)$ has the form*

$$\mathcal{E}_f(V) = \min\left\{\frac{1}{2}\int_D |\nabla u|^2\,dx + \frac{1}{2}\int_D Vu^2\,dx - \int_S u\,d\mathcal{H}^{d-1} \;:\; u \in H_0^1(D)\right\}.$$

By the results of Section 9.1 the maximization problem

$$\max\left\{\mathcal{E}_f(\mu) \;:\; \mu \in \mathcal{M}_{cap}(D),\ \int_D d\mu = 1\right\},$$

admits a solution μ_{opt} which is a capacitary measure. Repeating the proof of Theorem 9.20 we obtain the auxiliary variational problem

$$\mathcal{E}_f(\mu_{opt}) = \min\left\{\frac{1}{2}\int_D |\nabla u|^2\,dx + \frac{1}{2}\|u\|_{L^\infty}^2 - \int_S u\,d\mathcal{H}^{d-1} \;:\; u \in H_0^1(D)\right\}.$$

Denoting by u its unique solution and by M the maximum of u, we obtain that the optimal capacitary measure μ_{opt} is supported by the set $\{u = M\}$, this is contained in S (since the function u is subharmonic on $D \setminus S$) and so μ_{opt} is singular with respect to the Lebesgue measure. Moreover μ_{opt} has the form

$$\mu_{opt} = \frac{1}{M}\mathcal{H}^{d-1}\lfloor\{u = M\}.$$

The result in the following Theorem was proved in [356] (see also [505, Theorem 8.2.4]). We present it in a slightly different form as a simple consequence of Proposition 9.19. We recall the notation $\lambda_1(V)$ introduced in (9.14) for the first eigenvalue related to the potential V.

Theorem 9.22. *Let $D \subset \mathbb{R}^d$ be a bounded open set. Then there exists a unique solution to the maximization problem*

$$\max\left\{\lambda_1(V) \;:\; V \geq 0,\ \int_D V\,dx \leq 1\right\}, \tag{9.27}$$

given by

$$V_1 = \lambda\chi_\omega,$$

where $\omega = \left\{u_1 = \|u_1\|_{L^\infty(D)}\right\}$ and $u_1 \in H_0^1(D)$ solves the auxiliary variational problem

$$\lambda = \min\left\{\int_D |\nabla u|^2\,dx + \|u\|_{L^\infty(D)}^2 \;:\; u \in H_0^1(D),\ \int_D u^2\,dx = 1\right\}. \tag{9.28}$$

Proof. We first notice that due to the compact inclusion $H_0^1(D) \subset L^2(D)$ and the semi-continuity of the norm of the gradient there is a solution $u_\lambda \in H_0^1(D)$ of the problem

(9.28). We now set $f = \lambda u_\lambda$. Since for every $u \in H_0^1(D) \setminus \{0\}$ we have that

$$\min_{t\in\mathbb{R}} \left\{ \frac{t^2}{2} \int_D |\nabla u|^2\,dx + \frac{t^2}{2} \|u\|_{L^\infty}^2 - t \int_D uf\,dx \right\} = -\frac{\frac{1}{2}\left(\int_D uf\,dx\right)^2}{\int_D |\nabla u|^2\,dx + \|u\|_{L^\infty}^2},$$

we obtain that the minimizer of the functional J_1 corresponding to the function f is also the minimizer of the functional

$$J(u) = \frac{\int_D |\nabla u|^2\,dx + \|u\|_{L^\infty}^2}{\left(\int_D uf\,dx\right)^2}.$$

On the other hand, for every $u \in H_0^1(D)$ we have

$$\begin{aligned}\frac{\int_D |\nabla u|^2\,dx + \|u\|_{L^\infty}^2}{\left(\int_D uf\,dx\right)^2} &\geq \frac{\int_D |\nabla u|^2\,dx + \|u\|_{L^\infty}^2}{\|u\|_{L^2}^2 \|f\|_{L^2}^2} \\ &\geq \frac{\int_D |\nabla u_\lambda|^2\,dx + \|u_\lambda\|_{L^\infty}^2}{\|u_\lambda\|_{L^2}^2 \|f\|_{L^2}^2} = J(u_\lambda),\end{aligned}$$

which proves that u_λ is the minimizer of J_1. Thus u_λ satisfies the equation

$$-\Delta u_\lambda + V_1 u_\lambda = \lambda u_\lambda, \qquad u_\lambda \in H_0^1(D), \qquad \int_D u_\lambda^2\,dx.$$

where V_1 is such that

$$V_1 \geq 0, \qquad \int_D V_1\,dx = 1 \qquad \text{and} \qquad \int_D V_1 u_\lambda^2\,dx = \|u_\lambda\|_{L^\infty}^2.$$

Thus we have that

$$\lambda_1(V_1) = \min\left\{ \int_D |\nabla u|^2\,dx + \|u\|_{L^\infty}^2 \;:\; u \in H_0^1(D),\ \int_D u^2\,dx = 1 \right\},$$

On the other hand for every $V \geq 0$ such that $\int_D V\,dx = 1$ we have

$$\int_D |\nabla u|^2\,dx + \int_D u^2 V\,dx \leq \int_D |\nabla u|^2\,dx + \|u\|_{L^\infty}^2, \quad \text{for every} \quad u \in H_0^1(D),$$

which after taking the minimum with respect to u gives

$$\lambda_1(V) \leq \lambda = \lambda_1(V_1),$$

which proves that V_1 is a solution of (9.27).

In order to prove the uniqueness of the solution it is sufficient to check that there is a unique solution to the problem (9.28). In fact suppose that u_1 and u_2 are two distinct solutions of (9.28) and denote $M_i = \|u_i\|_{L^\infty}$, $\omega_i = \{u_i = M_i\}$ and $V_i = \lambda\chi_{\omega_i}$, for $i = 1, 2$. We consider now the potential $V = \dfrac{V_1 + V_2}{2}$. Since the function $V \to \lambda_1(V)$ is the infimum of a family of linear functions we know that it is concave and so, V is also a solution of (9.27). Now since V is optimal, we have that for every $A, B \subset \omega_1 \cup \omega_2$ with $|A| = |B|$,

$$\frac{d}{d\varepsilon}\Big|_{\varepsilon=0}\lambda_1\big(V + \varepsilon(\chi_A - \chi_B)\big) = 0.$$

Since the first eigenvalue is simple and the family of operators $-\Delta + V + \varepsilon(\chi_A - \chi_B)$ is analytic with respect to ε, we have that the functions $\varepsilon \mapsto \lambda_1\big(V + \varepsilon(\chi_A - \chi_B)\big)$ and $\varepsilon \mapsto u_\varepsilon$, where u_ε is the solution of

$$\begin{cases} -\Delta u_\varepsilon + \big(V + \varepsilon(\chi_A - \chi_B)\big)u_\varepsilon = \lambda_1\big(V + \varepsilon(\chi_A - \chi_B)\big)u_\varepsilon \\ u_\varepsilon \in H_0^1(D), \quad \int_D u_\varepsilon^2\,dx = 1 \end{cases}$$

are analytic. Taking the derivatives in ε at $\varepsilon = 0$ we obtain

$$\frac{d}{d\varepsilon}u_\varepsilon = u'$$

with

$$\begin{cases} -\Delta u' + Vu' + (\chi_A - \chi_B)u_0 = \lambda_1'(V)u_0 + \lambda_1(V)u' \\ u' \in H_0^1(D), \quad \int_D u'u_0\,dx = 0. \end{cases}$$

Multiplying both sides by u_0 and integrating by parts we get

$$\frac{d}{d\varepsilon}\Big|_{\varepsilon=0}\lambda_1\big(V + \varepsilon(\chi_A - \chi_B)\big) = \int_A u_0^2\,dx - \int_B u_0^2\,dx.$$

Since A and B are arbitrary we get that u_0 is a (positive, by the maximum principle) constant on $\omega_1 \cup \omega_2$ and since $u_0 \in H^2_{loc}(D)$ we obtain that

$$Vu_0 = -\Delta u_0 + V_0u_0 = \lambda u_0 \quad \text{on} \quad \omega_1 \cup \omega_2,$$

and as a consequence $V = \lambda$ on $\omega_1 \cup \omega_2$ which gives that $\omega_1 = \omega_2$, $V_1 = V_2$ and $u_1 = u_2$. □

Remark 9.23. *The proof above is constructed for the maximization of the first eigenvalue $\lambda_1(V)$ on the class*

$$\Big\{V \geq 0, \int_D V\,dx \leq 1\Big\}.$$

It would be interesting to consider the analogous maximization problem for $\lambda_k(V)$ on the same class of potentials

9.3 Existence results for confining potentials

In this section we consider the potential optimization problem

$$\min\left\{F(V)\ :\ V\in\mathcal{V}\right\} \tag{9.29}$$

where the functional F is as in the sections above and the admissible class $\mathcal{V}$ is given by

$$\mathcal{V}=\left\{V:D\to[0,+\infty]\ :\ V \text{ Lebesgue measurable, } \int_D \Psi(V)\,dx\le 1\right\} \tag{9.30}$$

and depends on a function $\Psi:[0,+\infty]\to[0,+\infty]$. On the function Ψ we make the following assumptions:

a) $\Psi:[0,+\infty]\to[0,+\infty]$ is an injective function;
b) there exist $p>1$ such that the function $s\mapsto\Psi^{-1}(s^p)$ is convex.

The assumptions above on the function Ψ are for instance satisfied by the following functions:

- $\Psi(s)=s^{-p}$, for any $p>0$;
- $\Psi(s)=e^{-\alpha s}$, for any $\alpha>0$.

and justify the terminology *"confining potentials"* we used. Indeed, large potentials turn out to be admissible.

The result showing the existence of an optimal potential in this case is as follows.

Theorem 9.24. *Let $D\subset\mathbb{R}^d$ be a bounded open set and let $\Psi:[0,+\infty]\to[0,+\infty]$ be a function satisfying the conditions a) and b) above. Let $F:\mathcal{M}_{cap}(D)\to\overline{\mathbb{R}}$ be a cost functional such that:*
i) F is lower semicontinuous with respect to the γ-convergence;
ii) F is increasing, that is

$$F(\mu_1)\le F(\mu_2) \text{ whenever } \mu_1\le\mu_2.$$

Then the optimization problem (9.29) has a solution, where the admissible class $\mathcal{V}$ is given by (9.30).

Proof. Let $V_n\in\mathcal{V}$ be a minimizing sequence for problem (9.29). Then the functions $v_n:=\left(\Psi(V_n)\right)^{1/p}$ are bounded in $L^p(D)$ and so, up to a subsequence, we may assume that v_n converges weakly in $L^p(D)$ to some function v. We will prove that the potential $V:=\Psi^{-1}(v^p)$ is optimal for the problem (9.29). Since v_n converges to v weakly in $L^p(D)$ we have

$$\int_D \Psi(V)\,dx=\int_D v^p\,dx\le\liminf_n\int_D v_n^p\,dx=\liminf_n\int_D\Psi(V_n)\,dx\le 1,$$

which shows that $V \in \mathcal{V}$. It remains to prove that

$$F(V) \le \liminf_{n} F(V_n).$$

By the compactness of the γ-convergence on the class $\mathcal{M}_{cap}(D)$, we can suppose that, up to a subsequence, V_n γ-converges to some capacitary measure $\mu \in \mathcal{M}_{cap}(D)$. Since F is assumed γ-lower semicontinuous, we have

$$F(\mu) \le \liminf_{n\to\infty} F(V_n). \tag{9.31}$$

We will show that $F(V) \le F(\mu)$, which, together with (9.31) will conclude the proof. By the definition of γ-convergence, we have that for any $u \in H_0^1(D)$, there is a sequence $u_n \in H_0^1(D)$ which converges to u in $L^2(D)$ and is such that

$$\begin{aligned}\int_D |\nabla u|^2\,dx + \int_D u^2\,d\mu &= \lim_{n\to\infty} \int_D |\nabla u_n|^2\,dx + \int_D u_n^2 V_n\,dx \\ &= \lim_{n\to\infty} \int_D |\nabla u_n|^2\,dx + \int_D u_n^2 \Psi^{-1}(v_n^p)\,dx \\ &\ge \int_D |\nabla u|^2\,dx + \int_D u^2 \Psi^{-1}(v^p)\,dx \\ &= \int_D |\nabla u|^2\,dx + \int_D u^2 V\,dx.\end{aligned} \tag{9.32}$$

The inequality in (9.32) is due to the $L^2(D)$ lower semicontinuity of the Dirichlet integral and to the strong-weak lower semicontinuity of integral functionals (see for instance [235]), which follows by the assumption b) on the function Ψ. Thus, for any $u \in H_0^1(D)$, we have

$$\int_D u^2\,d\mu \ge \int_D u^2 V\,dx,$$

which implies $V \le \mu$. Since F was assumed to increase monotonically, we obtain $F(V) \le F(\mu)$, which concludes the proof. □

Just like in the previous section, in some special cases, the solution to the optimization problem (9.29) can be computed explicitly through the solution to some auxiliary variational problem. This occurs for instance when $F(V) = \lambda_1(V)$ or when $F(V) = \mathcal{E}_f(V)$, with $f \in L^2(D)$. In fact, by the variational formulation

$$\lambda_1(V) = \min\left\{\int_D |\nabla u|^2 dx + \int_D u^2 V\,dx \;:\; u \in H_0^1(D),\ \int_D u^2\,dx = 1\right\},$$

we can rewrite the optimization problem (9.29) for $F(V) = \lambda_1(V)$ as

$$\begin{aligned}&\min_{V\in\mathcal{V}} \min_{\|u\|_2=1} \left\{\int_D |\nabla u|^2\,dx + \int_D u^2 V\,dx\right\} \\ &= \min_{\|u\|_2=1} \min_{V\in\mathcal{V}} \left\{\int_D |\nabla u|^2\,dx + \int_D u^2 V\,dx\right\}.\end{aligned} \tag{9.33}$$

The minimization with respect to V is easy to compute; in fact, if Ψ is differentiable with Ψ' invertible, then the minimum with respect to V in (9.33) is achieved for

$$V = (\Psi')^{-1}(\Lambda_u u^2), \tag{9.34}$$

where Λ_u is a constant such that

$$\int_D \Psi\left((\Psi')^{-1}(\Lambda_u u^2)\right)\,dx = 1.$$

Thus, the solution to the problem on the right hand side of (9.33) is given by the solution to the auxiliary variational problem

$$\min\left\{\int_D |\nabla u|^2\,dx + \int_D u^2(\Psi')^{-1}(\Lambda_u u^2)\,dx \;:\; u \in H_0^1(D),\ \int_D u^2\,dx = 1\right\}. \tag{9.35}$$

Analogously, in the case of the Dirichlet energy $F(V) = \mathcal{E}_f(V)$, we obtain that the optimal potential is given by (9.34), where this time u is a solution to the auxiliary variational problem

$$\min\left\{\int_D \frac{1}{2}|\nabla u|^2\,dx + \int_D \frac{1}{2}u^2(\Psi')^{-1}(\Lambda_u u^2)\,dx - \int_D fu\,dx \;:\; u \in H_0^1(D)\right\}. \tag{9.36}$$

Example 9.25. *Consider the case $\Psi(s) = s^{-p}$ with $p > 0$. Following the argument illustrated above we may conclude that the optimal potentials for the functionals $F(V) = \lambda_1(V)$ and $F(V) = \mathcal{E}_f(V)$ are given by*

$$V = \left(\int_D |u|^{2p/(p+1)}\,dx\right)^{1/p} u^{-2/(p+1)},$$

where u is the minimizer of the auxiliary variational problems (9.35) and (9.36) respectively. We also note that, in this case

$$\int_D u^2(\Psi')^{-1}(\Lambda_u u^2)\,dx = \left(\int_D |u|^{2p/(p+1)}\,dx\right)^{(1+p)/p}$$

and so the auxiliary variational problems (9.35) and (9.36) give rise to the nonlinear PDEs

$$-\Delta u + C_1(p,u)|u|^{-2/(p+1)}u = \lambda u \qquad u \in H_0^1(D)$$

$$-\Delta u + C_1(p,u)|u|^{-2/(p+1)}u = f \qquad u \in H_0^1(D)$$

respectively, where the constant $C(p,u)$ is given by

$$C(p,u) = \left(\int_D |u|^{2p/(p+1)}\,dx\right)^{1/p}.$$

Example 9.26. *Consider the case $\Psi(x) = e^{-\alpha x}$ with $\alpha > 0$. Again, the same argument we used above shows that the optimal potentials for the functionals $F(V) = \lambda_1(V)$ and $F(V) = \mathcal{E}_f(V)$ are given by*

$$V = \frac{1}{\alpha}\left(\log\left(\int_D u^2\,dx\right) - \log\left(u^2\right)\right),$$

where u is the minimizer of the auxiliary variational problems (9.35) and (9.36) respectively. We also note that, in this case

$$\int_D u^2(\Psi')^{-1}(\Lambda_u u^2)\,dx = \frac{1}{\alpha}\left(\int_D u^2\,dx \int_D \log\left(u^2\right)\,dx - \int_D u^2\log\left(u^2\right)\,dx\right)$$

and so the auxiliary variational problems (9.35) and (9.36) give rise to the nonlinear PDEs

$$-\Delta u + \frac{1}{\alpha}\left(C_2(u) + C_3(u)\frac{1}{u^2} - 2\log|u| - 1\right)u = \lambda u \qquad u \in H_0^1(D)$$

$$-\Delta u + \frac{1}{\alpha}\left(C_2(u) + C_3(u)\frac{1}{u^2} - 2\log|u| - 1\right)u = f \qquad u \in H_0^1(D)$$

respectively, where the constants $C_2(u)$ and $C_3(u)$ are given by

$$C_2(u) = 2\int_D \log|u|\,dx \qquad C_3(u) = \int_D u^2\,dx.$$

The function $\Psi(s) = e^{-\alpha s}$ in the constraint (9.30) can be used to simulate and approximate a volume constraint in a shape optimization problem of the form

$$\min\left\{F(\Omega)\ :\ \Omega \subset D,\ |\Omega| \le 1\right\}$$

in which the main unknown is a domain $\Omega \subset D$, or equivalently a potential of the form $V = \infty_{D\setminus\Omega}$, written as a capacitary measure. Taking the cost functional $F(V) = \mathcal{E}_f(V)$ and replacing the constraint (9.30) by the addition of a Lagrange multiplier term, we obtain the problem

$$\min\left\{\mathcal{E}_f(V) + \Lambda\int_D e^{-\alpha V}\,dx\ :\ V \ge 0\right\}, \tag{9.37}$$

where Λ is a Lagrange multiplier and the potential V now varies among the nonnegative Borel measurable functions on D. As before, we note that the problem (9.37) is equivalent to

$$\min\left\{\int_D \left(\frac{1}{2}|\nabla u|^2 + \frac{1}{2}Vu^2 - fu + \Lambda e^{-\alpha V}\right)dx\ :\ u \in H_0^1(D),\ V \ge 0\right\}. \tag{9.38}$$

Fixing $u \in H_0^1(D)$ and minimizing with respect to V leads to the problem

$$\min\left\{\int_D Vu^2\,dx + \Lambda\int_D e^{-\alpha V}\,dx\ :\ V \ge 0\right\},$$

whose solution V can be obtained through the relation

$$u^2 - \Lambda\alpha e^{-\alpha V} = 0 \qquad \text{on } \{V(x) > 0\}.$$

We note that on the set where $u^2 \ge \Lambda\alpha$ we necessarily have that $V = 0$. On the other hand, if $u^2 < \Lambda\alpha$, then by the optimality of V, we have that $V > 0$. Finally, the optimal potential V can be identified in terms of u by

$$V(x) = 0 \vee \left(-\frac{1}{\alpha}\log\frac{u^2}{\Lambda\alpha}\right). \tag{9.39}$$

Replacing the expression above in (9.38), we obtain the auxiliary problem

$$\min_{u\in H_0^1(D)}\Big\{\frac{1}{2}\int_D|\nabla u|^2\,dx-\frac{1}{2\alpha}\int_{\{u^2<\Lambda\alpha\}}u^2\log\Big(\frac{u^2}{\Lambda\alpha}\Big)\,dx$$
$$-\int_D fu\,dx+\Lambda|\{u^2\geq\Lambda\alpha\}|+\frac{1}{\alpha}\int_{\{u^2<\Lambda\alpha\}}u^2\,dx\Big\},$$

which can be equivalently written as

$$\min_{u\in H_0^1(D)}\Big\{\frac{1}{2}\int_D|\nabla u|^2\,dx-\frac{1}{2\alpha}\int_{\{u^2\leq\Lambda\alpha\}}u^2\log\Big(\frac{u^2}{\Lambda\alpha}\Big)\,dx$$
$$-\int_D fu\,dx+\Lambda|\{u^2>\Lambda\alpha\}|+\frac{1}{\alpha}\int_{\{u^2\leq\Lambda\alpha\}}u^2\,dx\Big\}. \tag{9.40}$$

Note that the function

$$h_\alpha(s)=\begin{cases}\frac{s^2}{2\alpha}\Big(2-\log\Big(\frac{s^2}{\Lambda\alpha}\Big)\Big) & \text{if } s^2\leq\Lambda\alpha\\ \Lambda & \text{if } s^2>\Lambda\alpha\end{cases}$$

is lower semicontinuous and nonnegative, which provides the necessary lower semicontinuity and coercivity to apply the direct methods of the calculus of variations and conclude that the auxiliary problem (9.40) has a solution $u_\alpha\in H_0^1(D)$. Moreover, on the quasi-open set $\{u^2>\Lambda\alpha\}$, we have $-\Delta u=f$. Let us denote by J_α the cost functional appearing in (9.40), that is

$$J_\alpha(u)=\frac{1}{2}\int_D|\nabla u|^2\,dx+\frac{1}{2\alpha}\int_{\{u^2\leq\Lambda\alpha\}}u^2\Big[2-\log\Big(\frac{u^2}{\Lambda\alpha}\Big)\Big]\,dx+\Lambda|\{u^2>\Lambda\alpha\}|$$
$$=\frac{1}{2}\int_D|\nabla u|^2\,dx+\int_D h_\alpha(u)\,dx.$$

As $\alpha\to0$ the functions h_α increase and converge to the function

$$h(s)=\begin{cases}\Lambda & \text{if } s>0\\ 0 & \text{if } s=0\end{cases}$$

Then the functionals J_α Γ-converges in $L^2(D)$, as $\alpha\to0$, to the functional

$$J(u)=\frac{1}{2}\int_D|\nabla u|^2\,dx+\int_D h(u)\,dx=\frac{1}{2}\int_D|\nabla u|^2\,dx+\Lambda|\{u\neq0\}|.$$

By the properties of the Γ-convergence this implies the convergence of the solutions u_α of (9.40) and hence, thanks to the relation (9.39), of the optimal potentials V_α for (9.37) to a limit potential of the form

$$V(x)=\begin{cases}+\infty & \text{if } u(x)=0\\ 0 & \text{if } u(x)\neq0,\end{cases}$$

where u is a solution to the limit problem

$$\min\left\{\frac{1}{2}\int_D |\nabla u|^2\,dx - \int_D fu\,dx + \Lambda|\{u \neq 0\}| \;:\; u \in H_0^1(D)\right\}.$$

This limit problem is indeed a shape optimization problem written in terms of the state function u; several results on the regularity of the optimal domains are known (see for instance [25], [187], [189], as well as Chapter 3 of the present book).

Acknowledgment: The work of Giuseppe Buttazzo has been supported by the Italian Ministry of Research and University through the Project 2010A2TFX2 "Calcolo delle Variazioni" and by the Gruppo Nazionale per l'Analisi Matematica, la Probabilitá e le loro Applicazioni (GNAMPA) of the Istituto Nazionale di Alta Matematica (INdAM). The work of Bozhidar Velichkov was supported by Université Grenoble Alpes through the project AGIR VARIFORM.

Virginie Bonnaillie-Noël and Bernard Helffer

10 Nodal and spectral minimal partitions – The state of the art in 2016 –

10.1 Introduction

We consider mainly the Dirichlet realization of the Laplacian operator in Ω, when Ω is a bounded domain in $\mathbb{R}^2$ with piecewise-C^1 boundary (domains with corners or cracks[10.1] permitted). This operator will be denoted by $H(\Omega)$. We would like to analyze the connections between the nodal domains of the eigenfunctions of $H(\Omega)$ and the partitions of Ω by k open sets D_i which are minimal in the sense that the maximum over the D_i's of the groundstate energy of the Dirichlet realization of the Laplacian $H(D_i)$ is minimal. This problem can be seen as a strong competition limit of segregating species in population dynamics (see [286, 289] and references therein).

Definition 10.1. *A* **partition** *(or k-partition for indicating the cardinality of the partition) of Ω is a family $\mathcal{D} = \{D_i\}_{1\le i\le k}$ of k mutually disjoint sets in Ω (with $k \ge 1$ an integer).*

We denote by $\mathfrak{O}_k = \mathfrak{O}_k(\Omega)$ the set of partitions of Ω where the D_i's are domains (i.e. open and connected). We now introduce the notion of the energy of a partition.

Definition 10.2. *For any integer $k \ge 1$, and for $\mathcal{D} = \{D_i\}_{1\le i\le k}$ in $\mathfrak{O}_k(\Omega)$, we introduce the energy of the partition:*

$$\Lambda(\mathcal{D}) = \max_{1\le i\le k} \lambda(D_i). \tag{10.1}$$

The optimal problem we are interested in is the determination, for any integer $k \ge 1$, of

$$\mathfrak{L}_k = \mathfrak{L}_k(\Omega) = \inf_{\mathcal{D}\in\mathfrak{O}_k(\Omega)} \Lambda(\mathcal{D}). \tag{10.2}$$

10.1 For example a square with a segment removed

Virginie Bonnaillie-Noël: Département de Mathématiques et Applications (DMA - UMR 8553), PSL Research University, CNRS, ENS Paris, 45 rue d'Ulm, F-75230 Paris cedex 05, France, E-mail: virginie.bonnaillie@ens.fr
Bernard Helffer: Laboratoire de Mathématique Jean Leray (UMR 6629), Université de Nantes, CNRS, 2 rue de la Houssinière, BP 92208, F-44322 Nantes cedex 3, France
and Laboratoire de Mathématiques (UMR 8628), Université Paris Sud, CNRS, Bât 425, F-91405 Orsay cedex, France, E-mail: bernard.helffer@univ-nantes.fr

We can also consider the case of a two-dimensional Riemannian manifold and the Laplacian is then the Laplace-Beltrami operator. We denote by $\{\lambda_j(\Omega), j \geq 1\}$ (or more simply λ_j if there is no ambiguity) the non decreasing sequence of its eigenvalues and by $\{u_j, j \geq 1\}$ some associated orthonormal basis of eigenfunctions. To be succinct, we often write $\lambda(\Omega)$ instead of $\lambda_1(\Omega)$. The groundstate u_1 can be chosen to be strictly positive in Ω, but the other excited eigenfunctions u_k must have zerosets. Here we recall that for $u \in C^0(\overline{\Omega})$, the nodal set (or zeroset) of u is defined by :

$$N(u) = \overline{\{\mathbf{x} \in \Omega \mid u(\mathbf{x}) = 0\}}\,. \tag{10.3}$$

In the case when u is an eigenfunction of the Laplacian, the $\mu(u)$ components of $\Omega \setminus N(u)$ are called the nodal domains of u and define naturally a partition of Ω by $\mu(u)$ open sets, which will be called a **nodal partition.**
Our main goal is to discuss the links between the partitions of Ω associated with these eigenfunctions and the minimal partitions of Ω.

10.2 Nodal partitions

10.2.1 Minimax characterization

Flexible criterion

We first give a flexible criterion for the determination of the bottom of the spectrum.

Theorem 10.3. *Let $\mathcal{H}$ be an Hilbert space of infinite dimension, $\|\cdot\|_{\mathcal{H}}$ be the endowed norm and P be a self-adjoint semibounded operator*[10.2] *of form domain $Q(P) \subset \mathcal{H}$ with compact injection. Let us introduce*

$$\mu_1(P) = \inf_{\phi \in Q(P)\setminus\{0\}} \frac{\langle P\phi \mid \phi\rangle_{\mathcal{H}}}{\|\phi\|^2_{\mathcal{H}}}\,, \tag{10.4}$$

and, for $n \geq 2$

$$\mu_n(P) = \sup_{\psi_1,\psi_2,\ldots,\psi_{n-1}\in Q(P)} \ \inf_{\substack{\phi\in[\mathrm{span}\,(\psi_1,\ldots,\psi_{n-1})]^{\perp}\\ \phi\in Q(P)\setminus\{0\}}} \frac{\langle P\phi \mid \phi\rangle_{\mathcal{H}}}{\|\phi\|^2_{\mathcal{H}}}\,. \tag{10.5}$$

Then $\mu_n(P)$ is the n-th eigenvalue when ordering the eigenvalues in increasing order (and counting their multiplicity).

Note that the proof involves the following proposition

10.2 The operator is associated with a coercive continuous symmetric sesquilinear form via Lax-Milgram's theorem. See for example [487].

Proposition 10.4. *Under the conditions of Theorem 10.3, suppose that there exist a constant a and a n-dimensional subspace $V \subset Q(P)$ such that,*

$$\langle P\phi\,,\,\phi\rangle_{\mathcal{H}} \le a\|\phi\|^2_{\mathcal{H}}, \qquad \forall\phi \in V.$$

Then $\mu_n(P) \le a$.

This could be applied when P is the Dirichlet Laplacian ($\mathcal{H} = L^2(\Omega)$ and $Q(P) = H^1_0(\Omega)$), the Neumann Laplacian ($\mathcal{H} = L^2(\Omega)$ and $Q(P) = H^1(\Omega)$) and the Harmonic oscillator ($\mathcal{H} = L^2(\mathbb{R}^n)$ and $Q(P) = B^1(\mathbb{R}^n) := \{u \in L^2(\mathbb{R}^n) : x_j u \in L^2(\mathbb{R}^n), \partial_{x_j} u \in L^2(\mathbb{R}^n)\}$).

An alternative characterization of λ_2

$\mathfrak{L}_2(\Omega)$ was introduced in (10.2). We now introduce another spectral sequence associated with the Dirichlet Laplacian.

Definition 10.5. *For any $k \ge 1$, we denote by $L_k(\Omega)$ (or L_k if there is no confusion) the smallest eigenvalue (if any) for which there exists an eigenfunction with k nodal domains. We set $L_k(\Omega) = +\infty$ if there is no eigenfunction with k nodal domains.*

Proposition 10.6. $\mathfrak{L}_2(\Omega) = \lambda_2(\Omega) = L_2(\Omega)$.

Proof. By definition of L_k, we have $\mathfrak{L}_2 \le L_2$.
The equality $\lambda_2 = L_2$ is a standard consequence of the fact that a second eigenfunction has exactly two nodal domains: the upper bound is a consequence of Courant and the lower bound is by orthogonality.
It remains to show that $\lambda_2 \le \mathfrak{L}_2$. This is a consequence of the min-max principle. For any $\varepsilon > 0$, there exists a 2-partition $\mathcal{D} = \{D_1, D_2\}$ of energy Λ such that $\Lambda < \mathfrak{L}_2 + \varepsilon$. We can construct a 2-dimensional space generated by the two ground states (extended by 0) u_1 and u_2 of energy less than Λ. This implies:

$$\lambda_2 \le \Lambda < \mathfrak{L}_2 + \varepsilon\,.$$

It is sufficient to take the limit $\varepsilon \to 0$ to conclude. □

10.2.2 On the local structure of nodal sets

We refer for this section to the survey of P. Bérard [123] or the book by I. Chavel [261]. We first mention a proposition (see [261, Lemma 1, p. 21-23] whose original proof is given in [133], Appendix D) which is implicitly used in many proofs and was overlooked in [292].

Proposition 10.7. *If u is an eigenfunction associated with λ and D is one of its nodal domains, then the restriction of u to D belongs to $H_0^1(D)$ and is an eigenfunction of the Dirichlet realization of the Laplacian in D. Moreover, λ is the ground state energy in D.*

Proposition 10.8. *Let f be a real valued eigenfunction of the Dirichlet-Laplacian on a two dimensional locally flat Riemannian manifold Ω with smooth boundary. Then $f \in C^\infty(\overline{\Omega})$. Furthermore, f has the following properties:*

1. *If f has a zero of order ℓ at a point $\mathbf{x}_0 \in \overline{\Omega}$, then the Taylor expansion of f is*

$$f(\mathbf{x}) = p_\ell(\mathbf{x} - \mathbf{x}_0) + \mathcal{O}(|\mathbf{x} - \mathbf{x}_0|^{\ell+1}), \tag{10.6}$$

   where p_ℓ is a real valued, non-zero, harmonic, homogeneous polynomial function of degree ℓ.
   Moreover if $\mathbf{x}_0 \in \partial\Omega$, the Dirichlet boundary conditions imply that

$$f(\mathbf{x}) = a\, r^\ell \sin \ell\omega + \mathcal{O}(r^{\ell+1}), \tag{10.7}$$

   for some non-zero $a \in \mathbb{R}$, where (r, ω) are polar coordinates of $\mathbf{x}$ around $\mathbf{x}_0$. The angle ω is chosen so that the tangent to the boundary at $\mathbf{x}_0$ is given by the equation $\sin \omega = 0$.
2. *The nodal set $\mathcal{N}(f)$ is the union of finitely many smoothly immersed circles in Ω, and smoothly immersed lines, with possible self-intersections, which connect points of $\partial\Omega$. Each of these immersions is called a nodal line. The connected components of $\Omega \setminus \mathcal{N}(f)$ are called nodal domains.*
3. *If f has a zero of order ℓ at $\mathbf{x}_0 \in \Omega$, then exactly ℓ segments of nodal lines pass through $\mathbf{x}_0$. The tangents to the nodal lines at $\mathbf{x}_0$ dissect the disk into 2ℓ equal angles. If f has a zero of order ℓ at $\mathbf{x}_0 \in \partial\Omega$, then exactly ℓ segments of nodal lines meet the boundary at $\mathbf{x}_0$. The tangents to the nodal lines at $\mathbf{x}_0$ are given by the equation $\sin \ell\omega = 0$, where ω is chosen as in (10.7).*

Proof. The proof that $f \in C^\infty(\overline{\Omega})$ can be found in [881, Theorem 20.4]. Moreover, the function f is analytic in Ω (property of the Laplacian). Hence, since f is not identically 0, Part *1* becomes trivial. See [139, 268] for the proof of the other parts. □

Remark 10.9.
- *In the case of Neumann condition Proposition 10.8 remains true if the Taylor expansion* (10.7) *for a zero of order ℓ at a point $\mathbf{x}_0 \in \partial\Omega$ is replaced by*

$$f(\mathbf{x}) = a\, r^\ell \cos \ell\omega + \mathcal{O}(r^{\ell+1}),$$

  where we have used the same polar coordinates (r, ω) centered at $\mathbf{x}_0$.
- *Proposition 10.8 remains true for polygonal domains, see [323] and for more general domains [477] (and references therein).*

From the above, we should remember that nodal sets are regular in the sense:

- The singular points $\mathbf{x}_0$ ($\ell > 1$) on the nodal lines are isolated.
- At the singular points, an even number of half-lines meet with equal angle.
- At the boundary, this is the same as adding the tangent line in the picture.

This will be made more precise later for general partitions in Subsection 10.4.2.

10.2.3 Weyl's theorem

We consider the Dirichlet realization of the Laplacian in a bounded regular set $\Omega \subset \mathbb{R}^n$, which will be denoted by $H(\Omega)$. For $\lambda \in \mathbb{R}$, we introduce the counting function $N(\lambda)$ by:

$$N(\lambda) := \sharp\{j \,:\, \lambda_j < \lambda\}\,. \tag{10.8}$$

We write $N(\lambda, \Omega)$ if we want to recall in which open set Ω the realization is considered. Weyl's theorem (established by H. Weyl in 1911) gives the asymptotic behavior of $N(\lambda)$ as $\lambda \to +\infty$.

Theorem 10.10 (Weyl). *As $\lambda \to +\infty$,*

$$N(\lambda) \sim \frac{\omega_n}{(2\pi)^n} |\Omega| \lambda^{\frac{n}{2}}\,, \tag{10.9}$$

where ω_n denotes the volume of a ball of radius 1 in $\mathbb{R}^n$ and $|\Omega|$ the volume of Ω.

In dimension $n = 2$, we find:

$$N(\lambda) \sim \frac{|\Omega|}{4\pi}\lambda\,. \tag{10.10}$$

Proof. The proof of Weyl's theorem can be found in [876], [292, p. 42] or in [261, p. 30-32]. We sketch here the so called Dirichlet-Neumann bracketing technique, which is presented in [292].

The idea is to use a suitable partition $\{D_i\}_i$ to prove lower and upper bounds according to $\sum_i N(\lambda, D_i)$. If the domains D_i are cubes, the eigenvalues of the Laplacian (with Dirichlet or Neumann conditions) are known explicitly and this gives explicit bounds for $N(\lambda, D_i)$ (see (10.26) for the case of the square). Let us provide details for the lower bound which is the most important for us. For any partition $\mathcal{D} = \{D_i\}_i$ in Ω, the min-max characterization of the spectrum implies

$$\sum_i N(\lambda, D_i) \leq N(\lambda, \Omega)\,. \tag{10.11}$$

Given $\varepsilon > 0$, we can find a partition $\{D_i\}_i$ of Ω by cubes such that $|\Omega \setminus \cup_i D_i| \leq \varepsilon|\Omega|$. Summing up in (10.11), and using Weyl's formula (lower bound) for each D_i, we obtain:

$$N(\lambda, \Omega) \geq (1-\varepsilon)\frac{\omega_n}{(2\pi)^n} |\Omega|\lambda^{\frac{n}{2}} + o_\varepsilon(\lambda^{\frac{n}{2}})\,.$$

Let us deal now with the upper bound. For any partition $\mathcal{D} = \{D_i\}_i$ in $\mathbb{R}^n$ such that $\Omega \subset \cup \overline{D_i}$ and ∂D_i of measure zero, we have, by comparison of form domains, the upper bound

$$N(\lambda, \Omega) \le \sum_i N(\lambda, -\Delta_{D_i}^{Neu}), \tag{10.12}$$

where $N(\lambda, \Delta_{D_i}^{Neu})$ denotes the number of eigenvalues, below λ, of the Neumann realization of the Laplacian in D_i. Then we choose a partition with suitable cubes for which the eigenvalues are known explicitly. □

Remark 10.11. *To improve the lower bound with a more explicit remainder, we can use more explicit lower bounds for the counting function for the cube (see Subsection 10.3.3 in the 2D case) and also consider cubes of size depending on λ. This will be also useful in Subsection 10.9.3.*

We do not need in this chapter improved Weyl's formulas with control of the remainder (see however in the analysis of Courant sharp cases (10.26) and (10.33)). We nevertheless mention a formula due to V. Ivrii in 1980 (cf [545, Chapter XXIX, Theorem 29.3.3 and Corollary 29.3.4]) which reads:

$$N(\lambda) = \frac{\omega_n}{(2\pi)^n}|\Omega|\lambda^{\frac{n}{2}} - \frac{1}{4}\frac{\omega_{n-1}}{(2\pi)^{n-1}}|\partial\Omega|\lambda^{\frac{n-1}{2}} + r(\lambda), \tag{10.13}$$

where $r(\lambda) = \mathcal{O}(\lambda^{\frac{n-1}{2}})$ in general but can also be shown to be $o(\lambda^{\frac{n-1}{2}})$ under some rather generic conditions about the geodesic billiards (the measure of periodic trajectories should be zero) and C^∞ boundary. This is only in this case that the second term is meaningful.
Formula (10.13) can also be established in the case of irrational rectangles as a very special case in [554], but more explicitly in [612] without any assumption of irrationality. This has also been extended in particular to specific triangles of interest (equilateral, right angled isosceles, hemiequilateral) by P. Bérard (see [125] and references therein).

Remark 10.12.
1. *The same asymptotics* (10.9) *are true for the Neumann realization of the Laplacian. The two-terms asymptotics* (10.13) *also holds but with the sign + before the second term.*
2. *For the harmonic oscillator (particular case of a Schrödinger operator $-\Delta + V$, with $V(\mathbf{x}) \to +\infty$ as $|\mathbf{x}| \to +\infty$) the situation is different. One can use either the fact that the spectrum is explicit or a pseudodifferential calculus. For the isotropic harmonic oscillator $-\Delta + |\mathbf{x}|^2$ in $\mathbb{R}^n$, the formula reads*

$$N(\lambda) \sim \frac{\omega_{2n-1}}{(2\pi)^n}\frac{\lambda^n}{2n}. \tag{10.14}$$

Note that the power of λ appearing in the asymptotics for the harmonic oscillator in $\mathbb{R}^n$ is, for a given n, the double of the one obtained for the Laplacian.

10.2.4 Courant's theorem and Courant sharp eigenvalues

This theorem was established by R. Courant [290] in 1923 for the Laplacian with Dirichlet or Neumann conditions.

Theorem 10.13 (Courant). *The number of nodal components of the k-th eigenfunction is not greater than k.*

Proof. The main arguments of the proof are already present in Courant-Hilbert [292, p. 453-454]. Suppose that u_k has $(k+1)$ nodal domains $\{D_i\}_{1\le i\le k+1}$. We also assume $\lambda_{k-1} < \lambda_k$. Considering k of these nodal domains and looking at $\Phi_a := \sum_{i=1}^{k} a_i\phi_i$ where ϕ_i is the ground state in each D_i, we can determine a_i such that Φ_a is orthogonal to the $(k-1)$ first eigenfunctions. On the other hand Φ_a has its energy bounded by λ_k. Hence it should be an eigenfunction for λ_k. But Φ_a vanishes in the open set D_{k+1} in contradiction with the property of an eigenfunction which cannot be flat at a point. □

On Courant's theorem with symmetry

Suppose that there exists an isometry g such that $g(\Omega) = \Omega$ and $g^2 = Id$. Then g acts naturally on $L^2(\Omega)$ by $gu(\mathbf{x}) = u(g^{-1}\mathbf{x})$, $\forall \mathbf{x} \in \Omega$, and one can naturally define an orthogonal decomposition of $L^2(\Omega)$

$$L^2(\Omega) = L^2_{odd} \oplus L^2_{even}\,,$$

where by definition $L^2_{odd} = \{u \in L^2\,,\ gu = -u\}$, resp. $L^2_{even} = \{u \in L^2\,,\ gu = u\}$. These two spaces are left invariant by the Laplacian and one can consider separately the spectrum of the two restrictions. Let us explain for the "odd case" what is the Courant theorem with symmetry. If u is an eigenfunction in L^2_{odd} associated with λ, we see immediately that the nodal domains appear in pairs (exchanged by g) and following the proof of the standard Courant theorem we see that, if $\lambda = \lambda_j^{odd}$ for some j (that is the j-th eigenvalue in the odd space), then the number $\mu(u)$ of nodal domains of u satisfies $\mu(u) \le 2j$.

We get a similar result for the "even" case (but in this case a nodal domain D is either g-invariant or $g(D)$ is a distinct nodal domain).

These remarks lead to improvement when each eigenspace has a specific symmetry. This will be the case for the sphere, the harmonic oscillator, the square (see (10.31)), the Aharonov-Bohm operator, ... where g can be the antipodal map, the map $(x, y) \mapsto (-x, -y)$, the map $(x, y) \mapsto (\pi - x, \pi - y)$, the deck map (as in Subsection 10.8.5), ...

Definition 10.14. *We say that (u, λ) is a spectral pair for $H(\Omega)$ if λ is an eigenvalue of the Dirichlet-Laplacian $H(\Omega)$ on Ω and $u \in E(\lambda)\setminus\{0\}$, where $E(\lambda)$ denotes the eigenspace attached to λ.*

Definition 10.15. *We say that a spectral pair (u, λ) is Courant sharp if $\lambda = \lambda_k$ and u has k nodal domains. We say that an eigenvalue λ_k is Courant sharp if there exists an eigenfunction u associated with λ_k such that (u, λ_k) is a Courant sharp spectral pair.*

If the Sturm-Liouville theory shows that in dimension 1 all the spectral pairs are Courant sharp, we will see below that when the dimension is ≥ 2, the Courant sharp situation can only occur for a finite number of eigenvalues.

The following property of transmission of the Courant sharp property to sub-partitions will be useful in the context of minimal partitions. Its proof can be found in [35].

Proposition 10.16.
1. *Let (u, λ) be a* **Courant sharp** *spectral pair for $H(\Omega)$ with $\lambda = \lambda_k$ and $\mu(u) = k$. Let $\mathcal{D}^{(k)} = \{D_i\}_{1\leq i\leq k}$ be the family of the nodal domains associated with u. Let L be a subset of $\{1, \dots, k\}$ with $\sharp L = \ell$ and let $\mathcal{D}_L$ be the subfamily $\{D_i\}_{i\in L}$. Let $\Omega_L = \mathrm{Int}\,(\overline{\cup_{i\in L} D_i}) \setminus \partial\Omega$. Then*
$$\lambda_\ell(\Omega_L) = \lambda_k, \tag{10.15}$$
where $\{\lambda_j(\Omega_L)\}_j$ are the eigenvalues of $H(\Omega_L)$.
2. *Moreover, when Ω_L is connected, $u\big|_{\Omega_L}$ is* **Courant sharp** *and $\lambda_\ell(\Omega_L)$ is simple.*

10.2.5 Pleijel's theorem

Motivated by Courant's Theorem, Pleijel's theorem (1956) says

Theorem 10.17 (Weak Pleijel's theorem). *If the dimension is ≥ 2, there is only a finite number of Courant sharp eigenvalues of the Dirichlet Laplacian.*

This theorem is the consequence of a more precise theorem which gives a link between Pleijel's theorem and partitions. For describing this result and its proof, we first recall the Faber-Krahn inequality:

Theorem 10.18 (Faber-Krahn inequality). *For any domain $D \subset \mathbb{R}^2$, we have*
$$|D|\,\lambda(D) \geq \lambda(\bigcirc)\,, \tag{10.16}$$
where $|D|$ denotes the area of D and $\bigcirc$ is the disk of unit area $\mathcal{B}\left(0, \frac{1}{\sqrt{\pi}}\right)$.

Remark 10.19. *Note that improvements can be useful when D is "far" from a disk. It is then interesting to have a lower bound for $|D|\,\lambda(D) - \lambda(\bigcirc)$. We refer for example to [181] and [447]. These ideas are behind recent improvements by Steinerberger [831], Bourgain [174] and Donnelly [340] of the strong Pleijel's theorem below. See also Subsection 10.9.1.*

By summation of Faber-Krahn's inequalities (10.16) applied to each D_i and having in mind Definition 10.2, we deduce:

Lemma 10.20. *For any open partition $\mathcal{D}$ in Ω we have*

$$|\Omega|\, \Lambda(\mathcal{D}) \geq \sharp(\mathcal{D})\, \lambda(\bigcirc)\,, \tag{10.17}$$

where $\sharp(\mathcal{D})$ denotes the number of elements of the partition.

Note that instead of using summation, we can prove the previous lemma by using the fact that there exists some D_i with $|D_i| \leq \frac{|\Omega|}{k}$ and apply Faber-Krahn's inequality for this D_i. There is no gain in the case considered in this chapter, but in other contexts (see for example Proposition 10.35), this could be useful, since Faber-Krahn's inequality holds only under a constraint on the area, which becomes satisfied for k large enough (see [133]).

Let us now give the strong form of Pleijel's theorem.

Theorem 10.21 (Strong Pleijel's theorem). *Let ϕ_n be an eigenfunction of $H(\Omega)$ associated with $\lambda_n(\Omega)$. Then*

$$\limsup_{n\to+\infty} \frac{\mu(\phi_n)}{n} \leq \frac{4\pi}{\lambda(\bigcirc)}\,, \tag{10.18}$$

where $\mu(\phi_n)$ is the cardinality of the connected components of $\Omega \setminus N(\phi_n)$.

Remark 10.22. *Of course, this implies Theorem 10.17. We have indeed*

$$\lambda(\bigcirc) = \pi \mathbf{j}^2\,,$$

and $\mathbf{j} \simeq 2.40$ is the smallest positive zero of the Bessel function of first kind. Hence

$$\frac{4\pi}{\lambda(\bigcirc)} = \left(\frac{2}{\mathbf{j}}\right)^2 < 1\,.$$

Proof. We start from the following identity

$$\frac{\mu(\phi_n)}{n}\,\frac{n}{\lambda_n}\,\frac{\lambda_n}{\mu(\phi_n)} = 1\,. \tag{10.19}$$

Applying Lemma 10.20 to the nodal partition of ϕ_n (which is associated with λ_n), we have

$$\frac{\lambda_n}{\mu(\phi_n)} \geq \frac{\lambda(\bigcirc)}{|\Omega|}\,.$$

Let us take a subsequence ϕ_{n_i} such that $\lim_{i\to+\infty} \frac{\mu(\phi_{n_i})}{n_i} = \limsup_{n\to+\infty} \frac{\mu(\phi_n)}{n}$, and implementing this equality in (10.19), we deduce:

$$\frac{\lambda(\bigcirc)}{|\Omega|}\,\limsup_{n\to+\infty} \frac{\mu(\phi_n)}{n}\,\liminf_{\lambda\to+\infty} \frac{N(\lambda)}{\lambda} \leq 1\,. \tag{10.20}$$

With Weyl's formula (10.10) in mind, we get (10.18). □

To finish this section, let us mention the particular case of irrational rectangles (see [152] and [763]).

Proposition 10.23. *Let us denote by $\mathcal{R}(a,b)$ the rectangle $(0,a\pi)\times(0,b\pi)$, with $a>0$ and $b>0$. We assume that b^2/a^2 is irrational. Then Theorem 10.21 is true for the rectangle $\mathcal{R}(a,b)$ with constant $4\pi/\lambda(\bigcirc)$ replaced by $4\pi/\lambda(\square)=2/\pi$, where $\square$ is a square of area 1. Moreover we have*

$$\limsup_{n\to+\infty}\frac{\mu(\phi_n)}{n}=\frac{2}{\pi}. \tag{10.21}$$

Proof. Since b^2/a^2 is irrational, the eigenvalues $\hat{\lambda}_{m,n}$ are simple and eigenpairs are given, for $m\geq 1$, $n\geq 1$, by

$$\hat{\lambda}_{m,n}=\frac{m^2}{a^2}+\frac{n^2}{b^2},\qquad \phi_{m,n}(x,y)=\sin\frac{mx}{a}\sin\frac{ny}{b}. \tag{10.22}$$

Without restriction, we can assume $a=1$. Thus we have $\mu(\phi_{m,n})=mn$. Applying Weyl asymptotics (10.10) with $\lambda=\hat{\lambda}_{m,n}$ gives

$$k(m,n):=\sharp\{(\tilde{m},\tilde{n}):\hat{\lambda}_{\tilde{m},\tilde{n}}(b)<\lambda\}=\frac{b\pi}{4}\left(m^2+\frac{n^2}{b^2}\right)+o(\lambda). \tag{10.23}$$

We have $\lambda_{k(m,n)+1}=\hat{\lambda}_{m,n}$. We observe that $\mu(\phi_{n,m})/k(n,m)$ is asymptotically given by

$$P(m,n;b):=\frac{4mn}{\pi\left(m^2b+\frac{n^2}{b}\right)}\leq\frac{2}{\pi}. \tag{10.24}$$

Taking a sequence (m_k,n_k) such that $b=\lim_{k\to\infty}\frac{n_k}{m_k}$ with $m_k\to+\infty$, we deduce

$$\lim_{k\to+\infty}P(m_k,n_k;b)=\frac{2}{\pi}, \tag{10.25}$$

which gives the proposition by using this sequence of eigenfunctions ϕ_{m_k,n_k}. □

Remark 10.24. *There is no hope in general to have a positive lower bound for* $\liminf\mu(\phi_n)/n$. *A. Stern for the square and the sphere (1925), H. Lewy for the sphere (1977), J. Leydold for the harmonic oscillator [656] (see [128, 129, 132] for the references, analysis of the proofs and new results) have constructed infinite sequences of eigenvalues such that a corresponding eigenfunctions have two or three nodal domains. On the contrary, it is conjectured in [500] that for the Neumann problem in the square this* lim inf *should be strictly positive.*
Coming back to the previous proof of Proposition 10.23, one immediately sees that

$$\lim_{m\to+\infty}P(m,1;b)=0.$$

Remark 10.25. *Inspired by computations of [152], it has been conjectured by Polterovich [763] that the constant $2/\pi=4\pi/\lambda(\square)$ is optimal for the validity of a strong*

Pleijel's theorem with a constant independent of the domain (see the discussion in [476]). A less ambitious conjecture is that Pleijel's theorem holds with the constant $4\pi/\lambda(\hexagon)$, where $\hexagon$ is the regular hexagon of area 1. This is directly related to the hexagonal conjecture which will be discussed in Section 10.9.

10.2.6 Notes

Pleijel's Theorem extends to bounded domains in $\mathbb{R}^n$, and more generally to compact n-manifolds with boundary, with a constant $\gamma(n) < 1$ replacing $4\pi/\lambda(\bigcirc)$ in the right-hand side of (10.18) (see Peetre [754], Bérard-Meyer [133]). It is also interesting to note that this constant is independent of the geometry. Pleijel's Theorem can be extended to the Robin problem as soon as the parameter α is large enough so that the ground state energy $\lambda^{Rob}(\bigcirc; \alpha)$ for the Robin problem in $\bigcirc$ is larger than 4π. We can indeed use the Bossel-Daners inequality (see [170, 319]) instead of the Faber-Krahn inequality for the nodal domains touching the boundary. Note also that the counting function $N(\lambda)$ for the Dirichlet problem is a lower bound for the counting function for the Robin problem. It is also true for the Neumann Laplacian in a piecewise analytic bounded domain in $\mathbb{R}^2$ (see [763] whose proof is based on a control of the asymptotics of the number of boundary points belonging to the nodal sets of the eigenfunction associated with λ_k as $k \to +\infty$, a difficult result proved by Toth-Zelditch [852]). In [130, 862], the authors determine an upper bound for Courant-sharp eigenvalues, expressed in terms of simple geometric invariants of Ω.

10.3 Courant sharp cases: examples

This section is devoted to determine the Courant sharp situation for some examples. This is a indeed a natural question in the theory but an addtional motivation is that it gives us examples of minimal partitions. This kind of analysis seems to have been initiated by Å. Pleijel. First, we recall that, according to Theorem 10.17, there are a finite number of Courant sharp eigenvalues. We will try to quantify this number or to find lower bounds or upper bounds for the largest integer n such that $\lambda_{n-1} < \lambda_n$ with λ_n Courant sharp.

10.3.1 Thin domains

This subsection is devoted to thin domains for which Léna proves in [648] that, under some geometrical assumptions, any eigenpair is Courant sharp as soon as the domain is thin enough.
Let us fix the framework. Let $a > 0$, $b > 0$ and $h \in C^\infty((-a, b), \mathbb{R}^+)$. We assume that h

has a unique maximum at 0 which is non degenerate. For $\varepsilon > 0$, we introduce

$$\Omega_\varepsilon = \left\{(x_1, x_2) \in \mathbb{R}^2\,,\ -a < x_1 < b \text{ and } -\varepsilon h(x_1) < x_2 < \varepsilon h(x_1)\right\}.$$

Theorem 10.26. *For any $k \geq 1$, there exists $\varepsilon_k > 0$ such that, if $0 < \varepsilon \leq \varepsilon_k$, the first k Dirichlet eigenvalues $\{\lambda_j(\Omega_\varepsilon), 1 \leq j \leq k\}$ are simple and Courant sharp.*

Sketch of the proof. The asymptotic behavior of the eigenvalues for a domain whose width is proportional to ε, as $\varepsilon \to 0$, was established by L. Friedlander and M. Solomyak [405] and the first terms of the expansion were given. An expansion at any order was proved by D. Borisov and P. Freitas for planar domains in [166]. The proof of Léna is based on a semi-classical approximation of the eigenpairs of the Schrödinger operator. He established some elliptic estimates with a control according to ε and applied some Sobolev imbeddings to prove the uniform convergence of the quasimodes and their derivative functions. The proof is achieved by adapting some arguments of [395] to localize the nodal sets. □

Remark 10.27.
- *The rectangle $\Omega_\varepsilon = (0, \pi) \times (0, \varepsilon\pi)$ does not fulfill the assumptions of Theorem 10.26 (see the assumptions done on h). Nevertheless (see Subsection 10.3.2), we have:*
  For any $k \geq 1$, there exists $\varepsilon_k > 0$ such that $\lambda_j(\Omega_\varepsilon)$ is Courant sharp for $1 \leq j \leq k$ and $0 < \varepsilon \leq \varepsilon_k$. Furthermore, when $0 < \varepsilon < \varepsilon_k$, the nodal partition of the corresponding eigenfunction u_j consists of j similar vertical strips.
- *In the case of the flat torus $\mathbb{T}(a, b) = (\mathbb{R}/a\mathbb{Z}) \times (\mathbb{R}/b\mathbb{Z})$, with $0 < b \leq a$, the first two positive eigenvalues are always Courant sharp. If $k \leq \frac{a}{b}$, the eigenvalues λ_{2j} for $1 \leq j < k$ are Courant sharp and if $k \leq \frac{a}{b} - 1$, the nodal partition of any corresponding eigenfunction consists of $2j$ similar strips (see Subsection 10.3.5).*

10.3.2 Irrational rectangles

The detailed analysis of the spectrum of the Dirichlet Laplacian in a rectangle is the example treated as the toy model in [761]. Let $\mathcal{R}(a, b) = (0, a\pi) \times (0, b\pi)$. We recall (10.22). If it is possible to determine the Courant sharp cases when b^2/a^2 is irrational (see for example [477]), it can become very difficult in general situation. If we assume that b^2/a^2 is irrational, then all the eigenvalues have multiplicity 1. For a given $\hat{\lambda}_{m,n}$, we know that the corresponding eigenfunction has mn nodal domains. As a result of a case by case analysis combined with Proposition 10.16, we obtain the following characterization[10.3] of the Courant sharp cases:

10.3 As observed by C. Léna, there was a mistake in [477] which is corrected here.

Theorem 10.28. *Let $a > b$ and $\frac{b^2}{a^2} \notin \mathbb{Q}$. Then the only cases when $\hat{\lambda}_{m,n}$ is a Courant sharp eigenvalue are the following:*

1. $(m, n) = (3, 2)$ *if* $\frac{8}{5} < \frac{a^2}{b^2} < \frac{7}{3}$;
2. $(m, n) = (2, 2)$ *if* $1 < \frac{a^2}{b^2} < \frac{5}{3}$;
3. $(m, 1)$ *if* $\frac{m^2-1}{3} < \frac{a^2}{b^2}$.

Remark 10.29. *Note that in Case 3 of Theorem 10.28, we can remove the assumption that $\frac{b^2}{a^2} \notin \mathbb{Q}$.*

10.3.3 Pleijel's reduction argument for the rectangle

The analysis of this subsection is independent of the arithmetic properties of b^2/a^2. Following (and improving) a remark in a course of R. Laugesen [631], one has a lower bound of $N(\lambda)$ in the case of the rectangle $\mathcal{R} = \mathcal{R}(a, b) := (0, a\pi) \times (0, b\pi)$, which can be expressed in terms of area and perimeter. One can indeed observe that the area of the intersection of the ellipse $\{\frac{(x+1)^2}{a^2} + \frac{(y+1)^2}{b^2} < \lambda\}$ with $\mathbb{R}^+ \times \mathbb{R}^+$ is a lower bound for $N(\lambda)$.

The formula (to compare with the two terms asymptotics (10.13)) reads for $\lambda \geq \frac{1}{a^2} + \frac{1}{b^2}$:

$$N(\lambda) > \frac{1}{4\pi}|\mathcal{R}|\lambda - \frac{1}{2\pi}|\partial\mathcal{R}|\sqrt{\lambda} + 1 . \tag{10.26}$$

We get, in the situation $\lambda_{n-1} < \lambda_n$,

$$n > \frac{\pi ab}{4}\lambda_n - (a + b)\sqrt{\lambda_n} + 2 . \tag{10.27}$$

On the other hand, if λ_n is Courant sharp, Lemma 10.20 gives the necessary condition

$$\frac{n}{\lambda_n} \leq \pi\frac{ab}{\mathbf{j}^2} . \tag{10.28}$$

Then, combining this last relation with (10.27), we get the inequality

$$\frac{a + b}{\sqrt{\lambda_n}} > \pi ab\left(\frac{1}{4} - \frac{1}{\mathbf{j}^2}\right) ,$$

and finally a Courant sharp eigenvalue λ_n should satisfy:

$$\lambda_n < \frac{1}{\pi^2}\left(\frac{1}{4} - \frac{1}{\mathbf{j}^2}\right)^{-2}\left(\frac{a + b}{ab}\right)^2 , \tag{10.29}$$

with

$$\frac{1}{\pi^2}\left(\frac{1}{4} - \frac{1}{\mathbf{j}^2}\right)^{-2} \simeq 17.36 .$$

Hence, using the expression (10.22) of the eigenvalues, we have just to look at the pairs $\ell, m \in \mathbb{N}^*$ such that

$$\frac{\ell^2}{a^2} + \frac{m^2}{b^2} < \frac{1}{\pi^2}\left(\frac{1}{4} - \frac{1}{\mathbf{j}}\right)^{-2}\left(\frac{a + b}{ab}\right)^2 .$$

Suppose that $a \geq b$. We can then normalize by taking $a = 1$. We get the condition:

$$\ell^2 + \frac{1}{b^2} m^2 < \frac{1}{\pi^2} \left(\frac{1}{4} - \frac{1}{\mathbf{j}^2} \right)^{-2} \left(\frac{1+b}{b} \right)^2 \simeq 17.36 \left(\frac{1+b}{b} \right)^2 \leq 69.44 \,.$$

This is compatible with the observation (see Subsections 10.3.1 and 10.3.2) that when b is small, the number of Courant sharp cases will increase. In any case, when $a = 1$, this number is $\geq [\frac{1}{b}]$ and using (10.28),

$$n \leq \lambda_n \frac{\pi a b}{\mathbf{j}^2} \leq \frac{1}{\pi} \left(\frac{\mathbf{j}}{4} - \frac{1}{\mathbf{j}} \right)^{-2} \frac{(1+b)^2}{b} \leq 9.27 \frac{(1+b)^2}{b} \,.$$

In the next subsection, we continue with a complete analysis of the square.

10.3.4 The square

We now take $a = b = 1$ and describe the Courant sharp cases.

Theorem 10.30. *In the case of the square, the Dirichlet eigenvalue λ_k is Courant sharp if and only if $k = 1, 2, 4$.*

Remark 10.31. *This result was obtained by Pleijel [761] who was nevertheless sketchy (see [132]) in his analysis of the eigenfunctions in the k-th eigenspace for $k = 5, 7, 9$ for which he only refers to pictures in Courant-Hilbert [292], actually reproduced from Pockel [762]. Details can be found in [132] or below.*

Proof. From the previous subsection, we know that it is enough to look at the eigenvalues which are less than 69 (actually 68 because 69 is not an eigenvalue). Looking at the necessary condition (10.28) eliminates most of the candidates associated with the remaining eigenvalues and we are left after computation with the analysis of the three cases $k = 5, 7, 9$. These three eigenvalues correspond respectively to the pairs $(m, n) = (1, 3)$, $(m, n) = (2, 3)$ and $(m, n) = (1, 4)$ and have multiplicity 2. Due to multiplicities, we have (at least) to consider the family of eigenfunctions $(x, y) \mapsto \Phi_{m,n}(x, y, \theta)$ defined by

$$(x, y) \mapsto \Phi_{m,n}(x, y, \theta) := \cos\theta\, \phi_{m,n}(x, y) + \sin\theta\, \phi_{n,m}(x, y) \,, \tag{10.30}$$

for $m, n \geq 1$, and $\theta \in [0, \pi)$.

Let us analyze each of the three cases $k = 5, 7, 9$. For λ_7 ($(m, n) = (2, 3)$) and λ_9 ($(m, n) = (1, 4)$), we can use some antisymmetry argument. We observe that

$$\phi_{m,n}(\pi - x, \pi - y, \theta) = (-1)^{m+n} \phi(x, y, \theta) \,. \tag{10.31}$$

Hence, when $m + n$ is odd, any eigenfunction corresponding to $m^2 + n^2$ has necessarily an even number of nodal domains. Hence λ_7 and λ_9 cannot be Courant sharp.

For the remaining eigenvalue λ_5 ($(m,n) = (1,3)$), we look at the zeroes of $\Phi_{1,3}(x,y,\theta)$ and consider the C^∞ change of variables $\cos x = u$, $\cos y = v$, which sends the square $(0,\pi)\times(0,\pi)$ onto $(-1,1)\times(-1,1)$. In these coordinates, the zero set of $\Phi_{1,3}(x,y,\theta)$ inside the square is given by: $\cos\theta\,(4v^2-1)+\sin\theta\,(4u^2-1)=0$.
Except the two easy cases when $\cos\theta = 0$ or $\sin\theta = 0$, which can be analyzed directly (product situation), we immediately get that the only possible singular point is $(u,v) = (0,0)$, i.e. $(x,y) = (\frac{\pi}{2},\frac{\pi}{2})$, and that this can only occur for $\cos\theta+\sin\theta = 0$, i.e. for $\theta = \frac{\pi}{4}$.
We can then conclude that the number of nodal domains is 2, 3 or 4.
This achieves the analysis of the Courant sharp cases for the Dirichlet-Laplacian in the square. □

Remark 10.32. *For an eigenvalue λ, let $\hat{\mu}_{max}(\lambda) = \max_{u\in E(\lambda)}\mu(u)$. For a given eigenvalue $\hat{\lambda}_{m,n}$ of the square with multiplicity ≥ 2, a natural question is to determine if*

$$\widehat{\mu}_{\max}(\hat{\lambda}_{m,n}) = \mu_{max}(m,n) \text{ with } \mu_{\max}(m,n) = \sup\{m_j n_j \,:\, m_j^2+n_j^2 = m^2+n^2\}.$$

The problem is not easy because one has to consider, in the case of degenerate eigenvalues, linear combinations of the canonical eigenfunctions associated with the $\hat{\lambda}_{m,n}$. Actually, as stated above, the answer is negative. As observed by Pleijel [761], the eigenfunction $\Phi_{1,3,\frac{3\pi}{4}}$ defined in (10.30) corresponds to the fifth eigenvalue and has four nodal domains delimited by the two diagonals, and $\mu_{max}(1,3) = 3$. One could think that this guess could hold for large enough eigenvalues but $u_k := \Phi_{1,3,\frac{3\pi}{4}}(2^k x, 2^k y)$ is an eigenfunction associated with the eigenvalue $\lambda_{n(k)} = \hat{\lambda}_{2^k,3\cdot 2^k} = 10\cdot 4^k$ with 4^{k+1} nodal domains. Using Weyl's asymptotics, we get that the corresponding quotient $\frac{\mu(u_k)}{n(k)}$ is asymptotic to $\frac{8}{5\pi}$. This does not contradict the Polterovich conjecture (see Remark 10.25).

10.3.5 Flat tori

Let $\mathbb{T}(a,b)$ be the torus $(\mathbb{R}/a\mathbb{Z})\times(\mathbb{R}/b\mathbb{Z})$, with $0 < b \leq a$. Then the eigenvalues of the Laplace-Beltrami operator on $\mathbb{T}(a,b)$ are

$$\lambda_{m,n}(a,b) = 4\pi^2\left(\frac{m^2}{a^2}+\frac{n^2}{b^2}\right), \qquad \text{with } m \geq 0,\ n \geq 0. \tag{10.32}$$

A basis of eigenfunctions is given by $\sin mx \sin ny$, $\cos mx \sin ny$, $\sin mx \cos ny$, $\cos mx \cos ny$, where we should eliminate the identically zero functions when $mn = 0$. The multiplicity can be 1 (when $m = n = 0$), 2 when $m = n$ (and no other pair gives the same eigenvalue), 4 for $m \neq n$ if no other pair gives the same eigenvalue, which can occur when $b^2/a^2 \in \mathbb{Q}$. Hence the multiplicity can be much higher than in the Dirichlet case.

Irrational tori

Theorem 10.33. *Suppose b^2/a^2 be irrational. If* $\min(m,n) \geq 1$, *then the eigenvalue* $\lambda_{m,n}(a,b)$ *is not Courant sharp.*

Proof. The proof given in [493] is based on two properties. The first one is to observe that, if $\lambda_{m,n}(a,b) = \lambda_{k(m,n)}$, then $k(m,n) \geq 4mn + 2m + 2n - 2$.
The second one is to prove (which needs some work) that for $m, n > 0$ any eigenvalue in $E(\lambda_{m,n}(a,b))$ has either $4mn$ nodal domains or $2D(m,n)$ nodal domains where $D(m,n)$ is the greatest common denominator of m and n. □

Hence we are reduced to the analysis of the case when $mn = 0$.
As mentioned in Remark 10.27, it is easy to see that, independently of the rationality or irationality of $\frac{b^2}{a^2}$, for $b \leq \frac{2}{k}$, the eigenvalues $\lambda_1 = 0$, and $\lambda_{2\ell}$ for $1 \leq \ell \leq k$ are Courant sharp.

The isotropic torus

In this case we can completely determine the cases where the eigenvalues are Courant sharp. The first eigenvalue has multiplicity 1 and the second eigenvalue has multiplicity 4. By the general theory we know that this is Courant sharp. C. Léna [649] has proven:

Theorem 10.34. *The only Courant sharp eigenvalues for the Laplacian on* $\mathbb{T}^2 := (\mathbb{R}/\mathbb{Z})^2$ *are the first and the second ones.*

Proof. The proof is based on a version of the Faber-Krahn inequality for the torus which reads:

Proposition 10.35. *If Ω is an open set in $\mathbb{T}^2$ of area $\leq \frac{1}{\pi}$, then the standard Faber-Krahn inequality is true.*

Combined with an explicit lower bound for the Weyl law, one gets

$$N(\lambda) \geq \frac{1}{4\pi}\lambda - \frac{2}{\pi}\sqrt{\lambda} - 3\,. \tag{10.33}$$

One can then proceed in a way similar to that of the rectangle case with the advantage here that the only remaining cases correspond to the first and second eigenvalues. □

10.3.6 The disk

Although the spectrum is explicitly computable, we are mainly interested in the ordering of the eigenvalues corresponding to different angular momenta. Consider the

Dirichlet realization in the unit disk $\mathcal{B}(0,1) \subset \mathbb{R}^2$ (where $\mathcal{B}(0,r)$ denotes the disk of radius r). We have in polar coordinates:

$$-\Delta = -\frac{\partial^2}{\partial r^2} - \frac{1}{r}\frac{\partial}{\partial r} - \frac{1}{r^2}\frac{\partial^2}{\partial \theta^2}\,.$$

The Dirichlet boundary conditions require that any eigenfunction u satisfies

$$u(1,\theta) = 0, \quad \text{for} \quad \theta \in [0, 2\pi)\,.$$

We analyze for any $\ell \in \mathbb{N}$ the eigenvalues $\tilde{\lambda}_{\ell,j}$ of

$$\Big(-\frac{d^2}{dr^2} - \frac{1}{r}\frac{d}{dr} + \frac{\ell^2}{r^2}\Big) f_{\ell,j} = \tilde{\lambda}_{\ell,j} f_{\ell,j}\,, \quad \text{in } (0,1)\,.$$

This operator is self adjoint for the scalar product in $L^2((0,1), r\,dr)$. The corresponding eigenfunctions of the eigenvalue problem take the form

$$u(r,\theta) = c\, f_{\ell,j}(r)\cos(\ell\theta + \theta_0), \qquad \text{with } c \neq 0\,,\ \theta_0 \in \mathbb{R}\,, \tag{10.34}$$

where the $f_{\ell,j}$ are suitable Bessel functions. For the corresponding $\tilde{\lambda}_{\ell,j}$'s, we find the following ordering

$$\begin{aligned} \lambda_1 = \tilde{\lambda}_{0,1} &< \lambda_2 = \lambda_3 = \tilde{\lambda}_{1,1} < \lambda_4 = \lambda_5 = \tilde{\lambda}_{2,1} < \lambda_6 = \tilde{\lambda}_{0,2} \\ &< \lambda_7 = \lambda_8 = \tilde{\lambda}_{3,1} < \lambda_9 = \tilde{\lambda}_{10} = \tilde{\lambda}_{1,2} < \lambda_{11} = \lambda_{12} = \tilde{\lambda}_{4,1} < \dots \end{aligned} \tag{10.35}$$

We recall that the zeros of the Bessel functions are related to the eigenvalues by the relation:

$$\tilde{\lambda}_{\ell,k} = (\mathbf{j}_{\ell,k})^2\,. \tag{10.36}$$

Moreover all the $\mathbf{j}_{\ell,k}$ are distinct (see Watson [870]). This goes back to deep results by C.L. Siegel [810] proving in 1929 a conjecture of J. Bourget (1866). The multiplicity is either 1 (if $\ell = 0$) or 2 if $\ell > 0$ and we have

$$\mu(u_1) = 1; \quad \mu(u) = 2\,,\ \forall u \in E(\lambda_2)\,; \quad \mu(u) = 4\,,\ \forall u \in E(\lambda_4)\,; \quad \mu(u_6) = 2\,,\ \cdots$$

Hence λ_1, λ_2 and λ_4 are Courant sharp and it is proven in [477] that we have finally:

Proposition 10.36. *Except the cases $k = 1$, 2 and 4, the eigenvalue λ_k of the Dirichlet Laplacian in the disk is never Courant sharp.*

Notice that the Neumann case can also be treated (see [501]) and that the result is the same. As observed in [85], Siegel's theorem also holds for the zeroes of the derivative of the above Bessel functions [808, 810].

10.3.7 Circular sectors

Let Σ_ω be a circular sector of opening ω. We are interested in finding the Courant sharp eigenvalues of Σ_ω for each ω. The eigenvalues $(\check{\lambda}_{m,n}(\omega), u^{\omega}_{m,n})$ of the Dirichlet Laplacian on the circular sector Σ_ω are given by

$$\check{\lambda}_{m,n}(\omega) = \mathbf{j}^2_{m\frac{\pi}{\omega},n} \qquad \text{and} \qquad u^{\omega}_{m,n}(\rho,\theta) = J_{m\frac{\pi}{\omega}}(\mathbf{j}_{m\frac{\pi}{\omega},n}\,\rho)\sin(m\pi(\tfrac{\theta}{\omega}+\tfrac{1}{2}))\,,$$

where $\mathbf{j}_{m\frac{\pi}{\omega},n}$ is the n-th positive zero of the Bessel function of the first kind $J_{m\frac{\pi}{\omega}}$. The Courant sharp situation is analyzed in [163] and can be summed up in the following proposition:

Proposition 10.37. *Let us define*

$$\omega^1_k = \inf\{\omega \in (0,2\pi] : \check{\lambda}_{1,k}(\omega) \geq \check{\lambda}_{2,1}(\omega)\}, \qquad \forall k \geq 2\,,$$

$$\omega^2_k = \inf\{\omega \in (0,2\pi] : \check{\lambda}_{k,1}(\omega) < \check{\lambda}_{1,2}(\omega)\}, \qquad \textit{for } 2 \leq k \leq 5\,.$$

If $2 \leq k \leq 5$, the eigenvalue λ_k is Courant sharp if and only if $\omega \in]0, \omega^1_k] \cup [\omega^2_k, 2\pi]$. If $k \geq 6$, the eigenvalue λ_k is Courant sharp for $\omega \leq \omega^1_k$.

10.3.8 Notes

Other domains have been analyzed:

- the square for the Neumann-Laplacian, by Helffer–Persson-Sundqvist [500],
- the annulus for the Neumann-Laplacian, by Helffer–Hoffmann-Ostenhof [492],
- the sphere by Leydold [657, 658] and Helffer-Hoffmann-Ostenhof–Terracini [495],
- the irrational torus by Helffer–Hoffmann-Ostenhof [493],
- the equilateral torus, the equilateral, hemi-equilateral and right angled isosceles triangles by Bérard-Helffer [131],
- the isotropic harmonic oscillator by Leydold [656], Bérard-Helffer [128] and Charron [258],
- the equilateral triangle, right isosceles triangle, hemi-equilateral triangle, and square in Chapter 6 where is described the known results and is plotted nodal domains for certain of the Courant-sharp cases.

In higher dimension, the cases of the cube [498], the ball [501] and the 3D-torus $\mathbb{T}^3$ [649] have been analyzed (see also [258] and [259] for Pleijel's theorem).

10.4 Introduction to minimal spectral partitions

This section is based on the paper [477], in which spectral minimal partitions were introduced.

10.4.1 Definition

With Definitions 10.1 and 10.2 in mind, we now introduce the notion of spectral minimal partitions.

Definition 10.38 (Minimal energy). *Minimizing the energy over all the k-partitions, we introduce:*

$$\mathfrak{L}_k(\Omega) = \inf_{\mathcal{D}\in\mathfrak{O}_k(\Omega)} \Lambda(\mathcal{D}). \tag{10.37}$$

We will say that $\mathcal{D} \in \mathfrak{O}_k(\Omega)$ is minimal if $\mathfrak{L}_k(\Omega) = \Lambda(\mathcal{D})$.

Sometimes (at least for the proofs), we have to relax this definition by considering quasi-open or measurable sets for the partitions. We will not discuss this point in detail (see [477]). We recall that, if $k = 2$, we have proved in Proposition 10.6 that $\mathfrak{L}_2(\Omega) = \lambda_2(\Omega)$.

More generally (see [477]), for any integer $k \geq 1$ and $p \in [1, +\infty[$, we define the p-energy of a k-partition $\mathcal{D} = \{D_i\}_{1\leq i\leq k}$ by

$$\Lambda_p(\mathcal{D}) = \left(\frac{1}{k}\sum_{i=1}^{k} \lambda(D_i)^p\right)^{\frac{1}{p}}. \tag{10.38}$$

The notion of p-minimal k-partition can be extended accordingly by minimizing $\Lambda_p(\mathcal{D})$. Then we can consider the optimization problem

$$\mathfrak{L}_{k,p}(\Omega) = \inf_{\mathcal{D}\in\mathfrak{O}_k} \Lambda_p(\mathcal{D}). \tag{10.39}$$

For $p = +\infty$, we write $\Lambda_\infty(\mathcal{D}) = \Lambda(\mathcal{D})$ and $\mathfrak{L}_{k,\infty}(\Omega) = \mathfrak{L}_k(\Omega)$.

10.4.2 Strong and regular partitions

The analysis of the properties of minimal partitions leads us to introduce two notions of regularity that we present briefly:

Definition 10.39. *A partition $\mathcal{D} = \{D_i\}_{1\leq i\leq k}$ of Ω in $\mathfrak{O}_k$ is called* **strong** *if*

$$\operatorname{Int}\left(\overline{\cup_i D_i}\right) \setminus \partial\Omega = \Omega. \tag{10.40}$$

We say that $\mathcal{D}$ is **nice** *if* $\operatorname{Int}(\overline{D_i}) = D_i$, *for any* $1 \leq i \leq k$.

For example, in Figure 10.12, only the fourth picture gives a nice partition. Attached to a strong partition, we associate a closed set in $\overline{\Omega}$:

Definition 10.40 (Boundary set).

$$\partial\mathcal{D} = \overline{\cup_i (\Omega \cap \partial D_i)} \,. \tag{10.41}$$

$\partial\mathcal{D}$ plays the role of the nodal set (in the case of a nodal partition). This leads us to introduce the set $\mathfrak{D}_k^{\mathrm{reg}}(\Omega)$ of **regular** partitions, which should satisfy the following properties :

(i) Except at finitely many distinct $\mathbf{x}_i \in \Omega \cap \partial\mathcal{D}$ in the neigborhood of which $\partial\mathcal{D}$ is the union of $\nu(\mathbf{x}_i)$ smooth curves ($\nu(\mathbf{x}_i) \geq 2$) with one end at $\mathbf{x}_i$, $\partial\mathcal{D}$ is locally diffeomorphic to a regular curve.
(ii) $\partial\Omega \cap \partial\mathcal{D}$ consists of a (possibly empty) finite set of points $\mathbf{y}_j$. Moreover $\partial\mathcal{D}$ is near $\mathbf{y}_j$ the union of $\rho(\mathbf{y}_j)$ distinct smooth half-curves which hit $\mathbf{y}_j$.
(iii) $\partial\mathcal{D}$ has the equal angle meeting property, that is the half curves cross with equal angle at each singular interior point of $\partial\mathcal{D}$ and also at the boundary together with the tangent to the boundary.

We denote by $X(\partial\mathcal{D})$ the set corresponding to the points $\mathbf{x}_i$ introduced in (i) and by $Y(\partial\mathcal{D})$ corresponding to the points $\mathbf{y}_i$ introduced in (ii).

Remark 10.41. *This notion of regularity for partitions is very close to what we have observed for the nodal partition of an eigenfunction in Proposition 10.8. The main difference is that, in the nodal case, there is always an even number of half-lines meeting at an interior singular point.*

Examples of regular partitions are given in Figure 10.1. More precisely, the partitions represented in Figures 10.1a are nodal (we have respectively some nodal partition associated with the double eigenvalue λ_2 and with λ_4 for the square, λ_{15} for the right angled isosceles triangle and the last two pictures are nodal partitions associated with the double eigenvalue λ_{12} on the equilateral triangle). On the contrary, all the partitions presented in Figures 10.1b are not nodal.

10.4.3 Bipartite partitions

Definition 10.42. *We say that two sets D_i, D_j of the partition $\mathcal{D}$ are neighbors and write $D_i \sim D_j$, if*

$$D_{ij} := \mathrm{Int}\,(\overline{D_i \cup D_j}) \setminus \partial\Omega$$

is connected. We say that a regular partition is bipartite if it can be colored by two colors (two neighbors having two different colors).

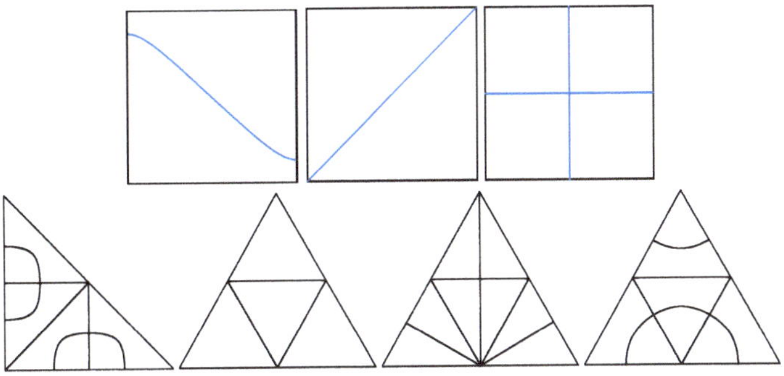

(a) Bipartite partitions.

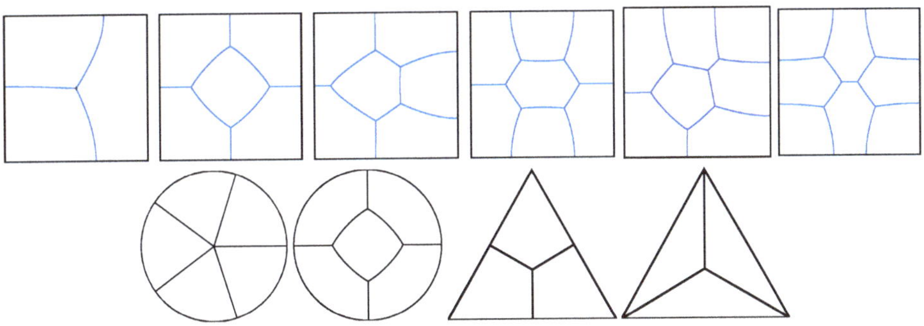

(b) Non bipartite partitions.

Fig. 10.1. Examples of partitions.

Nodal partitions are the main examples of bipartite partitions (see Figure 10.1a). Figure 10.1b gives examples of non bipartite partitions. Some examples can also be found in [307].

Note that in the case of a simply connected planar domain, we know by graph theory that, if for a regular partition all the $\nu(\mathbf{x}_i)$ are even, then the partition is bipartite. This is no more the case on an annulus or on a surface. See for example the third subfigure in Figure 10.4 for a counter-example on $\mathbb{T}^2$.

10.4.4 Main properties of minimal partitions

It has been proved by Conti-Terracini-Verzini [286, 288, 289] (existence) and Helffer–Hoffmann-Ostenhof–Terracini [477] (regularity) the following theorem:

Theorem 10.43. *For any k, there exists a minimal k-partition which is strong, nice and regular. Moreover any minimal k-partition has a strong, nice and regular representative*[10.4]*. The same result holds for the p-minimal k-partition problem with $p \in [1, +\infty)$.*

The proof is too involved to be presented here. We just give one statement used for the existence (see [288]) for $p \in [1, +\infty)$. The case $p = +\infty$ is harder.

Theorem 10.44.
Let $p \in [1, +\infty)$ and let $\mathcal{D} = \{D_i\}_{1 \le i \le k} \in \mathfrak{O}_k$ be a p-minimal k-partition associated with $\mathfrak{L}_{k,p}$ and let $(\phi_i)_i$ be any set of positive eigenfunctions normalized in L^2 corresponding to $(\lambda(D_i))_i$. Then, there exist $a_i > 0$, such that the functions $u_i = a_i \phi_i$ verify in Ω the variational inequalities
(I1) $-\Delta u_i \le \lambda(D_i) u_i$,
(I2) $-\Delta \left(u_i - \sum_{j \neq i} u_j \right) \ge \lambda(D_i) u_i - \sum_{j \neq i} \lambda(D_j) u_j$.

These inequalities imply that $U = (u_1, \dots, u_k)$ is in the class $\mathcal{S}^*$ as defined in [289] which ensures the Lipschitz continuity of the u_i's in Ω. Therefore we can choose a partition made of open representatives $D_i = \{u_i > 0\}$.
Other proofs of a somewhat weaker version of the existence statement have been given by Bucur-Buttazzo-Henrot [210], Caffarelli-Lin [249]. The minimal partition is shown to exist first in the class of quasi-open sets and it is then proved that a representative of the minimizer is open. Note that in some of these references these minimal partitions are also called **optimal** partitions.

When $p = +\infty$, minimal spectral partitions have two important properties.

Proposition 10.45. *If $\mathcal{D} = \{D_i\}_{1 \le i \le k}$ is a minimal k-partition, then*
1. *The minimal partition $\mathcal{D}$ is a* **spectral equipartition**, *that is satisfying:*
$$\lambda(D_i) = \lambda(D_j) , \qquad \textit{for any} \quad 1 \le i, j \le k .$$
2. *For any pair of neighbors $D_i \sim D_j$,*
$$\lambda_2(D_{ij}) = \mathfrak{L}_k(\Omega) . \tag{10.42}$$

Proof. For the first property, this can be understood, once the regularity is obtained by pushing the boundary and using the Hadamard formula [505] (see also Subsection 10.5.2). For the second property, we can observe that $\{D_i, D_j\}$ is necessarily a minimal 2-partition of D_{ij} and in this case, we know that $\mathfrak{L}_2(D_{ij}) = \lambda_2(D_{ij})$ by Proposition 10.6. Note that it is a stronger property than the claim in (10.42). □

10.4 possibly after a modification of the open sets of the partition by capacity 0 subsets.

Remark 10.46. *In the proof of Theorem 10.43, one obtains on the way an useful construction. Attached to each D_i, there is a distinguished ground state u_i such that $u_i > 0$ in D_i and such that for each pair of neighbors $\{D_i, D_j\}$, $u_i - u_j$ is the second eigenfunction of the Dirichlet Laplacian in D_{ij}.*

Let us now establish two important properties concerning the monotonicity (according to k or the domain Ω).

Proposition 10.47. *For any $k \geq 1$, we have*

$$\mathfrak{L}_k(\Omega) < \mathfrak{L}_{k+1}(\Omega)\,. \tag{10.43}$$

Proof. We take indeed a minimal $(k+1)$-partition of Ω. We have proved that this partition is regular. If we take any subpartition by k elements of the previous partitions, this cannot be a minimal k-partition (it does not have the "strong partition" property). The inequality in (10.43) is therefore strict. □

The second property concerns the **domain monotonicity**.

Proposition 10.48. *If $\Omega \subset \widetilde{\Omega}$, then*

$$\mathfrak{L}_k(\widetilde{\Omega}) \leq \mathfrak{L}_k(\Omega)\,, \quad \forall k \geq 1\,.$$

We observe indeed that each partition of Ω is a partition of $\widetilde{\Omega}$.

10.4.5 Minimal spectral partitions and Courant sharp property

A natural question is whether a minimal partition of Ω is a nodal partition. We have first the following converse theorem (see [475, 477]):

Theorem 10.49. *If the minimal partition is bipartite, this is a nodal partition.*

Proof. Combining the bipartite assumption and the pair compatibility condition mentioned in Remark 10.46, it is immediate to construct some $u \in H_0^1(\Omega)$ such that

$$u_{|D_i} = \pm u_i, \quad \forall 1 \leq i \leq k, \qquad \text{and} \qquad -\Delta u = \mathfrak{L}_k(\Omega) u \quad \text{in } \Omega \setminus X(\partial\mathcal{D}).$$

But $X(\partial\mathcal{D})$ consists of a finite set and $-\Delta u - \mathfrak{L}_k(\Omega)u$ belongs to $H^{-1}(\Omega)$. This implies that $-\Delta u = \mathfrak{L}_k(\Omega)u$ in Ω and hence u is an eigenfunction of $H(\Omega)$ whose nodal set is $\partial\mathcal{D}$. □

The next question is then to determine how general the previous situation is. Surprisingly, this only occurs in the so called Courant sharp situation. For any integer $k \geq 1$, we recall that $L_k(\Omega)$ was introduced in Definition 10.5. In general, one can show, as an easy consequence of the max-min characterization of the eigenvalues, that

$$\lambda_k(\Omega) \leq \mathfrak{L}_k(\Omega) \leq L_k(\Omega)\,. \tag{10.44}$$

The last result (due to [477]) gives the full picture of the equality cases:

Theorem 10.50. *Suppose $\Omega \subset \mathbb{R}^2$ is regular. If $\mathfrak{L}_k(\Omega) = L_k(\Omega)$ or $\mathfrak{L}_k(\Omega) = \lambda_k(\Omega)$, then*

$$\lambda_k(\Omega) = \mathfrak{L}_k(\Omega) = L_k(\Omega)\,. \tag{10.45}$$

In addition, there exists a Courant sharp eigenfunction associated with $\lambda_k(\Omega)$.

This answers a question informally mentioned in [229, Section 7].
Sketch of the proof. It is easy to see using a variation of the proof of Courant's theorem that the equality $\lambda_k = \mathfrak{L}_k$ implies (10.45). Hence the difficult part is to get (10.45) from the assumption that $L_k = \mathfrak{L}_k = \lambda_{m(k)}$, that is to prove that $m(k) = k$. This involves a construction of an exhaustive family $\{\Omega(t),\ t \in (0,1)\}$, interpolating between $\Omega(0) := \Omega \setminus \mathcal{N}(\phi_k)$ and $\Omega(1) := \Omega$, where ϕ_k is an eigenfunction corresponding to L_k such that its nodal partition is a minimal k-partition. This family is obtained by cutting small intervals in each regular component of $\mathcal{N}(\phi_k)$. L_k is an eigenvalue common to all $H(\Omega(t))$, but its labelling changes between $t = 0$ and $t = 1$ at some t_0 where the multiplicity of L_k should increase. By a tricky argument which is not detailed here, we get a contradiction. □

10.4.6 On subpartitions of minimal partitions

Starting with a given strong k-partition, one can consider subpartitions by considering a subfamily of D_i's such that $\mathrm{Int}\,(\cup\overline{D_i})$ is connected, typically a pair of two neighbors. Of course a subpartition of a minimal partition should be minimal. If this was not the case, we should be able to decrease the energy by deformation of the boundary. The next proposition is reminiscent of Proposition 10.16.

Proposition 10.51. *Let $\mathcal{D} = \{D_i\}_{1\leq i\leq k}$ be a minimal k-partition for $\mathfrak{L}_k(\Omega)$. Then, for any subset $I \in \{1,\dots,k\}$, the associated subpartition $\mathcal{D}^I = \{D_i\}_{i\in I}$ satisfies*

$$\mathfrak{L}_k(\Omega) = \Lambda(\mathcal{D}^I) = \mathfrak{L}_{|I|}(\Omega^I)\,, \tag{10.46}$$

where $\Omega^I := \mathrm{Int}\,(\overline{\cup_{i\in I} D_i})$.

It is clear from the definition and the previous results that any subpartition $\mathcal{D}^I$ of a minimal partition $\mathcal{D}$ should be minimal on Ω^I and this proves the proposition. One can also observe that, if this subpartition is bipartite, then it is nodal and actually Courant sharp. In the same spirit, starting with a minimal regular k-partition $\mathcal{D}$ of a domain Ω, we can extract (in many ways) in Ω a connected domain $\widetilde{\Omega}$ such that $\mathcal{D}$

becomes a minimal bipartite k-partition of $\widetilde{\Omega}$. This can be achieved by removing from Ω a union of a finite number of regular arcs corresponding to pieces of boundaries between two neighbors of the partition.

Corollary 10.52. *If $\mathcal{D}$ is a minimal regular k-partition, then for any extracted connected open set $\widetilde{\Omega}$ associated with $\mathcal{D}$, we have*

$$\lambda_k(\widetilde{\Omega}) = \mathfrak{L}_k(\widetilde{\Omega}) \,. \tag{10.47}$$

This last criterion has been analyzed in [162] for gluing of triangles, squares and hexagons as a test of minimality in connection with the hexagonal conjecture (see Subsection 10.9.1).

10.4.7 Notes

Similar results hold in the case of compact Riemannian surfaces when considering the Laplace-Beltrami operator (see [261]). Typical cases are analyzed like $\mathbb{S}^2$ (see [495]) and $\mathbb{T}^2$ (see [649]). In the case of dimension 3, let us mention that Theorem 10.50 is proved in [494]. The complete analysis of the properties of minimal partitions was not achieved in [477] but its completion is announced in the introduction of [782].

10.5 On p-minimal k-partitions

The notion of p-minimal k-partition has been already defined in Subsection 10.4.4. We would like in this section to analyze the dependence on p of these minimal partitions.

10.5.1 Main properties

Inequality (10.44) is replaced by the following one (see [495] for $p = \infty$ and [490] for general p)

$$\left(\frac{1}{k}\sum_{j=1}^{k} \lambda_j(\Omega)^p\right)^{\frac{1}{p}} \le \mathfrak{L}_{k,p}(\Omega) \,. \tag{10.48}$$

This is optimal for the disjoint union of k-disks with different (but close) radii.

10.5.2 Comparison between different p's

Proposition 10.53. *For any $k \ge 1$ and any $p \in [1, +\infty)$, there holds*

$$\frac{1}{k^{1/p}}\mathfrak{L}_k(\Omega) \le \mathfrak{L}_{k,p}(\Omega) \le \mathfrak{L}_k(\Omega) \,, \tag{10.49}$$

$$\mathfrak{L}_{k,p}(\Omega) \leq \mathfrak{L}_{k,q}(\Omega)\,, \qquad \text{if } p \leq q\,. \tag{10.50}$$

Let us notice that (10.49) implies that

$$\lim_{p \to +\infty} \mathfrak{L}_{k,p}(\Omega) = \mathfrak{L}_k(\Omega)\,,$$

and this can be useful in the numerical approach for the determination of $\mathfrak{L}_k(\Omega)$. Notice also the inequalities can be strict! This is the case if Ω is a disjoint union of two disks, possibly related by a thin channel (see [210, 495] for details).
In the case of the disk $\mathcal{B} \subset \mathbb{R}^2$, we do not know if the equality $\mathfrak{L}_{2,1}(\mathcal{B}) = \mathfrak{L}_{2,\infty}(\mathcal{B})$ is satisfied or not. Other aspects of this question will be discussed in Section 10.9.

Coming back to open sets in $\mathbb{R}^2$, it was established recently in [490] that the inequality

$$\mathfrak{L}_{2,1}(\Omega) < \mathfrak{L}_{2,\infty}(\Omega) \tag{10.51}$$

was satisfied in a generic sense. Moreover, we can give explicit examples (equilateral triangle) of convex domains for which this is true. This answers (by the negative) some question in [210]. The proof (see [490]) is based on the following proposition:

Proposition 10.54. *Let Ω be a domain in $\mathbb{R}^2$ and $k \geq 2$. Let $\mathcal{D}$ be a minimal k-partition for $\mathfrak{L}_k(\Omega)$ and suppose that, there is a pair of neighbors $\{D_i, D_j\}$ such that for the second eigenfunction ϕ_{ij} of $H(D_{ij})$ having D_i and D_j as nodal domains, we have*

$$\int_{D_i} |\phi_{ij}(x,y)|^2 \,dxdy \neq \int_{D_j} |\phi_{ij}(x,y)|^2 \,dxdy\,. \tag{10.52}$$

Then

$$\mathfrak{L}_{k,1}(\Omega) < \mathfrak{L}_{k,\infty}(\Omega)\,. \tag{10.53}$$

The proof involves the Hadamard formula (see [505]) which concerns the variation of some simple eigenvalue of the Dirichlet Laplacian by deformation of the boundary. Here we can make the deformation in $\partial D_i \cap \partial D_j$.

We recall from Proposition 10.45 that the ∞-minimal k-partition (we write simply minimal k-partition in this case) is a spectral equipartition. This is not necessarily the case for a p-minimal k-partition. Nevertheless we have the following property:

Proposition 10.55. *Let $\mathcal{D}$ be a p-minimal k-partition. If $\mathcal{D}$ is a spectral equipartition, then this k-partition is q-minimal for any $q \in [p, +\infty]$.*

This leads us to define a real $p(k, \Omega)$ as the infimum over $p \geq 1$ such that there exists a p-minimal k-equipartition.

10.5.3 Examples

Bourdin-Bucur-Oudet [173] have proposed an iterative method to exhibit numerically candidates for the 1-minimal k-partition. Their algorithm can be generalized to the case of the p-norm with $p < +\infty$ and this method has been implemented for several geometries like the square (see [155]) or the torus (see [159]). For any $k \geq 2$ and $p \geq 1$, we denote by $\mathfrak{D}^{k,p}$ the partition obtained numerically. Some examples of $\mathfrak{D}^{k,p}$ are given in Figures 10.2 for the square. For each partition $\mathfrak{D}^{k,p} = \{D_i^{k,p}\}_{1\leq i\leq k}$, we represent in Figures 10.3 the eigenvalues $(\lambda(D_i^{k,p}))_{1\leq i\leq k}$, the energies $\Lambda_p(\mathfrak{D}^{k,p})$ and $\Lambda_\infty(\mathfrak{D}^{k,p})$. We observe that for the case of the square, if $k \notin \{1, 2, 4\}$ (that is to say, if we are not in the Courant sharp situation), then the partitions obtained numerically are not spectral equipartitions for any $p < +\infty$. In the case $k = 3$, the first picture of Figure 10.2 suggests that the triple point (which is not at the center for $p = 1$) moves to the center as p tends to $+\infty$. Consequently, the p-minimal k-partition cannot be optimal for $p = +\infty$ and

$$p(k, \square) = +\infty \quad \text{for } k \in \{3, 5, 6, 7, 8\}.$$

Conversely, for any p, the algorithm produces a nodal partition when $k = 2$ and $k = 4$. This suggests

$$p(k, \square) = 1 \quad \text{for } k \in \{1, 2, 4\}.$$

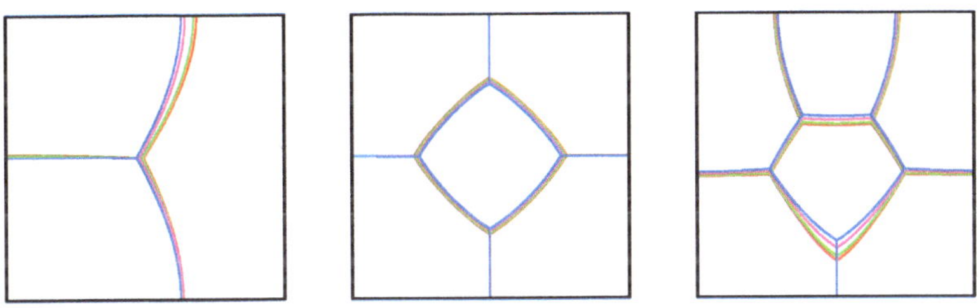

Fig. 10.2. Candidates $\mathfrak{D}^{k,p}$ for the p-minimal k-partition on the square, $p = 1, 2, 5, 10$ (in blue, magenta, green and red respectively), $k = 3, 5, 6$.

In the case of the isotropic torus $(\mathbb{R}/\mathbb{Z})^2$, numerical simulations in [159] for $3 \leq k \leq 6$ suggest that the p-minimal k-partition is a spectral equipartition for any $p \geq 1$ and thus

$$p(k, (\mathbb{R}/\mathbb{Z})^2) = 1 \quad \text{for } k = 3, 4, 5, 6.$$

Candidates are given in Figure 10.4 where we color two neighbors with two different colors, using the minimal number of colors.

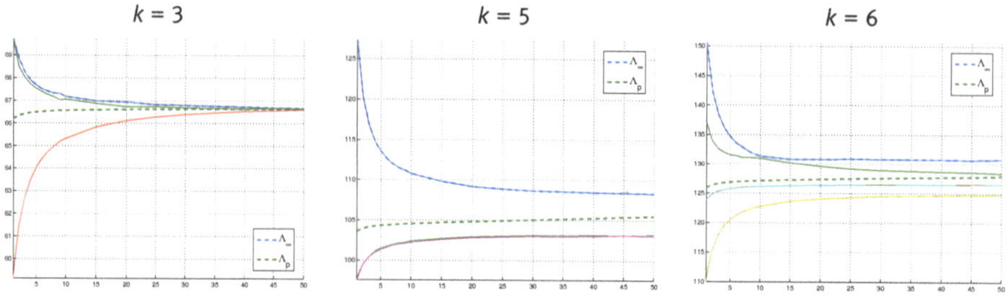

Fig. 10.3. Energies of $\underline{\mathfrak{D}}^{k,p}$ according to p, $k = 3, 5, 6$ on the square.

Fig. 10.4. Candidates for isotropic torus, $k = 3, 4, 5, 6$.

10.5.4 Notes

In the case of the sphere $\mathbb{S}^2$, it was proved that (10.51) is an equality (see [150, 402] and the references in [495]). But this is the only known case for which the equality is proved. For $k = 3$, it is a conjecture reinforced by the numerical simulations of Elliott-Ranner [363] or more recently of B. Bogosel [154]. For $k = 4$, simulations produced by Elliott-Ranner [363] for $\mathfrak{L}_{4,1}$ suggest that the spherical tetrahedron is a good candidate for a 1-minimal 4-partition. For $k > 6$, it seems that the candidates for the 1-minimal k-partitions are not spectral equipartitions.

10.6 Topology of regular partitions

10.6.1 Euler's formula for regular partitions

In the case of planar domains (see [475]), we will use the following result.

Proposition 10.56. *Let Ω be an open set in $\mathbb{R}^2$ with piecewise $C^{1,+}$ boundary and $\mathcal{D}$ be a k-partition with $\partial\mathcal{D}$ the boundary set (see Definition 10.40 and notation therein). Let b_0 be the number of components of $\partial\Omega$ and b_1 be the number of components of $\partial\mathcal{D} \cup \partial\Omega$. Denote by $\nu(\mathbf{x}_i)$ and $\rho(\mathbf{y}_i)$ the numbers of curves ending at $\mathbf{x}_i \in X(\partial\mathcal{D})$, respectively $\mathbf{y}_i \in Y(\partial\mathcal{D})$. Then*

$$k = 1 + b_1 - b_0 + \sum_{\mathbf{x}_i \in X(\partial\mathcal{D})} \left(\frac{\nu(\mathbf{x}_i)}{2} - 1\right) + \frac{1}{2} \sum_{\mathbf{y}_i \in Y(\partial\mathcal{D})} \rho(\mathbf{y}_i) . \tag{10.54}$$

This can be applied, together with other arguments to determine upper bounds for the number of singular points of minimal partitions. This version of the Euler's formula appears in [531] and can be recovered by the Gauss-Bonnet formula (see for example [124]). There is a corresponding result for compact manifolds involving the Euler characteristics.

Proposition 10.57. *Let M be a flat compact surface M without boundary. Then the Euler's formula for a partition $\mathcal{D} = \{D_i\}_{1\le i\le k}$ reads*

$$\sum_{i=1}^{k} \chi(D_i) = \chi(M) + \sum_{\mathbf{x}_i \in X(\partial\mathcal{D})} \left(\frac{\nu(\mathbf{x}_i)}{2} - 1\right),$$

where χ denotes the Euler characteristics.

It is well known that $\chi(\mathbb{S}^2) = 2$, $\chi(\mathbb{T}^2) = 0$ and that for open sets in $\mathbb{R}^2$ the Euler characteristic is 1 for the disk and 0 for the annulus.

10.6.2 Application to regular 3-partitions

Following [489], we describe here the possible "topological" types of non bipartite minimal 3-partitions for a general domain Ω in $\mathbb{R}^2$.

Proposition 10.58. *Let Ω be a simply-connected domain in $\mathbb{R}^2$ and consider a minimal 3-partition $\mathcal{D} = \{D_1, D_2, D_3\}$ associated with $\mathfrak{L}_3(\Omega)$ and suppose that it is not bipartite. Then the boundary set $\partial\mathcal{D}$ has one of the following properties:*

[a] *one interior singular point $\mathbf{x}_0 \in \Omega$ with $\nu(\mathbf{x}_0) = 3$, three points $\{\mathbf{y}_i\}_{1\le i\le 3}$ on the boundary $\partial\Omega$ with $\rho(\mathbf{y}_i) = 1$;*

[b] *two interior singular points* $\mathbf{x}_0, \mathbf{x}_1 \in \Omega$ *with* $\nu(\mathbf{x}_0) = \nu(\mathbf{x}_1) = 3$ *and two boundary singular points* $\mathbf{y}_1, \mathbf{y}_2 \in \partial\Omega$ *with* $\rho(\mathbf{y}_1) = 1 = \rho(\mathbf{y}_2)$*;*
[c] *two interior singular points* $\mathbf{x}_0, \mathbf{x}_1 \in \Omega$ *with* $\nu(\mathbf{x}_0) = \nu(\mathbf{x}_1) = 3$ *and no singular point on the boundary.*

The three types are described in Figure 10.5.

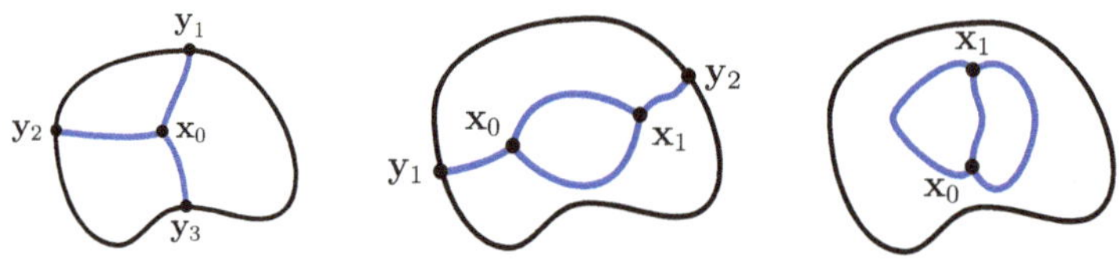

Fig. 10.5. Three topological types : [a], [b] and [c].

The proof of Proposition 10.58 essentially relies on the Euler formula. This leads (with some success) to analyze the minimal 3-partitions with some topological type. We actually do not know any example where the minimal 3-partitions are of type [b] and [c]. Numerical computations never produce candidates of type [b] or [c] (see [162] for the square and the disk, [163] for circular sectors and [161] for complements for the disk).
Note also that we do not know about results claiming that in a domain with a given symmetry there exists a minimal 3-partition with this symmetry. We know in the case of the disk (see [489, Proposition 1.6]) that a minimal 3-partition cannot keep all the symmetries of the disk.
In the case of circular sectors, it has been proved in [163] that a minimal 3-partition cannot be symmetric for some range of ω's.

10.6.3 Upper bound for the number of singular points

Proposition 10.59. *Let* $\mathcal{D}$ *be a minimal* k*-partition of a simply connected domain* Ω *with* $k \geq 2$*. Let* $X^{odd}(\partial\mathcal{D})$ *be the subset among the interior singular points* $X(\partial\mathcal{D})$ *for which* $\nu(\mathbf{x}_i)$ *is odd (see Definition 10.40). Then the cardinality of* $X^{odd}(\partial\mathcal{D})$ *satisfies*

$$\sharp X^{odd}(\partial\mathcal{D}) \leq 2k - 4\,. \tag{10.55}$$

Proof. Euler's formula implies that, for a minimal k-partition $\mathcal{D}$ of a simply connected domain Ω, the cardinality of $X^{odd}(\partial\mathcal{D})$ satisfies

$$\sharp X^{odd}(\partial\mathcal{D}) \leq 2k - 2\,. \tag{10.56}$$

Note that, if $b_1 = b_0$, we necessarily have a singular point in the boundary. If we use in addition the property that the open sets of the partitions are nice, we can exclude the case when there is only one point on the boundary. Hence, we obtain

$$b_1 - b_0 + \frac{1}{2} \sum_i \rho(\mathbf{y}_i) \geq 1 \,,$$

which implies (10.55). □

10.6.4 Notes

In the case of $\mathbb{S}^2$, one can prove that a minimal 3-partition is not nodal (the second eigenvalue has multiplicity 3), and as a step towards a characterization, one can show that non-nodal minimal partitions have necessarily two singular *triple points* (i.e. with $\nu(\mathbf{x}) = 3$).

If we assume, for some $k \geq 12$, that a minimal k-partition has only singular triple points and consists only of (spherical) pentagons and hexagons, then Euler's formula in its historical version for convex polyedra $V - E + F = \chi(\mathbb{S}^2) = 2$ (where F is the number of faces, E the number of edges and V the number of vertices) implies that the number of pentagons is 12. This is what is used for example for the soccer ball (12 pentagons and 20 hexagons). We refer to [363] for enlightening pictures.

More recently, it has been proved by Soave-Terracini [821, Theorem 1.12] that

$$\mathfrak{L}_3(\mathbb{S}^n) = \frac{3}{2}\left(n + \frac{1}{2}\right).$$

10.7 Examples of minimal k-partitions

10.7.1 The disk

In the case of the disk, Proposition 10.36 tells us that the minimal k-partition are nodal only for $k = 1, 2, 4$. Illustrations are given in Figure 10.6a. For other k's, the question is open. Numerical simulations in [156, 161] permit to exhibit candidates for the $\mathfrak{L}_k(\mathcal{B}_1)$ for $k = 3, 5$ (see Figure 10.6). Nevertheless we have no proof that the minimal 3-partition is the "Mercedes star" (see Figure 10.6b), except if we assume that the center belongs to the boundary set of the minimal partition [489] or if we assume that the minimal 3-partition is of type [a] (see [161, Proposition 1.4]).

10.7.2 The square

When Ω is a square, the only cases which are completely solved are $k = 1, 2, 4$ as mentioned in Theorem 10.30 and the minimal k-partitions for $k = 2, 4$ are presented

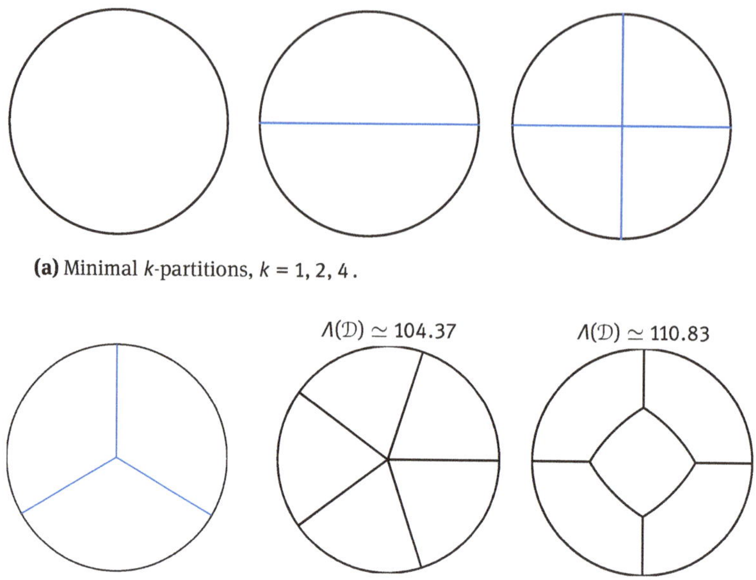

(a) Minimal k-partitions, $k = 1, 2, 4$.

(b) 3-partition.

(c) Energy for two 5-partitions.

Fig. 10.6. Candidates for the disk.

in the Figure 10.1a. Let us now discuss the 3-partitions. It is not difficult to see that $\mathfrak{L}_3$ is strictly less than L_3. We observe indeed that λ_4 is Courant sharp, so $\mathfrak{L}_4 = \lambda_4$, and there is no eigenfunction corresponding to $\lambda_2 = \lambda_3$ with three nodal domains (by Courant's Theorem). Restricting to the half-square and assuming that there is a minimal partition which is symmetric with one of the perpendicular bisectors of one side of the square or with one diagonal line, one needs only to consider a family of problems with mixed conditions on the symmetry axis (Dirichlet-Neumann, Dirichlet-Neumann-Dirichlet or Neumann-Dirichlet-Neumann according to the type of the configuration ([a], [b] or [c] respectively). Numerical computations[10.5] in [162] produce natural candidates for a symmetric minimal 3-partition. Two candidates $\mathfrak{D}^{perp}$ and $\mathfrak{D}^{diag}$ are obtained numerically by choosing the symmetry axis (perpendicular bisector or diagonal line) and represented in Figure 10.7. Numerics suggests that there is no candidate of type [b] or [c], that the two candidates $\mathfrak{D}^{perp}$ and $\mathfrak{D}^{diag}$ have the same energy $\Lambda(\mathfrak{D}^{perp}) \simeq \Lambda(\mathfrak{D}^{diag})$ and that the center is the unique singular point of the partition inside the square. Once this last property is accepted, one can perform the spectral analysis of an Aharonov-Bohm operator (see Section 10.8) with a pole at the center. This point of view is explored numerically in a rather systematic way by Bonnaillie-Noël–Helffer [156] and theoretically by Noris-Terracini [729] (see also

10.5 see http://w3.bretagne.ens-cachan.fr/math/simulations/MinimalPartitions/

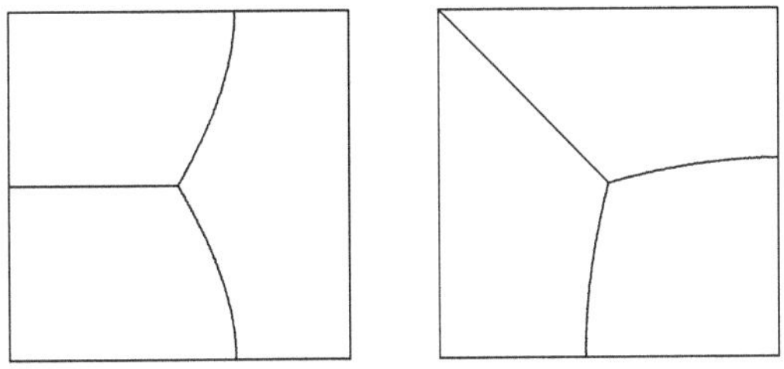

Fig. 10.7. Candidates $\mathfrak{D}^{perp}$ and $\mathfrak{D}^{diag}$ for the square.

[164]). In particular, it was proved that, if the singular point is at the center, the mixed Dirichlet-Neumann problems on the two half-squares (a rectangle and a right angled isosceles triangle depending on the considered symmetry) are isospectral with the Aharonov-Bohm operator. This explains why the two partitions $\mathfrak{D}^{perp}$ and $\mathfrak{D}^{diag}$ have the same energy. So this strongly suggests that there is a continuous family of minimal

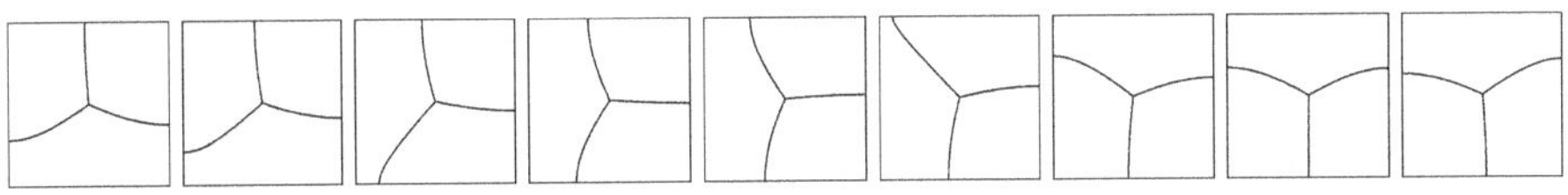

Fig. 10.8. A continuous family of 3-partitions with the same energy.

3-partitions of the square. This is analyzed indeed numerically in [156] and illustrated in Figure 10.8. This can be explained in the formalism of the Aharonov-Bohm operator presented in Section 10.8, observing that this operator has an eigenvalue of multiplicity 2 when the pole is at the center. This is reminiscent of the argument of isospectrality of Jakobson-Levitin-Nadirashvili-Polterovich [557] and Levitin-Parnovski-Polterovich [654]. We refer to [156, 158] for this discussion and more references therein.

Figure 10.9 gives some 5-partitions obtained with several approaches: Aharonov-Bohm approach (see Section 10.8), mixed conditions on one eighth of the square (with Dirichlet condition on the boundary of the square, Neumann condition on one of the other part and mixed Dirichlet-Neumann condition on the last boundary). The first 5-partition corresponds with what we got by minimizing over configurations with one interior singular point. The second 5-partition $\mathfrak{D}^{perp}$ (which has four interior singular points) gives the best known candidate to be minimal.

10.7.3 Flat tori

In the case of thin tori, we have a similar result to Subsection 10.3.1 for minimal partitions.

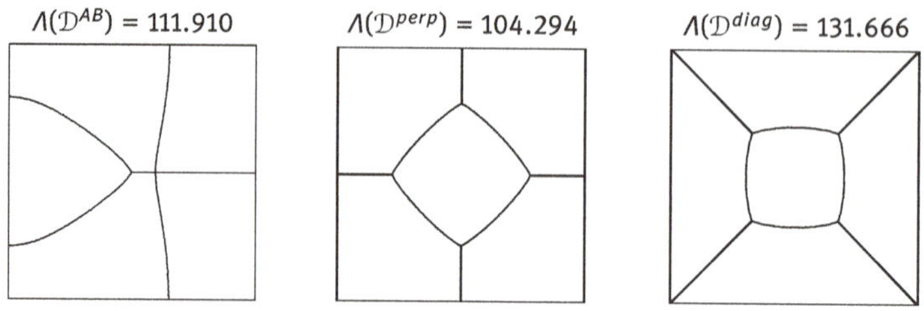

Fig. 10.9. Three candidates for the 5-partition of the square.

Theorem 10.60. *There exists $b_k > 0$ such that, if $b < b_k$, then $\mathfrak{L}_k(T(1,b)) = k^2\pi^2$ and the corresponding minimal k-partition $\mathfrak{D}_k = \{D_i\}_{1\le i\le k}$ is represented in $\mathcal{R}(1,b)$ by*

$$D_i = \left(\frac{i-1}{k}, \frac{i}{k}\right) \times [0, b), \quad \text{for } i = 1, \dots, k. \tag{10.57}$$

Moreover $b_k \ge \frac{1}{k}$ for k even and $b_k \ge \min(\frac{1}{k}, \frac{j^2}{k^2\pi})$ for k odd.

This result extends Remark 10.27 to odd k's, for which the minimal k-partitions are not nodal. We can also notice that the boundaries of the D_i in $T(1,b)$ are just k circles.

In the case of isotropic flat tori, we have seen in Subsection 10.3.5, that minimal partitions are not nodal for $k > 2$. Following [159], some candidates are given in Figure 10.4 for $k = 3, 4, 5, 6$.

10.7.4 Circular sectors

Figure 10.10 gives some symmetric and non symmetric examples for circular sectors. Note that the energy of the first partition in the second line is lower than any symmetric 3-partition. This proves that the minimal k-partition of a symmetric domain is not necessarily symmetric.

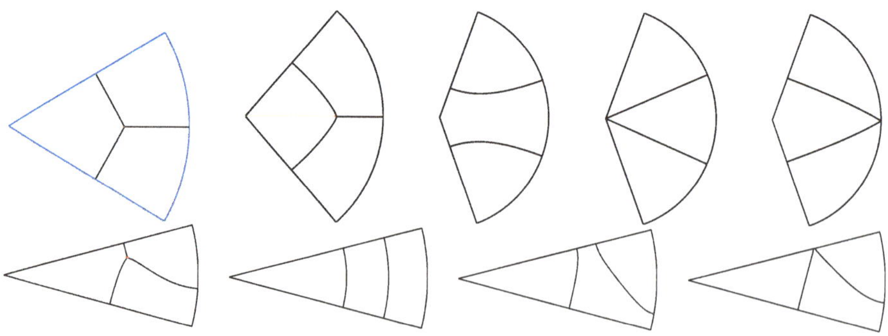

Fig. 10.10. Candidates for circular sectors.

10.7.5 Notes

The minimal 3-partitions for the sphere $\mathbb{S}^2$ have been determined mathematically in [495] (see Figure 10.11). This is an open question known as the Bishop conjecture [150] that the same partition is a 1-minimal 3-partition. The case of a thin annulus is treated in [492] for Neumann conditions and remains open for Dirichlet conditions.

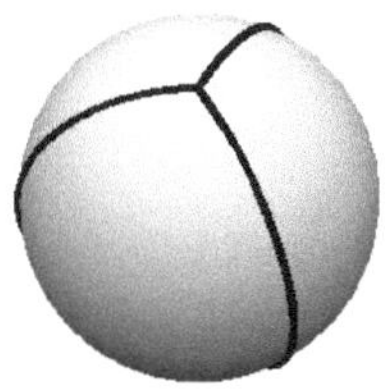

Fig. 10.11. Minimal 3-partition of the sphere.

10.8 Aharonov-Bohm approach

The introduction of Aharonov-Bohm operators in this context is an example of "physical mathematics". There is no magnetic field in our problem and it is introduced artificially. But the idea comes from [474], which was motivated by a problem in superconductivity in non simply connected domains.

10.8.1 Aharonov-Bohm operators

Let Ω be a planar domain and $\mathbf{p} = (p_1, p_2) \in \Omega$. Let us consider the Aharonov-Bohm Laplacian in a punctured domain $\dot{\Omega}_{\mathbf{p}} := \Omega \setminus \{\mathbf{p}\}$ with a singular magnetic potential and normalized flux α. We first introduce

$$\mathbf{A}^{\mathbf{p}}(\mathbf{x}) = (A_1^{\mathbf{p}}(\mathbf{x}), A_2^{\mathbf{p}}(\mathbf{x})) = \frac{(\mathbf{x}-\mathbf{p})^{\perp}}{|\mathbf{x}-\mathbf{p}|^2}, \qquad \text{with} \quad \mathbf{y}^{\perp} = (-y_2, y_1).$$

This magnetic potential satisfies

$$\operatorname{Curl} \mathbf{A}^{\mathbf{p}}(\mathbf{x}) = 0 \quad \text{in } \dot{\Omega}_{\mathbf{p}}.$$

If $\mathbf{p} \in \Omega$, its circulation along a path of index 1 around $\mathbf{p}$ is 2π (or the flux created by $\mathbf{p}$). If $\mathbf{p} \notin \Omega$, $\mathbf{A}^{\mathbf{p}}$ is a gradient and the circulation along any path in Ω is zero. From now on, we renormalize the flux by dividing the flux by 2π.

The Aharonov-Bohm Hamiltonian with singularity $\mathbf{p}$ and flux α (written for brevity

$H^{AB}(\dot{\Omega}_{\mathbf{p}}, \alpha))$ is defined by considering the Friedrichs extension starting from $C_0^\infty(\dot{\Omega}_{\mathbf{p}})$ and the associated differential operator is

$$-\Delta_{\alpha\mathbf{A}^{\mathbf{p}}} := (D_{x_1} - \alpha A_1^{\mathbf{p}})^2 + (D_{x_2} - \alpha A_2^{\mathbf{p}})^2 \qquad \text{with} \qquad D_{x_j} = -i\partial_{x_j}. \tag{10.58}$$

This construction can be extended to the case of a configuration with ℓ distinct points $\mathbf{p}_1, \ldots, \mathbf{p}_\ell$ (putting a flux α_j at each of these points). We just take as magnetic potential

$$\mathbf{A}_\alpha^{\mathbf{P}} = \sum_{j=1}^{\ell} \alpha_j \mathbf{A}^{\mathbf{p}_j}, \qquad \text{where} \quad \mathbf{P} = (\mathbf{p}_1, \ldots, \mathbf{p}_\ell) \quad \text{and} \quad \boldsymbol{\alpha} = (\alpha_1, \ldots, \alpha_\ell).$$

Let us point out that the $\mathbf{p}_j$'s can be in $\mathbb{R}^2 \setminus \Omega$, and in particular in $\partial\Omega$. It is important to observe the following

Proposition 10.61. *If $\boldsymbol{\alpha} = \boldsymbol{\alpha}'$ modulo $\mathbb{Z}^\ell$, then $H^{AB}(\dot{\Omega}_{\mathbf{p}}, \boldsymbol{\alpha})$ and $H^{AB}(\dot{\Omega}_{\mathbf{p}}, \boldsymbol{\alpha}')$ are unitary equivalent.*

10.8.2 The case when the fluxes are 1/2

Let us assume for the moment that there is a unique pole $\ell = 1$ and suppose that the flux α is $\frac{1}{2}$. For brevity, we omit α in the notation when it equals 1/2. Let $K_{\mathbf{p}}$ be the antilinear operator $K_{\mathbf{p}} = e^{i\theta_{\mathbf{p}}}\,\Gamma$, where Γ is the complex conjugation operator $\Gamma u = \bar{u}$ and

$$(x_1 - p_1) + i(x_2 - p_2) = \sqrt{|x_1 - p_1|^2 + |x_2 - p_2|^2}\, e^{i\theta_{\mathbf{p}}},$$

$\theta_{\mathbf{p}}$ such that

$$d\theta_{\mathbf{p}} = 2\mathbf{A}^{\mathbf{p}}.$$

Here we note that because the (normalized) flux of $2\mathbf{A}^{\mathbf{p}}$ belongs to $\mathbb{Z}$ for any path in $\dot{\Omega}_{\mathbf{p}}$, then $\mathbf{x} \mapsto \exp i\theta_{\mathbf{p}}(\mathbf{x})$ is a C^∞ function (this is indeed the variable θ in polar coordinates centered at $\mathbf{p}$).
A function u is called $K_{\mathbf{p}}$-real, if $K_{\mathbf{p}}u = u$. The operator $H^{AB}(\dot{\Omega}_{\mathbf{p}}) = H^{AB}(\dot{\Omega}_{\mathbf{p}}, \frac{1}{2})$ is preserving the $K_{\mathbf{p}}$-real functions. Therefore we can consider a basis of $K_{\mathbf{p}}$-real eigenfunctions. Hence we only analyze the restriction of $H^{AB}(\dot{\Omega}_{\mathbf{p}}, \frac{1}{2})$ to the $K_{\mathbf{p}}$-real space $L^2_{K_{\mathbf{p}}}$ where

$$L^2_{K_{\mathbf{p}}}(\dot{\Omega}_{\mathbf{p}}) = \{u \in L^2(\dot{\Omega}_{\mathbf{p}}) \; : \; K_{\mathbf{p}}\, u = u\}.$$

If there are several poles ($\ell > 1$) and $\boldsymbol{\alpha} = (\frac{1}{2}, \ldots, \frac{1}{2})$, we can also construct the antilinear operator $K_{\mathbf{P}}$, where $\theta_{\mathbf{p}}$ is replaced by

$$\Theta_{\mathbf{P}} = \sum_{j=1}^{\ell} \theta_{\mathbf{p}_j}. \tag{10.59}$$

10.8.3 Nodal sets of K-real eigenfunctions

As mentioned previously, we can find a basis of $K_{\mathbf{p}}$-real eigenfunctions. It was shown in [474] and [30] that the $K_{\mathbf{p}}$-real eigenfunctions have a regular nodal set (like the eigenfunctions of the Dirichlet Laplacian) with the exception that, at each singular point $\mathbf{p}_j$ ($j = 1, \dots, \ell$), an odd number $\nu(\mathbf{p}_j)$ of half-lines meet. So the only difference with the notion of regularity introduced in Subsection 10.4.2 is that some $\nu(p_j)$ can be equal to 1.

Proposition 10.62. *The zero set of a $K_{\mathbf{p}}$-real eigenfunction of $H^{AB}(\dot{\Omega}^{\mathbf{P}})$ is the boundary set of a regular partition if and only if $\nu(\mathbf{p}_j) \geq 2$ for $j = 1, \dots, \ell$.*

Let us illustrate the case of the square with one singular point. Figure 10.12 gives the nodal lines of some eigenfunctions of the Aharonov-Bohm operator. In these examples, there are always one or three lines ending at the singular point (represented by a red point). Note that only the fourth picture gives a regular and nice partition.

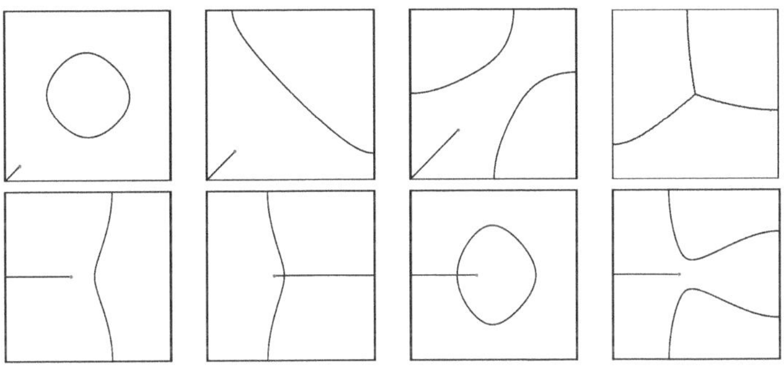

Fig. 10.12. Nodal lines of some Aharonov-Bohm eigenfunctions on the square.

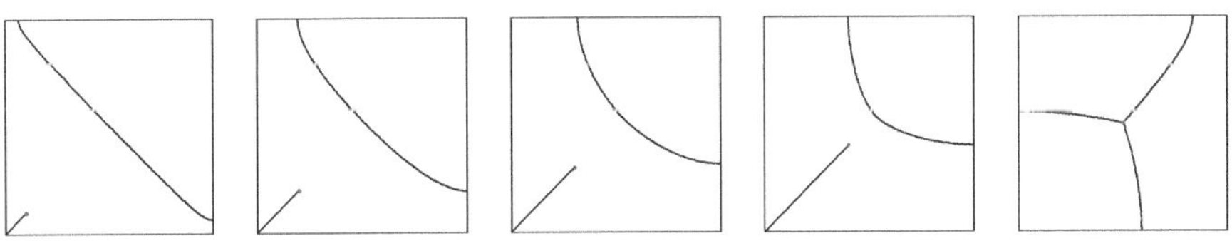

Fig. 10.13. Nodal lines for the third Aharonov-Bohm eigenfunction in function of **p** on the diagonal.

Our guess for the punctured square (**p** at the center) is that any nodal partition of a third $K_{\mathbf{p}}$-real eigenfunction gives a minimal 3-partition. Numerics shows that this is only true if the square is punctured at the center (see Figure 10.13 and [156] for a systematic study). Moreover the third eigenvalue is maximal there and has multiplicity two (see Figure 10.14).

10.8.4 Continuity with respect to the poles

In the case of a unique singular point, [729], [164, Theorem 1.1] establishes the continuity with respect to the singular point up to the boundary.

Theorem 10.63. *Let Ω be simply connected, $\alpha \in [0,1)$, and $\lambda_k^{AB}(\mathbf{p},\alpha)$ be the k-th eigenvalue of $H^{AB}(\dot{\Omega}_{\mathbf{p}},\alpha)$. Then the function $\mathbf{p} \in \Omega \mapsto \lambda_k^{AB}(\mathbf{p},\alpha)$ admits a continuous extension on $\overline{\Omega}$ and*

$$\lim_{\mathbf{p}\to\partial\Omega} \lambda_k^{AB}(\mathbf{p},\alpha) = \lambda_k\,, \qquad \forall k \geq 1\,, \tag{10.60}$$

where λ_k is the k-th eigenvalue of $H(\Omega)$.

The theorem implies that the function $\mathbf{p} \mapsto \lambda_k^{AB}(\mathbf{p},\alpha)$ has an extremal point in $\overline{\Omega}$. Note also that $\lambda_k^{AB}(\mathbf{p},\alpha)$ is well defined for $\mathbf{p} \notin \Omega$ and is equal to $\lambda_k(\Omega)$. One can indeed find a solution ϕ in Ω satisfying $d\phi = \mathbf{A}_{\mathbf{p}}$, and $u \mapsto \exp(i\alpha\phi)\,u$ defines the unitary transform intertwining $H(\Omega)$ and $H^{AB}(\dot{\Omega}_{\mathbf{p}},\alpha)$.

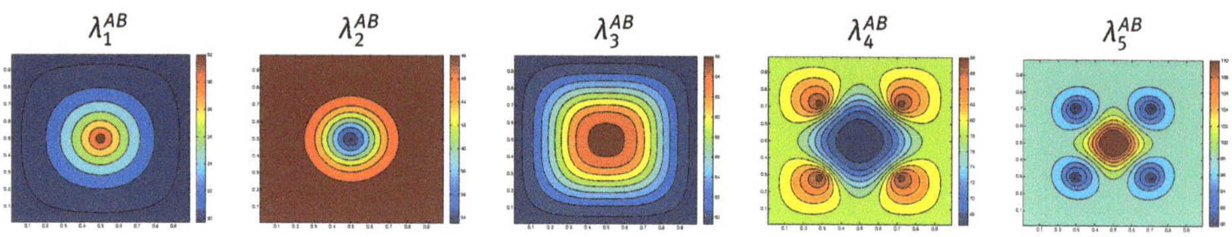

(a) $\mathbf{p} \mapsto \lambda_k^{AB}(\mathbf{p})$, $\mathbf{p} \in \Omega$, $1 \leq k \leq 5$.

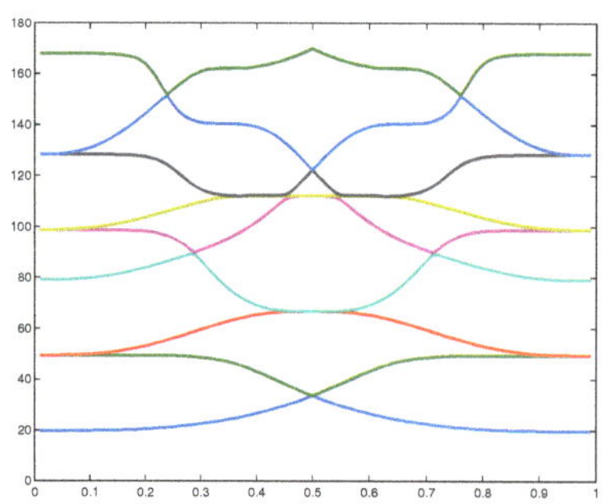

(b) $\mathbf{p} \mapsto \lambda_k^{AB}(\mathbf{p})$, $\mathbf{p} = (p,p)$, for $1 \leq k \leq 5$.

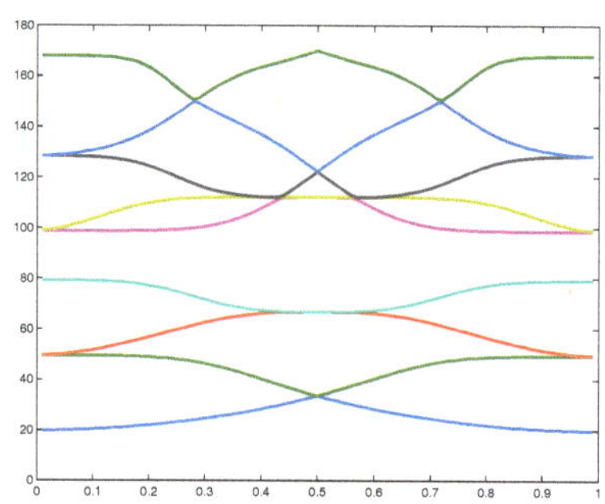

(c) $\mathbf{p} \mapsto \lambda_k^{AB}(\mathbf{p})$, $\mathbf{p} = (p,\frac{1}{2})$, for $1 \leq k \leq 5$.

Fig. 10.14. Aharonov-Bohm eigenvalues on the square as functions of the pole.

Figures 10.14–10.16 give some illustrations (see also [156, 164]) in the case of a square or of a circular sector with opening $\pi/3$ or $\pi/4$ and with a flux $\alpha = 1/2$. Figure 10.14 gives the first eigenvalues of $H^{AB}(\dot{\Omega}_{\mathbf{p}})$ in function of $\mathbf{p}$ in the square $\Omega = [0, 1]^2$ and demonstrates (10.60). When $\mathbf{p} = (1/2, 1/2)$, the eigenvalue is extremal and always double (see in particular Figures 10.14b and 10.14c which represent the first eigenvalues when the pole is either on a diagonal line or on a bisector line).

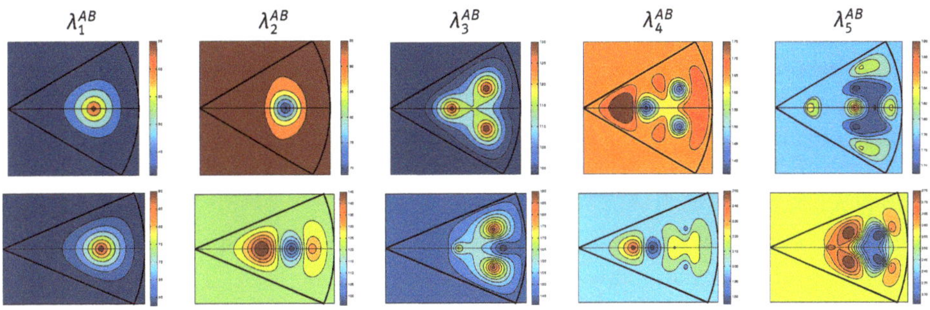

Fig. 10.15. Aharonov-Bohm eigenvalues for an circular sector Σ_ω of opening $\omega = \frac{\pi}{3}, \frac{\pi}{4}$.

Figure 10.15 gives the first five eigenvalues of $H^{AB}(\dot{\Sigma}_{\omega,\mathbf{p}})$ in function of $\mathbf{p}$, when Ω is a circular sector Σ_ω of opening $\omega = \pi/3$ (first line) and $\pi/4$ (second line). The k-th column gives $\lambda_k^{AB}(\mathbf{p})$. We recover (10.60) (we see that for $p \notin \Sigma_\omega$, $\lambda_k^{AB}(\mathbf{p}) = \lambda_k$) and observe that there always exists an extremal point on the symmetry axis. Figure 10.16 gives the eigenvalues of $H^{AB}(\dot{\Sigma}_{\frac{\pi}{4},\mathbf{p}})$ when $\mathbf{p}$ belongs to the bisector line of $\Sigma_{\pi/4}$.

Let us analyze what can happen at an extremal point (see [729, Theorem 1.1], [164,

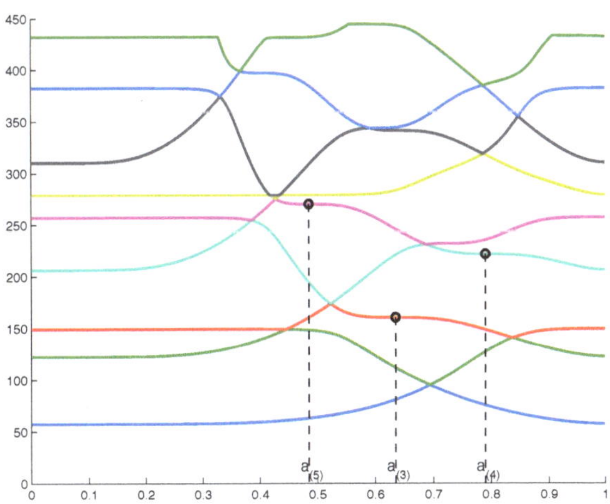

Fig. 10.16. $\mathbf{p} \mapsto \lambda_k^{AB}(\mathbf{p})$, $\mathbf{p} \in (0, 1) \times \{0\}$, $1 \le k \le 9$, on $\Sigma_{\pi/4}$.

Theorem 1.5]).

Theorem 10.64. *Suppose $\alpha = 1/2$. For any $k \geq 1$ and $\mathbf{p} \in \Omega$, we denote by $\varphi_k^{AB,\mathbf{p}}$ an eigenfunction associated with $\lambda_k^{AB}(\mathbf{p})$.*

- *If $\varphi_k^{AB,\mathbf{p}}$ has a zero of order $1/2$ at $\mathbf{p} \in \Omega$, then either $\lambda_k^{AB}(\mathbf{p})$ has multiplicity at least 2, or $\mathbf{p}$ is not an extremal point of the map $\mathbf{x} \mapsto \lambda_k^{AB}(\mathbf{x})$.*
- *If $\mathbf{p} \in \Omega$ is an extremal point of $\mathbf{x} \mapsto \lambda_k^{AB}(\mathbf{x})$, then either $\lambda_k^{AB}(\mathbf{p})$ has multiplicity at least 2, or $\varphi_k^{AB,\mathbf{p}}$ has a zero of order $m/2$ at $\mathbf{p}$, $m \geq 3$ odd.*

This theorem gives an interesting necessary condition for candidates to be minimal partitions. Indeed, knowing the behavior of the eigenvalues of Aharonov-Bohm operator, we can localize the position of the critical point for which the associated eigenfunction can produce a nice partition (with singular point where an odd number of lines end).

For the case of the square, we observe in Figure 10.14 that the eigenvalue is never simple at an extremal point. When Ω is the circular sector $\Sigma_{\pi/4}$ (see Figures 10.15 and 10.16), the only critical points of $\mathbf{x} \mapsto \lambda_k^{AB}(\mathbf{x})$ which correspond to simple eigenvalues are inflexion points located on the bisector line. Their abscissa are denoted $\mathrm{a}_{(\mathrm{k})}$, $\mathrm{k} = 3, 4, 5$ in Figure 10.16. Let $\mathbf{p}_{(k)} = (\mathrm{a}_{(\mathrm{k})}, 0)$. Figure 10.17 gives the nodal partitions associated with $\lambda_k^{AB}(\mathbf{p}_{(k)})$. We observe that there are always three lines ending at the singular point $\mathbf{p}_{(k)}$. In Figure 10.18 are represented the nodal partitions for singular points near $\mathbf{p}_{(3)}$. When $\mathbf{p} \neq \mathbf{p}_{(3)}$, there is just one line ending at $\mathbf{p}$.

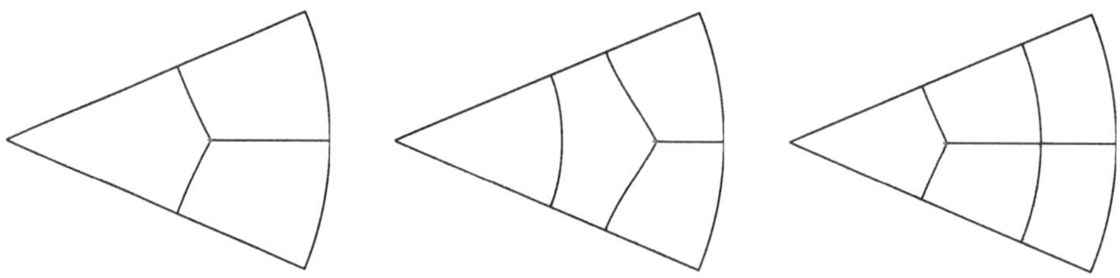

(a) $\lambda_3^{AB}(\mathbf{p}_{(3)})$ (b) $\lambda_4^{AB}(\mathbf{p}_{(4)})$ (c) $\lambda_5^{AB}(\mathbf{p}_{(5)})$

Fig. 10.17. Nodal lines of an eigenfunction associated with $\lambda_k(\mathbf{p}_{(k)})$, $k = 3, 4, 5$.

When there are several poles, the continuity result of Theorem 10.63 still holds. We will briefly address this result (see [650] for the proof and more details). This is rather clear in $\Omega^\ell \setminus \mathcal{C}$, where $\mathcal{C}$ denotes the $\mathbf{P}$'s such that $\mathbf{p}_i \neq \mathbf{p}_j$ when $i \neq j$. It is then convenient to extend the function $\mathbf{P} \mapsto \lambda_k^{AB}(\mathbf{P}, \boldsymbol{\alpha})$ to $(\mathbb{R}^2)^\ell$. We define $\lambda_k^{AB}(\mathbf{P}, \boldsymbol{\alpha})$ as the k-th eigenvalue of $H^{AB}(\dot{\Omega}_{\tilde{\mathbf{P}}}, \tilde{\boldsymbol{\alpha}})$, where the m-tuple $\tilde{\mathbf{P}} = (\tilde{\mathbf{p}}_1, \dots, \tilde{\mathbf{p}}_m)$ contains once, and only once, each point appearing in $\mathbf{P} = (\mathbf{p}_1, \dots, \mathbf{p}_\ell)$ and where $\tilde{\boldsymbol{\alpha}} = (\tilde{\alpha}_1, \dots, \tilde{\alpha}_M)$ with $\tilde{\alpha}_k = \sum_{j,\, \mathbf{p}_j = \tilde{\mathbf{p}}_k} \alpha_j$, for $1 \leq k \leq m$.

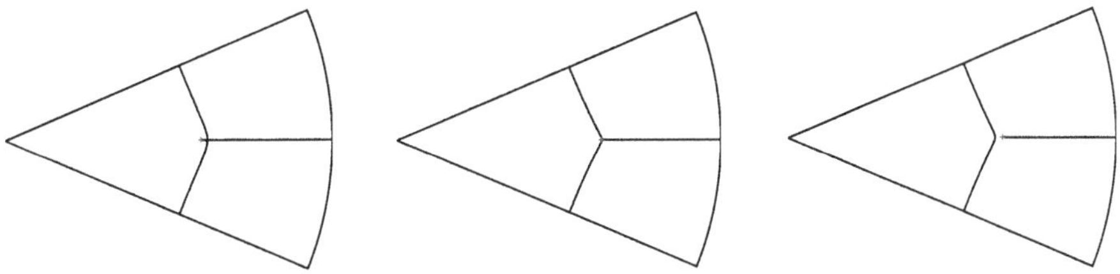

(a) $\lambda_3^{AB}(\mathbf{p})$, $\mathbf{p} = (0.60, 0)$ (b) $\lambda_3^{AB}(\mathbf{p})$, $\mathbf{p} \simeq (a_{(3)}, 0)$ (c) $\lambda_3^{AB}(\mathbf{p})$, $\mathbf{p} = (0.65, 0)$

Fig. 10.18. Nodal lines of an eigenfunction associated with $\lambda_3^{AB}(\mathbf{p})$.

Theorem 10.65. *If $k \geq 1$ and $\boldsymbol{\alpha} \in \mathbb{R}^\ell$, then the function $\mathbf{P} \mapsto \lambda_k^{AB}(\mathbf{P}, \boldsymbol{\alpha})$ is continuous in $\mathbb{R}^{2\ell}$.*

This result generalizes Theorems 10.63 and 10.64. It implies in particular continuity of the eigenvalues when one point tends to $\partial\Omega$, or in the case of coalescing points. For example, take $\ell = 2$, $\alpha_1 = \alpha_2 = \frac{1}{2}$, $\mathbf{P} = (\mathbf{p}_1, \mathbf{p}_2)$ and suppose that $\mathbf{p}_1$ and $\mathbf{p}_2$ tend to some $\mathbf{p}$ in Ω. Together with Proposition 10.61 we obtain in this case that $\lambda_k^{AB}(\mathbf{P}, \boldsymbol{\alpha})$ tends to $\lambda_k(\Omega)$.

10.8.5 Notes

More results on the Aharonov-Bohm eigenvalues as function of the poles can be found in [2, 156, 164, 650, 729]. We have only emphasized in this section the results which have direct applications to the research of candidates for minimal partitions.
In many of the papers analyzing minimal partitions, the authors refer to a double covering argument. Although this point of view (which appears first in [474] in the case of domains with holes) is essentially equivalent to the Aharonov approach, it has a more geometrical flavor. One can, in an abstract way, construct a double covering manifold $\dot{\Omega}_{\mathcal{R}}^{\mathbf{P}}$ above $\dot{\Omega}^{\mathbf{P}}$. One can then lift the initial spectral problem to one for the Laplace operator on this new (singular) manifold $\dot{\Omega}_{\mathcal{R}}^{\mathbf{P}}$. In this lifting, the $K^{\mathbf{P}}$-real eigenfunctions become eigenfunctions which are real and antisymmetric with respect to the deck map (exchanging two points having the same projection on $\dot{\Omega}^{\mathbf{P}}$). It appears that nodal components of antisymmetric Courant sharp eigenfunctions on $\dot{\Omega}_{\mathcal{R}}^{\mathbf{P}}$ (say with $2k$ nodal domains) give good candidates (by projection) for minimal k-partitions. The difficulty is of course with the choice of $\mathbf{P}$.
In the case of the disk, the construction is equivalent to considering $\theta \in (0, 4\pi)$ and the deck map corresponding to the translation by 2π. The nodal set of the 6-th eigenfunction gives by projection the Mercedes star and the 11-th eigenvalue (which is the 5-th in the space of antiperiodic functions) gives by projection the candidate presented in Figures 10.6b and 10.6c.

10.9 On the asymptotic behavior of minimal k-partitions

The hexagon has fascinating properties and appears naturally in many contexts (for example the honeycomb). If we consider polygons generating a tiling, the ground state energy $\lambda(\hexagon)$ gives the smallest value (at least in comparison with the square, the rectangle and the equilateral triangle). In this section we analyze the asymptotic behavior of minimal k-partitions as $k \to +\infty$.

10.9.1 The hexagonal conjecture

Conjecture 10.66. *The limit of $\mathfrak{L}_k(\Omega)/k$ as $k \to +\infty$ exists and*

$$|\Omega| \lim_{k\to+\infty} \frac{\mathfrak{L}_k(\Omega)}{k} = \lambda(\hexagon)\,.$$

Similarly, one has

Conjecture 10.67. *The limit of $\mathfrak{L}_{k,1}(\Omega)/k$ as $k \to +\infty$ exists and*

$$|\Omega| \lim_{k\to+\infty} \frac{\mathfrak{L}_{k,1}(\Omega)}{k} = \lambda(\hexagon)\,. \tag{10.61}$$

These conjectures, that we learned from M. Van den Berg in 2006 and which are also mentioned in Caffarelli-Lin [249] for $\mathfrak{L}_{k,1}$, imply in particular that the limit is independent of Ω.

Of course the optimality of the regular hexagonal tiling appears in various contexts in physics. It is easy to show by keeping the hexagons belonging to the intersection of Ω with the hexagonal tiling, the upper bound in Conjecture 10.66,

$$|\Omega| \limsup_{k\to+\infty} \frac{\mathfrak{L}_k(\Omega)}{k} \le \lambda(\hexagon)\,. \tag{10.62}$$

We recall that the Faber-Krahn inequality (10.16) gives a weaker lower bound

$$|\Omega| \frac{\mathfrak{L}_k(\Omega)}{k} \ge |\Omega| \frac{\mathfrak{L}_{k,1}(\Omega)}{k} \ge \lambda(\bigcirc)\,. \tag{10.63}$$

Note that Bourgain [174] and Steinerberger [831] have recently improved the lower bound by using an improved Faber-Krahn inequality together with considerations on packing property by disks (see Remark 10.19).

The inequality $\mathfrak{L}_{k,1}(\Omega) \le \mathfrak{L}_k(\Omega)$ together with the upper bound (10.62) shows that the second conjecture implies the first one.

Conjecture 10.66 has been explored in [162] by checking numerically non trivial consequences of this conjecture (see Corollary 10.52). Other recent numerical computations devoted to $\lim_{k\to+\infty} \frac{1}{k}\mathfrak{L}_{k,1}(\Omega)$ and to the asymptotic structure of the minimal partitions by Bourdin-Bucur-Oudet [173] are very enlightening.

10.9.2 Lower bounds for the length

We refer to [124] and references therein for proofs and more results. Let $\mathcal{D} = \{D_i\}_{1\le i\le k}$ be a regular spectral equipartition with energy $\Lambda = \Lambda(\mathcal{D})$. We define the length of the boundary set $\partial\mathcal{D}$ by the formula,

$$|\partial\mathcal{D}| := \frac{1}{2}\sum_{i=1}^{k} |\partial D_i| \,. \tag{10.64}$$

Proposition 10.68. *Let Ω be a bounded open set in $\mathbb{R}^2$, and let $\mathcal{D}$ be a regular spectral equipartition of Ω. The length $|\partial\mathcal{D}|$ of the boundary set of $\mathcal{D}$ is bounded from below in terms of the energy $\Lambda(\mathcal{D})$. More precisely,*

$$\frac{|\Omega|}{2\mathbf{j}}\sqrt{\Lambda(\mathcal{D})} + \frac{\pi\mathbf{j}}{2\sqrt{\Lambda(\mathcal{D})}}\left(\chi(\Omega) + \frac{1}{2}\sigma(\mathcal{D})\right) \le |\partial\mathcal{D}| \,. \tag{10.65}$$

Here

$$\sigma(\mathcal{D}) := \sum_{\mathbf{x}_i\in X(\partial\mathcal{D})}\left(\frac{\nu(\mathbf{x}_i)}{2} - 1\right) + \frac{1}{2}\sum_{\mathbf{y}_i\in Y(\partial\mathcal{D})}\rho(\mathbf{y}_i) \,,$$

which is the quantity appearing in Euler's formula (10.54).

The proof of [124] is obtained by combining techniques developed by Brüning-Gromes [200] together with ideas of A. Savo [798].

The hexagonal conjecture leads to a natural corresponding hexagonal conjecture for the length of the boundary set, namely

Conjecture 10.69.

$$\lim_{k\to+\infty}\frac{|\partial\mathcal{D}_k|}{\sqrt{k}} = \frac{1}{2}\ell(\hexagon)\sqrt{|\Omega|} \,, \tag{10.66}$$

where $\ell(\hexagon) = 2\sqrt{2\sqrt{3}}$ is the length of the boundary of $\hexagon$.

For regular spectral equipartitions $\mathcal{D}$ of the domain Ω, inequality (10.65) and Faber-Krahn's inequality yield,

$$\liminf_{\sharp(\mathcal{D})\to\infty}\frac{|\partial\mathcal{D}|}{\sqrt{\sharp(\mathcal{D})}} \ge \frac{\sqrt{\pi}}{2}\sqrt{|\Omega|} \,. \tag{10.67}$$

Assuming that $\chi(\Omega) \ge 0$, we have the uniform lower bound,

$$\frac{|\partial\mathcal{D}|}{\sqrt{\sharp(\mathcal{D})}} \ge \frac{\sqrt{\pi}}{2}\sqrt{|\Omega|} \,. \tag{10.68}$$

The following statement can be deduced from a particular case of Theorem 1-B established by T.C. Hales [443] in his proof of Lord Kelvin's honeycomb conjecture (see also [124]) which states that in $\mathbb{R}^2$ regular hexagons provide a perimeter-minimizing partition of the plane into unit areas.

Theorem 10.70. *For any regular partition $\mathcal{D}$ of a bounded open subset Ω of $\mathbb{R}^2$,*

$$|\partial\mathcal{D}| + \frac{1}{2}|\partial\Omega| \geq (12)^{\frac{1}{4}} (\min_i |D_i|)^{\frac{1}{2}} \sharp(\mathcal{D}) . \tag{10.69}$$

Theorem 10.71. *Let Ω be a regular bounded domain in $\mathbb{R}^2$. For $k \geq 1$, let $\mathcal{D}_k$ be a minimal regular k-partition of Ω. Then,*

$$\liminf_{k\to+\infty} \frac{|\partial\mathcal{D}_k|}{\sqrt{k}} \geq \frac{1}{2}\ell(\hexagon) \left(\frac{\pi\mathbf{j}^2}{\lambda(\hexagon)}\right)^{\frac{1}{2}} |\Omega|^{\frac{1}{2}} . \tag{10.70}$$

Proofs. Let $\mathcal{D} = \{D_i\}_{1\leq i\leq k}$ be a regular equipartition of Ω. Combining Faber-Krahn's inequality (10.16) for some D_i of minimal area with (10.69), we obtain

$$|\partial\mathcal{D}| + \frac{1}{2}|\partial\Omega| \geq (12)^{\frac{1}{4}} (\pi\mathbf{j}^2)^{\frac{1}{2}} \frac{\sharp(\mathcal{D})}{\sqrt{\Lambda(\mathcal{D})}} . \tag{10.71}$$

Let $\mathcal{D}_k$ be a minimal regular k-partition of Ω. Using (10.62) in (10.71) gives (10.70). □

To see the efficiency of each approach, we give the approximate value of the different constants:

$$\frac{1}{2}\ell(\hexagon) \simeq 1.8612 , \qquad \frac{1}{2}\ell(\hexagon) \left(\frac{\pi\mathbf{j}^2}{\lambda(\hexagon)}\right)^{\frac{1}{2}} \simeq 1.8407 , \qquad \frac{\sqrt{\pi}}{2} \simeq 0.8862 .$$

Assume now that $\widehat{\mathcal{D}_k} := \mathcal{D}(u_k)$ is the nodal partition of some k-th eigenfunction u_k of $H(\Omega)$. Assume furthermore that $\chi(\Omega) \geq 0$. Combining (10.65) with Weyl's theorem leads to:

$$\liminf_{k\to+\infty} \frac{|\partial\widehat{\mathcal{D}_k}|}{\sqrt{k}} \geq \frac{\sqrt{\pi}}{\mathbf{j}} \sqrt{|\Omega|} . \tag{10.72}$$

In the case of a compact manifold, this kind of lower bound appears first in [199], see also the celebrated work by Donnelly-Feffermann [341, 342] around a conjecture by Yau and the recent breakthrough by A. Logunov [678].

10.9.3 Magnetic characterization and lower bounds for the number of singular points

Helffer–Hoffmann-Ostenhof prove a magnetic characterization of minimal k-partitions (see [491, Theorem 5.1]):

Theorem 10.72. *Let Ω be simply connected and $\mathcal{D}$ be a minimal k-partition of Ω. Then $\mathcal{D}$ is the nodal partition of some k-th $K_{\mathbf{P}}$-real eigenfunction of $H^{AB}(\dot{\Omega}_{\mathbf{P}})$ with $\{\mathbf{p}_1, \ldots, \mathbf{p}_\ell\} = X^{odd}(\partial\mathcal{D})$.*

Proof. We return to the proof that a bipartite minimal partition is nodal for the Laplacian. Using the u_j whose existence was recalled for minimal partitions, we can find a sequence $\varepsilon_j = \pm 1$ such that $\sum_j \varepsilon_j \exp(\frac{i}{2}\Theta_{\mathbf{P}}(\mathbf{x}))\, u_j(\mathbf{x})$ is an eigenfunction of $H^{AB}(\dot{\Omega}_{\mathbf{P}})$, where $\Theta_{\mathbf{P}}$ was defined in (10.59). □

The next theorem of [488] improves a weaker version proved in [476].

Theorem 10.73. *For any sequence $(\mathfrak{D}_k)_{k\in\mathbb{N}}$ of regular minimal k-partitions, we have*

$$\liminf_{k\to\infty} \frac{\sharp X^{odd}(\partial\mathfrak{D}_k)}{k} > 0\,. \tag{10.73}$$

Although inspired by the proof of Pleijel's theorem, this proof includes (for any k) a lower bound in the Weyl's formula (for the eigenvalue $\mathfrak{L}_k$) for the Aharonov-Bohm operator $H^{AB}(\dot{\Omega}_{\mathbf{P}})$ associated with the odd singular points of $\mathfrak{D}_k$. The proof gives an explicit but very small lower bound in (10.73) which is independent of the sequence. This is to compare with the upper bound proven in Subsection 10.6.3 which gives

$$\limsup_{k\to\infty} \frac{\sharp X^{odd}(\partial\mathfrak{D}_k)}{k} \leq 2\,.$$

10.9.4 Notes

The hexagonal conjecture in the case of a compact Riemannian manifold is the same. We refer to [124] for the details, the idea being that, for k large, it is the local structure of the manifold which plays the main role, like for Pleijel's formula (see [133]). In [363] the authors analyze numerically the validity of the hexagonal conjecture in the case of the sphere (for $\mathfrak{L}_{k,1}$). As mentioned in Subsection 10.6.4, one can add in the hexagonal conjecture that there are $(k-12)$ hexagons and 12 pentagons for k large enough. In the case of a planar domain one expects curvilinear hexagons inside Ω, around $\sqrt{k}$ pentagons close to the boundary (see [173]) and a few number of other polygons.

Acknowledgment: We would like to thank particularly our first collaborators T. Hoffmann-Ostenhof and S. Terracini, and also P. Bérard, B. Bogosel, C. Léna, B. Noris, M. Nys, M. Persson Sundqvist and G. Vial who join us for the continuation of this programme devoted to minimal partitions.
We would also like to thank P. Charron, D. Bucur, T. Deheuvels, A. Henrot, D. Jakobson, J. Lamboley, J. Leydold, É. Oudet, M. Pierre, I. Polterovich, T. Ranner...for their interest, their help and for useful discussions.
The authors are supported by the ANR (Agence Nationale de la Recherche), project OPTIFORM n° ANR-12-BS01-0007-02 and by the Centre Henri Lebesgue (program "Investissements d'avenir" – n° ANR-11-LABX-0020-01). During the writing of this work, the second author was Simons foundation visiting fellow at the Isaac Newton Institute for Mathematical Sciences in Cambridge. He would like to thank the foundation for its support and the institute for its hospitality during the programme PESP supported by EPSRC Grant Number EP/K032208/1.

Pedro R. S. Antunes and Edouard Oudet

11 Numerical results for extremal problem for eigenvalues of the Laplacian

We consider in this chapter shape optimization problems for Dirichlet and Neumann eigenvalues,

$$\lambda_i^\star := \min\left\{\lambda_i(\Omega),\ \Omega \subset \mathbb{R}^d,\ |\Omega| = 1\right\},\ i = 1, 2, \ldots \tag{11.1}$$

and

$$\mu_i^\star := \max\left\{\mu_i(\Omega),\ \Omega \subset \mathbb{R}^d,\ |\Omega| = 1\right\},\ i = 2, 3, \ldots \tag{11.2}$$

Some of these shape optimizations have already been solved. The first Dirichlet eigenvalue is minimized by the ball, as proven by Faber and Krahn [377, 603]. The second Dirichlet eigenvalue is minimized by two balls of the same volume [604]. In two dimensional case, it has long been conjectured that the ball minimizes $\lambda_3(\Omega)$, but there has not been much progress in this direction. For higher eigenvalues, not much is known and even the existence of minimizers among quasi-open sets has only been proven quite recently (see Chapter 2) and [206, 700]. It is worth mentioning the work by Berger [137] who proved that for $i > 4$, the i-th Dirichlet eigenvalue is not minimized by any union of balls.

For the Neumann problem, we have $\mu_1(\Omega) = 0$. The second eigenvalue, $\mu_2(\Omega)$, is maximized by the ball. The result had been conjectured by Kornhauser and Stakgold in [599] and was proved by Szegö in [838] for Lipschitz simply connected planar domains and generalized by Weinberger in [871] to arbitrary domains, and any dimension. More recently, Girouard, Nadirashvili and Polterovich proved that the maximum of $\mu_3(\Omega)$ among simply connected bounded planar domains is attained by two disjoint balls of equal area [428].

Recently, many works have addressed numerical approaches that propose candidates for the optimizers for these and related spectral problems, and to suggest conjectures about their qualitative properties [42, 153, 734–737].

In the next two sections we describe briefly two of these approaches which have been successful for spectral problems. We first introduce some global optimization tools to provide a good initial guess of the optimal profile. This step does not require any topological information on the set but is restricted to a small class of shapes. Moreover, since this approach relies only on the parametrization of the space of shapes, any global algorithm can be used as a black box solver to find a starting candidate for the

Pedro R. S. Antunes: Grupo de Física Matemática da Universidade de Lisboa, Portugal, E-mail: prantunes@fc.ul.pt
Edouard Oudet: Laboratoire Jean Kuntzmann (LJK), Université Joseph Fourier et CNRS, Tour IRMA, BP 53, 51 rue des Mathématiques, 38041 Grenoble Cedex 9 - France, E-mail: edouard.oudet@imag.fr

local procedure. We then describe the method of fundamental solutions which is able in a second stage to both identify and evaluate precisely shapes which are locally optimal.

11.1 Some tools for global numerical optimization in spectral theory

Global optimality is perhaps one of the most challenging aspects in numerical shape optimization. Many approaches have been developed to tackle this difficulty: stochastic algorithms, multi scale methods, relaxation, homogenization, etc. In this section we first recall briefly an historical method developed to apply a genetic algorithm in the framework of shape optimization. It has the benefit of dramatically reducing the number of degrees of freedom, which makes the global optimization more efficient, it has the drawback of parametrizing non smooth profiles.

In the following we introduce a very naive approach based on implicit representation which makes it possible both to reduce the number of unknowns and to generate smooth shapes. In a second step we describe a new simple idea to restrict the search space in spectral optimization to one where homogeneous functionals are frequently involved.

11.1.1 An historical approach: Genetic algorithm and Voronoi cells

Consider a given grid covering a search domain in which we look for an optimal shape Ω. A purely discrete numerical approach consists of associating to every node of the grid a boolean value which expresses the fact that the node is or is not in the set Ω.

In the 90's, Allaire et al. [21] introduced a discretization framework in shape optimization based on Voronoi cells associated to a set of points independent of the grid. In this setting a grid point of the search space is considered to be a part of Ω if the seed of its Voronoi cell has a True boolean value. The interesting part of this method is that the complexity of the approach is not anymore related to the grid size but only on the number of seed points. This crucial distinction makes it possible to compute state solutions of partial differential equations with a reasonable precision whereas the number of unknowns is not too large. A drawback of the method is the non smoothness of Voronoi cells. Due to its polygonal faces, an large number of Voronoi seeds may be required to approximate smooth shapes. In the specific context of eigenvalues where some smoothness is expected, this kind of discretization is not optimal.

11.1.2 Smooth profiles with few parameters

In this section related to global spectral optimization we consider the minimization of functional of the type

$$G(\Omega) = F\left(|\Omega|, |\partial\Omega|, \{\lambda_i(\Omega)\}_{1\le i\le k}, \{\mu_i(\Omega)\}_{1\le i\le l}\right) \tag{11.3}$$

where F is some smooth fixed finite dimensional function. We address the optimization problem

$$\min_{\Omega} G(\Omega). \tag{11.4}$$

Depending on the context we may consider additional geometrical constraints imposed on the set Ω like a fixed measure or boundary measure, connectivity, convexity, etc.

Identifying the global optimal solution of a non convex and sometimes non smooth cost function is in almost all cases an untenable task. In order to decrease the complexity of this problem we introduce a reduction of the number mp of parameters which still allows a precise computation of the cost function. Since we use black box stochastic algorithm we need to reduce the number of parameters to $mp \le 100$ for instance. More precisely we develop this dimension reduction by introducing the two following steps.

In the spirit of level set methods, we first parameterize the set of shape as the level sets of truncated Fourier series. Contrary to standard boundary parametrization where the number of degrees of freedom is related to the number of boundary points in the mesh, our parametrization has the complexity of the number of terms in the Fourier series. Moreover, the very basic and crucial observation is the fact that this complexity is not related to the precision of the eigenvalue approximation.

Our second improvement is related to the reduction of the size of the search space. We reduce the complexity of the optimization process by substituting the cost evaluation of a given shape by the optimal value associated to the best homothetic connected components. Essentially it relies on the two following classical properties:

Proposition 11.1 (Homogeneity). *Let $\alpha > 0$ be a real. Then for all integers j,*

$$\lambda_j(\alpha\Omega) = \alpha^{-2}\lambda_j(\Omega), \quad \mu_j(\alpha\Omega) = \alpha^{-2}\mu_j(\Omega), \quad |\alpha\Omega| = \alpha^d|\Omega| \quad \textit{and} \quad |\partial(\alpha\Omega)| = \alpha^{d-1}|\partial\Omega|. \tag{11.5}$$

Notice that homogeneity can be used to a transform constrained problem like

$$\min\{\lambda_i(\Omega),\ |\Omega| = 1\},\ i = 1, 2, \dots \tag{11.6}$$

into an unconstrained one

$$\min\{\lambda_i(\Omega)|\Omega|\},\ i = 1, 2, \dots. \tag{11.7}$$

The second property makes it possible to compute more efficiently optimal profiles with multiply connected components:

Proposition 11.2 (Multiply connected components). *Let $\Omega_1, \dots, \Omega_m$ be the connected components of Ω. For $i = 1, \dots, m$, let Λ_i be the set of the eigenvalues of $-\Delta$ on Ω_i for Dirichlet or Neumann condition. Then the set of the eigenvalues of Δ is $\Lambda = \bigcup_{i=1}^m \Lambda_i$.*

We now describe our discretization of the search space for $d = 2$. The generalization to higher dimensions is straightforward. Let us consider coefficients $a_{i,j}$ such that $\sharp\{a_{i,j}\} = n$ (the number of coefficients). Then, define the Fourier series

$$\Phi_{\{a_{i,j}\}}(x) = \sum_{i,j} a_{i,j} \sin(\pi i x_1) \sin(\pi j x_2) + 1, \quad \text{where } x = (x_1, x_2) \in [0, 1]^2. \tag{11.8}$$

Notice that we add the constant value 1 to previous sum so that the function is non-zero on $\partial\Omega$. That is the level set domain does not intersect the boundary. We now define $\mathcal{F} : \mathbb{R}^n \to \mathcal{P}(\mathbb{R}^2)$ by

$$\mathcal{F}(\{a_{i,j}\}) = \{x \in [0, 1]^2,\ \Phi_{\{a_{i,j}\}}(x) \le 0\}. \tag{11.9}$$

Finally we build the sets

$$\Omega_{\{a_{i,j}\}} = \mathcal{F}(\{a_{i,j}\}). \tag{11.10}$$

Notice that the topology or more specifically the number of connected components of $\Omega_{\{a_{i,j}\}}$ is not imposed by the algorithm.

In practice, $B = [0, 1]^2$ is meshed by a Cartesian grid. $\Phi_{\{a_{i,j}\}}$ is evaluated at every point of the mesh and a linear interpolation is carried out to approximate $\mathcal{F}(\{a_{i,j}\})$. Through this discretization we associate a polygon $\Omega^{pol}_{\{a_{i,j}\}}$ to every $\Omega_{\{a_{i,j}\}}$. We then define the cost function associated to the parameters $a_{i,j}$ by

$$F(\{a_{i,j}\}) := G(\Omega^{pol}_{\{a_{i,j}\}}) \simeq G(\Omega_{\{a_{i,j}\}}) = G\left(\mathcal{F}(\{a_{i,j}\})\right).$$

Where the $\simeq$ symbol expresses the fact we do not optimize the true eigenvalues of the polygons int this process but rather the finite element approximation of these eigenvalues. Finally, it is standard to approximate the cost function $G(\Omega^{pol}_{\{a_{i,j}\}})$ by classical Finite Element Methods. Notice that every evaluation requires us to construct a new mesh adapted to the polygon $\Omega^{pol}_{\{a_{i,j}\}}$ since linear interpolation generates meshes of very bad quality. In all our experiments we fixed approximately the number of simplices per evaluation to obtain comparable results.

11.1.3 A fundamental complexity reduction: optimal connected components

We detailed in previous section a way to parametrize multi-connected shapes with few parameters. As it has been explained, every cost evaluation requires to mesh the new domain and to solve the associated discrete spectral optimization problem. This step can be very time-consuming especially in the case of 3 dimension computations. To tackle this difficulty, we would like to use the homogeneity of eigenmodes to investigate homothetical components in one single cost evaluation. Actually, due to the homogeneity of eigenmodes the computation of these modes associated to one geometrical configuration can be used to deduce the cost function of any domain made of homothetic components.

From properties 11.1 and 11.2 we obtain that if $\Omega = \alpha_1\Omega_1 \cup \alpha_2\Omega_2$ (disjoint union) then

$$\begin{aligned} G(\alpha_1\Omega_1 \cup \alpha_2\Omega_2) &= F\left(|\Omega|, |\partial\Omega|, \{\lambda_i(\Omega)\}_{1\le i\le k}, \{\mu_i(\Omega)\}_{1\le i\le l}\right) \\ &= F(\alpha_1^d|\Omega_1| + \alpha_2^d|\Omega_2|, \alpha_1^{d-1}|\partial\Omega_1| + \alpha_2^{d-1}|\partial\Omega_2|, \dots \\ &\quad \{\lambda_i(\alpha_1\Omega_1 \cup \alpha_2\Omega_2)\}_{1\le i\le k}, \{\mu_i(\alpha_1\Omega_1 \cup \alpha_2\Omega_2)\}_{1\le i\le l}) \end{aligned}$$

The crucial fact observation is that the computation of $\lambda_j(\alpha_1\Omega_1 \cup \alpha_2\Omega_2)$ and $\mu_j(\alpha_1\Omega_1 \cup \alpha_2\Omega_2)$ is equivalent to the sorting operation of the union of the two sets

$$\left\{\alpha_1^{-2}\lambda_i(\Omega_1);\ 1 \le j \le m < k\right\} \cup \left\{\alpha_2^{-2}\lambda_i(\Omega_2);\ 1 \le j \le k - m\right\}$$

and

$$\left\{\alpha_1^{-2}\mu_j(\Omega_1);\ 1 \le j \le m' < l\right\} \cup \left\{\alpha_2^{-2}\mu_j(\Omega_2);\ 1 \le j \le l - m'\right\}.$$

Let Ω be a fixed set with m connected components. Let us associate to Ω a new cost which is the best value obtained with respect to its homothetical connected components. More precisely we define:

$$G(\Omega^{pol}_{\{a_{i,j}\}}) = \min_{(\alpha_1,\dots,\alpha_m)\ge 0} G\left(\alpha_1\Omega_1^{pol} \cup \cdots \cup \alpha_m\Omega_m^{pol}\right) \tag{11.11}$$

where

$$\Omega^{pol}_{\{a_{i,j}\}} = \bigcup_{i=1}^{m} \Omega_i^{pol}. \tag{11.12}$$

Notice that due to the translation invariance of the problem we can always assume that every connected components are disjoint. As a consequence, we can associate to a fixed geometrical configuration with m connected components a new cost defined by (11.11). This small scale global problem (11.11) can be solved very efficiently by using global algorithm like Lipschitz optimization (see for instance [565]). Moreover, since the number of unknowns is small (the number of expected connected components) and the cost evaluation is pretty fast, this global optimization problem can be solved very quickly with respect to a finite element evaluation.

11.2 Numerical approach using the method of fundamental solutions

The eigenvalue problem for the Laplace operator is equivalent to obtaining the resonant frequencies $0 \le \kappa_1 \le \kappa_2 \le \cdots \le \kappa_p \le \cdots$ that lead to non trivial solutions of the

Helmholtz equation $u_P \not\equiv 0$:

$$\begin{cases} \Delta u_P + \kappa_P^2 u_P = 0 & \text{in } \Omega, \\ u_P = 0 \text{ or } \partial_n u = 0 & \text{on } \Gamma. \end{cases} \tag{11.13}$$

Among other numerical approaches, these eigenvalue problems can be solved by the Method of Fundamental Solutions (MFS) [29, 570]. The MFS is a Trefftz type method, where the particular solutions are point sources centered outside the domain. More precisely, denoting by $\|.\|$ the Euclidean norm in $\mathbb{R}^d$, we take a fundamental solution of the Helmholtz equation

$$\Phi_\kappa(x) = \frac{i}{4} H_0^{(1)}(\kappa \|x\|) \tag{11.14}$$

for the two-dimensional case, where $H_0^{(1)}$ is the first Hankel function and

$$\Phi_\kappa(x) = \frac{e^{i\kappa\|x\|}}{4\pi \|x\|} \tag{11.15}$$

in the three-dimensional case. We have $(\Delta + \kappa^2)\Phi_\kappa = -\delta$, where δ is the Dirac delta distribution. For a given frequency κ, we consider a basis built with point sources

$$\phi_j = \Phi_\kappa(\cdot - y_j) \tag{11.16}$$

where $y_j \notin \bar{\Omega}$. By $\hat{\Gamma} = \partial\hat{\Omega}$, we will denote an admissible source set, for instance, the boundary of a bounded open set $\hat{\Omega} \supset \bar{\Omega}$, with $\hat{\Gamma}$ surrounding $\partial\Omega$.

The MFS approximation is a linear combination

$$\sum_{j=1}^{m} \alpha_j \Phi_\kappa(\cdot - y_j), \tag{11.17}$$

where the source points y_j are placed on an admissible source set. The approximation of an eigenfunction by a MFS linear combination can be justified by density results (e.g. [29]),

Theorem 11.3. *Consider $\hat{\Gamma} = \partial\hat{\Omega}$, an admissible source set. Then,*

$$S(\hat{\Gamma}) = span\{\Phi_\kappa(\cdot - y)|_\Omega : y \subset \hat{\Gamma}\}$$

is dense in $\mathcal{H}_\kappa(\Omega) = \{v \in H^1(\Omega) : (\Delta + \kappa^2)v = 0\}$, with the $H^1(\Omega)$ topology.

Next we give a brief description of the application of the MFS for determining the eigensolutions for a given shape Ω. For details, see [28] and [39] respectively for the two and three-dimensional cases. The eigensolutions are obtained in two steps. First, we calculate an approximate eigenfrequency $\tilde{\kappa}$ and then, for that frequency, we obtain the approximation for the eigenfunction.

We define m collocation points x_i almost uniformly distributed on the boundary $\partial\Omega$ and for each of those points we define a corresponding source point,

$$y_i = x_i + \alpha n_i,$$

where α is a positive parameter and n_i is the unitary outward normal vector at the point x_i. Imposing the Dirichlet boundary conditions at the boundary points we obtain the system

$$\sum_{j=1}^{m} \alpha_j \Phi_\kappa(x_i - y_j) = 0. \tag{11.18}$$

A straightforward procedure for calculating the eigenfrequencies is to find the values κ for which the $m \times m$ matrix

$$A(\kappa) = \left[\Phi_\kappa(x_i - y_j)\right]_{m\times m} \tag{11.19}$$

is singular. The Neumann case is similar.

To obtain an eigenfunction associated with a certain resonant frequency κ we use a collocation method on $n+1$ points, with $x_1, \cdots, x_n$ on $\partial\Omega$ and a point $x_{n+1} \in \Omega$. The eigenfunction is approximated by an MFS approximation,

$$\tilde{u}(x) = \sum_{j=1}^{n+1} \alpha_j \Phi_\kappa(x - y_j) \tag{11.20}$$

and to exclude the trivial solution $\tilde{u}(x) \equiv 0$, the coefficients α_j are determined by solving the system

$$\tilde{u}(x_i) = \delta_{i,n+1}, \quad i = 1, \ldots, n+1$$

where $\delta_{i,j}$ is the Kronecker delta. An advantage of the Method of Fundamental Solutions approach with respect to Finite Element Method approach is the fact that we can calculate rigorous bounds for the errors associated to approximate eigenvalues. In the Dirichlet case, the error can be estimated by using an *a posteriori bound* due to Fox, Henrici and Moler (cf.[389]). This result provides upper bounds for the errors of the approximations obtained in several methods of particular solutions and was also used in rigorous proofs (eg. [28, 143, 642]).

Next, we define the class of admissible domains for the shape optimization. Note that if for some $i = 1, 2, \ldots,$ the optimizer of the problems (11.1) or (11.2) is disconnected, by Wolf-Keller Theorem (cf. [882]), each of the connected components are optimizers of a lower eigenvalue. Thus, we will focus on the numerical solution of the shape optimization problem among connected domains and then compare this optimal value against the optimal value obtained for disconnected sets by using Wolf-Keller theorem.

We consider the functions

$$\gamma_1(t) = a_0^{(1)} + \sum_{j=1}^{P} a_j^{(1)} \cos(jt) + \sum_{j=1}^{P} b_j^{(1)} \sin(jt)$$

and

$$\gamma_2(t) = a_0^{(2)} + \sum_{j=1}^{P} a_j^{(2)} \cos(jt) + \sum_{j=1}^{P} b_j^{(2)} \sin(jt),$$

for some $P \in \mathbb{N}$ and the vector $\mathcal{C} \in \mathbb{R}^{4P+2}$ with all the coefficients of these expansions,

$$\mathcal{C} = \left(a_0^{(1)}, a_1^{(1)}, \ldots, a_P^{(1)}, b_1^{(1)}, \ldots, b_P^{(1)}, a_0^{(2)}, a_1^{(2)}, \ldots, a_P^{(2)}, b_1^{(2)}, \ldots, b_P^{(2)}\right).$$

The class of planar admissible domains is the set

$$\mathcal{V} = \left\{V \subset \mathbb{R}^2 : \partial V = (\gamma_1(t), \gamma_2(t)) : t \in [0, 2\pi) \text{ is a Jordan curve}\right\}.$$

For the three dimensional case, we assume that Ω is star-shaped and its boundary can be parametrized by

$$\partial\Omega = \{r(\theta, \phi)\,(\sin(\theta)\cos(\phi), \sin(\theta)\sin(\phi), \cos(\theta))\,,\ \phi \in [0, 2\pi),\ \theta \in [0, \pi]\},$$

where r is expanded in terms of spherical harmonics

$$r(\theta, \phi) = \sum_{l=0}^{N} \sum_{m=-l}^{l} a_{l,m} y_l^m(\theta, \phi),$$

where

$$y_l^m(\theta, \phi) = \begin{cases} \sqrt{2} k_l^m \cos(m\phi) P_l^m(\cos(\theta)) & \text{if } m > 0, \\ k_l^0 P_l^0(\cos(\theta)) & \text{if } m = 0, \\ \sqrt{2} k_l^m \sin(-m\phi) P_l^{-m}(\cos(\theta)) & \text{if } m < 0, \end{cases}$$

P_l^m is an associated Legendre polynomial and

$$k_l^m = \sqrt{\frac{(2l+1)(l-|m|)!}{4\pi(l+|m|)!}}.$$

Then, we collect all the coefficients $a_{l,m}$ in a single vector

$$\mathcal{C} = (a_{0,0}, a_{1,-1}, a_{1,0}, a_{1,1}, a_{2,-2}, a_{2,-1}, a_{2,0}, a_{2,1}, a_{2,2}, \ldots).$$

The shape optimization is solved by searching for optimal vectors $\mathcal{C}$ using a gradient type method. In this context, a key ingredient is the Hadamard formula of derivation with respect to the domain (e.g. [505]). Consider an application $\Psi(t)$ such that

$$\Psi : t \in [0, T) \to W^{1,\infty}(\mathbb{R}^N, \mathbb{R}^N) \text{ is differentiable at 0 with } \Psi(0) = I,\ \Psi'(0) = V,$$

where $W^{1,\infty}(\mathbb{R}^N, \mathbb{R}^N)$ is the set of bounded Lipschitz maps from $\mathbb{R}^N$ into itself, I is the identity and V is a deformation field. We denote by $\Omega_t = \Psi(t)(\Omega)$, $\lambda_k(t) = \lambda_k(\Omega_t)$, and by u an associated normalized eigenfunction. If we assume that Ω is of class C^2 and $\lambda_k(\Omega)$ is simple, then

$$\lambda_k{}'(0) = -\int_{\partial\Omega} \left(\frac{\partial u}{\partial n}\right)^2 V.n d\sigma. \tag{11.21}$$

For the Neumann case, assuming that Ω is of class C^3, μ_k is simple and u is the associated normalized eigenfunction, we have

$$\mu_k{}'(\Omega)(0) = \int_{\partial\Omega} \left(|\nabla u|^2 - \mu_k u^2\right) V.nd\sigma. \tag{11.22}$$

11.3 The menagerie of the spectrum

In this section we present the main numerical results that we gathered for the solution of the shape optimization problems (11.1) and (11.2). In Figure 11.1 we plot the minimizers of the first 15 Dirichlet eigenvalues.

Table 11.1 shows the optimal Dirichlet eigenvalues, together with the corresponding multiplicity of each optimal eigenvalue.

Table 11.1. *The optimal Dirichlet eigenvalues λ_i^*, for $i = 1, 2, \ldots, 15$ and the multiplicity of the optimal eigenvalue.*

i	multiplicity	λ_i^*
1	1	18.17
2	2	36.34
3	3	46.13
4	3	64.30
5	2	78.15
6	3	88.48
7	3	106.12
8	3	118.88
9	3	132.34
10	4	142.69
11	4	159.40
12	4	172.88
13	4	186.91
14	4	198.94
15	5	209.62

In Figure 11.2 we plot the maximizers of the first 10 (non trivial) Neumann eigenvalues in the class of unions of simply connected domains.

Table 11.2 shows the optimal Neumann eigenvalues and the corresponding optimal multiplicity.

Next, we present some numerical results for the shape optimization problems (11.1) and (11.2) with three-dimensional domains. In Figure 11.3 we plot the 3D minimizers of the first 10 Dirichlet eigenvalues. Table 11.3 shows the optimal 3D Dirichlet

Fig. 11.1. The minimizers of the first 15 Dirichlet eigenvalues.

eigenvalues and the corresponding multiplicity. Figure 11.4 and Table 11.4 show similar results for Neumann eigenvalues.

Fig. 11.2. The maximizers of the first 10 (non trivial) Neumann eigenvalues.

11.4 Open problems

The numerical results that we obtained suggest some conjectures

Open problem 11.4. *Prove that the d-dimensional ball minimizes λ_{d+1} among all d-dimensional sets of a fixed volume.*

Open problem 11.5. *Prove that the d-dimensional minimizer of λ_{d+2} among all d-dimensional sets of a fixed volume is the union of two balls whose radii are in the ratio*

$$\frac{j_{\frac{d}{2},1}}{j_{\frac{d}{2}-1,1}},$$

where $j_{n,k}$ is the k-th zero of the Bessel function J_n.

Table 11.2. *The optimal Neumann eigenvalues μ_i^*, for $i = 1, 2, \ldots, 11$ and the corresponding multiplicity.*

i	multiplicity	μ_i^*
2	2	10.66
3	4	21.28
4	3	32.90
5	3	43.86
6	3	55.17
7	4	67.33
8	6	77.99
9	4	89.38
10	4	101.83
11	5	114.16

Table 11.3. *The optimal 3D Dirichlet eigenvalues λ_i^*, for $i = 1, 2, \ldots, 10$ and the corresponding multiplicity.*

i	multiplicity	λ_i^*
1	1	25.65
2	2	40.72
3	2	49.17
4	3	52.47
5	4	63.83
6	3	73.05
7	4	78.35
8	6	83.29
9	5	86.32
10	3	92.33

Acknowledgment: The research of the first author was partially supported by FCT, Portugal, through the program “Investigador FCT” with reference IF/00177/2013 and the scientific project PTDC/MAT-CAL/4334/2014.

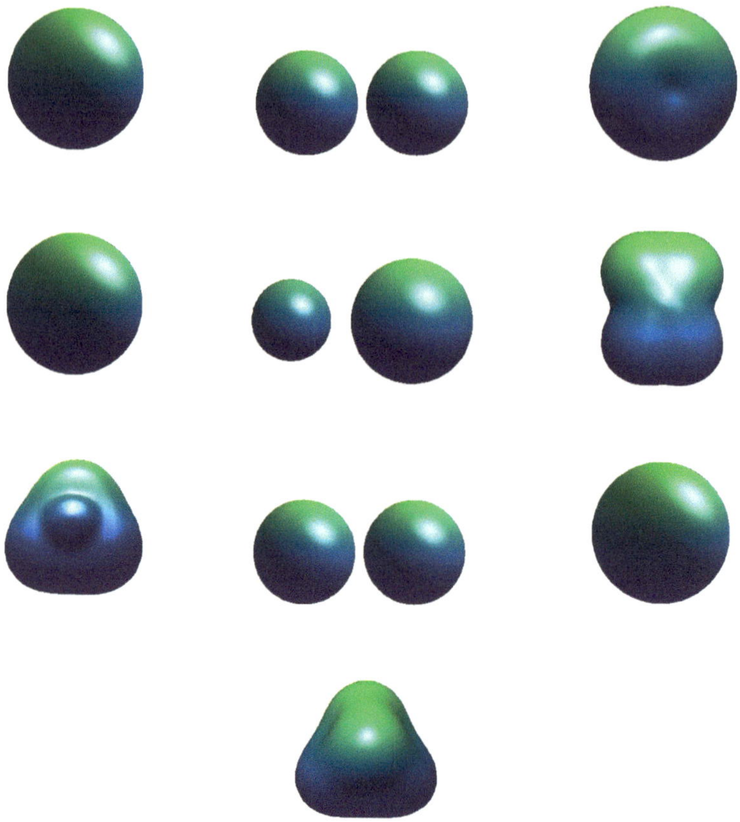

Fig. 11.3. The 3D minimizers of the first 10 Dirichlet eigenvalues.

Table 11.4. *The optimal 3D Neumann eigenvalues and the corresponding multiplicity.*

i	multiplicity	μ_i^*
2	3	11.25
3	6	18.87
4	4	23.52
5	5	29.02
6	5	33.55
7	4	37.83
8	3	41.18
9	3	46.63
10	4	52.49
11	7	55.91

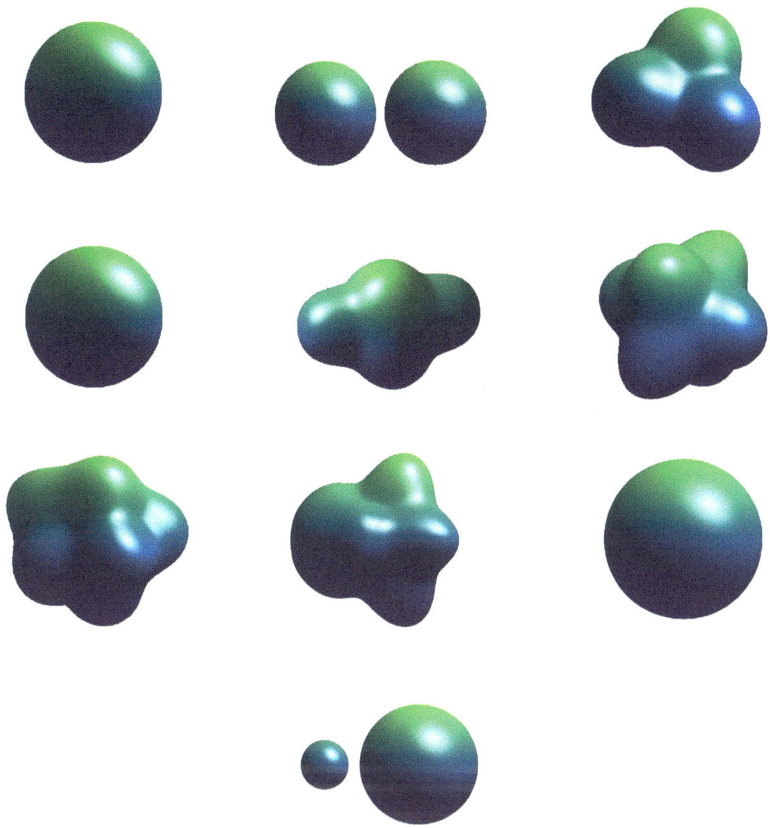

Fig. 11.4. The 3D maximizers of the first 10 (non trivial) Neumann eigenvalues.

Bibliography

[1] *Low eigenvalues of Laplace and Schrödinger operators*, Oberwolfach Rep. **6** (2009), *no. 1*, 355–427, Abstracts from the workshop held February 8–14, 2009, Organized by Mark Ashbaugh, Rafael Benguria, Richard Laugesen and Timo Weidl, Oberwolfach Reports. Vol. 6, no. 1. MR2604061 10.4171/OWR/2009/06 http://dx.doi.org/10.4171/OWR/2009/06

[2] **L. Abatangelo and V. Felli**, *Sharp asymptotic estimates for eigenvalues of Aharonov-Bohm operators with varying poles*, Calc. Var. Partial Differential Equations **54** (2015), *no. 4*, 3857–3903. MR3426097 10.1007/s00526-015-0924-0 http://dx.doi.org/10.1007/s00526-015-0924-0

[3] **S. Abramovich**, *The gap between the first two eigenvalues of a one-dimensional Schrödinger operator with symmetric potential*, Proc. Amer. Math. Soc. **111** (1991), *no. 2*, 451–453. MR1036981 10.2307/2048335

[4] **M. S. Abramowitz and I. A. Stegun**, *Handbook of mathematical functions*, Dover, New York, 1965, eds.

[5] **M. Abramowitz and I. A. Stegun**, *Handbook of mathematical functions with formulas, graphs, and mathematical tables*, National Bureau of Standards Applied Mathematics Series, vol. 55, For sale by the Superintendent of Documents, U.S. Government Printing Office, Washington, D.C., 1964. MR0167642

[6] **E. Acerbi, N. Fusco, and M. Morini**, *Minimality via second variation for a nonlocal isoperimetric problem*, Comm. Math. Phys. **322** (2013), *no. 2*, 515–557. MR3077924 10.1007/s00220-013-1733-y http://dx.doi.org/10.1007/s00220-013-1733-y

[7] **D. R. Adams and L. I. Hedberg**, *Function spaces and potential theory*, Grundlehren der Mathematischen Wissenschaften [Fundamental Principles of Mathematical Sciences], vol. 314, Springer-Verlag, Berlin, 1996. MR1411441 10.1007/978-3-662-03282-4

[8] **R. A. Adams**, *Sobolev spaces*, Academic Press [A subsidiary of Harcourt Brace Jovanovich, Publishers], New York-London, 1975, Pure and Applied Mathematics, Vol. 65. MR0450957

[9] **S. Agmon**, *On the eigenfunctions and on the eigenvalues of general elliptic boundary value problems*, Comm. Pure Appl. Math. **15** (1962), 119–147. MR0147774

[10] **M. S. Agranovich**, *Some asymptotic formulas for elliptic pseudodifferential operators*, Funktsional. Anal. i Prilozhen. **21** (1987), *no. 1*, 63–65. MR888015

[11] **M. S. Agranovich**, *On a mixed Poincaré-Steklov type spectral problem in a Lipschitz domain*, Russ. J. Math. Phys. **13** (2006), *no. 3*, 239–244. MR2262827 10.1134/S1061920806030010

[12] **M. S. Agranovich and B. A. Amosov**, *Estimates for s-numbers, and spectral asymptotics for integral operators of potential type on nonsmooth surfaces*, Funktsional. Anal. i Prilozhen. **30** (1996), *no. 2*, 1–18, 96. MR1402078 10.1007/BF02509448

[13] **N. Aguilera, H. W. Alt, and L. A. Caffarelli**, *An optimization problem with volume constraint*, SIAM J. Control Optim. **24** (1986), *no. 2*, 191–198. MR826512 10.1137/0324011 http://dx.doi.org/10.1137/0324011

[14] **L. V. Ahlfors**, *Conformal invariants*, AMS Chelsea Publishing, Providence, RI, 2010, Topics in geometric function theory, Reprint of the 1973 original, With a foreword by Peter Duren, F. W. Gehring and Brad Osgood. MR2730573

[15] **M. Aizenman and E. H. Lieb**, *On semiclassical bounds for eigenvalues of Schrödinger operators*, Phys. Lett. A **66** (1978), *no. 6*, 427–429. MR598768 10.1016/0375-9601(78)90385-7 http://dx.doi.org/10.1016/0375-9601(78)90385-7

[16] **E. Akhmetgaliyev, C.-Y. Kao, and B. Osting**, *Computational methods for extremal steklov problems*. ArXiv:1601.00605

[17] **J. H. Albert**, *Generic properties of eigenfunctions of elliptic partial differential operators*, Trans. Amer. Math. Soc. **238** (1978), 341–354. MR0471000 (57 #10743)

[18] **A. D. Aleksandrov**, *Uniqueness theorems for surfaces in the large. I*, Vestnik Leningrad. Univ. **11** (1956), *no. 19*, 5–17. MR0086338

[19] **G. Alessandrini and R. Magnanini**, *Elliptic equations in divergence form, geometric critical points of solutions, and Stekloff eigenfunctions*, SIAM J. Math. Anal. **25** (1994), *no. 5*, 1259–1268. MR1289138 10.1137/S0036141093249080

[20] **G. Alessandrini**, *Nodal lines of eigenfunctions of the fixed membrane problem in general convex domains*, Comment. Math. Helv. **69** (1994), *no. 1*, 142–154. MR1259610 10.1007/BF02564478

[21] **G. Allaire**, *Shape optimization by the homogenization method*, Applied Mathematical Sciences, vol. 146, Springer-Verlag, New York, 2002. MR1859696 10.1007/978-1-4684-9286-6

[22] **G. Allaire**, *Conception optimale de structures*, Mathématiques & Applications (Berlin) [Mathematics & Applications], vol. 58, Springer-Verlag, Berlin, 2007, With the collaboration of Marc Schoenauer (INRIA) in the writing of Chapter 8. MR2270119

[23] **W. Allegretto**, *Lower bounds on the number of points in the lower spectrum of elliptic operators*, Canad. J. Math. **31** (1979), *no. 2*, 419–426. MR528821 10.4153/CJM-1979-045-0 http://dx.doi.org/10.4153/CJM-1979-045-0

[24] **F. Almgren, J. E. Taylor, and L. Wang**, *Curvature-driven flows: a variational approach*, SIAM J. Control Optim. **31** (1993), *no. 2*, 387–438. MR1205983 10.1137/0331020 http://dx.doi.org/10.1137/0331020

[25] **H. W. Alt and L. A. Caffarelli**, *Existence and regularity for a minimum problem with free boundary*, J. Reine Angew. Math. **325** (1981), 105–144. MR618549

[26] **H. W. Alt, L. A. Caffarelli, and A. Friedman**, *Variational problems with two phases and their free boundaries*, Trans. Amer. Math. Soc. **282** (1984), *no. 2*, 431–461. MR732100 10.2307/1999245 http://dx.doi.org/10.2307/1999245

[27] **F. Alter and V. Caselles**, *Uniqueness of the Cheeger set of a convex body*, Nonlinear Anal. **70** (2009), *no. 1*, 32–44. MR2468216 10.1016/j.na.2007.11.032

[28] **C. J. S. Alves and P. R. S. Antunes**, *The method of fundamental solutions applied to the calculation of eigenfrequencies and eigenmodes of 2d simply connected shapes*, Computers, Materials & Continua **2** (2005), 251–266.

[29] **C. J. S. Alves and P. R. S. Antunes**, *The method of fundamental solutions applied to some inverse eigenproblems*, SIAM J. Sci. Comput. **35** (2013), *no. 3*, A1689–A1708. MR3070537 10.1137/110860380 http://dx.doi.org/10.1137/110860380

[30] **B. Alziary, J. Fleckinger-Pellé, and P. Takáč**, *Eigenfunctions and Hardy inequalities for a magnetic Schrödinger operator in $\mathbb{R}^2$*, Math. Methods Appl. Sci. **26** (2003), *no. 13*, 1093–1136. MR1994668 10.1002/mma.402

[31] **L. Ambrosio and A. Braides**, *Functionals defined on partitions in sets of finite perimeter. I. Integral representation and Γ-convergence*, J. Math. Pures Appl. (9) **69** (1990), *no. 3*, 285–305. MR1070481

[32] **L. Ambrosio and G. Buttazzo**, *An optimal design problem with perimeter penalization*, Calc. Var. Partial Differential Equations **1** (1993), *no. 1*, 55–69. MR1261717 10.1007/BF02163264

[33] **L. Ambrosio, N. Fusco, and D. Pallara**, *Functions of bounded variation and free discontinuity problems*, Oxford Mathematical Monographs, The Clarendon Press, Oxford University Press, New York, 2000. MR1857292

[34] **L. Ambrosio, N. Gigli, and G. Savaré**, *Gradient flows in metric spaces and in the space of probability measures*, Lectures in Mathematics ETH Zürich, Birkhäuser Verlag, Basel, 2005. MR2129498

[35] **A. Ancona, B. Helffer, and T. Hoffmann-Ostenhof,** *Nodal domain theorems à la Courant*, Doc. Math. **9** (2004), 283–299. MR2117417

[36] **B. Andrews and J. Clutterbuck,** *Proof of the fundamental gap conjecture*, J. Amer. Math. Soc. **24** (2011), *no. 3*, 899–916. MR2784332 10.1090/S0894-0347-2011-00699-1

[37] **L. C. Andrews,** *Special functions of mathematics for engineers*, second ed., McGraw-Hill, Inc., New York, 1992. MR1254207

[38] **P. R. S. Antunes, P. Freitas, and D. Krejčiřík,** *Bounds and extremal domains for robin eigenvalues with negative boundary parameter*, Preprint, 2015.

[39] **P. R. S. Antunes,** *Numerical calculation of eigensolutions of 3D shapes using the method of fundamental solutions*, Numer. Methods Partial Differential Equations **27** (2011), *no. 6*, 1525–1550. MR2838306 10.1002/num.20594 http://dx.doi.org/10.1002/num.20594

[40] **P. R. S. Antunes and P. Freitas,** *A numerical study of the spectral gap*, J. Phys. A **41** (2008), *no. 5*, 055201, 19. MR2433425 10.1088/1751-8113/41/5/055201

[41] **P. R. S. Antunes and P. Freitas,** *On the inverse spectral problem for Euclidean triangles*, Proc. R. Soc. Lond. Ser. A Math. Phys. Eng. Sci. **467** (2011), *no. 2130*, 1546–1562. MR2795790 10.1098/rspa.2010.0540

[42] **P. R. S. Antunes and P. Freitas,** *Numerical optimization of low eigenvalues of the Dirichlet and Neumann Laplacians*, J. Optim. Theory Appl. **154** (2012), *no. 1*, 235–257. MR2931377 10.1007/s10957-011-9983-3 http://dx.doi.org/10.1007/s10957-011-9983-3

[43] **P. R. S. Antunes and P. Freitas,** *Optimal spectral rectangles and lattice ellipses*, Proc. R. Soc. Lond. Ser. A Math. Phys. Eng. Sci. **469** (2013), *no. 2150*, 20120492, 15. MR3001382 10.1098/rspa.2012.0492

[44] **P. R. S. Antunes, P. Freitas, and D. Krejčiřík,** *Bounds and extremal domains for Robin eigenvalues with negative boundary parameter.* ArXiv:1605.08161

[45] **P. R. S. Antunes and A. Henrot,** *On the range of the first two Dirichlet and Neumann eigenvalues of the Laplacian*, Proc. R. Soc. Lond. Ser. A Math. Phys. Eng. Sci. **467** (2011), *no. 2130*, 1577–1603. MR2795792 10.1098/rspa.2010.0444

[46] **P. R. S. Antunes, P. Freitas, and J. B. Kennedy,** *Asymptotic behaviour and numerical approximation of optimal eigenvalues of the Robin Laplacian*, ESAIM: Control, Optimisation and Calculus of Variations 19, (2013), 438–459. MR3049718 10.1051/cocv/2012016 http://dx.doi.org/10.1051/cocv/2012016

[47] **P. R. Antunes and P. Freitas,** *New bounds for the principal Dirichlet eigenvalue of planar regions*, Experiment. Math. **15** (2006), *no. 3*, 333–342. MR2264470 http://projecteuclid.org/euclid.em/1175789762

[48] **W. Arendt, R. Nittka, W. Peter, and F. Steiner,** *Weyl's law: Spectral properties of the laplacian in mathematics and physics*, Mathematical Analysis of Evolution, Information, and Complexity (W. Arendt and S. W. P., eds.), Wiley-VCH Verlag, Weinheim, 2009, pp. 1–71.

[49] **W. Arendt and A. F. M. ter Elst,** *The Dirichlet-to-Neumann operator on rough domains*, J. Differential Equations **251** (2011), *no. 8*, 2100–2124. MR2823661 10.1016/j.jde.2011.06.017 http://dx.doi.org/10.1016/j.jde.2011.06.017

[50] **W. Arendt and R. Mazzeo,** *Friedlander's eigenvalue inequalities and the Dirichlet-to-Neumann semigroup*, Commun. Pure Appl. Anal. **11** (2012), *no. 6*, 2201–2212. MR2912743 10.3934/cpaa.2012.11.2201

[51] **W. Arendt and M. Warma,** *The Laplacian with Robin boundary conditions on arbitrary domains*, Potential Anal. **19** (2003), *no. 4*, 341–363. MR1988110 10.1023/A:1024181608863 http://dx.doi.org/10.1023/A:1024181608863

[52] **M. G. Armentano and R. G. Durán,** *Asymptotic lower bounds for eigenvalues by nonconforming finite element methods*, Electron. Trans. Numer. Anal. **17** (2004), 93–101 (electronic). MR2040799

[53] **J. M. Arrieta**, *Neumann eigenvalue problems on exterior perturbations of the domain*, J. Differential Equations **118** (1995), *no. 1*, 54–103. MR1329403 10.1006/jdeq.1995.1067

[54] **M. S. Ashbaugh and R. Benguria**, *Optimal lower bounds for eigenvalue gaps for Schrödinger operators with symmetric single-well potentials and related results*, Maximum principles and eigenvalue problems in partial differential equations (Knoxville, TN, 1987), Pitman Res. Notes Math. Ser., vol. 175, Longman Sci. Tech., Harlow, 1988, pp. 134–145. MR963464

[55] **M. S. Ashbaugh**, *Isoperimetric and universal inequalities for eigenvalues*, Spectral theory and geometry (Edinburgh, 1998), London Math. Soc. Lecture Note Ser., vol. 273, Cambridge Univ. Press, Cambridge, 1999, pp. 95–139. MR1736867 10.1017/CBO9780511566165.007 http://dx.doi.org/10.1017/CBO9780511566165.007

[56] **M. S. Ashbaugh**, *Open problems on eigenvalues of the Laplacian*, Analytic and geometric inequalities and applications, Math. Appl., vol. 478, Kluwer Acad. Publ., Dordrecht, 1999, pp. 13–28. MR1785859

[57] **M. S. Ashbaugh**, *The universal eigenvalue bounds of Payne-Pólya-Weinberger, Hile-Protter, and H. C. Yang*, Proc. Indian Acad. Sci. Math. Sci. **112** (2002), *no. 1*, 3–30, Spectral and inverse spectral theory (Goa, 2000). MR1894540 10.1007/BF02829638 http://dx.doi.org/10.1007/BF02829638

[58] **M. S. Ashbaugh**, *On universal inequalities for the low eigenvalues of the buckling problem*, Partial differential equations and inverse problems, Contemp. Math., vol. 362, Amer. Math. Soc., Providence, RI, 2004, pp. 13–31. MR2091488 10.1090/conm/362/06602 http://dx.doi.org/10.1090/conm/362/06602

[59] **M. S. Ashbaugh and R. Benguria**, *Best constant for the ratio of the first two eigenvalues of one-dimensional Schrödinger operators with positive potentials*, Proc. Amer. Math. Soc. **99** (1987), *no. 3*, 598–599. MR875408 10.2307/2046373

[60] **M. S. Ashbaugh and R. Benguria**, *Optimal lower bound for the gap between the first two eigenvalues of one-dimensional Schrödinger operators with symmetric single-well potentials*, Proc. Amer. Math. Soc. **105** (1989), *no. 2*, 419–424. MR942630 10.2307/2046959

[61] **M. S. Ashbaugh and R. D. Benguria**, *Optimal bounds for ratios of eigenvalues of one-dimensional Schrödinger operators with Dirichlet boundary conditions and positive potentials*, Comm. Math. Phys. **124** (1989), *no. 3*, 403–415. MR1012632 http://projecteuclid.org/euclid.cmp/1104179208

[62] **M. S. Ashbaugh and R. D. Benguria**, *Proof of the Payne-Pólya-Weinberger conjecture*, Bull. Amer. Math. Soc. (N.S.) **25** (1991), *no. 1*, 19–29. MR1085824 10.1090/S0273-0979-1991-16016-7 http://dx.doi.org/10.1090/S0273-0979-1991-16016-7

[63] **M. S. Ashbaugh and R. D. Benguria**, *A second proof of the Payne-Pólya-Weinberger conjecture*, Comm. Math. Phys. **147** (1992), *no. 1*, 181–190. MR1171765 http://projecteuclid.org/euclid.cmp/1104250531

[64] **M. S. Ashbaugh and R. D. Benguria**, *A sharp bound for the ratio of the first two eigenvalues of Dirichlet Laplacians and extensions*, Ann. of Math. (2) **135** (1992), *no. 3*, 601–628. MR1166646 10.2307/2946578 http://dx.doi.org/10.2307/2946578

[65] **M. S. Ashbaugh and R. D. Benguria**, *Eigenvalue ratios for Sturm-Liouville operators*, J. Differential Equations **103** (1993), *no. 1*, 205–219. MR1218744 10.1006/jdeq.1993.1047

[66] **M. S. Ashbaugh and R. D. Benguria**, *More bounds on eigenvalue ratios for Dirichlet Laplacians in n dimensions*, SIAM J. Math. Anal. **24** (1993), *no. 6*, 1622–1651. MR1241161 10.1137/0524091 http://dx.doi.org/10.1137/0524091

[67] **M. S. Ashbaugh and R. D. Benguria**, *Universal bounds for the low eigenvalues of Neumann Laplacians in n dimensions*, SIAM J. Math. Anal. **24** (1993), *no. 3*, 557–570. MR1215424 10.1137/0524034 http://dx.doi.org/10.1137/0524034

[68] **M. S. Ashbaugh and R. D. Benguria,** *Isoperimetric inequalities for eigenvalue ratios*, Partial differential equations of elliptic type (Cortona, 1992), Sympos. Math., XXXV, Cambridge Univ. Press, Cambridge, 1994, pp. 1–36. MR1297771

[69] **M. S. Ashbaugh and R. D. Benguria,** *The range of values of λ_2/λ_1 and λ_3/λ_1 for the fixed membrane problem*, Rev. Math. Phys. **6** (1994), *no. 5A*, 999–1009, Special issue dedicated to Elliott H. Lieb. MR1301363 10.1142/S0129055X9400033X http://dx.doi.org/10.1142/S0129055X9400033X

[70] **M. S. Ashbaugh and R. D. Benguria,** *On Rayleigh's conjecture for the clamped plate and its generalization to three dimensions*, Duke Math. J. **78** (1995), *no. 1*, 1–17. MR1328749 10.1215/S0012-7094-95-07801-6

[71] **M. S. Ashbaugh and R. D. Benguria,** *Bounds for ratios of the first, second, and third membrane eigenvalues*, Nonlinear problems in applied mathematics, SIAM, Philadelphia, PA, 1996, pp. 30–42. MR2410595

[72] **M. S. Ashbaugh and R. D. Benguria,** *A sharp bound for the ratio of the first two Dirichlet eigenvalues of a domain in a hemisphere of S^n*, Trans. Amer. Math. Soc. **353** (2001), *no. 3*, 1055–1087. MR1707696 10.1090/S0002-9947-00-02605-2 http://dx.doi.org/10.1090/S0002-9947-00-02605-2

[73] **M. S. Ashbaugh and R. D. Benguria,** *Isoperimetric inequalities for eigenvalues of the Laplacian*, Spectral theory and mathematical physics: a Festschrift in honor of Barry Simon's 60th birthday, Proc. Sympos. Pure Math., vol. 76, Amer. Math. Soc., Providence, RI, 2007, pp. 105–139. MR2310200 10.1090/pspum/076.1/2310200 http://dx.doi.org/10.1090/pspum/076.1/2310200

[74] **M. S. Ashbaugh and D. Bucur,** *On the isoperimetric inequality for the buckling of a clamped plate*, Z. Angew. Math. Phys. **54** (2003), *no. 5*, 756–770, Special issue dedicated to Lawrence E. Payne. MR2019178 10.1007/s00033-003-3204-3

[75] **M. S. Ashbaugh and F. Chiacchio,** *On low eigenvalues of the Laplacian with mixed boundary conditions*, J. Differential Equations **250** (2011), *no. 5*, 2544–2566. MR2756075 10.1016/j.jde.2010.11.003

[76] **M. S. Ashbaugh, F. Gesztesy, A. Laptev, M. Mitrea, and S. Sukhtaiev,** *A bound for the eigenvalue counting function for Krein–von Neumann and Friedrichs extensions*. ArXiv:1605.01170

[77] **M. S. Ashbaugh, F. Gesztesy, M. Mitrea, R. Shterenberg, and G. Teschl,** *The Krein-von Neumann extension and its connection to an abstract buckling problem*, Math. Nachr. **283** (2010), *no. 2*, 165–179. MR2604115 10.1002/mana.200910067 http://dx.doi.org/10.1002/mana.200910067

[78] **M. S. Ashbaugh, F. Gesztesy, M. Mitrea, R. Shterenberg, and G. Teschl,** *A survey on the Krein–von Neumann extension, the corresponding abstract buckling problem, and Weyl-type spectral asymptotics for perturbed Krein Laplacians in nonsmooth domains*, Mathematical physics, spectral theory and stochastic analysis, Oper. Theory Adv. Appl., vol. 232, Birkhäuser/Springer Basel AG, Basel, 2013, pp. 1–106. MR3077276 10.1007/978-3-0348-0591-9_1 http://dx.doi.org/10.1007/978-3-0348-0591-9_1

[79] **M. S. Ashbaugh, F. Gesztesy, M. Mitrea, and G. Teschl,** *Spectral theory for perturbed Krein Laplacians in nonsmooth domains*, Adv. Math. **223** (2010), *no. 4*, 1372–1467. MR2581375 10.1016/j.aim.2009.10.006 http://dx.doi.org/10.1016/j.aim.2009.10.006

[80] **M. S. Ashbaugh and E. M. Harrell, II,** *Maximal and minimal eigenvalues and their associated nonlinear equations*, J. Math. Phys. **28** (1987), *no. 8*, 1770–1786. MR899179 10.1063/1.527488

[81] **M. S. Ashbaugh, E. M. Harrell, II, and R. Svirsky,** *On minimal and maximal eigenvalue gaps and their causes*, Pacific J. Math. **147** (1991), *no. 1*, 1–24. MR1081670 http://projecteuclid.org/euclid.pjm/1102645038

[82] **M. S. Ashbaugh and L. Hermi,** *On Harrell-Stubbe Type Inequalities for the Discrete Spectrum of a Self-Adjoint Operator.* ArXiv:0712.4396

[83] **M. S. Ashbaugh and L. Hermi,** *A unified approach to universal inequalities for eigenvalues of elliptic operators*, Pacific J. Math. **217** (2004), *no. 2*, 201–219. MR2109931 10.2140/pjm.2004.217.201 http://dx.doi.org/10.2140/pjm.2004.217.201

[84] **M. Ashbaugh, R. Benguria, R. Laugesen, and T. Weidl,** *Low eigenvalues of laplace and schrodinger operators*, 2009, Mathematisches Forschungsinstitut Oberwolfach. Report No. 06/2009.

[85] **M. Ashu,** *Some properties of Bessel functions with applications to the Neumann eigenvalues on the disc*, Bachelor's thesis at Lund university (Adviser E. Wahlen), 2013.

[86] **R. Atar and K. Burdzy,** *On Neumann eigenfunctions in lip domains*, J. Amer. Math. Soc. **17** (2004), *no. 2*, 243–265 (electronic). MR2051611 10.1090/S0894-0347-04-00453-9

[87] **I. Athanasopoulos, L. A. Caffarelli, C. Kenig, and S. Salsa,** *An area-Dirichlet integral minimization problem*, Comm. Pure Appl. Math. **54** (2001), *no. 4*, 479–499. MR1808651 10.1002/1097-0312(200104)54:4<479::AID-CPA3>3.3.CO;2-U

[88] **H. Attouch,** *Variational convergence for functions and operators*, Applicable Mathematics Series, Pitman (Advanced Publishing Program), Boston, MA, 1984. MR773850

[89] **H. Attouch, G. Buttazzo, and G. Michaille,** *Variational analysis in Sobolev and BV spaces*, MPS/SIAM Series on Optimization, vol. 6, Society for Industrial and Applied Mathematics (SIAM), Philadelphia, PA; Mathematical Programming Society (MPS), Philadelphia, PA, 2006, Applications to PDEs and optimization. MR2192832

[90] **E. Aubry, J. Bertrand, and B. Colbois,** *Eigenvalue pinching on convex domains in space forms*, Trans. Amer. Math. Soc. **361** (2009), *no. 1*, 1–18. MR2439395 10.1090/S0002-9947-08-04775-2 http://dx.doi.org/10.1090/S0002-9947-08-04775-2

[91] **A. I. Ávila,** *Stability results for the first eigenvalue of the Laplacian on domains in space forms*, J. Math. Anal. Appl. **267** (2002), *no. 2*, 760–774. MR1888036 10.1006/jmaa.2001.7831 http://dx.doi.org/10.1006/jmaa.2001.7831

[92] **A. Baernstein II and B. A. Taylor,** *Spherical rearrangements, subharmonic functions, and *-functions in n-space*, Duke Math. J. **43** (1976), *no. 2*, 245–268. MR0402083

[93] **K. Ball,** *Volume ratios and a reverse isoperimetric inequality*, J. London Math. Soc. (2) **44** (1991), *no. 2*, 351–359. MR1136445 10.1112/jlms/s2-44.2.351 http://dx.doi.org/10.1112/jlms/s2-44.2.351

[94] **H. P. Baltes and E. R. Hilf,** *Spectra of finite systems*, Bibliographisches Institut, Mannheim-Vienna-Zurich, 1976, A review of Weyl's problem: the eigenvalue distribution of the wave equation for finite domains and its applications on the physics of small systems. MR0435624

[95] **R. Band, G. Berkolaiko, H. Raz, and U. Smilansky,** *The number of nodal domains on quantum graphs as a stability index of graph partitions*, Comm. Math. Phys. **311** (2012), *no. 3*, 815–838. MR2909765 10.1007/s00220-011-1384-9

[96] **C. Bandle and M. Flucher,** *Harmonic radius and concentration of energy; hyperbolic radius and Liouville's equations $\Delta U = e^U$ and $\Delta U = U^{(n+2)/(n-2)}$*, SIAM Rev. **38** (1996), *no. 2*, 191–238. MR1391227 10.1137/1038039 http://dx.doi.org/10.1137/1038039

[97] **C. Bandle,** *Isoperimetric inequalities and applications*, Monographs and Studies in Mathematics, vol. 7, Pitman (Advanced Publishing Program), Boston, Mass.-London, 1980. MR572958

[98] **C. Bandle,** *Extremal problems for eigenvalues of the Sturm-Liouville type*, General inequalities, 5 (Oberwolfach, 1986), Internat. Schriftenreihe Numer. Math., vol. 80, Birkhäuser, Basel, 1987, pp. 319–336. MR1018157 10.1007/978-3-0348-7192-1_26

[99] **C. Bandle and A. Wagner**, *Second domain variation for problems with Robin boundary conditions*, J. Optim. Theory Appl. **167** (2015), *no. 2*, 430–463. MR3412445 10.1007/s10957-015-0801-1 http://dx.doi.org/10.1007/s10957-015-0801-1

[100] **D. O. Banks**, *Bounds for the eigenvalues of nonhomogeneous hinged vibrating rods*, J. Math. Mech. **16** (1967), 949–966. MR0208057

[101] **D. Banks**, *Bounds for the eigenvalues of some vibrating systems*, Pacific J. Math. **10** (1960), 439–474. MR0117378

[102] **D. Banks**, *Upper bounds for the eigenvalues of some vibrating systems*, Pacific J. Math. **11** (1961), 1183–1203. MR0145146

[103] **D. O. Banks**, *Lower bounds for the eigenvalues of a vibrating string whose density satisfies a Lipschitz condition*, Pacific J. Math. **20** (1967), 393–410. MR0208058

[104] **R. Bañuelos and K. Burdzy**, *On the "hot spots" conjecture of J. Rauch*, J. Funct. Anal. **164** (1999), *no. 1*, 1–33. MR1694534 10.1006/jfan.1999.3397

[105] **R. Bañuelos, T. Kulczycki, I. Polterovich, and B. A. Siudeja**, *Eigenvalue inequalities for mixed Steklov problems*, Operator theory and its applications, Amer. Math. Soc. Transl. Ser. 2, vol. 231, Amer. Math. Soc., Providence, RI, 2010, pp. 19–34. MR2758960

[106] **R. Bañuelos, R. Latała, and P. J. Méndez-Hernández**, *A Brascamp-Lieb-Luttinger-type inequality and applications to symmetric stable processes*, Proc. Amer. Math. Soc. **129** (2001), *no. 10*, 2997–3008 (electronic). MR1840105 10.1090/S0002-9939-01-06137-8

[107] **L. Barbosa and P. Bérard**, *Eigenvalue and "twisted" eigenvalue problems, applications to CMC surfaces*, J. Math. Pures Appl. (9) **79** (2000), *no. 5*, 427–450. MR1759435 10.1016/S0021-7824(00)00160-4

[108] **M. Barchiesi, F. Cagnetti, and N. Fusco**, *Stability of the Steiner symmetrization of convex sets*, J. Eur. Math. Soc. (JEMS) **15** (2013), *no. 4*, 1245–1278. MR3055761 10.4171/JEMS/391

[109] **M. Barchiesi, G. M. Capriani, N. Fusco, and G. Pisante**, *Stability of Pólya-Szegő inequality for log-concave functions*, J. Funct. Anal. **267** (2014), *no. 7*, 2264–2297. MR3250365 10.1016/j.jfa.2014.03.015 http://dx.doi.org/10.1016/j.jfa.2014.03.015

[110] **M. Bareket**, *On an isoperimetric inequality for the first eigenvalue of a boundary value problem*, SIAM J. Math. Anal. **8** (1977), *no. 2*, 280–287. MR0430552

[111] **D. C. Barnes**, *Some isoperimetric inequalities for the eigenvalues of vibrating strings*, Pacific J. Math. **29** (1969), 43–61. MR0244545

[112] **D. C. Barnes**, *Extremal problems for eigenvalues with applications to buckling, vibration and sloshing*, SIAM J. Math. Anal. **16** (1985), *no. 2*, 341–357. MR777472 10.1137/0516025

[113] **E. R. Barnes**, *The shape of the strongest column and some related extremal eigenvalue problems*, Quart. Appl. Math. **34** (1976/77), *no. 4*, 393–409. MR0493674

[114] **J. Baxter, G. Dal Maso, and U. Mosco**, *Stopping times and Γ-convergence*, Trans. Amer. Math. Soc. **303** (1987), *no. 1*, 1–38. MR896006 10.2307/2000777

[115] **E. F. Beckenbach and R. Bellman**, *Inequalities*, Second revised printing. Ergebnisse der Mathematik und ihrer Grenzgebiete. Neue Folge, Band 30, Springer-Verlag, New York, Inc., 1965. MR0192009

[116] **E. Bednarczuk, M. Pierre, E. Rouy, and J. Sokolowski**, *Tangent sets in some functional spaces*, Nonlinear Anal. **42** (2000), *no. 5, Ser. A: Theory Methods*, 871–886. MR1776928 10.1016/S0362-546X(99)00134-0

[117] **P. R. Beesack**, *Isoperimetric inequalities for the nonhomogeneous clamped rod and plate*, J. Math. Mech. **8** (1959), 471–482. MR0107410

[118] **Z. Belhachmi, D. Bucur, G. Buttazzo, and J.-M. Sac-Epée**, *Shape optimization problems for eigenvalues of elliptic operators*, ZAMM Z. Angew. Math. Mech. **86** (2006), *no. 3*, 171–184. MR2205641 10.1002/zamm.200510259

[119] **K. Bellova and F. Lin**, *Nodal Sets of Steklov Eigenfunctions*. ArXiv:1402.4323

[120] **M. P. Bendsøe and O. Sigmund**, *Topology optimization*, Springer-Verlag, Berlin, 2003, Theory, methods and applications. MR2008524

[121] **R. D. Benguria and H. Linde**, *A second eigenvalue bound for the Dirichlet Laplacian in hyperbolic space*, Duke Math. J. **140** (2007), *no. 2*, 245–279. MR2359820 10.1215/S0012-7094-07-14022-5 http://dx.doi.org/10.1215/S0012-7094-07-14022-5

[122] **C. Bennewitz and E. J. M. Veling**, *Optimal bounds for the spectrum of a one-dimensional Schrödinger operator*, General inequalities, 6 (Oberwolfach, 1990), Internat. Ser. Numer. Math., vol. 103, Birkhäuser, Basel, 1992, pp. 257–268. MR1213012 10.1007/978-3-0348-7565-3_22

[123] **P. Bérard**, *Inégalités isopérimétriques et applications. Domaines nodaux des fonctions propres*, Goulaouic-Meyer-Schwartz Seminar, 1981/1982, École Polytech., Palaiseau, 1982, Exp. No. XI, 10. p. MR671608

[124] **P. Bérard and B. Helffer**, *Remarks on the boundary set of spectral equipartitions*, Philos. Trans. R. Soc. Lond. Ser. A Math. Phys. Eng. Sci. **372** (2014), *no. 2007*, 20120492, 15. MR3151084 10.1098/rsta.2012.0492

[125] **P. Bérard**, *Remarques sur la conjecture de Weyl*, Compositio Math. **48** (1983), *no. 1*, 35–53. MR700579 http://www.numdam.org/item?id=CM_1983__48_1_35_0

[126] **P. Bérard**, *Volume des ensembles nodaux des fonctions propres du laplacien*, Séminaire de Théorie Spectrale et Géométrie, Année 1984–1985, Univ. Grenoble I, Saint-Martin-d'Hères, 1985, pp. IV.1–IV.9. MR1046051 10.1080/07481188508252495

[127] **P. Bérard**, *Transplantation et isospectralité. I*, Math. Ann. **292** (1992), *no. 3*, 547–559. MR1152950 10.1007/BF01444635

[128] **P. Bérard and B. Helffer**, *On the number of nodal domains of the 2D isotropic quantum harmonic oscillator – an extension of results of A. Stern –*. ArXiv:1409.2333

[129] **P. Bérard and B. Helffer**, *A. Stern's analysis of the nodal sets of some families of spherical harmonics revisited*, Monatshefte für Mathematik (2015), 1–34, to appear. 10.1007/s00605-015-0788-6 ArXiv:1407.5564 http://dx.doi.org/10.1007/s00605-015-0788-6

[130] **P. Bérard and B. Helffer**, *The weak Pleijel theorem with geometric control*, 2015. ArXiv:1512.07089

[131] **P. Bérard and B. Helffer**, *Courant-sharp eigenvalues for the equilateral torus, and for the equilateral triangle*, Lett. Math. Phys. (To the memory of Louis Boutet de Monvel) (2016). 10.1007/s11005-016-0819-9 ArXiv:1503.00117 http://dx.doi.org/10.1007/s11005-016-0819-9

[132] **P. Bérard and B. Helffer**, *Dirichlet eigenfunctions of the square membrane: Courant's property, and A. Stern's and A. Pleijel's analyses*, Springer Proceedings in Mathematics & Statistics (2015), MIMS-GGTM conference in memory of M. S. Baouendi. Ali Baklouti, Aziz El Kacimi, Sadok Kallel, and Nordine Mir Editors. ArXiv:1402.6054

[133] **P. Bérard and D. Meyer**, *Inégalités isopérimétriques et applications*, Ann. Sci. École Norm. Sup. (4) **15** (1982), *no. 3*, 513–541. MR690651 http://www.numdam.org/item?id=ASENS_1982_4_15_3_513_0

[134] **P. H. Bérard**, *Spectres et groupes cristallographiques. I. Domaines euclidiens*, Invent. Math. **58** (1980), *no. 2*, 179–199. MR570879 10.1007/BF01403168

[135] **F. A. Berezin**, *Convex functions of operators*, Mat. Sb. (N.S.) **88(130)** (1972), 268–276. MR0300121

[136] **F. A. Berezin**, *Covariant and contravariant symbols of operators*, Izv. Akad. Nauk SSSR Ser. Mat. **36** (1972), 1134–1167. MR0350504

[137] **A. Berger**, *The eigenvalues of the Laplacian with Dirichlet boundary condition in $\mathbb{R}^2$ are almost never minimized by disks*, Ann. Global Anal. Geom. **47** (2015), *no. 3*, 285–304. MR3318833 10.1007/s10455-014-9446-9 http://dx.doi.org/10.1007/s10455-014-9446-9

[138] **J. Berger and J. Rubinstein**, *On the zero set of the wave function in superconductivity*, Comm. Math. Phys. **202** (1999), *no. 3*, 621–628. MR1690956 10.1007/s002200050598

[139] **L. Bers**, *Local behavior of solutions of general linear elliptic equations*, Comm. Pure Appl. Math. **8** (1955), 473–496. MR0075416

[140] **D. P. Bertsekas**, *Necessary and sufficient condition for a penalty method to be exact*, Math. Programming **9** (1975), *no. 1*, 87–99. MR0384144

[141] **A. L. Besse**, *Manifolds all of whose geodesics are closed*, Ergebnisse der Mathematik und ihrer Grenzgebiete [Results in Mathematics and Related Areas], vol. 93, Springer-Verlag, Berlin-New York, 1978, With appendices by D. B. A. Epstein, J.-P. Bourguignon, L. Bérard-Bergery, M. Berger and J. L. Kazdan. MR496885

[142] **G. Besson**, *Sur la multiplicité de la première valeur propre des surfaces riemanniennes*, Ann. Inst. Fourier (Grenoble) **30** (1980), *no. 1*, x, 109–128. MR576075 http://www.numdam.org/item?id=AIF_1980__30_1_109_0

[143] **T. Betcke and L. N. Trefethen**, *Reviving the method of particular solutions*, SIAM Rev. **47** (2005), *no. 3*, 469–491 (electronic). MR2178637 10.1137/S0036144503437336 http://dx.doi.org/10.1137/S0036144503437336

[144] **H. A. Bethe**, *Intermediate quantum mechanics*, Notes by R. W. Jackiw, W. A. Benjamin, Inc., New York-Amsterdam, 1964. MR0161579

[145] **M. F. Betta, F. Brock, A. Mercaldo, and M. R. Posteraro**, *A weighted isoperimetric inequality and applications to symmetrization*, J. Inequal. Appl. **4** (1999), *no. 3*, 215–240. MR1734159 10.1155/S1025583499000375

[146] **T. Bhattacharya**, *Some observations on the first eigenvalue of the p-Laplacian and its connections with asymmetry*, Electron. J. Differential Equations (2001), No. 35, 15 pp. (electronic). MR1836803

[147] **T. Bhattacharya and A. Weitsman**, *Estimates for Green's function in terms of asymmetry*, Applied analysis (Baton Rouge, LA, 1996), Contemp. Math., vol. 221, Amer. Math. Soc., Providence, RI, 1999, pp. 31–58. MR1647193 10.1090/conm/221/03117 http://dx.doi.org/10.1090/conm/221/03117

[148] **C. Bianchini and A. Henrot**, *Optimal sets for a class of minimization problems with convex constraints*, J. Convex Anal. **19** (2012), *no. 3*, 725–758. MR3013757

[149] **Binoy and G. Santhanam**, *Sharp upperbound and a comparison theorem for the first nonzero Steklov eigenvalue*, J. Ramanujan Math. Soc. **29** (2014), *no. 2*, 133–154. MR3237730

[150] **C. J. Bishop**, *Some questions concerning harmonic measure*, Partial differential equations with minimal smoothness and applications (Chicago, IL, 1990), IMA Vol. Math. Appl., vol. 42, Springer, New York, 1992, pp. 89–97. MR1155854 10.1007/978-1-4612-2898-1_7

[151] **W. Blaschke**, *Vorlesungen über Differentialgeometrie und geometrische Grundlagen von Einsteins Relativita'tstheorie*, vol. 2, Springer Verlag, 1923.

[152] **G. Blum, S. Gnutzmann, and U. Smilansky**, *Nodal domain statistics: A criterion for quantum chaos*, Phys. Rev. Lett. **88** (2002), 114101–114104.

[153] **B. Bogosel**, *The method of fundamental solutions applied to boundary eigenvalue problems*, J. Comput. Appl. Math. **306** (2016), 265–285. MR3505897

[154] **B. Bogosel**, *Shape optimization and spectral problems*, Ph.D. thesis, Université de Grenoble, 2015.

[155] **B. Bogosel and V. Bonnaillie-Noël**, *p-minimal k-partitions for eigenvalues*, In preparation, 2016.

[156] **V. Bonnaillie-Noël and B. Helffer**, *Numerical analysis of nodal sets for eigenvalues of Aharonov-Bohm Hamiltonians on the square with application to minimal partitions*, Exp. Math. **20** (2011), *no. 3*, 304–322. MR2836255 10.1080/10586458.2011.565240

[157] **V. Bonnaillie-Noël and B. Helffer**, *À la découverte des partitions minimales*, Images des Mathématiques **CNRS** (2012). http://images.math.cnrs.fr/A-la-decouverte-des-partitions.html

[158] **V. Bonnaillie-Noël, B. Helffer, and T. Hoffmann-Ostenhof**, *Aharonov-Bohm Hamiltonians, isospectrality and minimal partitions*, J. Phys. A **42** (2009), *no. 18*, 185203, 20. MR2591197 10.1088/1751-8113/42/18/185203

[159] **V. Bonnaillie-Noël and C. Léna**, *Spectral minimal partitions for a family of tori*, Exp. Math. ArXiv:1503.04545

[160] **V. Bonnaillie-Noël and G. Vial**, *Computations for nodal domains and spectral minimal partitions*, (2007). http://www.bretagne.ens-cachan.fr/math/simulations/MinimalPartitions

[161] **V. Bonnaillie-Noël and B. Helffer**, *On spectral minimal partitions: the disk revisited*, Ann. Univ. Buchar. Math. Ser. **4(LXII)** (2013), *no. 1*, 321–342. MR3093548

[162] **V. Bonnaillie-Noël, B. Helffer, and G. Vial**, *Numerical simulations for nodal domains and spectral minimal partitions*, ESAIM Control Optim. Calc. Var. **16** (2010), *no. 1*, 221–246. MR2598097 10.1051/cocv:2008074

[163] **V. Bonnaillie-Noël and C. Léna**, *Spectral minimal partitions of a sector*, Discrete Contin. Dyn. Syst. Ser. B **19** (2014), *no. 1*, 27–53. MR3245081 10.3934/dcdsb.2014.19.27

[164] **V. Bonnaillie-Noël, B. Noris, M. Nys, and S. Terracini**, *On the eigenvalues of Aharonov-Bohm operators with varying poles*, Anal. PDE **7** (2014), *no. 6*, 1365–1395. MR3270167 10.2140/apde.2014.7.1365

[165] **T. Bonnesen**, *Über das isoperimetrische Defizit ebener Figuren*, Math. Ann. **91** (1924), *no. 3-4*, 252–268. MR1512192 10.1007/BF01556082 http://dx.doi.org/10.1007/BF01556082

[166] **D. Borisov and P. Freitas**, *Singular asymptotic expansions for Dirichlet eigenvalues and eigenfunctions of the Laplacian on thin planar domains*, Ann. Inst. H. Poincaré Anal. Non Linéaire **26** (2009), *no. 2*, 547–560. MR2504043 10.1016/j.anihpc.2007.12.001

[167] **D. Borisov and P. Freitas**, *Asymptotics of Dirichlet eigenvalues and eigenfunctions of the Laplacian on thin domains in $\mathbb{R}^d$*, J. Funct. Anal. **258** (2010), *no. 3*, 893–912. MR2558181 10.1016/j.jfa.2009.07.014

[168] **M.-H. Bossel**, *Longueurs extrémales et fonctionnelles de domaine*, Complex Variables Theory Appl. **6** (1986), *no. 2-4*, 203–234. MR871731

[169] **M.-H. Bossel**, *Membranes élastiquement liées: extension du théorème de Rayleigh-Faber-Krahn et de l'inégalité de Cheeger*, C. R. Acad. Sci. Paris Sér. I Math. **302** (1986), *no. 1*, 47–50. MR827106

[170] **M.-H. Bossel**, *Membranes élastiquement liées inhomogènes ou sur une surface: une nouvelle extension du théorème isopérimétrique de Rayleigh-Faber-Krahn*, Z. Angew. Math. Phys. **39** (1988), *no. 5*, 733–742. MR963641 10.1007/BF00948733 http://dx.doi.org/10.1007/BF00948733

[171] **G. Bouchitté and G. Buttazzo**, *Optimal design problems for Schrödinger operators with noncompact resolvents*, **Preprint Dipartimento di Matematica, Universita di Pisa** (2015). http://cvgmt.sns.it. ArXiv:1502.00091

[172] **L. Boulton and M. Levitin**, *Trends and tricks in spectral theory*, Ediciones IVIC (2007). http://evm.ivic.gob.ve/boulton_levitin_final.pdf

[173] **B. Bourdin, D. Bucur, and É. Oudet**, *Optimal partitions for eigenvalues*, SIAM J. Sci. Comput. **31** (2009/10), *no. 6*, 4100–4114. MR2566585 10.1137/090747087

[174] **J. Bourgain**, *On Pleijel's nodal domain theorem*, Int. Math. Res. Not. IMRN (2015), *no. 6*, 1601–1612. MR3340367

[175] **A. Braides**, *Γ-convergence for beginners*, Oxford Lecture Series in Mathematics and its Applications, vol. 22, Oxford University Press, Oxford, 2002. MR1968440 10.1093/acprof:oso/9780198507840.001.0001

[176] **B. Brandolini, F. Chiacchio, and C. Trombetti**, *Sharp estimates for eigenfunctions of a Neumann problem*, Comm. Partial Differential Equations **34** (2009), *no. 10-12*, 1317–1337. MR2581975 10.1080/03605300903089859

[177] **J. J. A. M. Brands**, *Bounds for the ratios of the first three membrane eigenvalues*, Arch. Rational Mech. Anal. **16** (1964), 265–268. MR0163068

[178] **L. Brasco and A. Pratelli**, *Sharp stability of some spectral inequalities*, Geom. Funct. Anal. **22** (2012), *no. 1*, 107–135. MR2899684 10.1007/s00039-012-0148-9 http://dx.doi.org.acces.bibl.ulaval.ca/10.1007/s00039-012-0148-9

[179] **L. Brasco**, *On torsional rigidity and principal frequencies: an invitation to the Kohler-Jobin rearrangement technique*, ESAIM Control Optim. Calc. Var. **20** (2014), *no. 2*, 315–338. MR3264206 10.1051/cocv/2013065 http://dx.doi.org/10.1051/cocv/2013065

[180] **L. Brasco, G. De Philippis, and B. Ruffini**, *Spectral optimization for the Stekloff-Laplacian: the stability issue*, J. Funct. Anal. **262** (2012), *no. 11*, 4675–4710. MR2913683 10.1016/j.jfa.2012.03.017

[181] **L. Brasco, G. De Philippis, and B. Velichkov**, *Faber-Krahn inequalities in sharp quantitative form*, Duke Math. J. **164** (2015), *no. 9*, 1777–1831. MR3357184 10.1215/00127094-3120167 http://dx.doi.org/10.1215/00127094-3120167

[182] **L. Brasco and G. Franzina**, *An anisotropic eigenvalue problem of Stekloff type and weighted Wulff inequalities*, NoDEA Nonlinear Differential Equations Appl. **20** (2013), *no. 6*, 1795–1830. MR3128695 10.1007/s00030-013-0231-4 http://dx.doi.org/10.1007/s00030-013-0231-4

[183] **L. Brasco and R. Magnanini**, *The heart of a convex body*, Geometric properties for parabolic and elliptic PDE's, Springer INdAM Ser., vol. 2, Springer, Milan, 2013, pp. 49–66. MR3050226 10.1007/978-88-470-2841-8_4 http://dx.doi.org/10.1007/978-88-470-2841-8_4

[184] **L. Brasco, C. Nitsch, and A. Pratelli**, *On the boundary of the attainable set of the Dirichlet spectrum*, Z. Angew. Math. Phys. **64** (2013), *no. 3*, 591–597. MR3068840 10.1007/s00033-012-0250-8 http://dx.doi.org/10.1007/s00033-012-0250-8

[185] **H. Brezis**, *Analyse fonctionnelle*, Collection Mathématiques Appliquées pour la Maîtrise. [Collection of Applied Mathematics for the Master's Degree], Masson, Paris, 1983, Théorie et applications. [Theory and applications]. MR697382

[186] **H. Brezis and E. H. Lieb**, *Sobolev inequalities with remainder terms*, J. Funct. Anal. **62** (1985), *no. 1*, 73–86. MR790771 10.1016/0022-1236(85)90020-5 http://dx.doi.org/10.1016/0022-1236(85)90020-5

[187] **T. Briançon**, *Regularity of optimal shapes for the Dirichlet's energy with volume constraint*, ESAIM Control Optim. Calc. Var. **10** (2004), *no. 1*, 99–122 (electronic). MR2084257 10.1051/cocv:2003038

[188] **T. Briançon, M. Hayouni, and M. Pierre**, *Lipschitz continuity of state functions in some optimal shaping*, Calc. Var. Partial Differential Equations **23** (2005), *no. 1*, 13–32. MR2133659 10.1007/s00526-004-0286-5

[189] **T. Briançon and J. Lamboley**, *Regularity of the optimal shape for the first eigenvalue of the Laplacian with volume and inclusion constraints*, Ann. Inst. H. Poincaré Anal. Non Linéaire **26** (2009), *no. 4*, 1149–1163. MR2542718 10.1016/j.anihpc.2008.07.003

[190] **P. Briet, F. Germinet, and G. Raikov**, *Spectral and scattering theory for quantum magnetic systems*, Contemporary Mathematics (2009), Proceedings of the conference held in Luminy, July 2008, American Mathematical Society, Providence, RI, xiii+186, MR2640520 10.1090/conm/500

[191] **F. Brock**, *Continuous polarization and symmetry of solutions of variational problems with potentials*, Calculus of variations, applications and computations (Pont-à-Mousson, 1994), Pitman Res. Notes Math. Ser., vol. 326, Longman Sci. Tech., Harlow, 1995, pp. 25–34. MR1419331 10.1002/mana.19951720104

[192] **F. Brock**, *An isoperimetric inequality for eigenvalues of the Stekloff problem*, ZAMM Z. Angew. Math. Mech. **81** (2001), *no. 1*, 69–71. MR1808500

[193] **F. Brock and D. Daners**, *Conjecture concerning a Faber-Krahn inequality for robin problems*, 2007, Open Problem in Mini-Workshop: Shape Analysis for Eigenvalues (Organized by: D. Bucur, G. Buttazzo, A. Henrot), pp. 1022–1023.

[194] **F. Brock**, *Continuous Steiner-symmetrization*, Math. Nachr. **172** (1995), 25–48. MR1330619 10.1002/mana.19951720104

[195] **F. Brock**, *Continuous rearrangement and symmetry of solutions of elliptic problems*, Proc. Indian Acad. Sci. Math. Sci. **110** (2000), *no. 2*, 157–204. MR1758811 10.1007/BF02829490

[196] **F. Brock and A. Y. Solynin**, *An approach to symmetrization via polarization*, Trans. Amer. Math. Soc. **352** (2000), *no. 4*, 1759–1796. MR1695019 10.1090/S0002-9947-99-02558-1

[197] **V. Bruneau, K. Pankrashkin, and N. Popoff**, *On the accumulation of Robin eigenvalues in conical domains*, 2016. ArXiv:1602.07448

[198] **V. Bruneau and N. Popoff**, *On the negative spectrum of the Robin Laplacian in corner domains*, Anal. PDE 9 (2016), *no. 5*, 1259–1283. MR3531372

[199] **J. Brüning**, *Über Knoten von Eigenfunktionen des Laplace-Beltrami-Operators*, Math. Z. **158** (1978), *no. 1*, 15–21. MR0478247

[200] **J. Brüning and D. Gromes**, *Über die Länge der Knotenlinien schwingender Membranen*, Math. Z. **124** (1972), 79–82. MR0287202

[201] **D. Bucur, G. Buttazzo, and C. Nitsch**, *Symmetry breaking for a problem in optimal insulation*, 2016. ArXiv:1601.02146

[202] **D. Bucur, A. Henrot, J. Sokolowski, and A. Żochowski**, *Continuity of the elasticity system solutions with respect to the geometrical domain variations*, Adv. Math. Sci. Appl. **11** (2001), *no. 1*, 57–73. MR1841560

[203] **D. Bucur**, *Concentration-compacité et γ-convergence*, C. R. Acad. Sci. Paris Sér. I Math. **327** (1998), *no. 3*, 255–258. MR1650269 10.1016/S0764-4442(98)80142-0

[204] **D. Bucur**, *Uniform concentration-compactness for Sobolev spaces on variable domains*, J. Differential Equations **162** (2000), *no. 2*, 427–450. MR1751712 10.1006/jdeq.1999.3726 http://dx.doi.org/10.1006/jdeq.1999.3726

[205] **D. Bucur**, *Regularity of optimal convex shapes*, J. Convex Anal. **10** (2003), *no. 2*, 501–516. MR2044433

[206] **D. Bucur**, *Minimization of the k-th eigenvalue of the Dirichlet Laplacian*, Arch. Ration. Mech. Anal. **206** (2012), *no. 3*, 1073–1083. MR2989451 10.1007/s00205-012-0561-0 http://dx.doi.org/10.1007/s00205-012-0561-0

[207] **D. Bucur and G. Buttazzo**, *Variational methods in shape optimization problems*, Progress in Nonlinear Differential Equations and their Applications, 65, Birkhäuser Boston, Inc., Boston, MA, 2005. MR2150214

[208] **D. Bucur and G. Buttazzo**, *On the characterization of the compact embedding of Sobolev spaces*, Calc. Var. Partial Differential Equations **44** (2012), *no. 3-4*, 455–475. MR2915329 10.1007/s00526-011-0441-8

[209] **D. Bucur, G. Buttazzo, and I. Figueiredo**, *On the attainable eigenvalues of the Laplace operator*, SIAM J. Math. Anal. **30** (1999), *no. 3*, 527–536. MR1677942 10.1137/S0036141097329135

[210] **D. Bucur, G. Buttazzo, and A. Henrot**, *Existence results for some optimal partition problems*, Adv. Math. Sci. Appl. **8** (1998), *no. 2*, 571–579. MR1657219

[211] **D. Bucur, G. Buttazzo, and A. Henrot**, *Minimization of $\lambda_2(\Omega)$ with a perimeter constraint*, Indiana Univ. Math. J. **58** (2009), *no. 6*, 2709–2728. MR2603765 10.1512/iumj.2009.58.3768

[212] **D. Bucur, G. Buttazzo, and A. Lux**, *Quasistatic evolution in debonding problems via capacitary methods*, Arch. Ration. Mech. Anal. **190** (2008), *no. 2*, 281–306. MR2448320 10.1007/s00205-008-0166-9

[213] **D. Bucur and D. Daners,** *An alternative approach to the Faber-Krahn inequality for Robin problems*, Calc. Var. Partial Differential Equations **37** (2010), *no. 1-2*, 75–86. MR2564398 10.1007/s00526-009-0252-3

[214] **D. Bucur and I. Fragalà,** *Blaschke-Santaló and Mahler inequalities for the first eigenvalue of the Dirichlet Laplacian*, Proc. Lond. Math. Soc. (3) 113 (2016), *no. 3*, 387–417. MR3551851

[215] **D. Bucur, I. Fragalà, and J. Lamboley,** *Optimal convex shapes for concave functionals*, ESAIM Control Optim. Calc. Var. **18** (2012), *no. 3*, 693–711. MR3041661 10.1051/cocv/2011167 http://dx.doi.org/10.1051/cocv/2011167

[216] **D. Bucur and A. Giacomini,** *A variational approach to the isoperimetric inequality for the Robin eigenvalue problem*, Arch. Ration. Mech. Anal. **198** (2010), *no. 3*, 927–961. MR2729323 10.1007/s00205-010-0298-6

[217] **D. Bucur and A. Giacomini,** *Faber-Krahn inequalities for the Robin-Laplacian: a free discontinuity approach*, Arch. Ration. Mech. Anal. **218** (2015), *no. 2*, 757–824. MR3375539 10.1007/s00205-015-0872-z http://dx.doi.org/10.1007/s00205-015-0872-z

[218] **D. Bucur and A. Giacomini,** *The Saint-Venant inequality for the Laplace operator with Robin boundary conditions*, Milan J. Math. **83** (2015), *no. 2*, 327–343. MR3412286 10.1007/s00032-015-0243-0 http://dx.doi.org/10.1007/s00032-015-0243-0

[219] **D. Bucur and A. Henrot,** *Minimization of the third eigenvalue of the Dirichlet Laplacian*, R. Soc. Lond. Proc. Ser. A Math. Phys. Eng. Sci. **456** (2000), *no. 1996*, 985–996. MR1805088 10.1098/rspa.2000.0546 http://dx.doi.org/10.1098/rspa.2000.0546

[220] **D. Bucur and A. Henrot,** *Stability for the Dirichlet problem under continuous Steiner symmetrization*, Potential Anal. **13** (2000), *no. 2*, 127–145. MR1782253 10.1023/A:1008753703185

[221] **D. Bucur and D. Mazzoleni,** *A surgery result for the spectrum of the Dirichlet Laplacian*, SIAM J. Math. Anal. **47** (2015), *no. 6*, 4451–4466. MR3427044

[222] **D. Bucur, D. Mazzoleni, A. Pratelli, and B. Velichkov,** *Lipschitz regularity of the eigenfunctions on optimal domains*, Arch. Ration. Mech. Anal. **216** (2015), *no. 1*, 117–151. MR3305655 10.1007/s00205-014-0801-6 http://dx.doi.org/10.1007/s00205-014-0801-6

[223] **D. Bucur and P. Trebeschi,** *Shape optimisation problems governed by nonlinear state equations*, Proc. Roy. Soc. Edinburgh Sect. A **128** (1998), *no. 5*, 945–963. MR1642112 10.1017/S0308210500030006

[224] **D. Bucur and B. Velichkov,** *Multiphase shape optimization problems*, SIAM J. Control Optim. **52** (2014), *no. 6*, 3556–3591. MR3274808 10.1137/130917272 http://dx.doi.org/10.1137/130917272

[225] **D. Bucur and J.-P. Zolésio,** *N-dimensional shape optimization under capacitary constraint*, J. Differential Equations **123** (1995), *no. 2*, 504–522. MR1362884 10.1006/jdeq.1995.1171

[226] **A. Burchard and M. Fortier,** *Random polarizations*, Adv. Math. **234** (2013), 550–573. MR3003937 10.1016/j.aim.2012.10.010

[227] **A. Burchard and H. Hajaiej,** *Rearrangement inequalities for functionals with monotone integrands*, J. Funct. Anal. **233** (2006), *no. 2*, 561–582. MR2214588 10.1016/j.jfa.2005.08.010

[228] **K. Burdzy,** *The hot spots problem in planar domains with one hole*, Duke Math. J. **129** (2005), *no. 3*, 481–502. MR2169871 10.1215/S0012-7094-05-12932-5

[229] **K. Burdzy, R. Holyst, D. Ingerman, and P. March,** *Configurational transition in a Fleming-Viot-type model and probabilistic interpretation of Laplacian eigenfunctions*, J. Phys. A, Math. Gen. **29** (1996), *no. 11*, 2633–2642 (English). Zbl 0901.60054 10.1088/0305-4470/29/11/004

[230] **K. Burdzy and W. Werner,** *A counterexample to the "hot spots" conjecture*, Ann. of Math. (2) **149** (1999), *no. 1*, 309–317. MR1680567 10.2307/121027

[231] **V. I. Burenkov and E. B. Davies,** *Spectral stability of the Neumann Laplacian*, J. Differential Equations **186** (2002), *no. 2*, 485–508. MR1942219 10.1016/S0022-0396(02)00033-5

[232] **P. Buser**, *Isospectral Riemann surfaces*, Ann. Inst. Fourier (Grenoble) **36** (1986), *no. 2*, 167–192. MR850750 http://www.numdam.org/item?id=AIF_1986__36_2_167_0

[233] **P. Buser, J. Conway, P. Doyle, and K.-D. Semmler**, *Some planar isospectral domains*, Internat. Math. Res. Notices (1994), *no. 9*, 391ff., approx. 9 pp. (electronic). MR1301439 10.1155/S1073792894000437

[234] **G. Buttazzo and A. Wagner**, *On some rescaled shape optimization problems*, Adv. Calc. Var. **3** (2010), *no. 2*, 213–232. MR2609348 10.1515/ACV.2010.013

[235] **G. Buttazzo**, *Semicontinuity, relaxation and integral representation in the calculus of variations*, Pitman Research Notes in Mathematics Series, vol. 207, Longman Scientific & Technical, Harlow; copublished in the United States with John Wiley & Sons, Inc., New York, 1989. MR1020296

[236] **G. Buttazzo and G. Dal Maso**, *Shape optimization for Dirichlet problems: relaxed solutions and optimality conditions*, Bull. Amer. Math. Soc. (N.S.) **23** (1990), *no. 2*, 531–535. MR1027896 10.1090/S0273-0979-1990-15971-3

[237] **G. Buttazzo and G. Dal Maso**, *Shape optimization for Dirichlet problems: relaxed formulation and optimality conditions*, Appl. Math. Optim. **23** (1991), *no. 1*, 17–49. MR1076053 10.1007/BF01442391

[238] **G. Buttazzo and G. Dal Maso**, *An existence result for a class of shape optimization problems*, Arch. Rational Mech. Anal. **122** (1993), *no. 2*, 183–195. MR1217590 10.1007/BF00378167 http://dx.doi.org/10.1007/BF00378167

[239] **G. Buttazzo, M. Giaquinta, and S. Hildebrandt**, *One-dimensional variational problems*, Oxford Lecture Series in Mathematics and its Applications, vol. 15, The Clarendon Press, Oxford University Press, New York, 1998, An introduction. MR1694383

[240] **G. Buttazzo and C. Timofte**, *On the relaxation of some optimal partition problems*, Adv. Math. Sci. Appl. **12** (2002), *no. 2*, 509–520. MR1943977

[241] **G. Buttazzo and N. Varchon**, *On the optimal reinforcement of an elastic membrane*, Riv. Mat. Univ. Parma (7) **4*** (2005), 115–125. MR2197483

[242] **L. A. Caffarelli**, *A Harnack inequality approach to the regularity of free boundaries. I. Lipschitz free boundaries are $C^{1,\alpha}$*, Rev. Mat. Iberoamericana **3** (1987), *no. 2*, 139–162. MR990856 10.4171/RMI/47 http://dx.doi.org/10.4171/RMI/47

[243] **L. A. Caffarelli**, *A Harnack inequality approach to the regularity of free boundaries. III. Existence theory, compactness, and dependence on X*, Ann. Scuola Norm. Sup. Pisa Cl. Sci. (4) **15** (1988), *no. 4*, 583–602 (1989). MR1029856 http://www.numdam.org/item?id=ASNSP_1988_4_15_4_583_0

[244] **L. A. Caffarelli**, *A Harnack inequality approach to the regularity of free boundaries. II. Flat free boundaries are Lipschitz*, Comm. Pure Appl. Math. **42** (1989), *no. 1*, 55–78. MR973745 10.1002/cpa.3160420105 http://dx.doi.org/10.1002/cpa.3160420105

[245] **L. A. Caffarelli, D. Jerison, and C. E. Kenig**, *Some new monotonicity theorems with applications to free boundary problems*, Ann. of Math. (2) **155** (2002), *no. 2*, 369–404. MR1906591 10.2307/3062121 http://dx.doi.org/10.2307/3062121

[246] **L. A. Caffarelli, D. Jerison, and C. E. Kenig**, *Global energy minimizers for free boundary problems and full regularity in three dimensions*, Noncompact problems at the intersection of geometry, analysis, and topology, Contemp. Math., vol. 350, Amer. Math. Soc., Providence, RI, 2004, pp. 83–97. MR2082392 10.1090/conm/350/06339 http://dx.doi.org/10.1090/conm/350/06339

[247] **L. A. Caffarelli and F. H. Lin**, *Analysis on the junctions of domain walls*, Discrete Contin. Dyn. Syst. **28** (2010), *no. 3*, 915–929. MR2644773 10.3934/dcds.2010.28.915

[248] **L. A. Caffarelli and S. Salsa**, *A geometric approach to free boundary problems*, Graduate Studies in Mathematics, vol. 68, American Mathematical Society, Providence, RI, 2005. MR2145284 10.1090/gsm/068 http://dx.doi.org/10.1090/gsm/068

[249] **L. A. Cafferelli and F. H. Lin**, *An optimal partition problem for eigenvalues*, J. Sci. Comput. **31** (2007), *no. 1-2*, 5–18. MR2304268 10.1007/s10915-006-9114-8

[250] **F. Cakoni, N. Chaulet, and H. Haddar**, *On the asymptotics of a Robin eigenvalue problem*, C. R. Math. Acad. Sci. Paris **351** (2013), *no. 13-14*, 517–521. MR3095098 10.1016/j.crma.2013.07.022 http://dx.doi.org/10.1016/j.crma.2013.07.022

[251] **G. Carlier and T. Lachand-Robert**, *Convex bodies of optimal shape*, J. Convex Anal. **10** (2003), *no. 1*, 265–273. MR1999913

[252] **C. Carstensen and J. Gedicke**, *Guaranteed lower bounds for eigenvalues*, Math. Comp. **83** (2014), *no. 290*, 2605–2629. MR3246802 10.1090/S0025-5718-2014-02833-0

[253] **V. Caselles, A. Chambolle, and M. Novaga**, *Uniqueness of the Cheeger set of a convex body*, Pacific J. Math. **232** (2007), *no. 1*, 77–90. MR2358032 10.2140/pjm.2007.232.77

[254] **J. Céa and K. Malanowski**, *An example of a max-min problem in partial differential equations*, SIAM J. Control **8** (1970), 305–316. MR0274915

[255] **A. Chambolle and F. Doveri**, *Continuity of Neumann linear elliptic problems on varying two-dimensional bounded open sets*, Comm. Partial Differential Equations **22** (1997), *no. 5-6*, 811–840. MR1452169 10.1080/03605309708821285

[256] **K. Chandrasekharan and S. Minakshisundaram**, *Typical means*, Oxford University Press, 1952. MR0055458

[257] **P.-K. Chang and D. DeTurck**, *On hearing the shape of a triangle*, Proc. Amer. Math. Soc. **105** (1989), *no. 4*, 1033–1038. MR953738 10.2307/2047071

[258] **P. Charron**, *A Pleijel type theorem for the quantum harmonic oscillator*, J. Spect. Theory, To appear. ArXiv:1512.07880

[259] **P. Charron, B. Helffer, and T. Hoffmann-Ostenhof**, *Pleijel's theorem for Schrödinger operators with radial potentials*, To appear in Annales mathématiques du Québec , arXiv:1604.08372

[260] **T. Chatelain and M. Choulli**, *Clarke generalized gradient for eigenvalues*, Commun. Appl. Anal. **1** (1997), *no. 4*, 443–454. MR1470812

[261] **I. Chavel**, *Eigenvalues in Riemannian geometry*, Pure and Applied Mathematics, vol. 115, Academic Press, Inc., Orlando, FL, 1984, Including a chapter by Burton Randol, With an appendix by Jozef Dodziuk. MR768584

[262] **D. Chen and T. Zheng**, *Bounds for ratios of the membrane eigenvalues*, J. Differential Equations **250** (2011), *no. 3*, 1575–1590. MR2737218 10.1016/j.jde.2010.10.009 http://dx.doi.org/10.1016/j.jde.2010.10.009

[263] **D. Chen, T. Zheng, and H. Yang**, *Estimates of the gaps between consecutive eigenvalues of Laplacian*, Pacific J. Math. **282** (2016), *no. 2*, 293–311. MR3478937 10.2140/pjm.2016.282.293 http://dx.doi.org/10.2140/pjm.2016.282.293

[264] **D. Chenais**, *On the existence of a solution in a domain identification problem*, J. Math. Anal. Appl. **52** (1975), *no. 2*, 189–219. MR0385666

[265] **Q.-M. Cheng and H. Yang**, *Estimates on eigenvalues of Laplacian*, Math. Ann. **331** (2005), *no. 2*, 445–460. MR2115463 10.1007/s00208-004-0589-z http://dx.doi.org/10.1007/s00208-004-0589-z

[266] **Q.-M. Cheng and H. Yang**, *Bounds on eigenvalues of Dirichlet Laplacian*, Math. Ann. **337** (2007), *no. 1*, 159–175. MR2262780 10.1007/s00208-006-0030-x http://dx.doi.org/10.1007/s00208-006-0030-x

[267] **S. Y. Cheng**, *Eigenvalue comparison theorems and its geometric applications*, Math. Z. **143** (1975), *no. 3*, 289–297. MR0378001

[268] **S. Y. Cheng**, *Eigenfunctions and nodal sets*, Comment. Math. Helv. **51** (1976), *no. 1*, 43–55. MR0397805

[269] **M. Chipot and G. Dal Maso**, *Relaxed shape optimization: the case of nonnegative data for the Dirichlet problem*, Adv. Math. Sci. Appl. **1** (1992), *no. 1*, 47–81. MR1161483

[270] **G. Chiti**, *An isoperimetric inequality for the eigenfunctions of linear second order elliptic operators*, Boll. Un. Mat. Ital. A (6) **1** (1982), *no. 1*, 145–151. MR652376

[271] **G. Chiti**, *A reverse Hölder inequality for the eigenfunctions of linear second order elliptic operators*, Z. Angew. Math. Phys. **33** (1982), *no. 1*, 143–148. MR652928 10.1007/BF00948319

[272] **G. Chiti**, *A bound for the ratio of the first two eigenvalues of a membrane*, SIAM J. Math. Anal. **14** (1983), *no. 6*, 1163–1167. MR718816 10.1137/0514090 http://dx.doi.org/10.1137/0514090

[273] **A. Cianchi, L. Esposito, N. Fusco, and C. Trombetti**, *A quantitative Pólya-Szegö principle*, J. Reine Angew. Math. **614** (2008), 153–189. MR2376285 10.1515/CRELLE.2008.005 http://dx.doi.org/10.1515/CRELLE.2008.005

[274] **A. Cianchi and N. Fusco**, *Functions of bounded variation and rearrangements*, Arch. Ration. Mech. Anal. **165** (2002), *no. 1*, 1–40. MR1947097 10.1007/s00205-002-0214-9

[275] **M. Cicalese and G. P. Leonardi**, *A selection principle for the sharp quantitative isoperimetric inequality*, Arch. Ration. Mech. Anal. **206** (2012), *no. 2*, 617–643. MR2980529 10.1007/s00205-012-0544-1 http://dx.doi.org/10.1007/s00205-012-0544-1

[276] **D. Cioranescu and F. Murat**, *Un terme étrange venu d'ailleurs*, Nonlinear partial differential equations and their applications. Collège de France Seminar, Vol. II (Paris, 1979/1980), Res. Notes in Math., vol. 60, Pitman, Boston, Mass.-London, 1982, pp. 98–138, 389–390. MR652509

[277] **C. Clark**, *The asymptotic distribution of eigenvalues and eigenfunctions for elliptic boundary value problems*, SIAM Rev. **9** (1967), 627–646. MR0510064

[278] **B. Colbois and A. El Soufi**, *Extremal eigenvalues of the Laplacian on Euclidean domains and closed surfaces*, Math. Z. **278** (2014), *no. 1-2*, 529–546. MR3267589 10.1007/s00209-014-1325-3 http://dx.doi.org/10.1007/s00209-014-1325-3

[279] **B. Colbois, A. El Soufi, and A. Girouard**, *Isoperimetric control of the Steklov spectrum*, J. Funct. Anal. **261** (2011), *no. 5*, 1384–1399. MR2807105 10.1016/j.jfa.2011.05.006

[280] **B. Colbois and A. Girouard**, *The spectral gap of graphs and Steklov eigenvalues on surfaces*, Electron. Res. Announc. Math. Sci. **21** (2014), 19–27. MR3181630 10.3934/era.2014.21.19

[281] **B. Colbois and D. Maerten**, *Eigenvalues estimate for the Neumann problem of a bounded domain*, J. Geom. Anal. **18** (2008), *no. 4*, 1022–1032. MR2438909 10.1007/s12220-008-9041-z

[282] **Y. Colin de Verdière**, *Construction de laplaciens dont une partie finie du spectre est donnée*, Ann. Sci. École Norm. Sup. (4) **20** (1987), *no. 4*, 599–615. MR932800 http://www.numdam.org/item?id=ASENS_1987_4_20_4_599_0

[283] **Y. Colin de Verdière**, *Magnetic interpretation of the nodal defect on graphs*, Anal. PDE **6** (2013), *no. 5*, 1235–1242. MR3125555 10.2140/apde.2013.6.1235

[284] **E. Colorado and J. García-Melián**, *The behavior of the principal eigenvalue of a mixed elliptic problem with respect to a parameter*, J. Math. Anal. Appl. **377** (2011), *no. 1*, 53–69. MR2754808 10.1016/j.jmaa.2010.10.015 http://dx.doi.org/10.1016/j.jmaa.2010.10.015

[285] **R. Cominetti and J.-P. Penot**, *Tangent sets to unilateral convex sets*, C. R. Acad. Sci. Paris Sér. I Math. **321** (1995), *no. 12*, 1631–1636. MR1367820

[286] **M. Conti, S. Terracini, and G. Verzini**, *An optimal partition problem related to nonlinear eigenvalues*, J. Funct. Anal. **198** (2003), *no. 1*, 160–196. MR1962357 10.1016/S0022-1236(02)00105-2

[287] **M. Conti, S. Terracini, and G. Verzini,** *Asymptotic estimates for the spatial segregation of competitive systems*, Adv. Math. **195** (2005), *no. 2*, 524–560. MR2146353 10.1016/j.aim.2004.08.006

[288] **M. Conti, S. Terracini, and G. Verzini,** *On a class of optimal partition problems related to the Fučík spectrum and to the monotonicity formulae*, Calc. Var. Partial Differential Equations **22** (2005), *no. 1*, 45–72. MR2105968 10.1007/s00526-004-0266-9

[289] **M. Conti, S. Terracini, and G. Verzini,** *A variational problem for the spatial segregation of reaction-diffusion systems*, Indiana Univ. Math. J. **54** (2005), *no. 3*, 779–815. MR2151234 10.1512/iumj.2005.54.2506

[290] **R. Courant,** *Ein allgemeiner Satz zur Theorie der Eigenfunktionen selbstadjungierter Differentialausdrücke*, Nachr. Ges. Göttingen (1923), 81–84.

[291] **R. Courant,** *Dirichlet's Principle, Conformal Mapping, and Minimal Surfaces*, Interscience Publishers, Inc., New York, N.Y., 1950, Appendix by M. Schiffer. MR0036317

[292] **R. Courant and D. Hilbert,** *Methods of mathematical physics. Vol. I*, Interscience Publishers, Inc., New York, N.Y., 1953. MR0065391

[293] **R. Courant and D. Hilbert,** *Methods of mathematical physics. Vol. II: Partial differential equations*, (Vol. II by R. Courant.), Interscience Publishers (a division of John Wiley & Sons), New York-Lon don, 1962. MR0140802

[294] **S. J. Cox, B. Kawohl, and P. X. Uhlig,** *On the optimal insulation of conductors*, J. Optim. Theory Appl. **100** (1999), *no. 2*, 253–263. MR1673432 10.1023/A:1021773901158

[295] **S. Cox and R. Lipton,** *Extremal eigenvalue problems for two-phase conductors*, Arch. Rational Mech. Anal. **136** (1996), *no. 2*, 101–117. MR1423004 10.1007/BF02316974

[296] **S. J. Cox,** *The two phase drum with the deepest bass note*, Japan J. Indust. Appl. Math. **8** (1991), *no. 3*, 345–355. MR1137646 10.1007/BF03167141

[297] **S. J. Cox,** *The shape of the ideal column*, Math. Intelligencer **14** (1992), *no. 1*, 16–24. MR1147343 10.1007/BF03024137

[298] **S. J. Cox,** *Extremal eigenvalue problems for the Laplacian*, Recent advances in partial differential equations (El Escorial, 1992), RAM Res. Appl. Math., vol. 30, Masson, Paris, 1994, pp. 37–53. MR1266201 10.2307/25305835

[299] **S. J. Cox,** *The generalized gradient at a multiple eigenvalue*, J. Funct. Anal. **133** (1995), *no. 1*, 30–40. MR1351641 10.1006/jfan.1995.1117

[300] **S. J. Cox and J. R. McLaughlin,** *Extremal eigenvalue problems for composite membranes. I, II*, Appl. Math. Optim. **22** (1990), *no. 2*, 153–167, 169–187. MR1055658 10.1007/BF01447325

[301] **S. J. Cox and M. L. Overton,** *On the optimal design of columns against buckling*, SIAM J. Math. Anal. **23** (1992), *no. 2*, 287–325. MR1147865 10.1137/0523015

[302] **S. J. Cox and M. Ross,** *Extremal eigenvalue problems for starlike planar domains*, J. Differential Equations **120** (1995), *no. 1*, 174–197. MR1339673 10.1006/jdeq.1995.1109

[303] **S. J. Cox and P. X. Uhlig,** *Where best to hold a drum fast*, SIAM J. Optim. **9** (1999), *no. 4*, 948–964 (electronic), Dedicated to John E. Dennis, Jr., on his 60th birthday. MR1724771 10.1137/S1052623497326083

[304] **S. J. Cox and P. X. Uhlig,** *Where best to hold a drum fast*, SIAM Rev. **45** (2003), *no. 1*, 75–92 (electronic), Reprint [SIAM J. Optim. **9** (1999), no. 4, 948–964 [MR1724771 (2000j:35208)]. MR1981255 10.1137/S003614450241427 http://dx.doi.org/10.1137/S003614450241427

[305] **M. Crouzeix,** *Variational approach of a magnetic shaping problem*, European J. Mech. B Fluids **10** (1991), *no. 5*, 527–536. MR1139607

[306] **M. Crouzeix and P.-A. Raviart,** *Conforming and nonconforming finite element methods for solving the stationary Stokes equations. I*, Rev. Française Automat. Informat. Recherche Opérationnelle Sér. Rouge **7** (1973), *no. R-3*, 33–75. MR0343661

[307] **O. Cybulski, V. Babin, and R. Hołyst**, *Minimization of the Renyi entropy production in the space-partitioning process*, Phys. Rev. E (3) **71** (2005), *no. 4*, 046130, 10. MR2139992 10.1103/PhysRevE.71.046130
[308] **H. L. Cycon, R. G. Froese, W. Kirsch, and B. Simon**, *Schrödinger operators with application to quantum mechanics and global geometry*, study ed., Texts and Monographs in Physics, Springer-Verlag, Berlin, 1987. MR883643
[309] **B. Dacorogna**, *Weak continuity and weak lower semicontinuity of nonlinear functionals*, Lecture Notes in Mathematics, vol. 922, Springer-Verlag, Berlin-New York, 1982. MR658130
[310] **B. Dacorogna**, *Direct methods in the calculus of variations*, Applied Mathematical Sciences, vol. 78, Springer-Verlag, Berlin, 1989. MR990890 10.1007/978-3-642-51440-1
[311] **Q.-y. Dai and Y.-x. Fu**, *Faber-Krahn inequality for Robin problems involving p-Laplacian*, Acta Math. Appl. Sin. Engl. Ser. **27** (2011), *no. 1*, 13–28. MR2754757 10.1007/s10255-011-0036-3 http://dx.doi.org/10.1007/s10255-011-0036-3
[312] **G. Dal Maso**, *Γ-convergence and μ-capacities*, Ann. Scuola Norm. Sup. Pisa Cl. Sci. (4) **14** (1987), *no. 3*, 423–464 (1988). MR951228 http://www.numdam.org/item?id=ASNSP_1987_4_14_3_423_0
[313] **G. Dal Maso**, *An introduction to Γ-convergence*, Progress in Nonlinear Differential Equations and their Applications, 8, Birkhäuser Boston, Inc., Boston, MA, 1993. MR1201152 10.1007/978-1-4612-0327-8
[314] **G. Dal Maso and U. Mosco**, *Wiener's criterion and Γ-convergence*, Appl. Math. Optim. **15** (1987), *no. 1*, 15–63. MR866165 10.1007/BF01442645 http://dx.doi.org/10.1007/BF01442645
[315] **G. Dal Maso and F. Murat**, *Asymptotic behaviour and correctors for Dirichlet problems in perforated domains with homogeneous monotone operators*, Ann. Scuola Norm. Sup. Pisa Cl. Sci. (4) **24** (1997), *no. 2*, 239–290. MR1487956 http://www.numdam.org/item?id=ASNSP_1997_4_24_2_239_0
[316] **M. Dambrine**, *On variations of the shape Hessian and sufficient conditions for the stability of critical shapes*, RACSAM. Rev. R. Acad. Cienc. Exactas Fís. Nat. Ser. A Mat. **96** (2002), *no. 1*, 95–121. MR1915674
[317] **E. N. Dancer and D. Daners**, *Domain perturbation for elliptic equations subject to Robin boundary conditions*, J. Differential Equations **138** (1997), *no. 1*, 86–132. MR1458457 10.1006/jdeq.1997.3256 http://dx.doi.org/10.1006/jdeq.1997.3256
[318] **D. Daners**, *Robin boundary value problems on arbitrary domains*, Trans. Amer. Math. Soc. **352** (2000), *no. 9*, 4207–4236. MR1650081 10.1090/S0002-9947-00-02444-2 http://dx.doi.org/10.1090/S0002-9947-00-02444-2
[319] **D. Daners**, *A Faber-Krahn inequality for Robin problems in any space dimension*, Math. Ann. **335** (2006), *no. 4*, 767–785. MR2232016 10.1007/s00208-006-0753-8
[320] **D. Daners**, *Inverse positivity for general Robin problems on Lipschitz domains*, Arch. Math. (Basel) **92** (2009), *no. 1*, 57–69. MR2471988 10.1007/s00013-008-2918-z http://dx.doi.org/10.1007/s00013-008-2918-z
[321] **D. Daners and J. B. Kennedy**, *Uniqueness in the Faber-Krahn inequality for Robin problems*, SIAM J. Math. Anal. **39** (2007/08), *no. 4*, 1191–1207. MR2368899 10.1137/060675629 http://dx.doi.org/10.1137/060675629
[322] **D. Daners and J. B. Kennedy**, *On the asymptotic behaviour of the eigenvalues of a Robin problem*, Differential Integral Equations **23** (2010), *no. 7-8*, 659–669. MR2654263
[323] **M. Dauge**, *Elliptic boundary value problems on corner domains*, Lecture Notes in Mathematics, vol. 1341, Springer-Verlag, Berlin, 1988, Smoothness and asymptotics of solutions. MR961439
[324] **R. Dautray and J.-L. Lions**, *Analyse mathématique et calcul numérique pour les sciences et les techniques. Tome 1*, Collection du Commissariat à l'Énergie Atomique: Série Scientifique.

[Collection of the Atomic Energy Commission: Science Series], Masson, Paris, 1984, With the collaboration of Michel Artola, Marc Authier, Philippe Bénilan, Michel Cessenat, Jean-Michel Combes, André Gervat, Hélène Lanchon, Bertrand Mercier, Claude Wild and Claude Zuily. MR792484

[325] **R. Dautray and J.-L. Lions**, *Analyse mathématique et calcul numérique pour les sciences et les techniques. Tome 2*, Collection du Commissariat à l'Énergie Atomique: Série Scientifique. [Collection of the Atomic Energy Commission: Science Series], Masson, Paris, 1985, With the collaboration of Michel Artola, Philippe Bénilan, Michel Bernadou, Michel Cessenat, Jean-Claude Nédélec, Jacques Planchard and Bruno Scheurer. MR902801

[326] **R. Dautray and J.-L. Lions**, *Mathematical analysis and numerical methods for science and technology. Vol. 1*, Springer-Verlag, Berlin, 1990, Physical origins and classical methods, With the collaboration of Philippe Bénilan, Michel Cessenat, André Gervat, Alain Kavenoky and Hélène Lanchon, Translated from the French by Ian N. Sneddon, With a preface by Jean Teillac. MR1036731

[327] **E. De Giorgi**, *New problems on minimizing movements*, Boundary value problems for partial differential equations and applications, RMA Res. Notes Appl. Math., vol. 29, Masson, Paris, 1993, pp. 81–98. MR1260440

[328] **G. De Philippis and F. Maggi**, *Sharp stability inequalities for the Plateau problem*, J. Differential Geom. **96** (2014), *no. 3*, 399–456. MR3189461 http://projecteuclid.org/euclid.jdg/1395321846

[329] **G. De Philippis, J. Lamboley, M. Pierre, and B. Velichkov**, *Regularity of the minimizers of shape optimization problems involving the perimeter*, Preprint, 2016. ArXiv:1605.06294

[330] **G. De Philippis and B. Velichkov**, *Existence and regularity of minimizers for some spectral functionals with perimeter constraint*, Appl. Math. Optim. **69** (2014), *no. 2*, 199–231. MR3175194 10.1007/s00245-013-9222-4 http://dx.doi.org/10.1007/s00245-013-9222-4

[331] **D. De Silva**, *Free boundary regularity for a problem with right hand side*, Interfaces Free Bound. **13** (2011), *no. 2*, 223–238. MR2813524 10.4171/IFB/255 http://dx.doi.org/10.4171/IFB/255

[332] **D. De Silva and D. Jerison**, *A singular energy minimizing free boundary*, J. Reine Angew. Math. **635** (2009), 1–21. MR2572253 10.1515/CRELLE.2009.074 http://dx.doi.org/10.1515/CRELLE.2009.074

[333] **H. L. de Vries**, *On the upper bound for the ratio of the first two membrane eigenvalues*, Z. Naturforsch. **22a** (1967), 152–153. MR0209664

[334] **P. A. Deift**, *Applications of a commutation formula*, Duke Math. J. **45** (1978), *no. 2*, 267–310. MR495676 http://projecteuclid.org/euclid.dmj/1077312819

[335] **F. Della Pietra and N. Gavitone**, *Faber-Krahn inequality for anisotropic eigenvalue problems with Robin boundary conditions*, Potential Anal. **41** (2014), *no. 4*, 1147–1166. MR3269718 10.1007/s11118-014-9412-y http://dx.doi.org/10.1007/s11118-014-9412-y

[336] **J. Deng**, *Some results on the Schiffer conjecture in R^2*, J. Differential Equations **253** (2012), *no. 8*, 2515–2526. MR2950461 10.1016/j.jde.2012.06.002

[337] **B. Dittmar**, *Sums of reciprocal eigenvalues of the Laplacian*, Math. Nachr. **237** (2002), 45–61. MR1894352 10.1002/1522-2616(200211)245:1<45::AID-MANA45>3.0.CO;2-L

[338] **B. Dittmar**, *Sums of reciprocal Stekloff eigenvalues*, Math. Nachr. **268** (2004), 44–49. MR2054531 10.1002/mana.200310158

[339] **B. Dittmar**, *Sums of free membrane eigenvalues*, J. Anal. Math. **95** (2005), 323–332. MR2145568 10.1007/BF02791506

[340] **H. Donnelly**, *Counting nodal domains in Riemannian manifolds*, Ann. Global Anal. Geom. **46** (2014), *no. 1*, 57–61. MR3205801 10.1007/s10455-013-9408-7

[341] **H. Donnelly and C. Fefferman**, *Nodal sets of eigenfunctions on Riemannian manifolds*, Invent. Math. **93** (1988), *no. 1*, 161–183. MR943927 10.1007/BF01393691

[342] **H. Donnelly and C. Fefferman**, *Nodal sets for eigenfunctions of the Laplacian on surfaces*, J. Amer. Math. Soc. **3** (1990), *no. 2*, 333–353. MR1035413 10.2307/1990956

[343] **J. L. Doob**, *Classical potential theory and its probabilistic counterpart*, Grundlehren der Mathematischen Wissenschaften [Fundamental Principles of Mathematical Sciences], vol. 262, Springer-Verlag, New York, 1984. MR731258 10.1007/978-1-4612-5208-5

[344] **P. G. Doyle and J. P. Rossetti**, *Laplace-isospectral hyperbolic 2-orbifolds are representation-equivalent*. ArXiv:1103.4372

[345] **C. Draghici**, *Polarization and rearrangement inequalities for multiple integrals*, ProQuest LLC, Ann Arbor, MI, 2003, Thesis (Ph.D.)–Washington University in St. Louis. MR2705198 http://gateway.proquest.com/openurl?url_ver=Z39.88-2004&rft_val_fmt=info:ofi/fmt:kev:mtx:dissertation&res_dat=xri:pqdiss&rft_dat=xri:pqdiss:3105946

[346] **V. N. Dubinin**, *Capacities and geometric transformations of subsets in n-space*, Geom. Funct. Anal. **3** (1993), *no. 4*, 342–369. MR1223435 10.1007/BF01896260

[347] **R. J. Duffin**, *Nodal lines of a vibrating plate*, J. Math. Physics **31** (1953), 294–299. MR0052293

[348] **J. J. Duistermaat and V. W. Guillemin**, *The spectrum of positive elliptic operators and periodic bicharacteristics*, Invent. Math. **29** (1975), *no. 1*, 39–79. MR0405514

[349] **N. Dunford and J. T. Schwartz**, *Linear operators. Part II*, Wiley Classics Library, John Wiley & Sons, Inc., New York, 1988, Spectral theory. Selfadjoint operators in Hilbert space, With the assistance of William G. Bade and Robert G. Bartle, Reprint of the 1963 original, A Wiley-Interscience Publication. MR1009163

[350] **C. Durso**, *Solution of the inverse spectral problem for triangles*, 1990, Thesis (Ph.D.)–Massachusetts Institute of Technology.

[351] **M. S. P. Eastham and H. Kalf**, *Schrödinger-type operators with continuous spectra*, Research Notes in Mathematics, vol. 65, Pitman (Advanced Publishing Program), Boston, Mass.-London, 1982. MR667015

[352] **J. Edward**, *Pre-compactness of isospectral sets for the Neumann operator on planar domains*, Comm. Partial Differential Equations **18** (1993), *no. 7-8*, 1249–1270. MR1233194 (94h:35177) 10.1080/03605309308820973 http://dx.doi.org.acces.bibl.ulaval.ca/10.1080/03605309308820973

[353] **J. Edward and S. Wu**, *Determinant of the Neumann operator on smooth Jordan curves*, Proc. Amer. Math. Soc. **111** (1991), *no. 2*, 357–363. MR1031662 (91f:58094) 10.2307/2048324 http://dx.doi.org.acces.bibl.ulaval.ca/10.2307/2048324

[354] **J. Edward**, *An inverse spectral result for the Neumann operator on planar domains*, J. Funct. Anal. **111** (1993), *no. 2*, 312–322. MR1203456 10.1006/jfan.1993.1015

[355] **J. Edward**, *An inequality for Steklov eigenvalues for planar domains*, Z. Angew. Math. Phys. **45** (1994), *no. 3*, 493–496. MR1278688 10.1007/BF00945933

[356] **H. Egnell**, *Extremal properties of the first eigenvalue of a class of elliptic eigenvalue problems*, Ann. Scuola Norm. Sup. Pisa Cl. Sci. (4) **14** (1987), *no. 1*, 1–48. MR937535 http://www.numdam.org/item?id=ASNSP_1987_4_14_1_1_0

[357] **Y. Egorov and V. Kondratiev**, *On spectral theory of elliptic operators*, Operator Theory: Advances and Applications, vol. 89, Birkhäuser Verlag, Basel, 1996. MR1409364 10.1007/978-3-0348-9029-8

[358] **Y. V. Egorov and S. Karaa**, *Optimisation de la première valeur propre de l'opérateur de Sturm-Liouville*, C. R. Acad. Sci. Paris Sér. I Math. **319** (1994), *no. 8*, 793–798. MR1300944

[359] **I. Ekeland and R. Temam**, *Convex analysis and variational problems*, North-Holland Publishing Co., Amsterdam-Oxford; American Elsevier Publishing Co., Inc., New York, 1976, Translated from the French, Studies in Mathematics and its Applications, Vol. 1. MR0463994

[360] **A. El Soufi, E. Harrell, S. Ilias, and J. Stubbe,** *On sums of eigenvalues of elliptic operators on manifolds*, J. Spectr. Theory, to appear. ArXiv:1507.02632

[361] **A. El Soufi and S. Ilias,** *Riemannian manifolds admitting isometric immersions by their first eigenfunctions*, Pacific J. Math. **195** (2000), *no. 1*, 91–99. MR1781616 10.2140/pjm.2000.195.91

[362] **A. El Soufi and S. Ilias,** *Laplacian eigenvalue functionals and metric deformations on compact manifolds*, J. Geom. Phys. **58** (2008), *no. 1*, 89–104. MR2378458 10.1016/j.geomphys.2007.09.008

[363] **C. M. Elliott and T. Ranner,** *A computational approach to an optimal partition problem on surfaces*, Interfaces Free Bound. **17** (2015), *no. 3*, 353–379. MR3421911

[364] **C. Enache and G. A. Philippin,** *Some inequalities involving eigenvalues of the Neumann Laplacian*, Math. Methods Appl. Sci. **36** (2013), *no. 16*, 2145–2153. MR3124783 10.1002/mma.2743 http://dx.doi.org/10.1002/mma.2743

[365] **J. F. Escobar,** *The geometry of the first non-zero Stekloff eigenvalue*, J. Funct. Anal. **150** (1997), *no. 2*, 544–556. MR1479552 10.1006/jfan.1997.3116

[366] **J. F. Escobar,** *An isoperimetric inequality and the first Steklov eigenvalue*, J. Funct. Anal. **165** (1999), *no. 1*, 101–116. MR1696453 10.1006/jfan.1999.3402

[367] **J. F. Escobar,** *A comparison theorem for the first non-zero Steklov eigenvalue*, J. Funct. Anal. **178** (2000), *no. 1*, 143–155. MR1800793 10.1006/jfan.2000.3662

[368] **L. Esposito, N. Fusco, and C. Trombetti,** *A quantitative version of the isoperimetric inequality: the anisotropic case*, Ann. Sc. Norm. Super. Pisa Cl. Sci. (5) **4** (2005), *no. 4*, 619–651. MR2207737

[369] **M. Essén,** *On estimating eigenvalues of a second order linear differential operator*, General inequalities, 5 (Oberwolfach, 1986), Internat. Schriftenreihe Numer. Math., vol. 80, Birkhäuser, Basel, 1987, pp. 347–366. MR1018159

[370] **L. C. Evans and R. F. Gariepy,** *Measure theory and fine properties of functions*, revised ed., Textbooks in Mathematics, CRC Press, Boca Raton, FL, 2015. MR3409135

[371] **P. Exner,** *Leaky quantum graphs: a review*, Analysis on graphs and its applications, Proc. Sympos. Pure Math., vol. 77, Amer. Math. Soc., Providence, RI, 2008, pp. 523–564. MR2459890 10.1090/pspum/077/2459890 http://dx.doi.org/10.1090/pspum/077/2459890

[372] **P. Exner, E. M. Harrell, and M. Loss,** *Optimal eigenvalues for some Laplacians and Schrödinger operators depending on curvature*, Mathematical results in quantum mechanics (Prague, 1998), Oper. Theory Adv. Appl., vol. 108, Birkhäuser, Basel, 1999, pp. 47–58. MR1708787

[373] **P. Exner and A. Minakov,** *Curvature-induced bound states in Robin waveguides and their asymptotical properties*, J. Math. Phys. **55** (2014), *no. 12*, 122101, 19. MR3390531 10.1063/1.4903184 http://dx.doi.org/10.1063/1.4903184

[374] **P. Exner, A. Minakov, and L. Parnovski,** *Asymptotic eigenvalue estimates for a Robin problem with a large parameter*, Port. Math. **71** (2014), *no. 2*, 141–156. MR3229039 10.4171/PM/1945 http://dx.doi.org/10.4171/PM/1945

[375] **P. Exner and K. Pankrashkin,** *Strong coupling asymptotics for a singular Schrödinger operator with an interaction supported by an open arc*, Comm. Partial Differential Equations **39** (2014), *no. 2*, 193–212. MR3169783 10.1080/03605302.2013.851213 http://dx.doi.org/10.1080/03605302.2013.851213

[376] **P. Exner and K. Yoshitomi,** *Asymptotics of eigenvalues of the Schrödinger operator with a strong δ-interaction on a loop*, J. Geom. Phys. **41** (2002), *no. 4*, 344–358. MR1888470 10.1016/S0393-0440(01)00071-7 http://dx.doi.org/10.1016/S0393-0440(01)00071-7

[377] **G. Faber**, *Beweis, dass unter allen homogenen membranen von gleicher fläche und gleicher spannung die kreisförmige den tiefsten Grundton gibt*, Sitz. bayer. Akad. Wiss. (1923), 169–172.
[378] **X.-Q. Fan, L.-F. Tam, and C. Yu**, *Extremal problems for Steklov eigenvalues on annuli*. Calc. Var. Partial Differential Equations 54 (2015), *no. 1*, 1043–1059. MR3385191
[379] **H. Federer**, *Geometric measure theory*, Die Grundlehren der mathematischen Wissenschaften, Band 153, Springer-Verlag New York Inc., New York, 1969. MR0257325
[380] **V. Ferone, C. Nitsch, and C. Trombetti**, *On a conjectured reverse Faber-Krahn inequality for a Steklov-type Laplacian eigenvalue*, Commun. Pure Appl. Anal. **14** (2015), *no. 1*, 63–82. MR3299025 10.3934/cpaa.2015.14.63 http://dx.doi.org/10.3934/cpaa.2015.14.63
[381] **A. Figalli, F. Maggi, and A. Pratelli**, *A mass transportation approach to quantitative isoperimetric inequalities*, Invent. Math. **182** (2010), *no. 1*, 167–211. MR2672283 10.1007/s00222-010-0261-z http://dx.doi.org/10.1007/s00222-010-0261-z
[382] **A. Filinovskiy**, *On the asymptotic behavior of the first eigenvalue of Robin problem with large parameter*, J. Elliptic Parabol. Equ. **1** (2015), 123–135. MR3403415
[383] **A. V. Filinovskiy**, *On the estimates of eigenvalues of the boundary value problem with large parameter*, Tatra Mt. Math. Publ. **63** (2015), 101–113. MR3411438 10.1515/tmmp-2015-0023 http://dx.doi.org/10.1515/tmmp-2015-0023
[384] **A. V. Filinovskiy**, *Estimates of eigenvalues of a boundary value problem with a parameter*, Math. Commun. **19** (2014), *no. 3*, 531–543. MR3284409
[385] **A. Filinovskiy**, *Asymptotic behavior of the first eigenvalue of the robin problem*, Differ. Equ. **47** (2011), 1681–1682, On the seminar on qualitative theory of differential equations at Moscow State University, pp. 1680–1696.
[386] **N. Filonov**, *On an inequality for the eigenvalues of the Dirichlet and Neumann problems for the Laplace operator*, Algebra i Analiz **16** (2004), *no. 2*, 172–176. MR2068346 10.1090/S1061-0022-05-00857-5
[387] **M. Flucher**, *Approximation of Dirichlet eigenvalues on domains with small holes*, J. Math. Anal. Appl. **193** (1995), *no. 1*, 169–199. MR1338506 10.1006/jmaa.1995.1228
[388] **D. W. Fox and J. R. Kuttler**, *Sloshing frequencies*, Z. Angew. Math. Phys. **34** (1983), *no. 5*, 668–696. MR723140 10.1007/BF00948809
[389] **L. Fox, P. Henrici, and C. Moler**, *Approximations and bounds for eigenvalues of elliptic operators*, SIAM J. Numer. Anal. **4** (1967), 89–102. MR0215542
[390] **I. Fragalà, F. Gazzola, J. Lamboley, and M. Pierre**, *Counterexamples to symmetry for partially overdetermined elliptic problems*, Analysis (Munich) **29** (2009), *no. 1*, 85–93. MR2524107 10.1524/anly.2009.1016 http://dx.doi.org/10.1524/anly.2009.1016
[391] **R. L. Frank, A. Laptev, and S. Molchanov**, *Eigenvalue estimates for magnetic Schrödinger operators in domains*, Proc. Amer. Math. Soc. **136** (2008), *no. 12*, 4245–4255. MR2431037 10.1090/S0002-9939-08-09523-3 http://dx.doi.org/10.1090/S0002-9939-08-09523-3
[392] **A. Fraser and R. Schoen**, *Sharp eigenvalue bounds and minimal surfaces in the ball*. Invent. Math. 203 (2016), *no. 3*, 823–890. MR3461367
[393] **A. Fraser and R. Schoen**, *The first Steklov eigenvalue, conformal geometry, and minimal surfaces*, Adv. Math. **226** (2011), *no. 5*, 4011–4030. MR2770439 10.1016/j.aim.2010.11.007
[394] **A. Fraser and R. Schoen**, *Minimal surfaces and eigenvalue problems*, Geometric analysis, mathematical relativity, and nonlinear partial differential equations, Contemp. Math., vol. 599, Amer. Math. Soc., Providence, RI, 2013, pp. 105–121. MR3202476 10.1090/conm/599/11927
[395] **P. Freitas and D. Krejčiřík**, *Location of the nodal set for thin curved tubes*, Indiana Univ. Math. J. **57** (2008), *no. 1*, 343–375. MR2400260 10.1512/iumj.2008.57.3170

[396] **P. Freitas**, *On minimal eigenvalues of Schrödinger operators on manifolds*, Comm. Math. Phys. **217** (2001), *no. 2*, 375–382. MR1821228 10.1007/s002200100365

[397] **P. Freitas**, *Upper and lower bounds for the first Dirichlet eigenvalue of a triangle*, Proc. Amer. Math. Soc. **134** (2006), *no. 7*, 2083–2089 (electronic). MR2215778 10.1090/S0002-9939-06-08339-0

[398] **P. Freitas**, *Precise bounds and asymptotics for the first Dirichlet eigenvalue of triangles and rhombi*, J. Funct. Anal. **251** (2007), *no. 1*, 376–398. MR2353712 10.1016/j.jfa.2007.04.012

[399] **P. Freitas and A. Henrot**, *On the first twisted Dirichlet eigenvalue*, Comm. Anal. Geom. **12** (2004), *no. 5*, 1083–1103. MR2103312 http://projecteuclid.org/euclid.cag/1117666354

[400] **P. Freitas and D. Krejčiřík**, *The first Robin eigenvalue with negative boundary parameter*, Adv. Math. **280** (2015), 322–339. MR3350222 10.1016/j.aim.2015.04.023 http://dx.doi.org/10.1016/j.aim.2015.04.023

[401] **P. Freitas and B. A. Siudeja**, *Bounds for the first Dirichlet eigenvalue of triangles and quadrilaterals*, ESAIM Control Optim. Calc. Var. **16** (2010), *no. 3*, 648–676. MR2674631 10.1051/cocv/2009018

[402] **S. Friedland and W. K. Hayman**, *Eigenvalue inequalities for the Dirichlet problem on spheres and the growth of subharmonic functions*, Comment. Math. Helv. **51** (1976), *no. 2*, 133–161. MR0412442

[403] **S. Friedland**, *Extremal eigenvalue problems defined for certain classes of functions*, Arch. Rational Mech. Anal. **67** (1977), *no. 1*, 73–81. MR473319 10.1007/BF00280828

[404] **L. Friedlander**, *Some inequalities between Dirichlet and Neumann eigenvalues*, Arch. Rational Mech. Anal. **116** (1991), *no. 2*, 153–160. MR1143438 10.1007/BF00375590

[405] **L. Friedlander and M. Solomyak**, *On the spectrum of the Dirichlet Laplacian in a narrow strip*, Israel J. Math. **170** (2009), 337–354. MR2506330 10.1007/s11856-009-0032-y

[406] **A. Friedman**, *Reinforcement of the principal eigenvalue of an elliptic operator*, Arch. Rational Mech. Anal. **73** (1980), *no. 1*, 1–17. MR555579 10.1007/BF00283252

[407] **B. Fuglede**, *Finely harmonic functions*, Lecture Notes in Mathematics, Vol. 289, Springer-Verlag, Berlin-New York, 1972. MR0450590

[408] **B. Fuglede**, *Stability in the isoperimetric problem for convex or nearly spherical domains in* $\mathbf{R}^n$, Trans. Amer. Math. Soc. **314** (1989), *no. 2*, 619–638. MR942426 10.2307/2001401 http://dx.doi.org/10.2307/2001401

[409] **B. Fuglede**, *Bonnesen's inequality for the isoperimetric deficiency of closed curves in the plane*, Geom. Dedicata **38** (1991), *no. 3*, 283–300. MR1112666 10.1007/BF00181191 http://dx.doi.org/10.1007/BF00181191

[410] **K. Funano**, *Applications of the 'ham sandwich theorem' to eigenvalues of the Laplacian*, ArXiv:1609.05549v2

[411] **N. Fusco, F. Maggi, and A. Pratelli**, *The sharp quantitative isoperimetric inequality*, Ann. of Math. (2) **168** (2008), *no. 3*, 941–980. MR2456887 10.4007/annals.2008.168.941

[412] **N. Fusco**, *The quantitative isoperimetric inequality and related topics*, Bull. Math. Sci. **5** (2015), *no. 3*, 517–607. MR3404715 10.1007/s13373-015-0074-x http://dx.doi.org/10.1007/s13373-015-0074-x

[413] **N. Fusco, F. Maggi, and A. Pratelli**, *Stability estimates for certain Faber-Krahn, isocapacitary and Cheeger inequalities*, Ann. Sc. Norm. Super. Pisa Cl. Sci. (5) **8** (2009), *no. 1*, 51–71. MR2512200

[414] **A. Gabard**, *Sur la représentation conforme des surfaces de Riemann à bord et une caractérisation des courbes séparantes*, Comment. Math. Helv. **81** (2006), *no. 4*, 945–964. MR2271230 10.4171/CMH/82

[415] **S. Gallot, D. Hulin, and J. Lafontaine**, *Riemannian geometry*, second ed., Universitext, Springer-Verlag, Berlin, 1990. MR1083149 10.1007/978-3-642-97242-3

[416] **P. R. Garabedian and M. Schiffer**, *Variational problems in the theory of elliptic partial differential equations*, J. Rational Mech. Anal. **2** (1953), 137–171. MR0054819

[417] **R. D. Gentry and D. O. Banks**, *Bounds for functions of eigenvalues of vibrating systems*, J. Math. Anal. Appl. **51** (1975), 100–128. MR0372312

[418] **F. Gesztesy, M. Mitrea, S. Sukhtaiev, and A. Laptev**, *A bound for the eigenvalue counting function for higher-order Krein Laplacians on open sets*, Mathematical results in quantum mechanics, World Sci. Publ., Hackensack, NJ, 2015, pp. 3–29. MR3380497

[419] **F. Gesztesy and M. Mitrea**, *Nonlocal Robin Laplacians and some remarks on a paper by Filonov on eigenvalue inequalities*, J. Differential Equations **247** (2009), *no. 10*, 2871–2896. MR2568160 10.1016/j.jde.2009.07.007 http://dx.doi.org/10.1016/j.jde.2009.07.007

[420] **D. Gilbarg and N. S. Trudinger**, *Elliptic partial differential equations of second order*, second ed., Grundlehren der Mathematischen Wissenschaften [Fundamental Principles of Mathematical Sciences], vol. 224, Springer-Verlag, Berlin, 1983. MR737190 10.1007/978-3-642-61798-0

[421] **D. Gilbarg and N. S. Trudinger**, *Elliptic partial differential equations of second order*, Classics in Mathematics, Springer-Verlag, Berlin, 2001, Reprint of the 1998 edition. MR1814364

[422] **P. B. Gilkey**, *Asymptotic formulae in spectral geometry*, Studies in Advanced Mathematics, Chapman & Hall/CRC, Boca Raton, FL, 2004. MR2040963

[423] **P. B. Gilkey and G. Grubb**, *Logarithmic terms in asymptotic expansions of heat operator traces*, Comm. Partial Differential Equations **23** (1998), *no. 5-6*, 777–792. MR1632752 10.1080/03605309808821365

[424] **T. Giorgi and R. G. Smits**, *Monotonicity results for the principal eigenvalue of the generalized Robin problem*, Illinois J. Math. **49** (2005), *no. 4*, 1133–1143 (electronic). MR2210355 http://projecteuclid.org/euclid.ijm/1258138130

[425] **T. Giorgi and R. G. Smits**, *Eigenvalue estimates and critical temperature in zero fields for enhanced surface superconductivity*, Z. Angew. Math. Phys. **58** (2007), *no. 2*, 224–245. MR2305713 10.1007/s00033-005-0049-y http://dx.doi.org/10.1007/s00033-005-0049-y

[426] **T. Giorgi and R. G. Smits**, *Bounds and monotonicity for the generalized Robin problem*, Z. Angew. Math. Phys. **59** (2008), *no. 4*, 600–618. MR2417381 10.1007/s00033-007-6125-8 http://dx.doi.org/10.1007/s00033-007-6125-8

[427] **A. Girouard, R. S. Laugesen, and B. A. Siudeja**, *Steklov eigenvalues and quasiconformal maps of simply connected planar domains*, Arch. Ration. Mech. Anal. **219** (2016), *no. 2*, 903–936. MR3437867 10.1007/s00205-015-0912-8 http://dx.doi.org/10.1007/s00205-015-0912-8

[428] **A. Girouard, N. S. Nadirashvili, and I. Polterovich**, *Maximization of the second positive Neumann eigenvalue for planar domains*, J. Differential Geom. **83** (2009), *no. 3*, 637–661. MR2581359 http://projecteuclid.org/euclid.jdg/1264601037

[429] **A. Girouard, L. Parnovski, I. Polterovich, and D. A. Sher**, *The Steklov spectrum of surfaces: asymptotics and invariants*, Math. Proc. Cambridge Philos. Soc. **157** (2014), *no. 3*, 379–389. MR3286514 10.1017/S030500411400036X

[430] **A. Girouard and I. Polterovich**, *Shape optimization for low Neumann and Steklov eigenvalues*, Math. Methods Appl. Sci. **33** (2010), *no. 4*, 501–516. MR2641628 10.1002/mma.1222

[431] **A. Girouard and I. Polterovich**, *Upper bounds for Steklov eigenvalues on surfaces*, Electron. Res. Announc. Math. Sci. **19** (2012), 77–85. MR2970718 10.3934/era.2012.19.77

[432] **E. Giusti**, *Minimal surfaces and functions of bounded variation*, Monographs in Mathematics, vol. 80, Birkhäuser Verlag, Basel, 1984. MR775682 10.1007/978-1-4684-9486-0

[433] **E. Giusti**, *Metodi diretti nel calcolo delle variazioni*, Unione Matematica Italiana, Bologna, 1994. MR1707291

[434] **H. H. Goldstine**, *A history of the calculus of variations from the 17th through the 19th century*, Studies in the History of Mathematics and Physical Sciences, vol. 5, Springer-Verlag, New York-Berlin, 1980. MR601774

[435] **C. Gordon, P. Herbrich, and D. Webb**, *Steklov isospectral manifolds*, In preparation.

[436] **C. Gordon, P. Perry, and D. Schueth**, *Isospectral and isoscattering manifolds: a survey of techniques and examples*, Geometry, spectral theory, groups, and dynamics, Contemp. Math., vol. 387, Amer. Math. Soc., Providence, RI, 2005, pp. 157–179. MR2179793 10.1090/conm/387/07241

[437] **D. Grieser and S. Maronna**, *Hearing the shape of a triangle*, Notices Amer. Math. Soc. **60** (2013), *no. 11*, 1440–1447. MR3154630 10.1090/noti1063

[438] **A. Grigor′yan, Y. Netrusov, and S.-T. Yau**, *Eigenvalues of elliptic operators and geometric applications*, Surveys in differential geometry. Vol. IX, Surv. Differ. Geom., IX, Int. Press, Somerville, MA, 2004, pp. 147–217. MR2195408 10.4310/SDG.2004.v9.n1.a5

[439] **P. Grisvard**, *Elliptic problems in nonsmooth domains*, Monographs and Studies in Mathematics, vol. 24, Pitman (Advanced Publishing Program), Boston, MA, 1985. MR775683

[440] **G. Grubb and R. T. Seeley**, *Weakly parametric pseudodifferential operators and Atiyah-Patodi-Singer boundary problems*, Invent. Math. **121** (1995), *no. 3*, 481–529. MR1353307 10.1007/BF01884310

[441] **C. Guillarmou and L. Guillopé**, *The determinant of the Dirichlet-to-Neumann map for surfaces with boundary*, Int. Math. Res. Not. IMRN (2007), *no. 22*, Art. ID rnm099, 26. MR2376211 (2009b:58050) 10.1093/imrn/rnm099 http://dx.doi.org.acces.bibl.ulaval.ca/10.1093/imrn/rnm099

[442] **B. Gustafsson and H. Shahgholian**, *Existence and geometric properties of solutions of a free boundary problem in potential theory*, J. Reine Angew. Math. **473** (1996), 137–179. MR1390686

[443] **T. C. Hales**, *The honeycomb conjecture*, Discrete Comput. Geom. **25** (2001), *no. 1*, 1–22. MR1797293 10.1007/s004540010071

[444] **R. R. Hall, W. K. Hayman, and A. W. Weitsman**, *On asymmetry and capacity*, J. Analyse Math. **56** (1991), 87–123. MR1243100 10.1007/BF02820461 http://dx.doi.org/10.1007/BF02820461

[445] **Q. Han and F.-H. Lin**, *On the geometric measure of nodal sets of solutions*, J. Partial Differential Equations **7** (1994), *no. 2*, 111–131. MR1280175

[446] **B. Hanin, S. Zelditch, and P. Zhou**, *Nodal Sets of Random Eigenfunctions for the Isotropic Harmonic Oscillator*, Int. Math. Res. Not. IMRN, to appear. ArXiv:1310.4532

[447] **W. Hansen and N. Nadirashvili**, *Isoperimetric inequalities in potential theory*, Proceedings from the International Conference on Potential Theory (Amersfoort, 1991), vol. 3, 1994, pp. 1–14. MR1266215 10.1007/BF01047833

[448] **W. Hansen and N. S. Nadirashvili**, *Isoperimetric inequalities for capacities*, Harmonic analysis and discrete potential theory (Frascati, 1991), Plenum, New York, 1992, pp. 193–206. MR1222458

[449] **G. H. Hardy, J. E. Littlewood, and G. Pólya**, *Inequalities*, Cambridge, at the University Press, 1952, 2d ed. MR0046395

[450] **G. H. Hardy, J. E. Littlewood, and G. Pólya**, *Inequalities*, Cambridge Mathematical Library, Cambridge University Press, Cambridge, 1988, Reprint of the 1952 edition. MR944909

[451] **E. M. Harrell, A. Henrot, and J. Lamboley**, *On the local minimizers of the Mahler volume*, J. Convex Anal. **22** (2015), *no. 3*, 809–825. MR3400156

[452] **E. M. Harrell, II**, *General bounds for the eigenvalues of Schrödinger operators*, Maximum principles and eigenvalue problems in partial differential equations (Knoxville, TN, 1987), Pitman Res. Notes Math. Ser., vol. 175, Longman Sci. Tech., Harlow, 1988, pp. 146–166. MR963465 Zbl 0678.35075

[453] **E. M. Harrell**, *Geometric lower bounds for the spectrum of elliptic PDEs with Dirichlet conditions in part*, J. Comput. Appl. Math. **194** (2006), *no. 1*, 26–35. MR2230968 10.1016/j.cam.2005.06.012

[454] **E. M. Harrell and J. Stubbe**, *On sums of graph eigenvalues*, Linear Algebra Appl. **455** (2014), 168–186. MR3217405 10.1016/j.laa.2014.05.001

[455] **E. M. Harrell, II**, *Hamiltonian operators with maximal eigenvalues*, J. Math. Phys. **25** (1984), *no. 1*, 48–51. MR728885 10.1063/1.525996

[456] **E. M. Harrell, II**, *Some geometric bounds on eigenvalue gaps*, Comm. Partial Differential Equations **18** (1993), *no. 1-2*, 179–198. MR1211730 10.1080/03605309308820926 http://dx.doi.org/10.1080/03605309308820926

[457] **E. M. Harrell, II and L. Hermi**, *Differential inequalities for Riesz means and Weyl-type bounds for eigenvalues*, J. Funct. Anal. **254** (2008), *no. 12*, 3173–3191. MR2418623 10.1016/j.jfa.2008.02.016 http://dx.doi.org/10.1016/j.jfa.2008.02.016

[458] **E. M. Harrell, II and L. Hermi**, *On Riesz means of eigenvalues*, Comm. Partial Differential Equations **36** (2011), *no. 9*, 1521–1543. MR2825601 10.1080/03605302.2011.595865 http://dx.doi.org/10.1080/03605302.2011.595865

[459] **E. M. Harrell, II, P. Kröger, and K. Kurata**, *On the placement of an obstacle or a well so as to optimize the fundamental eigenvalue*, SIAM J. Math. Anal. **33** (2001), *no. 1*, 240–259 (electronic). MR1858877 10.1137/S0036141099357574

[460] **E. M. Harrell, II and P. L. Michel**, *Commutator bounds for eigenvalues, with applications to spectral geometry*, Comm. Partial Differential Equations **19** (1994), *no. 11-12*, 2037–2055. MR1301181 10.1080/03605309408821081 http://dx.doi.org/10.1080/03605309408821081

[461] **E. M. Harrell, II and P. L. Michel**, *Commutator bounds for eigenvalues of some differential operators*, Evolution equations (Baton Rouge, LA, 1992), Lecture Notes in Pure and Appl. Math., vol. 168, Dekker, New York, 1995, pp. 235–244. MR1300432

[462] **E. M. Harrell, II and P. L. Michel**, *Erratum to: "Commutator bounds for eigenvalues, with applications to spectral geometry" [Comm. Partial Differential Equations* **19** *(1994), no. 11-12; 2037–2055; MR1301181 (95i:58182)]*, Comm. Partial Differential Equations **20** (1995), *no. 7-8*, 1453. MR1335758

[463] **E. M. Harrell, II and J. Stubbe**, *On trace identities and universal eigenvalue estimates for some partial differential operators*, Trans. Amer. Math. Soc. **349** (1997), *no. 5*, 1797–1809. MR1401772 10.1090/S0002-9947-97-01846-1 http://dx.doi.org/10.1090/S0002-9947-97-01846-1

[464] **E. M. Harrell, II and J. Stubbe**, *Universal bounds and semiclassical estimates for eigenvalues of abstract Schrödinger operators*, SIAM J. Math. Anal. **42** (2010), *no. 5*, 2261–2274. MR2729439 10.1137/090763743 http://dx.doi.org/10.1137/090763743

[465] **E. M. Harrell, II and J. Stubbe**, *Trace identities for commutators, with applications to the distribution of eigenvalues*, Trans. Amer. Math. Soc. **363** (2011), *no. 12*, 6385–6405. MR2833559 10.1090/S0002-9947-2011-05252-9 http://dx.doi.org/10.1090/S0002-9947-2011-05252-9

[466] **A. Hasnaoui and L. Hermi**, *A sharp upper bound for the first Dirichlet eigenvalue of a class of wedge-like domains*, Z. Angew. Math. Phys. **66** (2015), *no. 5*, 2419–2440. MR3412305 Zbl 1328.35130 10.1007/s00033-015-0530-1 http://dx.doi.org/10.1007/s00033-015-0530-1

[467] **A. Hassannezhad**, *Conformal upper bounds for the eigenvalues of the Laplacian and Steklov problem*, J. Funct. Anal. **261** (2011), *no. 12*, 3419–3436. MR2838029 10.1016/j.jfa.2011.08.003

[468] **A. Hatcher**, *Algebraic topology*, Cambridge University Press, Cambridge, 2002. MR1867354

[469] **M. Hayouni**, *Sur la minimisation de la première valeur propre du laplacien*, C. R. Acad. Sci. Paris Sér. I Math. **330** (2000), *no. 7*, 551–556. MR1760437 10.1016/S0764-4442(00)00229-9

[470] **P. Hébrard and A. Henrot**, *A spillover phenomenon in the optimal location of actuators*, SIAM J. Control Optim. **44** (2005), *no. 1*, 349–366. MR2177160 10.1137/S0363012903436247

[471] **J. Heinonen, T. Kilpeläinen, and O. Martio**, *Nonlinear potential theory of degenerate elliptic equations*, Oxford Mathematical Monographs, The Clarendon Press, Oxford University Press, New York, 1993, Oxford Science Publications. MR1207810

[472] **J. Heinonen, T. Kilpeläinen, and O. Martio**, *Nonlinear potential theory of degenerate elliptic equations*, Dover Publications, Inc., Mineola, NY, 2006, Unabridged republication of the 1993 original. MR2305115

[473] **B. Helffer**, *On spectral minimal partitions: a survey*, Milan J. Math. **78** (2010), *no. 2*, 575–590. MR2781853 10.1007/s00032-010-0129-0

[474] **B. Helffer, M. Hoffmann-Ostenhof, T. Hoffmann-Ostenhof, and M. P. Owen**, *Nodal sets for groundstates of Schrödinger operators with zero magnetic field in non-simply connected domains*, Comm. Math. Phys. **202** (1999), *no. 3*, 629–649. MR1690957 10.1007/s002200050599

[475] **B. Helffer and T. Hoffmann-Ostenhof**, *Converse spectral problems for nodal domains*, Mosc. Math. J. **7** (2007), *no. 1*, 67–84, 167. MR2324557

[476] **B. Helffer and T. Hoffmann-Ostenhof**, *A review on large k minimal spectral k-partitions and Pleijel's theorem*, Spectral theory and partial differential equations, Contemp. Math., vol. 640, Amer. Math. Soc., Providence, RI, 2015, pp. 39–57. MR3381015 10.1090/conm/640/12841 http://dx.doi.org/10.1090/conm/640/12841

[477] **B. Helffer, T. Hoffmann-Ostenhof, and S. Terracini**, *Nodal domains and spectral minimal partitions*, Ann. Inst. H. Poincaré Anal. Non Linéaire **26** (2009), *no. 1*, 101–138. MR2483815 10.1016/j.anihpc.2007.07.004

[478] **B. Helffer and D. Robert**, *Riesz means of bound states and semiclassical limit connected with a Lieb-Thirring's conjecture*, Asymptotic Anal. **3** (1990), *no. 2*, 91–103. MR1061661

[479] **B. Helffer and D. Robert**, *Riesz means of bounded states and semi-classical limit connected with a Lieb-Thirring conjecture. II*, Ann. Inst. H. Poincaré Phys. Théor. **53** (1990), *no. 2*, 139–147. MR1079775 http://www.numdam.org/item?id=AIHPA_1990__53_2_139_0

[480] **B. Helffer and J. Sjöstrand**, *Multiple wells in the semiclassical limit. I*, Comm. Partial Differential Equations **9** (1984), *no. 4*, 337–408. MR740094 10.1080/03605308408820335 http://dx.doi.org/10.1080/03605308408820335

[481] **B. Helffer and J. Sjöstrand**, *Multiple wells in the semiclassical limit. I*, Comm. Partial Differential Equations **9** (1984), *no. 4*, 337–408. MR740094 10.1080/03605308408820335 http://dx.doi.org/10.1080/03605308408820335

[482] **B. Helffer and J. Sjöstrand**, *Puits multiples en limite semi-classique. II. Interaction moléculaire. Symétries. Perturbation*, Ann. Inst. H. Poincaré Phys. Théor. **42** (1985), *no. 2*, 127–212. MR798695 http://www.numdam.org/item?id=AIHPB_1985__42_2_127_0

[483] **B. Helffer and J. Sjöstrand**, *Effet tunnel pour l'équation de Schrödinger avec champ magnétique*, Ann. Scuola Norm. Sup. Pisa Cl. Sci. (4) **14** (1987), *no. 4*, 625–657 (1988). MR963493 http://www.numdam.org/item?id=ASNSP_1987_4_14_4_625_0

[484] **B. Helffer**, *Semi-classical analysis for the Schrödinger operator and applications*, Lecture Notes in Mathematics, vol. 1336, Springer-Verlag, Berlin, 1988. MR960278

[485] **B. Helffer**, *On spectral theory for Schrödinger operators with magnetic potentials*, Spectral and scattering theory and applications, Adv. Stud. Pure Math., vol. 23, Math. Soc. Japan, Tokyo, 1994, pp. 113–141. MR1275398

[486] **B. Helffer**, *Domaines nodaux et partitions spectrales minimales (d'après B. Helffer, T. Hoffmann-Ostenhof et S. Terracini)*, Séminaire: Équations aux Dérivées Partielles. 2006–2007, Sémin. Équ. Dériv. Partielles, École Polytech., Palaiseau, 2007, pp. Exp. No. VIII, 23. MR2385195

[487] **B. Helffer**, *Spectral theory and its applications*, Cambridge Studies in Advanced Mathematics, vol. 139, Cambridge University Press, Cambridge, 2013. MR3027462

[488] **B. Helffer**, *Lower bound for the number of critical points of minimal spectral k-partitions for k large*, Annales mathématiques du Québec (2016), 1–8. 10.1007/s40316-016-0058-6 ArXiv:1504.01015 http://dx.doi.org/10.1007/s40316-016-0058-6

[489] **B. Helffer and T. Hoffmann-Ostenhof**, *On minimal partitions: new properties and applications to the disk*, Spectrum and dynamics, CRM Proc. Lecture Notes, vol. 52, Amer. Math. Soc., Providence, RI, 2010, pp. 119–135. MR2743435

[490] **B. Helffer and T. Hoffmann-Ostenhof**, *Remarks on two notions of spectral minimal partitions*, Adv. Math. Sci. Appl. **20** (2010), *no. 1*, 249–263. MR2760728

[491] **B. Helffer and T. Hoffmann-Ostenhof**, *On a magnetic characterization of spectral minimal partitions*, J. Eur. Math. Soc. (JEMS) **15** (2013), *no. 6*, 2081–2092. MR3120736 10.4171/JEMS/415

[492] **B. Helffer and T. Hoffmann-Ostenhof**, *Spectral minimal partitions for a thin strip on a cylinder or a thin annulus like domain with Neumann condition*, Operator methods in mathematical physics, Oper. Theory Adv. Appl., vol. 227, Birkhäuser/Springer Basel AG, Basel, 2013, pp. 107–115. MR3050161 10.1007/978-3-0348-0531-5_5

[493] **B. Helffer and T. Hoffmann-Ostenhof**, *Minimal partitions for anisotropic tori*, J. Spectr. Theory **4** (2014), *no. 2*, 221–233. MR3232810 10.4171/JST/68

[494] **B. Helffer, T. Hoffmann-Ostenhof, and S. Terracini**, *Nodal minimal partitions in dimension 3*, Discrete Contin. Dyn. Syst. **28** (2010), *no. 2*, 617–635. MR2644760 10.3934/dcds.2010.28.617

[495] **B. Helffer, T. Hoffmann-Ostenhof, and S. Terracini**, *On spectral minimal partitions: the case of the sphere*, Around the research of Vladimir Maz'ya. III, Int. Math. Ser. (N. Y.), vol. 13, Springer, New York, 2010, pp. 153–178. MR2664708 10.1007/978-1-4419-1345-6_6

[496] **B. Helffer and A. Kachmar**, *Eigenvalues for the Robin Laplacian in domains with variable curvature*, to appear in Trans. Amer. Math. Soc., 2014. ArXiv:1411.2700

[497] **B. Helffer, A. Kachmar, and N. Raymond**, *Tunneling for the Robin Laplacian in smooth planar domains*, 2015. ArXiv:1509.03986

[498] **B. Helffer and R. Kiwan**, *Dirichlet eigenfunctions on the cube, sharpening the Courant nodal inequality*, Functional Analysis and Operator Theory for Quantum Physics. A Festschrift in Honor of Pavel Exner. (J. Dittrich, H. Kovarik, and A. Laptev, eds.), Europ. Math. Soc. Publ. House, to appear. ArXiv:1506.05733

[499] **B. Helffer and K. Pankrashkin**, *Tunneling between corners for Robin Laplacians*, J. Lond. Math. Soc. (2) **91** (2015), *no. 1*, 225–248. MR3338614 10.1112/jlms/jdu070 http://dx.doi.org/10.1112/jlms/jdu070

[500] **B. Helffer and M. Persson Sundqvist**, *Nodal domains in the square—the Neumann case*, Moscow Mathematical Journal, to appear. ArXiv:1410.6702

[501] **B. Helffer and M. Persson Sundqvist**, *On nodal domains in Euclidean balls*, Proc. Amer. Math. Soc. 10.1090/proc/13098 ArXiv:1506.04033

[502] **R. Hempel, L. A. Seco, and B. Simon**, *The essential spectrum of Neumann Laplacians on some bounded singular domains*, J. Funct. Anal. **102** (1991), *no. 2*, 448–483. MR1140635 10.1016/0022-1236(91)90130-W

[503] **A. Henrot, G. A. Philippin, and A. Safoui**, *Some isoperimetric inequalities with application to the Stekloff problem*, J. Convex Anal. **15** (2008), *no. 3*, 581–592. MR2431413

[504] **A. Henrot**, *Minimization problems for eigenvalues of the Laplacian*, J. Evol. Equ. **3** (2003), *no. 3*, 443–461, Dedicated to Philippe Bénilan. MR2019029 10.1007/s00028-003-0111-0

[505] **A. Henrot**, *Extremum problems for eigenvalues of elliptic operators*, Frontiers in Mathematics, Birkhäuser Verlag, Basel, 2006. MR2251558

[506] **A. Henrot and H. Maillot**, *Optimization of the shape and the location of the actuators in an internal control problem*, Boll. Unione Mat. Ital. Sez. B Artic. Ric. Mat. (8) **4** (2001), *no. 3*, 737–757. MR1859433

[507] **A. Henrot and E. Oudet**, *Le stade ne minimise pas λ_2 parmi les ouverts convexes du plan*, C. R. Acad. Sci. Paris Sér. I Math. **332** (2001), *no. 5*, 417–422. MR1826627 10.1016/S0764-4442(01)01864-X

[508] **A. Henrot and E. Oudet**, *Minimizing the second eigenvalue of the Laplace operator with Dirichlet boundary conditions*, Arch. Ration. Mech. Anal. **169** (2003), *no. 1*, 73–87. MR1996269 10.1007/s00205-003-0259-4

[509] **A. Henrot and M. Pierre**, *About existence of equilibria in electromagnetic casting*, Quart. Appl. Math. **49** (1991), *no. 3*, 563–575. MR1121687

[510] **A. Henrot and M. Pierre**, *Variation et optimisation de formes*, Mathématiques & Applications (Berlin) [Mathematics & Applications], vol. 48, Springer, Berlin, 2005, Une analyse géométrique. [A geometric analysis]. MR2512810 10.1007/3-540-37689-5 http://dx.doi.org/10.1007/3-540-37689-5

[511] **L. Hermi**, *On the spectrum of the Dirichlet Laplacian and other elliptic operators*, ProQuest LLC, Ann Arbor, MI, 1999, Thesis (Ph.D.)–University of Missouri - Columbia. MR2700115 http://gateway.proquest.com/openurl?url_ver=Z39.88-2004&rft_val_fmt=info:ofi/fmt:kev:mtx:dissertation&res_dat=xri:pqdiss&rft_dat=xri:pqdiss:9953864

[512] **L. Hermi**, *Two new Weyl-type bounds for the Dirichlet Laplacian*, Trans. Amer. Math. Soc. **360** (2008), *no. 3*, 1539–1558 (electronic). MR2357704 10.1090/S0002-9947-07-04254-7 http://dx.doi.org/10.1090/S0002-9947-07-04254-7

[513] **J. Hersch, L. E. Payne, and M. M. Schiffer**, *Some inequalities for Stekloff eigenvalues*, Arch. Rational Mech. Anal. **57** (1975), 99–114. MR0387837

[514] **J. Hersch**, *Sur la fréquence fondamentale d'une membrane vibrante: évaluations par défaut et principe de maximum*, Z. Angew. Math. Phys. **11** (1960), 387–413. MR0125319

[515] **J. Hersch**, *Contribution to the method of interior parallels applied to vibrating membranes*, Studies in mathematical analysis and related topics, Stanford Univ. Press, Stanford, Calif., 1962, pp. 132–139. MR0163492

[516] **J. Hersch**, *The method of interior parallels applied to polygonal or multiply connected membranes*, Pacific J. Math. **13** (1963), 1229–1238. MR0163493

[517] **J. Hersch**, *Contraintes rectilignes parallèles et valeurs propres de membranes vibrantes*, Z. Angew. Math. Phys. **17** (1966), 457–460. MR0208192

[518] **J. Hersch**, *Transplantation harmonique, transplantation par modules, et théorèmes isopérimétriques*, Comment. Math. Helv. **44** (1969), 354–366. MR0247572

[519] **J. Hersch**, *Quatre propriétés isopérimétriques de membranes sphériques homogènes*, C. R. Acad. Sci. Paris Sér. A-B **270** (1970), A1645–A1648. MR0292357

[520] **J. Hersch and M. Monkewitz**, *Une inégalité isopérimétrique renforçant celle de Szegö-Weinberger sur les membranes libres*, C. R. Acad. Sci. Paris Sér. A-B **273** (1971), A62–A64. MR0295172

[521] **J. Hersch and L. E. Payne**, *Extremal principles and isoperimetric inequalities for some mixed problems of Stekloff's type*, Z. Angew. Math. Phys. **19** (1968), 802–817. MR0243210

[522] **G. N. Hile and M. H. Protter**, *Inequalities for eigenvalues of the Laplacian*, Indiana Univ. Math. J. **29** (1980), *no. 4*, 523–538. MR578204 10.1512/iumj.1980.29.29040 http://dx.doi.org/10.1512/iumj.1980.29.29040

[523] **G. N. Hile and Z. Y. Xu**, *Inequalities for sums of reciprocals of eigenvalues*, J. Math. Anal. Appl. **180** (1993), *no. 2*, 412–430. MR1251868 10.1006/jmaa.1993.1409

[524] **G. N. Hile and R. Z. Yeh**, *Inequalities for eigenvalues of the biharmonic operator*, Pacific J. Math. **112** (1984), *no. 1*, 115–133. MR739143 http://projecteuclid.org/euclid.pjm/1102710102

[525] **L. Hillairet and C. Judge**, *Generic spectral simplicity of polygons*, Proc. Amer. Math. Soc. **137** (2009), *no. 6*, 2139–2145. MR2480296 10.1090/S0002-9939-09-09621-X

[526] **L. Hillairet and C. Judge**, *Spectral simplicity and asymptotic separation of variables*, Comm. Math. Phys. **302** (2011), *no. 2*, 291–344. MR2770015 10.1007/s00220-010-1185-6

[527] **L. Hillairet and C. Judge**, *Erratum to: Spectral simplicity and asymptotic separation of variables*, Comm. Math. Phys. **311** (2012), *no. 3*, 839–842. MR2909766 10.1007/s00220-012-1467-2

[528] **J.-B. Hiriart-Urruty and C. Lemaréchal**, *Convex analysis and minimization algorithms. I*, Grundlehren der Mathematischen Wissenschaften [Fundamental Principles of Mathematical Sciences], vol. 305, Springer-Verlag, Berlin, 1993, Fundamentals. MR1261420

[529] **P. D. Hislop and C. V. Lutzer**, *Spectral asymptotics of the Dirichlet-to-Neumann map on multiply connected domains in* $\mathbb{R}^d$, Inverse Problems **17** (2001), *no. 6*, 1717–1741. MR1872919 10.1088/0266-5611/17/6/313

[530] **M. Hoffmann-Ostenhof, T. Hoffmann-Ostenhof, and N. Nadirashvili**, *On the nodal line conjecture*, Advances in differential equations and mathematical physics (Atlanta, GA, 1997), Contemp. Math., vol. 217, Amer. Math. Soc., Providence, RI, 1998, pp. 33–48. MR1605269 10.1090/conm/217/02980

[531] **T. Hoffmann-Ostenhof, P. W. Michor, and N. Nadirashvili**, *Bounds on the multiplicity of eigenvalues for fixed membranes*, Geom. Funct. Anal. **9** (1999), *no. 6*, 1169–1188. MR1736932 10.1007/s000390050111

[532] **T. Hoffmann-Ostenhof**, *Eigenfunctions for 2-dimensional tori and for rectangles with Neumann boundary conditions*, Mosc. Math. J. **15** (2015), *no. 1*, 101–106, 182. MR3427414

[533] **G. Hong**, *The singular homogeneous solutions to one phase free boundary problem*, Proc. Amer. Math. Soc. **143** (2015), *no. 9*, 4009–4015. MR3359589 10.1090/S0002-9939-2015-12553-1 http://dx.doi.org/10.1090/S0002-9939-2015-12553-1

[534] **I. Hong**, *On an inequality concerning the eigenvalue problem of membrane*, Kōdai Math. Sem. Rep. **6** (1954), 113–114, {Volume numbers not printed on issues until Vol. **7** (1955).}. MR0070015

[535] **S. M. Hook**, *Domain-independent upper bounds for eigenvalues of elliptic operators*, Trans. Amer. Math. Soc. **318** (1990), *no. 2*, 615–642. MR994167 10.2307/2001323 http://dx.doi.org/10.2307/2001323

[536] **S. M. Hook**, *Inequalities for eigenvalues of selfadjoint operators*, Trans. Amer. Math. Soc. **318** (1990), *no. 1*, 237–259. MR943604 10.2307/2001237 http://dx.doi.org/10.2307/2001237

[537] **S. M. Hook**, *Bounds for the fundamental frequencies of an elastic medium*, Michigan Math. J. **38** (1991), *no. 1*, 3–12. MR1091505 10.1307/mmj/1029004257 http://dx.doi.org/10.1307/mmj/1029004257

[538] **S. M. Hook**, *INEQUALITIES FOR EIGENVALUES OF SELF-ADJOINT OPERATORS (LAPLACIAN, ELLIPTIC)*, ProQuest LLC, Ann Arbor, MI, 1986, Thesis (Ph.D.)–University of California, Berkeley. MR2635049 http://gateway.proquest.com/openurl?url_ver=Z39.88-2004&rft_val_fmt=info:ofi/fmt:kev:mtx:dissertation&res_dat=xri:pqdiss&rft_dat=xri:pqdiss:8624792

[539] **W. Hooker and M. H. Protter**, *Bounds for the first eigenvalue of a rhombic membrane*, J. Math. and Phys. **39** (1960/1961), 18–34. MR0127610

[540] **L. Hörmander**, *Inequalities between normal and tangential derivatives of harmonic functions*, unpublished, c. 1955.

[541] **L. Hörmander**, *The analysis of linear partial differential operators. IV*, Grundlehren der Mathematischen Wissenschaften [Fundamental Principles of Mathematical Sciences], vol. 275, Springer-Verlag, Berlin, 1985, Fourier integral operators. MR781537

[542] **L. Hörmander**, *The analysis of linear partial differential operators. I*, Classics in Mathematics, Springer-Verlag, Berlin, 2003, Distribution theory and Fourier analysis, Reprint of the second (1990) edition [Springer, Berlin; MR1065993 (91m:35001a)]. MR1996773

[543] **L. Hörmander,** *The analysis of linear partial differential operators. II*, Classics in Mathematics, Springer-Verlag, Berlin, 2005, Differential operators with constant coefficients, Reprint of the 1983 original. MR2108588

[544] **L. Hörmander,** *The analysis of linear partial differential operators. III*, Classics in Mathematics, Springer, Berlin, 2007, Pseudo-differential operators, Reprint of the 1994 edition. MR2304165

[545] **L. Hörmander,** *The analysis of linear partial differential operators. IV*, Classics in Mathematics, Springer-Verlag, Berlin, 2009, Fourier integral operators, Reprint of the 1994 edition. MR2512677 10.1007/978-3-642-00136-9

[546] **M. Horváth,** *On the first two eigenvalues of Sturm-Liouville operators*, Proc. Amer. Math. Soc. **131** (2003), *no. 4*, 1215–1224 (electronic). MR1948113 10.1090/S0002-9939-02-06637-6

[547] **M. Horváth and M. Kiss,** *A bound for ratios of eigenvalues of Schrödinger operators with single-well potentials*, Proc. Amer. Math. Soc. **134** (2006), *no. 5*, 1425–1434 (electronic). MR2199189 10.1090/S0002-9939-05-08100-1

[548] **H. Howards, M. Hutchings, and F. Morgan,** *The isoperimetric problem on surfaces*, Amer. Math. Monthly **106** (1999), *no. 5*, 430–439. MR1699261 10.2307/2589147

[549] **M.-J. Huang,** *On the eigenvalue ratio for vibrating strings*, Proc. Amer. Math. Soc. **127** (1999), *no. 6*, 1805–1813. MR1621941 10.1090/S0002-9939-99-05015-7

[550] **Y. L. Huang and C. K. Law,** *Eigenvalue ratios for the regular Sturm-Liouville system*, Proc. Amer. Math. Soc. **124** (1996), *no. 5*, 1427–1436. MR1328351 10.1090/S0002-9939-96-03396-5

[551] **D. Hundertmark, A. Laptev, and T. Weidl,** *New bounds on the Lieb-Thirring constants*, Invent. Math. **140** (2000), *no. 3*, 693–704. MR1760755 10.1007/s002220000077 http://dx.doi.org/10.1007/s002220000077

[552] **D. Hundertmark,** *Some bound state problems in quantum mechanics*, Spectral theory and mathematical physics: a Festschrift in honor of Barry Simon's 60th birthday, Proc. Sympos. Pure Math., vol. 76, Amer. Math. Soc., Providence, RI, 2007, pp. 463–496. MR2310215 10.1090/pspum/076.1/2310215 http://dx.doi.org/10.1090/pspum/076.1/2310215

[553] **S. Ilias and O. Makhoul,** *A Reilly inequality for the first Steklov eigenvalue*, Differential Geom. Appl. **29** (2011), *no. 5*, 699–708. MR2831826 10.1016/j.difgeo.2011.07.005

[554] **V. Y. Ivriĭ,** *Weyl's asymptotic formula for the Laplace-Beltrami operator in Riemannian polyhedra and in domains with conic singularities of the boundary*, Dokl. Akad. Nauk SSSR **288** (1986), *no. 1*, 35–38. MR841136

[555] **V. Ivriĭ,** *Precise spectral asymptotics for elliptic operators acting in fiberings over manifolds with boundary*, Lecture Notes in Mathematics, vol. 1100, Springer-Verlag, Berlin, 1984. MR771297

[556] **D. Jakobson, M. Levitin, N. S. Nadirashvili, N. Nigam, and I. Polterovich,** *How large can the first eigenvalue be on a surface of genus two?*, Int. Math. Res. Not. (2005), *no. 63*, 3967–3985. MR2202582 10.1155/IMRN.2005.3967

[557] **D. Jakobson, M. Levitin, N. S. Nadirashvili, and I. Polterovich,** *Spectral problems with mixed Dirichlet-Neumann boundary conditions: isospectrality and beyond*, J. Comput. Appl. Math. **194** (2006), *no. 1*, 141–155. MR2230975 10.1016/j.cam.2005.06.019

[558] **P. Jammes,** *Multiplicité du spectre de Steklov sur les surfaces et nombre chromatique*. ArXiv:1304.4559

[559] **P. Jammes,** *Une inégalité de Cheeger pour le spectre de Steklov, note=to appear*, Ann. inst. Fourier. ArXiv:1302.6540

[560] **P. Jammes,** *Prescription du spectre de Steklov dans une classe conforme*, Anal. PDE **7** (2014), *no. 3*, 529–549. MR3227426 10.2140/apde.2014.7.529

[561] **S. A. Janczewsky**, *Oscillation theorems for the differential boundary value problems of the fourth order*, Ann. of Math. (2) **29** (1927/28), *no. 1-4*, 521–542. MR1502859 10.2307/1968021

[562] **D. Jerison and N. S. Nadirashvili**, *The "hot spots" conjecture for domains with two axes of symmetry*, J. Amer. Math. Soc. **13** (2000), *no. 4*, 741–772. MR1775736 10.1090/S0894-0347-00-00346-5

[563] **D. Jerison and O. Savin**, *Some remarks on stability of cones for the one-phase free boundary problem*, Geom. Funct. Anal. **25** (2015), *no. 4*, 1240–1257. MR3385632 10.1007/s00039-015-0335-6 http://dx.doi.org/10.1007/s00039-015-0335-6

[564] **A. Jollivet and V. Sharafutdinov**, *On an inverse problem for the Steklov spectrum of a Riemannian surface*, Inverse problems and applications, Contemp. Math., vol. 615, Amer. Math. Soc., Providence, RI, 2014, pp. 165–191. MR3221604 10.1090/conm/615/12260

[565] **D. R. Jones, C. D. Perttunen, and B. E. Stuckman**, *Lipschitzian optimization without the Lipschitz constant*, J. Optim. Theory Appl. **79** (1993), *no. 1*, 157–181. MR1246501 10.1007/BF00941892 http://dx.doi.org/10.1007/BF00941892

[566] **C. Jouron**, *Sur un problème d'optimisation où la contrainte porte sur la fréquence fondamentale*, RAIRO Anal. Numér. **12** (1978), *no. 4*, 349–375, vi. MR519018

[567] **M. Kac**, *Can one hear the shape of a drum?*, Amer. Math. Monthly **73** (1966), *no. 4, part II*, 1–23. MR0201237

[568] **S. Karaa**, *Extremal eigenvalue gaps for the Schrödinger operator with Dirichlet boundary conditions*, J. Math. Phys. **39** (1998), *no. 4*, 2325–2332. MR1614731 10.1063/1.532290

[569] **S. Karaa**, *Sharp estimates for the eigenvalues of some differential equations*, SIAM J. Math. Anal. **29** (1998), *no. 5*, 1279–1300 (electronic). MR1628259 10.1137/S0036141096307849

[570] **A. Karageorghis**, *The method of fundamental solutions for the calculation of the eigenvalues of the Helmholtz equation*, Appl. Math. Lett. **14** (2001), *no. 7*, 837–842. MR1849238 10.1016/S0893-9659(01)00053-2 http://dx.doi.org/10.1016/S0893-9659(01)00053-2

[571] **S. Karlin**, *Some extremal problems for eigenvalues of certain matrix and integral operators*, Advances in Math. **9** (1972), 93–136. MR0322585

[572] **L. Karp and N. Peyerimhoff**, *Extremal properties of the principal Dirichlet eigenvalue for regular polygons in the hyperbolic plane*, Arch. Math. (Basel) **79** (2002), *no. 3*, 223–231. MR1933381 Zbl 1034.52013 10.1007/s00013-002-8308-z http://dx.doi.org/10.1007/s00013-002-8308-z

[573] **M. Karpukhin**, *Bounds between Laplace and Steklov eigenvalues on nonnegatively curved manifolds*. ArXiv:1512.09038

[574] **M. Karpukhin, G. Kokarev, and I. Polterovich**, *Multiplicity bounds for Steklov eigenvalues on Riemannian surfaces, journal=Ann. Inst. Fourier*, **64** (2014), 2481–2502. ArXiv:1209.4869

[575] **T. Kato**, *Perturbation theory for linear operators*, Die Grundlehren der mathematischen Wissenschaften, Band 132, Springer-Verlag New York, Inc., New York, 1966. MR0203473

[576] **T. Kato**, *Perturbation theory for linear operators*, Classics in Mathematics, Springer-Verlag, Berlin, 1995, Reprint of the 1980 edition. MR1335452

[577] **B. Kawohl and V. Fridman**, *Isoperimetric estimates for the first eigenvalue of the p-laplace operator and the cheeger constant*, Comment. Math. Univ. Carolin. **44** (2003), 659–667.

[578] **B. Kawohl, O. Pironneau, L. Tartar, and J.-P. Zolésio**, *Optimal shape design*, Lecture Notes in Mathematics, vol. 1740, Springer-Verlag, Berlin; Centro Internazionale Matematico Estivo (C.I.M.E.), Florence, 2000, Lectures given at the Joint C.I.M./C.I.M.E. Summer School held in Tróia, June 1–6, 1998, Edited by A. Cellina and A. Ornelas, Fondazione C.I.M.E.. [C.I.M.E. Foundation]. MR1804683 10.1007/BFb0106739

[579] **B. Kawohl**, *Rearrangements and convexity of level sets in PDE*, Lecture Notes in Mathematics, vol. 1150, Springer-Verlag, Berlin, 1985. MR810619

[580] **B. Kawohl**, *Remarks on some old and current eigenvalue problems*, Partial differential equations of elliptic type (Cortona, 1992), Sympos. Math., XXXV, Cambridge Univ. Press, Cambridge, 1994, pp. 165–183. MR1297778

[581] **J. B. Keller**, *The shape of the strongest column*, Arch. Rational Mech. Anal. **5** (1960), 275–285 (1960). MR0128160

[582] **J. B. Keller**, *The minimum ratio of two eigenvalues*, SIAM J. Appl. Math. **31** (1976), *no. 3*, 485–491. MR0422751

[583] **J. B. Kennedy**, *On the isoperimetric problem for the higher eigenvalues of the Robin and Wentzell Laplacians*, Z. Angew. Math. Phys. **61** (2010), *no. 5*, 781–792. MR2726626 10.1007/s00033-009-0052-9 http://dx.doi.org/10.1007/s00033-009-0052-9

[584] **J. B. Kennedy**, *The nodal line of the second eigenfunction of the Robin Laplacian in $\mathbb{R}^2$ can be closed*, J. Differential Equations **251** (2011), *no. 12*, 3606–3624. MR2837697 10.1016/j.jde.2011.08.012

[585] **J. B. Kennedy**, *Closed nodal surfaces for simply connected domains in higher dimensions*, Indiana Univ. Math. J. **62** (2013), *no. 3*, 785–798. MR3164844 10.1512/iumj.2013.62.4975

[586] **J. B. Kennedy**, *An isoperimetric inequality for the second eigenvalue of the Laplacian with Robin boundary conditions*, Proc. Amer. Math. Soc. **137** (2009), *no. 2*, 627–633. MR2448584 10.1090/S0002-9939-08-09704-9 http://dx.doi.org/10.1090/S0002-9939-08-09704-9

[587] **S. Kesavan**, *On two functionals connected to the Laplacian in a class of doubly connected domains*, Proc. Roy. Soc. Edinburgh Sect. A **133** (2003), *no. 3*, 617–624. MR1983689 10.1017/S0308210500002560

[588] **S. Kesavan**, *Symmetrization & applications*, Series in Analysis, vol. 3, World Scientific Publishing Co. Pte. Ltd., Hackensack, NJ, 2006. MR2238193 10.1142/9789812773937 http://dx.doi.org/10.1142/9789812773937

[589] **D. Kinderlehrer and L. Nirenberg**, *Regularity in free boundary problems*, Ann. Scuola Norm. Sup. Pisa Cl. Sci. (4) **4** (1977), *no. 2*, 373–391. MR0440187

[590] **D. Kinderlehrer and G. Stampacchia**, *An introduction to variational inequalities and their applications*, Pure and Applied Mathematics, vol. 88, Academic Press, Inc. [Harcourt Brace Jovanovich, Publishers], New York-London, 1980. MR567696

[591] **M.-T. Kohler-Jobin**, *Une méthode de comparaison isopérimétrique de fonctionnelles de domaines de la physique mathématique. I. Une démonstration de la conjecture isopérimétrique $P\lambda^2 \geq \pi j_0^4/2$ de Pólya et Szegő*, Z. Angew. Math. Phys. **29** (1978), *no. 5*, 757–766. MR511908 10.1007/BF01589287 http://dx.doi.org/10.1007/BF01589287

[592] **M.-T. Kohler-Jobin**, *Une méthode de comparaison isopérimétrique de fonctionnelles de domaines de la physique mathématique. II. Cas inhomogène: une inégalité isopérimétrique entre la fréquence fondamentale d'une membrane et l'énergie d'équilibre d'un problème de Poisson*, Z. Angew. Math. Phys. **29** (1978), *no. 5*, 767–776. MR511909 10.1007/BF01589288

[593] **M.-T. Kohler-Jobin**, *Symmetrization with equal Dirichlet integrals*, SIAM J. Math. Anal. **13** (1982), *no. 1*, 153–161. MR641547 10.1137/0513011 http://dx.doi.org/10.1137/0513011

[594] **G. Kokarev**, *Variational aspects of Laplace eigenvalues on Riemannian surfaces*, Adv. Math. **258** (2014), 191–239. MR3190427 10.1016/j.aim.2014.03.006

[595] **A. V. Kolesnikov and R. I. Zhdanov**, *On isoperimetric sets of radially symmetric measures*, Concentration, functional inequalities and isoperimetry, Contemp. Math., vol. 545, Amer. Math. Soc., Providence, RI, 2011, pp. 123–154. MR2858470 10.1090/conm/545/10769 http://dx.doi.org/10.1090/conm/545/10769

[596] **T. Kolokolnikov, M. S. Titcombe, and M. J. Ward**, *Optimizing the fundamental Neumann eigenvalue for the Laplacian in a domain with small traps*, European J. Appl. Math. **16** (2005), *no. 2*, 161–200. MR2155290 10.1017/S0956792505006145

[597] **V. Komkov**, *An optimal design problem. A nonexistence theorem*, Arch. Mech. (Arch. Mech. Stos.) **33** (1981), *no. 1*, 147–151. MR639776

[598] **N. Korevaar**, *Upper bounds for eigenvalues of conformal metrics*, J. Differential Geom. **37** (1993), *no. 1*, 73–93. MR1198600 http://projecteuclid.org/euclid.jdg/1214453423

[599] **E. T. Kornhauser and I. Stakgold**, *A variational theorem for $\nabla^2 u + \lambda u = 0$ and its applications*, J. Math. Physics **31** (1952), 45–54. MR0047236

[600] **H. Kovařík**, *On the lowest eigenvalue of Laplace operators with mixed boundary conditions*, J. Geom. Anal. **24** (2014), *no. 3*, 1509–1525. MR3223564 10.1007/s12220-012-9383-4 http://dx.doi.org/10.1007/s12220-012-9383-4

[601] **H. Kovařík and K. Pankrashkin**, *On the p-Laplacian with Robin boundary conditions and boundary trace theorems*. ArXiv:1603.01737

[602] **V. Kozlov and N. Kuznetsov**, *The ice-fishing problem: the fundamental sloshing frequency versus geometry of holes*, Math. Methods Appl. Sci. **27** (2004), *no. 3*, 289–312. MR2023011 10.1002/mma.442

[603] **E. Krahn**, *Über eine von Rayleigh formulierte Minimaleigenschaft des Kreises*, Math. Ann. **94** (1925), *no. 1*, 97–100. MR1512244 10.1007/BF01208645

[604] **E. Krahn**, *Uber minimaleigenschaften der krugel in drei un mehr dimensionen*, Acta Comm. Univ. Dorpat. **A9** (1926), 1–44.

[605] **M. G. Krein**, *On certain problems on the maximum and minimum of characteristic values and on the Lyapunov zones of stability*, Amer. Math. Soc. Transl. (2) **1** (1955), 163–187. MR0073776

[606] **P. Kuchment**, *The mathematics of photonic crystals*, Mathematical modeling in optical science, Frontiers Appl. Math., vol. 22, SIAM, Philadelphia, PA, 2001, pp. 207–272. MR1831334 (2002k:78002)

[607] **A. Kufner, O. John, and S. Fučík**, *Function spaces*, Noordhoff International Publishing, Leyden; Academia, Prague, 1977, Monographs and Textbooks on Mechanics of Solids and Fluids; Mechanics: Analysis. MR0482102

[608] **J. R. Kuttler and V. G. Sigillito**, *Lower bounds for Stekloff and free membrane eigen-values*, SIAM Rev. **10** (1968), 368–370. MR0233500

[609] **J. R. Kuttler and V. G. Sigillito**, *An inequality of a Stekloff eigenvalue by the method of defect*, Proc. Amer. Math. Soc. **20** (1969), 357–360. MR0235323

[610] **J. R. Kuttler and V. G. Sigillito**, *Eigenvalues of the Laplacian in two dimensions*, SIAM Rev. **26** (1984), *no. 2*, 163–193. MR738929 10.1137/1026033 http://dx.doi.org/10.1137/1026033

[611] **J. F. Kuzanek**, *Existence and uniqueness of solutions to a fourth order nonlinear eigenvalue problem*, SIAM J. Appl. Math. **27** (1974), 341–354. MR0365319

[612] **N. Kuznetsov**, *Asymptotic distribution of eigenfrequencies of a plane membrane in the case of separable variables*, Differencial′nye Uravnenija **2** (1966), 1385–1402. MR0206524

[613] **N. Kuznetsov, T. Kulczycki, M. Kwaśnicki, A. Nazarov, S. Poborchi, I. Polterovich, and B. A. Siudeja**, *The legacy of Vladimir Andreevich Steklov*, Notices Amer. Math. Soc. **61** (2014), *no. 1*, 9–22. MR3137253 10.1090/noti1073

[614] **A. A. Lacey, J. R. Ockendon, and J. Sabina**, *Multidimensional reaction diffusion equations with nonlinear boundary conditions*, SIAM J. Appl. Math. **58** (1998), *no. 5*, 1622–1647. MR1637882 10.1137/S0036139996308121 http://dx.doi.org/10.1137/S0036139996308121

[615] **T. Lachand-Robert and M. A. Peletier**, *An example of non-convex minimization and an application to Newton's problem of the body of least resistance*, Ann. Inst. H. Poincaré Anal. Non Linéaire **18** (2001), *no. 2*, 179–198. MR1808028 10.1016/S0294-1449(00)00062-7

[616] **T. Lachand-Robert and M. A. Peletier**, *Newton's problem of the body of minimal resistance in the class of convex developable functions*, Math. Nachr. **226** (2001), 153–176. MR1839407 10.1002/1522-2616(200106)226:1<153::AID-MANA153>3.3.CO;2-U http://dx.doi.org/10.1002/1522-2616(200106)226:1<153::AID-MANA153>3.3.CO;2-U

[617] **O. A. Ladyzhenskaya and N. N. Ural′tseva**, *Linear and quasilinear elliptic equations*, Translated from the Russian by Scripta Technica, Inc. Translation editor: Leon Ehrenpreis, Academic Press, New York-London, 1968. MR0244627

[618] **H. Lamb**, *Hydrodynamics*, sixth ed., Cambridge Mathematical Library, Cambridge University Press, Cambridge, 1993, With a foreword by R. A. Caflisch [Russel E. Caflisch]. MR1317348

[619] **P. D. Lamberti and L. Provenzano**, *Viewing the Steklov eigenvalues of the Laplace operator as critical Neumann eigenvalues*. Current trends in analysis and its applications, 171–178, Trends Math., Birkhäuser/Springer, Cham, 2015.MR3496508

[620] **J. Lamboley**, *Variations autour de formes irrégulières optimales*, PhD Thesis, https://tel.archives-ouvertes.fr/tel-00346316/.

[621] **J. Lamboley**, *About Hölder-regularity of the convex shape minimizing λ_2*, Appl. Anal. **90** (2011), *no. 2*, 263–278. MR2754044 10.1080/00036811.2010.496361 http://dx.doi.org/10.1080/00036811.2010.496361

[622] **J. Lamboley and A. Novruzi**, *Polygons as optimal shapes with convexity constraint*, SIAM J. Control Optim. **48** (2009/10), *no. 5*, 3003–3025. MR2599908 10.1137/080738581 http://dx.doi.org/10.1137/080738581

[623] **J. Lamboley, A. Novruzi, and M. Pierre**, *Regularity and singularities of optimal convex shapes in the plane*, Arch. Ration. Mech. Anal. **205** (2012), *no. 1*, 311–343. MR2927624 10.1007/s00205-012-0514-7 http://dx.doi.org/10.1007/s00205-012-0514-7

[624] **J. Lamboley, A. Novruzi, and M. Pierre**, *Estimates of first and second order shape derivatives in nonsmooth multidimensional domains and applications*, J. Funct. Anal. **270** (2016), *no. 7*, 2616–2652. MR3464052 10.1016/j.jfa.2016.02.013 http://dx.doi.org/10.1016/j.jfa.2016.02.013

[625] **S. Lang**, *Complex analysis*, fourth ed., Graduate Texts in Mathematics, vol. 103, Springer-Verlag, New York, 1999. MR1659317 10.1007/978-1-4757-3083-8 http://dx.doi.org/10.1007/978-1-4757-3083-8

[626] **A. Laptev**, *Dirichlet and Neumann eigenvalue problems on domains in Euclidean spaces*, J. Funct. Anal. **151** (1997), *no. 2*, 531–545. MR1491551 10.1006/jfan.1997.3155 http://dx.doi.org/10.1006/jfan.1997.3155

[627] **A. Laptev and Y. Safarov**, *A generalization of the Berezin-Lieb inequality*, Contemporary mathematical physics, Amer. Math. Soc. Transl. Ser. 2, vol. 175, Amer. Math. Soc., Providence, RI, 1996, pp. 69–79. MR1402917 Zbl 0868.47016

[628] **A. Laptev**, *On the Lieb-Thirring conjecture for a class of potentials*, The Maz′ya anniversary collection, Vol. 2 (Rostock, 1998), Oper. Theory Adv. Appl., vol. 110, Birkhäuser, Basel, 1999, pp. 227–234. MR1747896

[629] **A. Laptev and T. Weidl**, *Recent results on Lieb-Thirring inequalities*, Journées "Équations aux Dérivées Partielles" (La Chapelle sur Erdre, 2000), Univ. Nantes, Nantes, 2000, pp. Exp. No. XX, 14. MR1775696

[630] **A. Laptev and T. Weidl**, *Sharp Lieb-Thirring inequalities in high dimensions*, Acta Math. **184** (2000), *no. 1*, 87–111. MR1756570 10.1007/BF02392782 http://dx.doi.org/10.1007/BF02392782

[631] **R. S. Laugesen**, *Spectral theory of partial differential equations*, University of Illinois at Urbana-Champaign, 2011.

[632] **R. S. Laugesen**, *Tight frames and rotations: sharp bounds on eigenvalues of the laplacian*, Proceedings of the AMSI International Conference on Harmonic Analysis and Applications (Macquarie University, February 2011), vol. 45, 2011, Edited by X. Duong, J. Hogan, C. Meaney, A. Sikora, pp. 63–82.

[633] **R. S. Laugesen, Z. C. Pan, and S. S. Son**, *Neumann eigenvalue sums on triangles are (mostly) minimal for equilaterals*, Math. Inequal. Appl. **15** (2012), *no. 2*, 381–394. MR2962241 10.7153/mia-15-32

[634] **R. S. Laugesen and B. A. Siudeja**, *Maximizing Neumann fundamental tones of triangles*, J. Math. Phys. **50** (2009), *no. 11*, 112903, 18. MR2567204 10.1063/1.3246834

[635] **R. S. Laugesen and B. A. Siudeja**, *Minimizing Neumann fundamental tones of triangles: an optimal Poincaré inequality*, J. Differential Equations **249** (2010), *no. 1*, 118–135. MR2644129 10.1016/j.jde.2010.02.020

[636] **R. S. Laugesen and B. A. Siudeja**, *Sums of Laplace eigenvalues—rotationally symmetric maximizers in the plane*, J. Funct. Anal. **260** (2011), *no. 6*, 1795–1823. MR2754893 10.1016/j.jfa.2010.12.018

[637] **R. S. Laugesen and B. A. Siudeja**, *Sums of Laplace eigenvalues: rotations and tight frames in higher dimensions*, J. Math. Phys. **52** (2011), *no. 9*, 093703, 13. MR2867829 10.1063/1.3635379

[638] **R. S. Laugesen, J. Liang, and A. Roy**, *Sums of magnetic eigenvalues are maximal on rotationally symmetric domains*, Ann. Henri Poincaré **13** (2012), *no. 4*, 731–750. MR2913619 10.1007/s00023-011-0142-z

[639] **R. S. Laugesen and B. A. Siudeja**, *Sharp spectral bounds on starlike domains*, J. Spectr. Theory **4** (2014), *no. 2*, 309–347. MR3232813 10.4171/JST/71

[640] **R. S. Laugesen and B. A. Siudeja**, *Magnetic spectral bounds on starlike plane domains*, ESAIM Control Optim. Calc. Var. **21** (2015), *no. 3*, 670–689. 10.1051/cocv/2014043 http://dx.doi.org/10.1051/cocv/2014043

[641] **R. S. Laugesen and C. Morpurgo**, *Extremals for eigenvalues of Laplacians under conformal mapping*, J. Funct. Anal. **155** (1998), *no. 1*, 64–108. MR1622804 10.1006/jfan.1997.3222 http://dx.doi.org/10.1006/jfan.1997.3222

[642] **R. S. Laugesen and B. A. Siudeja**, *Dirichlet eigenvalue sums on triangles are minimal for equilaterals*, Comm. Anal. Geom. **19** (2011), *no. 5*, 855–885. MR2886710 10.4310/CAG.2011.v19.n5.a2

[643] **R. Lavine**, *The eigenvalue gap for one-dimensional convex potentials*, Proc. Amer. Math. Soc. **121** (1994), *no. 3*, 815–821. MR1185270 10.2307/2160281

[644] **P. D. Lax and R. S. Phillips**, *On the scattering frequencies of the Laplace operator for exterior domains*, Comm. Pure Appl. Math. **25** (1972), 85–101. MR0296471

[645] **N. N. Lebedev**, *Special functions and their applications*, Dover Publications, Inc., New York, 1972, Revised edition, translated from the Russian and edited by Richard A. Silverman, Unabridged and corrected republication. MR0350075

[646] **J. M. Lee and G. Uhlmann**, *Determining anisotropic real-analytic conductivities by boundary measurements*, Comm. Pure Appl. Math. **42** (1989), *no. 8*, 1097–1112. MR1029119 10.1002/cpa.3160420804

[647] **C. Léna**, *Courant-sharp eigenvalues of the three-dimensional square torus.* Proc. Amer. Math. Soc. 144 (2016), *no. 9*, 3949–3958. MR3513551

[648] **C. Léna**, *Contributions à l'étude des partitions spectrales minimales*, Ph.D. thesis, Université Paris Sud 11, 2013.

[649] **C. Léna**, *Courant-sharp eigenvalues of a two-dimensional torus*, C. R. Math. Acad. Sci. Paris **353** (2015), *no. 6*, 535–539. MR3348988 10.1016/j.crma.2015.03.014

[650] **C. Léna**, *Eigenvalues variations for Aharonov-Bohm operators*, J. Math. Phys. **56** (2015), *no. 1*, 011502, 18. MR3390812 10.1063/1.4905647 http://dx.doi.org/10.1063/1.4905647

[651] **H. A. Levine and H. F. Weinberger**, *Inequalities between Dirichlet and Neumann eigenvalues*, Arch. Rational Mech. Anal. **94** (1986), *no. 3*, 193–208. MR846060 10.1007/BF00279862

[652] **M. Levitin and L. Parnovski**, *Commutators, spectral trace identities, and universal estimates for eigenvalues*, J. Funct. Anal. **192** (2002), *no. 2*, 425–445. MR1923409 10.1006/jfan.2001.3913 http://dx.doi.org/10.1006/jfan.2001.3913

[653] **M. Levitin and L. Parnovski**, *On the principal eigenvalue of a Robin problem with a large parameter*, Math. Nachr. **281** (2008), *no. 2*, 272–281. MR2387365 10.1002/mana.200510600 http://dx.doi.org/10.1002/mana.200510600

[654] **M. Levitin, L. Parnovski, and I. Polterovich**, *Isospectral domains with mixed boundary conditions*, J. Phys. A **39** (2006), *no. 9*, 2073–2082. MR2211977 10.1088/0305-4470/39/9/006

[655] **M. Levitin and R. Yagudin**, *Range of the first three eigenvalues of the planar Dirichlet Laplacian*, LMS J. Comput. Math. **6** (2003), 1–17 (electronic). MR1971489 10.1112/S1461157000000346 http://dx.doi.org/10.1112/S1461157000000346

[656] **J. Leydold**, *Knotenlinien und Knotengebiete von Eigenfunktionen*, Diplom Arbeit, Universität Wien, 1989.

[657] **J. Leydold**, *Nodal properties of spherical harmonics*, ProQuest LLC, Ann Arbor, MI, 1993, Thesis (Dr.natw.)–Universitaet Wien (Austria). MR2714918 http://gateway.proquest.com/openurl?url_ver=Z39.88-2004&rft_val_fmt=info:ofi/fmt:kev:mtx:dissertation&res_dat=xri:pqdiss&rft_dat=xri:pqdiss:C488577

[658] **J. Leydold**, *On the number of nodal domains of spherical harmonics*, Topology **35** (1996), *no. 2*, 301–321. MR1380499 10.1016/0040-9383(95)00028-3

[659] **P. Li and S. T. Yau**, *A new conformal invariant and its applications to the Willmore conjecture and the first eigenvalue of compact surfaces*, Invent. Math. **69** (1982), *no. 2*, 269–291. MR674407 10.1007/BF01399507

[660] **P. Li and S. T. Yau**, *On the Schrödinger equation and the eigenvalue problem*, Comm. Math. Phys. **88** (1983), *no. 3*, 309–318. MR701919 http://projecteuclid.org/euclid.cmp/1103922378

[661] **P. Li and S. T. Yau**, *On the Schrödinger equation and the eigenvalue problem*, Comm. Math. Phys. **88** (1983), *no. 3*, 309–318. MR701919 http://projecteuclid.org/euclid.cmp/1103922378

[662] **E. H. Lieb**, *The number of bound states of one-body Schroedinger operators and the Weyl problem*, Geometry of the Laplace operator (Proc. Sympos. Pure Math., Univ. Hawaii, Honolulu, Hawaii, 1979), Proc. Sympos. Pure Math., XXXVI, Amer. Math. Soc., Providence, R.I., 1980, pp. 241–252. MR573436

[663] **E. H. Lieb and M. Loss**, *Analysis*, second ed., Graduate Studies in Mathematics, vol. 14, American Mathematical Society, Providence, RI, 2001. MR1817225 10.1090/gsm/014 http://dx.doi.org/10.1090/gsm/014

[664] **C. S. Lin**, *On the second eigenfunctions of the Laplacian in* $\mathbf{R}^2$, Comm. Math. Phys. **111** (1987), *no. 2*, 161–166. MR899848 http://projecteuclid.org/euclid.cmp/1104159536

[665] **J.-L. Lions and E. Magenes**, *Problemi ai limiti non omogenei. I*, Ann. Scuola Norm. Sup. Pisa (3) **14** (1960), 269–308. MR0133583

[666] **J.-L. Lions and E. Magenes**, *Problèmes aux limites non homogènes. II*, Ann. Inst. Fourier (Grenoble) **11** (1961), 137–178. MR0146525

[667] **J.-L. Lions and E. Magenes**, *Problèmes aux limites non homogènes. IV*, Ann. Scuola Norm. Sup. Pisa (3) **15** (1961), 311–326. MR0140938

[668] **J.-L. Lions and E. Magenes**, *Problèmes aux limites non homogénes. VI*, J. Analyse Math. **11** (1963), 165–188. MR0161021

[669] **J. L. Lions and E. Magenes**, *Problèmes aux limites non homogènes. VII*, Ann. Mat. Pura Appl. (4) **63** (1963), 201–224. MR0170228

[670] **J.-L. Lions and E. Magenes**, *Problèmes aux limites non homogènes et applications. Vol. 1*, Travaux et Recherches Mathématiques, No. 17, Dunod, Paris, 1968. MR0247243

[671] **J.-L. Lions and E. Magenes**, *Problèmes aux limites non homogènes et applications. Vol. 2*, Travaux et Recherches Mathématiques, No. 18, Dunod, Paris, 1968. MR0247244

[672] **J.-L. Lions and E. Magenes**, *Problèmes aux limites non homogènes et applications. Vol. 3*, Dunod, Paris, 1970, Travaux et Recherches Mathématiques, No. 20. MR0291887

[673] **P.-L. Lions**, *The concentration-compactness principle in the calculus of variations. The locally compact case. I*, Ann. Inst. H. Poincaré Anal. Non Linéaire **1** (1984), *no. 2*, 109–145. MR778970 http://www.numdam.org/item?id=AIHPC_1984__1_2_109_0

[674] **P.-L. Lions, F. Pacella, and M. Tricarico**, *Best constants in Sobolev inequalities for functions vanishing on some part of the boundary and related questions*, Indiana Univ. Math. J. **37** (1988), *no. 2*, 301–324. MR963504 10.1512/iumj.1988.37.37015

[675] **G. Liu**, *Heat invariants of the perturbed polyharmonic Steklov problem*. ArXiv:1405.3350

[676] **X. Liu and S. Oishi**, *Verified eigenvalue evaluation for the Laplacian over polygonal domains of arbitrary shape*, SIAM J. Numer. Anal. **51** (2013), *no. 3*, 1634–1654. MR3061473 10.1137/120878446

[677] A. Logg, K.-A. Mardal, and G. N. Wells (eds.), *Automated solution of differential equations by the finite element method*, Lecture Notes in Computational Science and Engineering, vol. 84, Springer, Heidelberg, 2012, The FEniCS book. MR3075806 10.1007/978-3-642-23099-8 http://dx.doi.org/10.1007/978-3-642-23099-8

[678] **A. Logunov**, *Nodal sets of laplace eigenfunctions: proof of nadirashvili's conjecture and of the lower bound in yau's conjecture*. ArXiv:1605.02589

[679] **Y. Lou and M. Zhu**, *A singularly perturbed linear eigenvalue problem in C^1 domains*, Pacific J. Math. **214** (2004), *no. 2*, 323–334. MR2042936 10.2140/pjm.2004.214.323 http://dx.doi.org/10.2140/pjm.2004.214.323

[680] **Z. Lu and J. Rowlett**, *The fundamental gap of simplices*, Comm. Math. Phys. **319** (2013), *no. 1*, 111–145. MR3034027 10.1007/s00220-013-1670-9

[681] **K. A. Lurie and A. V. Cherkaev**, *Effective characteristics of composite materials and the optimal design of structural elements [partial translation of MR0885713 (88e:73019)]*, Topics in the mathematical modelling of composite materials, Progr. Nonlinear Differential Equations Appl., vol. 31, Birkhäuser Boston, Boston, MA, 1997, pp. 175–258. MR1493043

[682] **J. M. Luttinger**, *Generalized isoperimetric inequalities*, J. Mathematical Phys. **14** (1973), 586–593. MR0337197

[683] **F. Maggi and C. Villani**, *Balls have the worst best Sobolev inequalities. II. Variants and extensions*, Calc. Var. Partial Differential Equations **31** (2008), *no. 1*, 47–74. MR2342614 10.1007/s00526-007-0105-x http://dx.doi.org/10.1007/s00526-007-0105-x

[684] **F. Maggi and C. Villani**, *Balls have the worst best Sobolev inequalities*, J. Geom. Anal. **15** (2005), *no. 1*, 83–121. MR2132267 10.1007/BF02921860 http://dx.doi.org/10.1007/BF02921860

[685] **T. J. Mahar and B. E. Willner**, *An extremal eigenvalue problem*, Comm. Pure Appl. Math. **29** (1976), *no. 5*, 517–529. MR0425244

[686] **T. J. Mahar and B. E. Willner**, *Sturm-Liouville eigenvalue problems in which the squares of the eigenfunctions are linearly dependent*, Comm. Pure Appl. Math. **33** (1980), *no. 4*, 567–578. MR575738 10.1002/cpa.3160330406

[687] **E. Makai**, *Bounds for the principal frequency of a membrane and the torsional rigidity of a beam*, Acta Sci. Math. Szeged **20** (1959), 33–35. MR0105222

[688] **E. Makai**, *On the principal frequency of a membrane and the torsional rigidity of a beam*, Studies in mathematical analysis and related topics, Stanford Univ. Press, Stanford, Calif., 1962, pp. 227–231. MR0167004

[689] **E. Makai**, *A proof of Saint-Venant's theorem on torsional rigidity*, Acta Math. Acad. Sci. Hungar **17** (1966), 419–422. MR0202053

[690] **E. Malkovich and V. Sharafutdinov**, *Zeta-invariants of the Steklov spectrum for a planar domain*. ArXiv:1404.2117

[691] **P. Marcellini**, *Bounds for the third membrane eigenvalue*, J. Differential Equations **37** (1980), *no. 3*, 438–443. MR590001 10.1016/0022-0396(80)90108-4 http://dx.doi.org/10.1016/0022-0396(80)90108-4

[692] **A. W. Marshall, I. Olkin, and B. C. Arnold**, *Inequalities: theory of majorization and its applications*, second ed., Springer Series in Statistics, Springer, New York, 2011. MR2759813 10.1007/978-0-387-68276-1

[693] **U. Massari and M. Miranda**, *Minimal surfaces of codimension one*, North-Holland Mathematics Studies, vol. 91, North-Holland Publishing Co., Amsterdam, 1984, Notas de Matemática [Mathematical Notes], 95. MR795963

[694] **E. F. Masur**, *Optimal structural design under multiple eigenvalue constraints*, Internat. J. Solids Structures **20** (1984), *no. 3*, 211–231. MR735699 10.1016/0020-7683(84)90034-9

[695] **J. Mathews and R. L. Walker**, *Mathematical methods of physics*, Second edition, W. A. Benjamin, Inc., New York, 1970. MR0261817

[696] **V. G. Maz'ja**, *Sobolev spaces*, Springer Series in Soviet Mathematics, Springer-Verlag, Berlin, 1985, Translated from the Russian by T. O. Shaposhnikova. MR817985 10.1007/978-3-662-09922-3

[697] **W. G. Mazja, S. A. Nasarow, and B. A. Plamenewski**, *Asymptotische Theorie elliptischer Randwertaufgaben in singulär gestörten Gebieten. I*, Mathematische Lehrbücher und Monographien, II. Abteilung: Mathematische Monographien [Mathematical Textbooks and Monographs, Part II: Mathematical Monographs], vol. 82, Akademie-Verlag, Berlin, 1991, Störungen isolierter Randsingularitäten. [Perturbations of isolated boundary singularities]. MR1101139

[698] **R. Mazzeo and J. Rowlett**, *A heat trace anomaly on polygons*, Math. Proc. Cambridge Philos. Soc. **159** (2015), *no. 2*, 303–319. MR3395373 10.1017/S0305004115000365 http://dx.doi.org/10.1017/S0305004115000365

[699] **D. Mazzoleni and D. Zucco**, *Convex combinations of low eigenvalues, fraenkel asymmetries and attainable sets*, (2015). ArXiv:1512.00993

[700] **D. Mazzoleni and A. Pratelli**, *Existence of minimizers for spectral problems*, J. Math. Pures Appl. (9) **100** (2013), *no. 3*, 433–453. MR3095209 10.1016/j.matpur.2013.01.008 http://dx.doi.org/10.1016/j.matpur.2013.01.008

[701] **B. J. McCartin**, *Laplacian eigenstructure of the equilateral triangle*, Hikari Ltd., Ruse, 2011. MR2918422

[702] **K. P. McHale and U. Ufuktepe**, *Inequalities for the vibrating clamped plate problem*, Turkish J. Math. **25** (2001), *no. 2*, 283–298. MR1840177

[703] **K. P. P. Mchale**, *Inequalities for vibration and buckling of a clamped plate*, ProQuest LLC, Ann Arbor, MI, 1997, Thesis (Ph.D.)–University of Missouri - Columbia. MR2697875 http://gateway.proquest.com/openurl?url_ver=Z39.88-2004&rft_val_fmt=info:ofi/fmt:kev:mtx:dissertation&res_dat=xri:pqdiss&rft_dat=xri:pqdiss:9842551

[704] **W. McLean**, *Strongly elliptic systems and boundary integral equations*, Cambridge University Press, Cambridge, 2000. MR1742312

[705] **A. D. Melas**, *On the nodal line of the second eigenfunction of the Laplacian in* $\mathbf{R}^2$, J. Differential Geom. **35** (1992), *no. 1*, 255–263. MR1152231 http://projecteuclid.org/euclid.jdg/1214447811

[706] **A. D. Melas**, *The stability of some eigenvalue estimates*, J. Differential Geom. **36** (1992), *no. 1*, 19–33. MR1168980 http://projecteuclid.org/euclid.jdg/1214448441

[707] **A. D. Melas**, *A lower bound for sums of eigenvalues of the Laplacian*, Proc. Amer. Math. Soc. **131** (2003), *no. 2*, 631–636 (electronic). MR1933356 10.1090/S0002-9939-02-06834-X http://dx.doi.org/10.1090/S0002-9939-02-06834-X

[708] **Y. Miyamoto**, *The "hot spots" conjecture for a certain class of planar convex domains*, J. Math. Phys. **50** (2009), *no. 10*, 103530, 7. MR2572703 10.1063/1.3251335
[709] **Y. Miyamoto**, *A planar convex domain with many isolated "hot spots" on the boundary*, Jpn. J. Ind. Appl. Math. **30** (2013), *no. 1*, 145–164. MR3022811 10.1007/s13160-012-0091-z
[710] **E. Mohr**, *Über die Rayleighsche Vermutung: unter allen Platten von gegebener Fläche und konstanter Dichte und Elastizität hat die kreisförmige den tiefsten Grundton*, Ann. Mat. Pura Appl. (4) **104** (1975), 85–122. MR0381462
[711] **C. B. Morrey, Jr.**, *Multiple integrals in the calculus of variations*, Die Grundlehren der mathematischen Wissenschaften, Band 130, Springer-Verlag New York, Inc., New York, 1966. MR0202511
[712] **J. Mossino**, *Inégalités isopérimétriques et applications en physique*, Travaux en Cours. [Works in Progress], Hermann, Paris, 1984. MR733257
[713] **C. Müller**, *Analysis of spherical symmetries in Euclidean spaces*, Applied Mathematical Sciences, vol. 129, Springer-Verlag, New York, 1998. MR1483320 10.1007/978-1-4612-0581-4 http://dx.doi.org/10.1007/978-1-4612-0581-4
[714] **F. Murat and L. Tartar**, *Calcul des variations et homogénéisation*, Homogenization methods: theory and applications in physics (Bréau-sans-Nappe, 1983), Collect. Dir. Études Rech. Élec. France, vol. 57, Eyrolles, Paris, 1985, pp. 319–369. MR844873
[715] **M. K. V. Murthy and G. Stampacchia**, *Boundary value problems for some degenerate-elliptic operators*, Ann. Mat. Pura Appl. (4) **80** (1968), 1–122. MR0249828
[716] **N. S. Nadirashvili**, *Multiplicity of eigenvalues of the Neumann problem*, Dokl. Akad. Nauk SSSR **286** (1986), *no. 6*, 1303–1305. MR830292
[717] **N. S. Nadirashvili**, *New isoperimetric inequalities in mathematical physics*, Partial differential equations of elliptic type (Cortona, 1992), Sympos. Math., XXXV, Cambridge Univ. Press, Cambridge, 1994, pp. 197–203. MR1297780
[718] **N. S. Nadirashvili**, *Rayleigh's conjecture on the principal frequency of the clamped plate*, Arch. Rational Mech. Anal. **129** (1995), *no. 1*, 1–10. MR1328469 10.1007/BF00375124
[719] **N. S. Nadirashvili**, *Conformal maps and isoperimetric inequalities for eigenvalues of the Neumann problem*, Proceedings of the Ashkelon Workshop on Complex Function Theory (1996), Israel Math. Conf. Proc., vol. 11, Bar-Ilan Univ., Ramat Gan, 1997, pp. 197–201. MR1476715
[720] **N. Nadirashvili**, *Berger's isoperimetric problem and minimal immersions of surfaces*, Geom. Funct. Anal. **6** (1996), *no. 5*, 877–897. MR1415764 10.1007/BF02246788
[721] **S. A. Nazarov and J. Taskinen**, *On the spectrum of the Steklov problem in a domain with a peak*, Vestnik St. Petersburg Univ. Math. **41** (2008), *no. 1*, 45–52. MR2406898 10.3103/S1063454108010081 http://dx.doi.org/10.3103/S1063454108010081
[722] **S. A. Nazarov and J. Taskinen**, *On the spectrum of the third boundary value problem in a domain with a peak*, Funktsional. Anal. i Prilozhen. **45** (2011), *no. 1*, 93–96. MR2848745 10.1007/s10688-011-0010-0 http://dx.doi.org/10.1007/s10688-011-0010-0
[723] **S. A. Nazarov and J. Taskinen**, *Spectral anomalies of the Robin Laplacian in non-Lipschitz domains*, J. Math. Sci. Univ. Tokyo **20** (2013), *no. 1*, 27–90. MR3112086
[724] **S. A. Nazarov and J. Sokołowski**, *Asymptotic analysis of shape functionals*, J. Math. Pures Appl. (9) **82** (2003), *no. 2*, 125–196. MR1976204 10.1016/S0021-7824(03)00004-7
[725] **J. Nečas**, *Direct methods in the theory of elliptic equations*, Springer Monographs in Mathematics, Springer, Heidelberg, 2012, Translated from the 1967 French original by Gerard Tronel and Alois Kufner, Editorial coordination and preface by Šárka Nečasová and a contribution by Christian G. Simader. MR3014461 10.1007/978-3-642-10455-8
[726] **N. Nigam, B. A. Siudeja, and B. Young**, *Nearly radial neumann modes on symmetric domains*, 2015. ArXiv:1508.07019

[727] **C. Nitsch**, *On the first Dirichlet Laplacian eigenvalue of regular polygons*, Kodai Math. J. **37** (2014), *no. 3*, 595–607. MR3273886 10.2996/kmj/1414674611

[728] **B. Noris, M. Nys, and S. Terracini**, *On the eigenvalues of Aharonov-Bohm operators with varying poles: pole approaching the boundary of the domain*. ArXiv:1411.5244

[729] **B. Noris and S. Terracini**, *Nodal sets of magnetic Schrödinger operators of Aharonov-Bohm type and energy minimizing partitions*, Indiana Univ. Math. J. **59** (2010), *no. 4*, 1361–1403. MR2815036 10.1512/iumj.2010.59.3964

[730] **P. Nowosad**, *Isoperimetric eigenvalue problems in algebras*, Comm. Pure Appl. Math. **21** (1968), 401–465. MR0238087

[731] **R. Osserman**, *The isoperimetric inequality*, Bull. Amer. Math. Soc. **84** (1978), *no. 6*, 1182–1238. MR0500557

[732] **R. Osserman**, *Bonnesen-style isoperimetric inequalities*, Amer. Math. Monthly **86** (1979), *no. 1*, 1–29. MR519520 10.2307/2320297 http://dx.doi.org/10.2307/2320297

[733] **R. Osserman**, *Isoperimetric inequalities and eigenvalues of the Laplacian*, Proceedings of the International Congress of Mathematicians (Helsinki, 1978), Acad. Sci. Fennica, Helsinki, 1980, pp. 435–442. MR562638

[734] **B. Osting**, *Optimization of spectral functions of Dirichlet-Laplacian eigenvalues*, J. Comput. Phys. **229** (2010), *no. 22*, 8578–8590. MR2719190 10.1016/j.jcp.2010.07.040 http://dx.doi.org/10.1016/j.jcp.2010.07.040

[735] **B. Osting and C.-Y. Kao**, *Minimal convex combinations of sequential Laplace-Dirichlet eigenvalues*, SIAM J. Sci. Comput. **35** (2013), *no. 3*, B731–B750. MR3070538 10.1137/120881865 http://dx.doi.org/10.1137/120881865

[736] **B. Osting and C.-Y. Kao**, *Minimal convex combinations of three sequential Laplace-Dirichlet eigenvalues*, Appl. Math. Optim. **69** (2014), *no. 1*, 123–139. MR3162497 10.1007/s00245-013-9219-z http://dx.doi.org/10.1007/s00245-013-9219-z

[737] **É. Oudet**, *Numerical minimization of eigenmodes of a membrane with respect to the domain*, ESAIM Control Optim. Calc. Var. **10** (2004), *no. 3*, 315–330 (electronic). MR2084326 10.1051/cocv:2004011 http://dx.doi.org/10.1051/cocv:2004011

[738] **S. Ozawa**, *Singular variation of domains and eigenvalues of the Laplacian*, Duke Math. J. **48** (1981), *no. 4*, 767–778. MR782576 10.1215/S0012-7094-81-04842-0

[739] **K. Pankrashkin**, *On the asymptotics of the principal eigenvalue for a robin problem with a large parameter in planar domains*, Nanosystems: Physics, Chemistry, Mathematics **4** (2013), 474–483.

[740] **K. Pankrashkin and N. Popoff**, *Mean curvature bounds and eigenvalues of Robin Laplacians*, Calc. Var. Partial Differential Equations **54** (2015), *no. 2*, 1947–1961. MR3396438 10.1007/s00526-015-0850-1 http://dx.doi.org/10.1007/s00526-015-0850-1

[741] **O. Parzanchevski**, *On G-sets and isospectrality*, Ann. Inst. Fourier (Grenoble) **63** (2013), *no. 6*, 2307–2329. MR3237449 http://aif.cedram.org/item?id=AIF_2013__63_6_2307_0

[742] **M. N. Pascu**, *Scaling coupling of reflecting Brownian motions and the hot spots problem*, Trans. Amer. Math. Soc. **354** (2002), *no. 11*, 4681–4702 (electronic). MR1926894 10.1090/S0002-9947-02-03020-9

[743] **L. E. Payne**, *Inequalities for eigenvalues of membranes and plates*, J. Rational Mech. Anal. **4** (1955), 517–529. MR0070834

[744] **L. E. Payne**, *Isoperimetric inequalities and their applications*, SIAM Rev. **9** (1967), 453–488. MR0218975

[745] **L. E. Payne**, *Some isoperimetric inequalities for harmonic functions*, SIAM J. Math. Anal. **1** (1970), 354–359. MR0437782

[746] **L. E. Payne, G. Pólya, and H. F. Weinberger**, *On the ratio of consecutive eigenvalues*, J. Math. and Phys. **35** (1956), 289–298. MR0084696

[747] **L. E. Payne and P. W. Schaefer**, *Eigenvalue and eigenfunction inequalities for the elastically supported membrane*, Z. Angew. Math. Phys. **52** (2001), *no. 5*, 888–895. MR1872006 10.1007/PL00001580

[748] **L. E. Payne and H. F. Weinberger**, *Lower bounds for vibration frequencies of elastically supported membranes and plates*, J. Soc. Indust. Appl. Math. **5** (1957), 171–182. MR0092431

[749] **L. E. Payne and H. F. Weinberger**, *An optimal Poincaré inequality for convex domains*, Arch. Rational Mech. Anal. **5** (1960), 286–292 (1960). MR0117419

[750] **L. E. Payne and H. F. Weinberger**, *Some isoperimetric inequalities for membrane frequencies and torsional rigidity*, J. Math. Anal. Appl. **2** (1961), 210–216. MR0149735

[751] **L. E. Payne, G. Pólya, and H. F. Weinberger**, *Sur le quotient de deux fréquences propres consécutives*, C. R. Acad. Sci. Paris **241** (1955), 917–919. MR0073046

[752] **L. E. Payne**, *On two conjectures in the fixed membrane eigenvalue problem*, Z. Angew. Math. Phys. **24** (1973), 721–729. MR0333487

[753] **L. E. Payne**, *Some comments on the past fifty years of isoperimetric inequalities*, Inequalities (Birmingham, 1987), Lecture Notes in Pure and Appl. Math., vol. 129, Dekker, New York, 1991, pp. 143–161. MR1112577

[754] **J. Peetre**, *A generalization of Courant's nodal domain theorem*, Math. Scand. **5** (1957), 15–20. MR0092917

[755] **R. Petrides**, *Bornes sur des valeurs propres et métriques extrémales*, Ph.D. thesis, Université de Lyon, 2015.

[756] **R. Petrides**, *Existence and regularity of maximal metrics for the first Laplace eigenvalue on surfaces*, Geom. Funct. Anal. **24** (2014), *no. 4*, 1336–1376. MR3248488 10.1007/s00039-014-0292-5

[757] **G. A. Philippin**, *Some remarks on the elastically supported membrane*, Z. Angew. Math. Phys. **29** (1978), *no. 2*, 306–314. MR0483914

[758] **E. Picard**, *Mémoire sur la théorie des équations aux dérivées partielles et la méthode des approximations successives*, J. Math. Pures Appl. **6** (1890), 145–210. http://sites.mathdoc.fr/JMPA/PDF/JMPA_1890_4_6_A3_0.pdf

[759] **M. A. Pinsky**, *The eigenvalues of an equilateral triangle*, SIAM J. Math. Anal. **11** (1980), *no. 5*, 819–827. MR586910 10.1137/0511073

[760] **O. Pironneau**, *Optimal shape design for elliptic systems*, Springer Series in Computational Physics, Springer-Verlag, New York, 1984. MR725856 10.1007/978-3-642-87722-3

[761] **Å. Pleijel**, *Remarks on Courant's nodal line theorem*, Comm. Pure Appl. Math. **9** (1956), 543–550. MR0080861

[762] **F. Pockels**, *über die partielle Differentialgleichung* $-\delta u - k^2 u = 0$ *and deren Auftreten in mathematischen Physik*, Historical Math. Monographs, Cornell University, 2013.

[763] **I. Polterovich**, *Pleijel's nodal domain theorem for free membranes*, Proc. Amer. Math. Soc. **137** (2009), *no. 3*, 1021–1024. MR2457442 10.1090/S0002-9939-08-09596-8

[764] **I. Polterovich and D. A. Sher**, *Heat invariants of the Steklov problem*, J. Geom. Anal. **25** (2015), *no. 2*, 924–950. MR3319956 10.1007/s12220-013-9451-4

[765] **I. Polterovich, D. A. Sher, and J. A. Toth**, *Nodal length of Steklov eigenfunctions on real-analytic Riemannian surfaces*. ArXiv:1506.07600

[766] **G. Pólya**, *On the characteristic frequencies of a symmetric membrane*, Math. Z. **63** (1955), 331–337. MR0073047

[767] **G. Pólya**, *Two more inequalities between physical and geometrical quantities*, J. Indian Math. Soc. (N.S.) **24** (1960), 413–419 (1961). MR0133059

[768] **G. Pólya**, *On the eigenvalues of vibrating membranes*, Proc. London Math. Soc. (3) **11** (1961), 419–433. MR0129219

[769] **G. Pólya and M. Schiffer**, *Convexity of functionals by transplantation*, J. Analyse Math. **3** (1954), 245–346, With an appendix by Heinz Helfenstein. MR0066530

[770] **G. Pólya and G. Szegö**, *Isoperimetric Inequalities in Mathematical Physics*, Annals of Mathematics Studies, no. 27, Princeton University Press, Princeton, N. J., 1951. MR0043486

[771] **G. Pólya**, *Collected papers. Vol. III*, Mathematicians of Our Time, vol. 21, MIT Press, Cambridge, MA, 1984, Analysis, Edited by Joseph Hersch, Gian-Carlo Rota, M. C. Reynolds and R. M. Shortt. MR758989

[772] **G. Pólya**, *Sur le rôle des domaines symétriques dans le calcul de certaines grandeurs physiques*, C. R. Acad. Sci. Paris **235** (1952), 1079–1081. MR0052477

[773] **C. Pommerenke**, *Boundary behaviour of conformal maps*, Grundlehren der Mathematischen Wissenschaften [Fundamental Principles of Mathematical Sciences], vol. 299, Springer-Verlag, Berlin, 1992. MR1217706 10.1007/978-3-662-02770-7 http://dx.doi.org/10.1007/978-3-662-02770-7

[774] **R. Ponge**, *Generic multiplicity of eigenvalues and resonances*, In preparation.

[775] **T. Povel**, *Confinement of Brownian motion among Poissonian obstacles in* $\mathbf{R}^d$, $d \geq 3$, Probab. Theory Related Fields **114** (1999), *no. 2*, 177–205. MR1701519 10.1007/s440-1999-8036-0 http://dx.doi.org/10.1007/s440-1999-8036-0

[776] **M. H. Protter**, *Can one hear the shape of a drum? revisited*, SIAM Rev. **29** (1987), *no. 2*, 185–197. MR889243 10.1137/1029041 http://dx.doi.org/10.1137/1029041

[777] **M. H. Protter**, *The asymptotics of eigenvalues*, Maximum principles and eigenvalue problems in partial differential equations (Knoxville, TN, 1987), Pitman Res. Notes Math. Ser., vol. 175, Longman Sci. Tech., Harlow, 1988, pp. 178–188. MR963467

[778] **M. H. Protter**, *Universal inequalities for eigenvalues*, Maximum principles and eigenvalue problems in partial differential equations (Knoxville, TN, 1987), Pitman Res. Notes Math. Ser., vol. 175, Longman Sci. Tech., Harlow, 1988, pp. 111–120. MR963462

[779] **P. Pucci and J. Serrin**, *A general variational identity*, Indiana Univ. Math. J. **35** (1986), *no. 3*, 681–703. MR855181 10.1512/iumj.1986.35.35036

[780] **R. Pütter**, *On the nodal lines of second eigenfunctions of the fixed membrane problem*, Comment. Math. Helv. **65** (1990), *no. 1*, 96–103. MR1036131 10.1007/BF02566596

[781] **A. G. Ramm and P. N. Shivakumar**, *Inequalities for the minimal eigenvalue of the Laplacian in an annulus*, Math. Inequal. Appl. **1** (1998), *no. 4*, 559–563. MR1646670 10.7153/mia-01-54

[782] **M. Ramos, H. Tavares, and S. Terracini**, *Existence and regularity of solutions to optimal partition problems involving Laplacian eigenvalues*. ArXiv:1403.6313

[783] **T. M. Rassias**, *The isoperimetric inequality and eigenvalues of the Laplacian*, Constantin Carathéodory: an international tribute, Vol. I, II, World Sci. Publ., Teaneck, NJ, 1991, pp. 1146–1163. MR1130882

[784] **J. Rauch**, *Five problems: an introduction to the qualitative theory of partial differential equations*, Partial differential equations and related topics (Program, Tulane Univ., New Orleans, La., 1974), Springer, Berlin, 1975, pp. 355–369. Lecture Notes in Math., Vol. 446. MR0509045

[785] **S. Raulot and A. Savo**, *On the first eigenvalue of the Dirichlet-to-Neumann operator on forms*, J. Funct. Anal. **262** (2012), *no. 3*, 889–914. MR2863852 10.1016/j.jfa.2011.10.008 http://dx.doi.org/10.1016/j.jfa.2011.10.008

[786] **S. Raulot and A. Savo**, *On the spectrum of the Dirichlet-to-Neumann operator acting on forms of a Euclidean domain*, J. Geom. Phys. **77** (2014), 1–12. MR3157898 10.1016/j.geomphys.2013.11.002 http://dx.doi.org/10.1016/j.geomphys.2013.11.002

[787] **J. W. S. L. Rayleigh**, *The theory of sound*, 1877.

[788] **J. W. S. Rayleigh, Baron**, *The Theory of Sound*, Dover Publications, New York, N. Y., 1945, 2d ed. MR0016009

[789] **M. Reed and B. Simon**, *Methods of modern mathematical physics. IV. Analysis of operators*, Academic Press [Harcourt Brace Jovanovich, Publishers], New York-London, 1978. MR0493421

[790] **X. Ren and J. Wei**, *On a two-dimensional elliptic problem with large exponent in nonlinearity*, Trans. Amer. Math. Soc. **343** (1994), *no. 2*, 749–763. MR1232190 10.2307/2154740 http://dx.doi.org/10.2307/2154740

[791] **J. Rohleder**, *Strict inequality of Robin eigenvalues for elliptic differential operators on Lipschitz domains*, J. Math. Anal. Appl. **418** (2014), *no. 2*, 978–984. 3206692 10.1016/j.jmaa.2014.04.036

[792] **B. Rousselet**, *Shape design sensitivity of a membrane*, J. Optim. Theory Appl. **40** (1983), *no. 4*, 595–623. MR717179 10.1007/BF00933973

[793] **J. Royo-Letelier**, *Segregation and symmetry breaking of strongly coupled two-component Bose-Einstein condensates in a harmonic trap*, Calc. Var. Partial Differential Equations **49** (2014), *no. 1-2*, 103–124. MR3148108 10.1007/s00526-012-0571-7

[794] **G. Rozenbljum**, *Asymptotic behavior of the eigenvalues for some two-dimensional spectral problems*, Selected translations, Birkhäuser Boston, Inc., Secaucus, NJ, 1986, Selecta Math. Soviet. 5 (1986), no. 3, pp. i–ii and 201–306. MR865338 (87h:00034)

[795] **Y. Safarov**, *Lower bounds for the generalized counting function*, The Maz′ya anniversary collection, Vol. 2 (Rostock, 1998), Oper. Theory Adv. Appl., vol. 110, Birkhäuser, Basel, 1999, pp. 275–293. MR1747899

[796] **Y. G. Safarov**, *Riesz means of a function of the distribution of the eigenvalues of an elliptic operator*, Zap. Nauchn. Sem. Leningrad. Otdel. Mat. Inst. Steklov. (LOMI) **163** (1987), *no. Kraev. Zadachi Mat. Fiz. i Smezhn. Vopr. Teor. Funktsii 19*, 143–145, 189–190. MR918948 10.1007/BF02208718 http://dx.doi.org/10.1007/BF02208718

[797] **J. J. Sakurai**, *Modern quantum mechanics (revised edition)*, Addison-Wesley, 1993.

[798] **A. Savo**, *Lower bounds for the nodal length of eigenfunctions of the Laplacian*, Ann. Global Anal. Geom. **19** (2001), *no. 2*, 133–151. MR1826398 10.1023/A:1010774905973

[799] **H. H. Schaefer and M. P. Wolff**, *Topological vector spaces*, second ed., Graduate Texts in Mathematics, vol. 3, Springer-Verlag, New York, 1999. MR1741419 10.1007/978-1-4612-1468-7

[800] **M. Schiffer**, *Hadamard's formula and variation of domain-functions*, Amer. J. Math. **68** (1946), 417–448. MR0018750

[801] **R. Schneider**, *Convex bodies: the Brunn-Minkowski theory*, expanded ed., Encyclopedia of Mathematics and its Applications, vol. 151, Cambridge University Press, Cambridge, 2014. MR3155183

[802] **R. Schoen and S.-T. Yau**, *Lectures on differential geometry*, Conference Proceedings and Lecture Notes in Geometry and Topology, I, International Press, Cambridge, MA, 1994, Lecture notes prepared by Wei Yue Ding, Kung Ching Chang [Gong Qing Zhang], Jia Qing Zhong and Yi Chao Xu. MR1333601

[803] **B. Schwarz**, *On the extrema of the frequencies of nonhomogeneous strings with equimeasurable density*, J. Math. Mech. **10** (1961), 401–422. MR0125271

[804] **B. Schwarz**, *Some results on the frequencies of nonhomogeneous rods*, J. Math. Anal. Appl. **5** (1962), 169–175. MR0148986

[805] **J. Serrin**, *A symmetry problem in potential theory*, Arch. Rational Mech. Anal. **43** (1971), 304–318. MR0333220

[806] **A. P. Seyranian and O. G. Privalova**, *The Lagrange problem on an optimal column: old and new results*, Struct. Multidiscip. Optim. **25** (2003), *no. 5-6*, 393–410, Special issue to honour Niels Olhoff (Nyborg, 2001). MR2038925 10.1007/s00158-003-0333-4

[807] **S. E. Shamma**, *Asymptotic behavior of Stekloff eigenvalues and eigenfunctions*, SIAM J. Appl. Math. **20** (1971), 482–490. MR0306697

[808] **A. B. Shidlovskii**, *Transcendental numbers*, de Gruyter Studies in Mathematics, vol. 12, Walter de Gruyter & Co., Berlin, 1989, Translated from the Russian by Neal Koblitz, With a foreword by W. Dale Brownawell. MR1033015 10.1515/9783110889055

[809] **N. Shimakura**, *La première valeur propre du laplacien pour le problème de Dirichlet*, J. Math. Pures Appl. (9) **62** (1983), *no. 2*, 129–152. MR713393

[810] **C. L. Siegel**, *Gesammelte Abhandlungen. Bände I, II, III*, Herausgegeben von K. Chandrasekharan und H. Maass, Springer-Verlag, Berlin-New York, 1966. MR0197270

[811] **J. Simon**, *Differentiation with respect to the domain in boundary value problems*, Numer. Funct. Anal. Optim. **2** (1980), *no. 7-8*, 649–687 (1981). MR619172 10.1080/01630563.1980.10120631

[812] **I. M. Singer, B. Wong, S.-T. Yau, and S. S.-T. Yau**, *An estimate of the gap of the first two eigenvalues in the Schrödinger operator*, Ann. Scuola Norm. Sup. Pisa Cl. Sci. (4) **12** (1985), *no. 2*, 319–333. MR829055 http://www.numdam.org/item?id=ASNSP_1985_4_12_2_319_0

[813] **B. A. Siudeja**, *Eigenvalue software*. http://github.com/siudej/Eigenvalues

[814] **B. A. Siudeja**, *Githhub repositories*. http://github.com/siudej

[815] **B. A. Siudeja**, *Nearly radial Neumann modes on symmetric domains*, J. Spectr. Theory, to appear.

[816] **B. A. Siudeja**, *Generalized tight p-frames and spectral bounds for Laplace-like operators*, Appl. Comput. Harmon. Anal., **42** (2017), *no. 2*, 167–198. MR3591237

[817] **B. A. Siudeja**, *Sharp bounds for eigenvalues of triangles*, Michigan Math. J. **55** (2007), *no. 2*, 243–254. MR2369934 10.1307/mmj/1187646992

[818] **B. A. Siudeja**, *Isoperimetric inequalities for eigenvalues of triangles*, Indiana Univ. Math. J. **59** (2010), *no. 3*, 1097–1120. MR2779073 10.1512/iumj.2010.59.3744

[819] **B. A. Siudeja**, *Hot spots conjecture for a class of acute triangles*, Math. Z. **280** (2015), *no. 3-4*, 783–806. MR3369351 10.1007/s00209-015-1448-1 http://dx.doi.org/10.1007/s00209-015-1448-1

[820] **B. A. Siudeja**, *On mixed Dirichlet-Neumann eigenvalues of triangles*, Proc. Amer. Math. Soc. **144** (2016), *no. 6*, 2479–2493. MR3477064 10.1090/proc/12888 http://dx.doi.org/10.1090/proc/12888

[821] **N. Soave and S. Terracini**, *Liouville theorems and 1-dimensional symmetry for solutions of an elliptic system modelling phase separation*, Adv. Math. **279** (2015), 29–66. MR3345178 10.1016/j.aim.2015.03.015

[822] **C. D. Sogge and S. Zelditch**, *Lower bounds on the Hausdorff measure of nodal sets*, Math. Res. Lett. **18** (2011), *no. 1*, 25–37. MR2770580 10.4310/MRL.2011.v18.n1.a3

[823] **J. Sokołowski and J.-P. Zolésio**, *Introduction to shape optimization*, Springer Series in Computational Mathematics, vol. 16, Springer-Verlag, Berlin, 1992, Shape sensitivity analysis. MR1215733 10.1007/978-3-642-58106-9

[824] **A. Y. Solynin**, *Continuous symmetrization of sets*, Zap. Nauchn. Sem. Leningrad. Otdel. Mat. Inst. Steklov. (LOMI) **185** (1990), *no. Anal. Teor. Chisel i Teor. Funktsii. 10*, 125–139, 186. MR1097593 10.1007/BF01374083

[825] **A. Y. Solynin**, *Isoperimetric inequalities for polygons and dissymetrization*, Algebra i Analiz **4** (1992), *no. 2*, 210–234. MR1182401

[826] **A. Y. Solynin**, *Polarization and functional inequalities*, Algebra i Analiz **8** (1996), *no. 6*, 148–185. MR1458141

[827] **A. Y. Solynin**, *Continuous symmetrization via polarization*, Algebra i Analiz **24** (2012), *no. 1*, 157–222. MR3013297 10.1090/S1061-0022-2012-01234-3

[828] **A. Y. Solynin and V. A. Zalgaller**, *The inradius, the first eigenvalue, and the torsional rigidity of curvilinear polygons*, Bull. Lond. Math. Soc. **42** (2010), *no. 5*, 765–783. MR2721739 10.1112/blms/bdq028

[829] **R. P. Sperb**, *Untere und obere Schranken für den tiefsten Eigenwert der elastisch gestützten Membran*, Z. Angew. Math. Phys. **23** (1972), 231–244. MR0312800

[830] **R. Sperb**, *Bounds for the first eigenvalue of the elastically supported membrane on convex domains*, Z. Angew. Math. Phys. **54** (2003), *no. 5*, 879–903, Special issue dedicated to Lawrence E. Payne. MR2019188 10.1007/s00033-003-3212-3 http://dx.doi.org/10.1007/s00033-003-3212-3

[831] **S. Steinerberger**, *A geometric uncertainty principle with an application to Pleijel's estimate*, Ann. Henri Poincaré **15** (2014), *no. 12*, 2299–2319. MR3272823 10.1007/s00023-013-0310-4

[832] **E. Stredulinsky and W. P. Ziemer**, *Area minimizing sets subject to a volume constraint in a convex set*, J. Geom. Anal. **7** (1997), *no. 4*, 653–677. MR1669207 10.1007/BF02921639 http://dx.doi.org/10.1007/BF02921639

[833] **J. Stubbe**, *Universal monotonicity of eigenvalue moments and sharp Lieb-Thirring inequalities*, J. Eur. Math. Soc. (JEMS) **12** (2010), *no. 6*, 1347–1353. MR2734344 10.4171/JEMS/233 http://dx.doi.org/10.4171/JEMS/233

[834] **T. Sunada**, *Riemannian coverings and isospectral manifolds*, Ann. of Math. (2) **121** (1985), *no. 1*, 169–186. MR782558 10.2307/1971195

[835] **V. Šverák**, *On optimal shape design*, J. Math. Pures Appl. (9) **72** (1993), *no. 6*, 537–551. MR1249408

[836] **B. Sz.-Nagy**, *Über Parallelmengen nichtkonvexer ebener Bereiche*, Acta Sci. Math. Szeged **20** (1959), 36–47. MR0107213

[837] **G. Szegő**, *On membranes and plates*, Proc. Nat. Acad. Sci. U. S. A. **36** (1950), 210–216. MR0035629

[838] **G. Szegő**, *Inequalities for certain eigenvalues of a membrane of given area*, J. Rational Mech. Anal. **3** (1954), 343–356. MR0061749

[839] **A.-S. Sznitman**, *Fluctuations of principal eigenvalues and random scales*, Comm. Math. Phys. **189** (1997), *no. 2*, 337–363. MR1480023 10.1007/s002200050206 http://dx.doi.org/10.1007/s002200050206

[840] **I. Tadjbakhsh and J. B. Keller**, *Strongest columns and isoperimetric inequalities for eigenvalues*, Trans. ASME Ser. E. J. Appl. Mech. **29** (1962), 159–164. MR0137381

[841] **R. Tahraoui**, *Quelques remarques sur le contrôle des valeurs propres*, Nonlinear partial differential equations and their applications. Collège de France seminar, Vol. VIII (Paris, 1984–1985), Pitman Res. Notes Math. Ser., vol. 166, Longman Sci. Tech., Harlow, 1988, pp. 176–213. MR946490

[842] **G. Talenti**, *Best constant in Sobolev inequality*, Ann. Mat. Pura Appl. (4) **110** (1976), 353–372. MR0463908

[843] **G. Talenti**, *Elliptic equations and rearrangements*, Ann. Scuola Norm. Sup. Pisa Cl. Sci. (4) **3** (1976), *no. 4*, 697–718. MR0601601

[844] **G. Talenti**, *On the first eigenvalue of the clamped plate*, Ann. Mat. Pura Appl. (4) **129** (1981), 265–280 (1982). MR648335 10.1007/BF01762146

[845] **G. Talenti**, *Estimates for eigenvalues of Sturm-Liouville problems*, General inequalities, 4 (Oberwolfach, 1983), Internat. Schriftenreihe Numer. Math., vol. 71, Birkhäuser, Basel, 1984, pp. 341–350. MR821811

[846] **G. Talenti**, *Linear elliptic p.d.e.'s: level sets, rearrangements and a priori estimates of solutions*, Boll. Un. Mat. Ital. B (6) **4** (1985), *no. 3*, 917–949. MR831299

[847] **G. Talenti**, *Rearrangements of functions and partial differential equations*, Nonlinear diffusion problems (Montecatini Terme, 1985), Lecture Notes in Math., vol. 1224, Springer, Berlin, 1986, pp. 153–178. MR877989 10.1007/BFb0072690

[848] **T. Tao et al.**, *Polymath7: the hot spots conjecture*. http://michaelnielsen.org/polymath1/index.php?title=The_hot_spots_conjecture

[849] **L. Tartar**, *Compensated compactness and applications to partial differential equations*, Nonlinear analysis and mechanics: Heriot-Watt Symposium, Vol. IV, Res. Notes in Math., vol. 39, Pitman, Boston, Mass.-London, 1979, pp. 136–212. MR584398

[850] **M. E. Taylor**, *Partial differential equations. II*, Applied Mathematical Sciences, vol. 116, Springer-Verlag, New York, 1996, Qualitative studies of linear equations. MR1395149 10.1007/978-1-4757-4187-2

[851] **C. J. Thompson**, *On the ratio of consecutive eigenvalues in N-dimensions*, Studies in Appl. Math. **48** (1969), 281–283. MR0257592

[852] **J. A. Toth and S. Zelditch**, *Counting nodal lines which touch the boundary of an analytic domain*, J. Differential Geom. **81** (2009), *no. 3*, 649–686. MR2487604 http://projecteuclid.org/euclid.jdg/1236604347

[853] **B. A. Troesch**, *An isoperimetric sloshing problem*, Comm. Pure Appl. Math. **18** (1965), 319–338. MR0183210

[854] **B. A. Troesch**, *Elliptical membranes with smallest second eigenvalue*, Math. Comp. **27** (1973), 767–772. MR0421277

[855] **N. S. Trudinger**, *Linear elliptic operators with measurable coefficients*, Ann. Scuola Norm. Sup. Pisa (3) **27** (1973), 265–308. MR0369884

[856] **Ü. Ufuktepe and K. P. Mchale**, *Inequalities for buckling of a clamped plate*, J. Interdiscip. Math. **5** (2002), *no. 1*, 41–47. MR1925334 10.1080/09720502.2002.10700305 http://dx.doi.org/10.1080/09720502.2002.10700305

[857] **K. Uhlenbeck**, *Generic properties of eigenfunctions*, Amer. J. Math. **98** (1976), *no. 4*, 1059–1078. MR0464332 (57 #4264)

[858] **G. Uhlmann**, *Inverse problems: seeing the unseen*, Bull. Math. Sci. **4** (2014), *no. 2*, 209–279. MR3228575 10.1007/s13373-014-0051-9

[859] **M. van den Berg and M. Iversen**, *On the minimization of Dirichlet eigenvalues of the Laplace operator*, J. Geom. Anal. **23** (2013), *no. 2*, 660–676. MR3023854 10.1007/s12220-011-9258-0

[860] **M. van den Berg**, *Estimates for the torsion function and Sobolev constants*, Potential Anal. **36** (2012), *no. 4*, 607–616. MR2904636 10.1007/s11118-011-9246-9 http://dx.doi.org/10.1007/s11118-011-9246-9

[861] **M. van den Berg, D. Bucur, and K. Gittins**, *Maximising Neumann eigenvalues on rectangles*, Bull. Lond. Math. Soc. 48 (2016), *no. 5*, 877–894. MR3556370

[862] **M. van den Berg and K. Gittins**, *On the number of Courant sharp Dirichlet eigenvalues*. ArXiv:1602.08376

[863] **B. Velichkov**, *A note on the monotonicity formula of Caffarelli-Jerison Kenig*, Atti Accad. Naz. Lincei Rend. Lincei Mat. Appl. **25** (2014), *no. 2*, 165–189. MR3210965 10.4171/RLM/673 http://dx.doi.org/10.4171/RLM/673

[864] **B. Velichkov**, *Existence and regularity results for some shape optimization problems*, Tesi. Scuola Normale Superiore di Pisa (Nuova Series) [Theses of Scuola Normale Superiore di Pisa (New Series)], vol. 19, Edizioni della Normale, Pisa, 2015. MR3309888 10.1007/978-88-7642-527-1 http://dx.doi.org/10.1007/978-88-7642-527-1

[865] **E. J. M. Veling**, *Optimal lower bounds for the spectrum of a second order linear differential equation with a p-integrable coefficient*, Proc. Roy. Soc. Edinburgh Sect. A **92** (1982), *no. 1-2*, 95–101. MR667128 10.1017/S030821050001996X

[866] **A. Wagner**, *Optimal shape problems for eigenvalues*, Comm. Partial Differential Equations **30** (2005), *no. 7-9*, 1039–1063. MR2180294 10.1081/PDE-200064443 http://dx.doi.org/10.1081/PDE-200064443
[867] **S. Wang**, *Generalization of the thomas-reiche-kuhn and the bethe sum rules*, Phys. Rev. A **60** (1999), 262–266. 10.1103/PhysRevA.60.262 http://link.aps.org/doi/10.1103/PhysRevA.60.262
[868] **X. Wang and J. Zhu**, *A lower bound for the nodal sets of Steklov eigenfunctions*. ArXiv:1411.0708
[869] **M. Warma**, *The Robin and Wentzell-Robin Laplacians on Lipschitz domains*, Semigroup Forum **73** (2006), *no. 1*, 10–30. MR2277314 10.1007/s00233-006-0617-2 http://dx.doi.org/10.1007/s00233-006-0617-2
[870] **G. N. Watson**, *A treatise on the theory of Bessel functions*, Cambridge Mathematical Library, Cambridge University Press, Cambridge, 1995, Reprint of the second (1944) edition. MR1349110
[871] **H. F. Weinberger**, *An isoperimetric inequality for the N-dimensional free membrane problem*, J. Rational Mech. Anal. **5** (1956), 633–636. MR0079286
[872] **H. F. Weinberger**, *An effectless cutting of a vibrating membrane*, Pacific J. Math. **13** (1963), 1239–1240. MR0156092
[873] **R. Weinstock**, *Inequalities for a classical eigenvalue problem*, J. Rational Mech. Anal. **3** (1954), 745–753. MR0064989
[874] **G. S. Weiss**, *Partial regularity for weak solutions of an elliptic free boundary problem*, Comm. Partial Differential Equations **23** (1998), *no. 3-4*, 439–455. MR1620644 10.1080/03605309808821352 http://dx.doi.org/10.1080/03605309808821352
[875] **G. S. Weiss**, *Partial regularity for a minimum problem with free boundary*, J. Geom. Anal. **9** (1999), *no. 2*, 317–326. MR1759450 10.1007/BF02921941 http://dx.doi.org/10.1007/BF02921941
[876] **H. Weyl**, *Über die asymptotische Verteilung der Eigenwerte*, Nachrichten der Königlichen Gesellschaft der Wissenschaften zu Göttingen, 110–117.
[877] **H. Weyl**, *Ramifications, old and new, of the eigenvalue problem*, Bull. Amer. Math. Soc. **56** (1950), 115–139. MR0034940
[878] **C. Wieners**, *A numerical existence proof of nodal lines for the first eigenfunction of the plate equation*, Arch. Math. (Basel) **66** (1996), *no. 5*, 420–427. MR1383907 10.1007/BF01781561
[879] **B. E. Willner and T. J. Mahar**, *The two-dimensional eigenvalue range and extremal eigenvalue problems*, SIAM J. Math. Anal. **13** (1982), *no. 4*, 621–631. MR661593 10.1137/0513040
[880] **B. E. Willner and T. J. Mahar**, *Extrema of functions of eigenvalues*, J. Math. Anal. Appl. **72** (1979), *no. 2*, 730–739. MR559401 10.1016/0022-247X(79)90260-9
[881] **J. Wloka**, *Partielle Differentialgleichungen*, B. G. Teubner, Stuttgart, 1982, Sobolevräume und Randwertaufgaben. [Sobolev spaces and boundary value problems], Mathematische Leitfäden. [Mathematical Textbooks]. MR652934
[882] **S. A. Wolf and J. B. Keller**, *Range of the first two eigenvalues of the Laplacian*, Proc. Roy. Soc. London Ser. A **447** (1994), *no. 1930*, 397–412. MR1312811 10.1098/rspa.1994.0147
[883] **V. Wolontis**, *Properties of conformal invariants*, Amer. J. Math. **74** (1952), 587–606. MR0048585
[884] **C. Xia**, *Eigenvalues on Riemannian manifolds*, Publicações Matemáticas do IMPA. [IMPA Mathematical Publications], Instituto Nacional de Matemática Pura e Aplicada (IMPA), Rio de Janeiro, 2013, 29o Colóquio Brasileiro de Matemática. MR3113924
[885] **Y. Xu**, *The first nonzero eigenvalue of Neumann problem on Riemannian manifolds*, J. Geom. Anal. **5** (1995), *no. 1*, 151–165. MR1315661 10.1007/BF02926446 http://dx.doi.org/10.1007/BF02926446

[886] **H. C. Yang**, *Estimates of the difference between consecutive eigenvalues*, revision of International Centre for Theoretical Physics preprint IC/91/60, Trieste, Italy, April, 1991, 1995.

[887] **Q. H. Yu and J. Q. Zhong**, *Lower bounds of the gap between the first and second eigenvalues of the Schrödinger operator*, Trans. Amer. Math. Soc. **294** (1986), *no. 1*, 341–349. MR819952 10.2307/2000135

[888] **L. Zalcman**, *A bibliographic survey of the Pompeiu problem*, Approximation by solutions of partial differential equations (Hanstholm, 1991), NATO Adv. Sci. Inst. Ser. C Math. Phys. Sci., vol. 365, Kluwer Acad. Publ., Dordrecht, 1992, pp. 185–194. MR1168719

[889] **S. Zelditch**, *Measure of nodal sets of analytic Steklov eigenfunctions*. ArXiv:1403.0647

[890] **S. Zelditch**, *Eigenfunctions and nodal sets*, Surveys in differential geometry. Geometry and topology, Surv. Differ. Geom., vol. 18, Int. Press, Somerville, MA, 2013, pp. 237–308. MR3087922 10.4310/SDG.2013.v18.n1.a7

[891] **S. Zelditch**, *Maximally degenerate Laplacians*, Ann. Inst. Fourier (Grenoble) **46** (1996), *no. 2*, 547–587. MR1393525 http://www.numdam.org/item?id=AIF_1996__46_2_547_0

[892] **A. Zhiruar and I. Polterovich**, *On the Hersch-Payne-Schiffer estimates for the eigenvalues of the Steklov problem*, Funktsional. Anal. i Prilozhen. **44** (2010), *no. 2*, 33–47. MR2681956 10.1007/s10688-010-0014-1

[893] **J. Zhu**, *Geometry and interior nodal sets of Steklov eigenfunctions*. ArXiv:1510.07300

[894] **J. Zhu**, *Interior nodal sets of Steklov eigenfunctions on surfaces*. ArXiv:1507.00621

[895] **W. P. Ziemer**, *Weakly differentiable functions*, Graduate Texts in Mathematics, vol. 120, Springer-Verlag, New York, 1989, Sobolev spaces and functions of bounded variation. MR1014685 10.1007/978-1-4612-1015-3

Index

Kolibri 360 GmbH
Ferdinand-Jühlke-Straße 7
99095 Erfurt, Deutschland
produktsicherheit@kolibri360.de